Methods in Enzymology

Volume 201
PROTEIN PHOSPHORYLATION
Part B
Analysis of Protein Phosphorylation,
Protein Kinase Inhibitors, and Protein Phosphatases

METHODS IN ENZYMOLOGY

EDITORS-IN-CHIEF

John N. Abelson   Melvin I. Simon

DIVISION OF BIOLOGY
CALIFORNIA INSTITUTE OF TECHNOLOGY
PASADENA, CALIFORNIA

FOUNDING EDITORS

Sidney P. Colowick and Nathan O. Kaplan

Methods in Enzymology

Volume 201

Protein Phosphorylation

Part B
Analysis of Protein Phosphorylation, Protein Kinase Inhibitors, and Protein Phosphatases

EDITED BY

Tony Hunter

Bartholomew M. Sefton

MOLECULAR BIOLOGY AND VIROLOGY LABORATORY
THE SALK INSTITUTE
SAN DIEGO, CALIFORNIA

ACADEMIC PRESS, INC.
Harcourt Brace Jovanovich, Publishers
San Diego   New York   Boston
London   Sydney   Tokyo   Toronto

Academic Press, Inc.
San Diego, California 92101

United Kingdom Edition published by
ACADEMIC PRESS LIMITED
24-28 Oval Road, London NW1 7DX

Library of Congress Catalog Card Number: 54-9110

ISBN   0-12-182102-1   (alk. paper)

PRINTED IN THE UNITED STATES OF AMERICA
91 92 93 94   9 8 7 6 5 4 3 2 1

Table of Contents

Section I. Analysis of Protein Phosphorylation

A. Phosphoamino Acid Determination

B. Antibodies to Phosphoamino Acids: Immunoprecipitation and Immunoblotting

Section II. Protein Kinase Inhibitors

A. Peptide Inhibitors

B. Chemical Inhibitors

Section III. Protein Phosphatases

Section IV. Protein Phosphatase Inhibitors

Contributors to Volume 201

Article numbers are in parentheses following the names of contributors.
Affiliations listed are current.

RUEDI H. AEBERSOLD (15), *The Biomedical Research Centre, University of British Columbia, Vancouver, British Columbia V6T 1W5, Canada*

TETSU AKIYAMA (30), *Department of Oncogene Research, Research Institute of Microbial Diseases, Osaka University, Osaka 565, Japan*

JONATHAN M. BACKER (7), *Research Division, Joslin Diabetes Center, Boston, Massachusetts 02215*

ROBERT M. BELL (26), *Department of Biochemistry, Duke University Medical Center, Durham, North Carolina 27710*

NORBERT BERNDT (35), *School of Medicine, University of Southern California, Childrens Hospital of Los Angeles, Los Angeles, California 90027*

WILLIAM J. BOYLE (11), *Department of Molecular and Cellular Biology, AMGEN, Thousand Oaks, California 91320*

JONATHAN CHEN (23), *Abbott Laboratories, Pharmaceutical Products Division, Abbott Park, Illinois 60064*

PATRICIA T. W. COHEN (34, 35), *Department of Biochemistry, Protein Phosphorylation Unit, University of Dundee, Dundee DD1 4HN, Scotland*

PHILIP COHEN (13, 33, 36), *Department of Biochemistry, MRC Protein Phosphorylation Unit, University of Dundee, Dundee DD1 4HN, Scotland*

JONATHAN A. COOPER (21), *Fred Hutchinson Cancer Research Center, Seattle, Washington 98104*

ALAIN J. COZZONE (2), *Institut de Biologie et Chimie des Protéines, C.N.R.S., Université de Lyon, 69622 Villeurbanne, Cedex, France*

ANDREW J. CZERNIK (23), *Laboratory of Molecular and Cellular Neuroscience, The Rockefeller University, New York, New York 10021*

CURTIS D. DILTZ (37, 38), *Department of Biochemistry, University of Washington, Seattle, Washington 98195*

ARIANNA DONELLA-DEANA (17), *Dipartimento di Chimica Biologica, Universita di Padova, I-35121 Padova, Italy*

BERTRAND DUCLOS (2), *Institut de Biologie et Chimie des Protéines, C.N.R.S., Université de Lyon, 69622 Villeurbanne, Cedex, France*

EDMOND H. FISCHER (37, 38), *Department of Biochemistry, University of Washington, Seattle, Washington 98195*

A. RAYMOND FRACKELTON, JR. (8), *Department of Medicine, Roger Williams Medical Center and Brown University, Providence, Rhode Island 02908*

MATI FRIDKIN (5), *Department of Organic Chemistry, The Weizmann Institute of Science, Rehovot 76100, Israel*

HIDESUKE FUKAZAWA (31), *Department of Antibiotics, National Institute of Health, Tokyo 141, Japan*

AVIV GAZIT (29), *Department of Biological Chemistry and Organic Chemistry, The Hebrew University of Jerusalem, Jerusalem 91904, Israel*

BRADFORD W. GIBSON (13), *Department of Pharmaceutical Chemistry, School of Pharmacology, University of California, San Francisco, San Francisco, California 94143*

CHAIM GILON (29), *Department of Organic Chemistry, The Hebrew University of Jerusalem, Jerusalem 91904, Israel*

JEAN-ANTOINE GIRAULT (23), *INSERM U114, College de France, 75235 Paris, France*

DAVID B. GLASS (25), *Department of Pharmacology and Biochemistry, School of Medicine, Emory University, Atlanta, Georgia 30322*

JOHN R. GLENNEY (9), *Department of Biochemistry, University of Kentucky, Lexington, Kentucky 40536*

JULIUS A. GORDON (41), *Pathology and Laboratory Medicine, University of Texas A&M School of Medicine, College Station, Texas 77843*

PAUL GREENGARD (23), *Laboratory of Molecular and Cellular Neuroscience, The Rockefeller University, New York, New York 10021*

YUSUF A. HANNUN (26), *Departments of Medicine and Cell Biology, Duke University Medical Center, Durham, North Carolina 27710*

D. GRAHAME HARDIE (40), *Protein Phosphorylation Group, Department of Biochemistry, Medical Sciences Institute, University of Dundee, Dundee DD1 4HN, Scotland*

TIMOTHY A. J. HAYSTEAD (40), *Department of Pharmacology, University of Washington, Seattle, Washington 98195*

DAPHNA HEFFETZ (5), *Department of Chemical Immunology, The Weizmann Institute of Science, Rehovot 76100, Israel*

LUDWIG M. G. HEILMEYER, JR. (14), *Abteilung Biochemie Supramolekularer Systeme, Ruhr-Universität Bochum, D-4630 Bochum 1, Germany*

HIROYOSHI HIDAKA (27), *Department of Pharmacology, Nagoya University School of Medicine, Showa-ku, Nagoya 466, Japan*

EDELTRAUT HOFFMANN-POSORSKE (14, 17), *Abteilung Biochemie Supramolekularer Systeme, Ruhr-Universität Bochum, D-4630 Bochum 1, Germany*

CHARLES F. B. HOLMES (13), *Department of Biochemistry, University of Alberta, Edmonton, Alberta T6G 2H7, Canada*

LEROY E. HOOD (15), *Division of Biology, California Institute of Technology, Pasadena, California 91125*

COLIN M. HOUSE (24), *St. Vincent's Institute of Medical Research, Fitzroy, Victoria 3065, Australia*

MICHAEL J. HUBBARD (36), *Department of Biochemistry, University of Otago, Dunedin, New Zealand*

TONY HUNTER (11), *Molecular Biology and Virology Laboratory, The Salk Institute, San Diego, California 92138*

TAMARA R. HURLEY (12), *Molecular Biology and Virology Laboratory, The Salk Institute, San Deigo, California 92138*

MASAYA IMOTO (32), *Department of Applied Chemistry, Faculty of Science and Technology, Keio University, Yokohama 223, Japan*

THOMAS S. INGEBRITSEN (39), *Department of Zoology and Genetics, Iowa State University, Ames, Iowa 50011*

MARK P. KAMPS (3, 10), *The Whitehead Institute for Biomedical Research, Cambridge, Massachusetts 02142*

B. KANNAN (8), *Georgetown University Hospital, Lombardi Cancer Research Center, Washington, D.C. 20007*

JOHN KEBABIAN (23), *Abbott Laboratories, Pharmaceutical Products Division, Abbott Park, Illinois 60064*

BRUCE E. KEMP (24), *St. Vincent's Institute of Medical Research, Fitzroy, Victoria 3065, Australia*

RYOJI KOBAYASHI (27), *Department of Pharmacology, Nagoya University School of Medicine, Showa-ku, Nagoya 466, Japan*

HORST KORTE (17), *Abteilung Biochemie Supramolekularer Systeme, Ruhr-Universität Bochum, D-4630 Bochum 1, Germany*

LYNN M. KOZMA (4), *Department of Microbiology, University of Virginia School of Medicine, Charlottesville, Virginia 22908*

ALEXANDER LEVITZKI (29), *Department of*

Biological Chemistry, The Hebrew University of Jerusalem, Jerusalem 91904, Israel

KUNXIN LUO (12), Molecular Biology and Virology Laboratory, The Salk Institute, San Diego, California 92138

SYLVIE MARCANDIER (2), Institut de Biologie et Chimie des Protéines, C.N.R.S., Université de Lyon, 69622 Villeurbanne, Cedex, France

F. MERMELSTEIN (8), Department of Biochemistry, University of Medicine and Dentistry of New Jersey, Robert Wood Johnson Medical School, Piscataway, New Jersey 08854

ALFRED H. MERRILL, JR. (26), Department of Biochemistry, Emory University, Atlanta, Georgia 30322

HELMUT E. MEYER (14, 17), Abteilung Biochemie Supramolekularer Systeme, Ruhr-Universität Bochum, D-4630 Bochum 1, Germany

GARY W. MOY (22), Marine Biology Research Division, Scripps Institution of Oceanography, University of California, San Diego, La Jolla, California 92093

ANGUS C. NAIRN (23), Laboratory of Molecular and Cellular Neuroscience, The Rockefeller University, New York, New York 10021

HIROSHI OGAWARA (30), Department of Biochemistry, Meiji College of Pharmacy, Tokyo 154, Japan

NIR OSHEROV (29), Department of Biological Chemistry, The Hebrew University of Jerusalem, Jerusalem 91904, Israel

RICHARD B. PEARSON (24), St. Vincent's Institute of Medical Research, Fitzroy, Victoria 3065, Australia

JOHN W. PERICH (18, 19), School of Chemistry, University of Melbourne, Parkville, Victoria 3052, Australia

ISRAEL POSNER (29), Department of Biological Chemistry, The Hebrew University of Jerusalem, Jerusalem 91904, Israel

M. POSNER (8), Department of Medicine,

New England Deaconess Hospital, Boston, Massachusetts 02215

DAVID P. RINGER (1), Biochemical Pharmacology Section, Biomedical Division, The Samuel Roberts Noble Foundation, Inc., Ardmore, Oklahoma 73402

PETER J. ROACH (16), Department of Biochemistry and Molecular Biology, Indiana University School of Medicine, Indianapolis, Indiana 46223

ANTHONY J. ROSSOMANDO (4), Department of Microbiology, University of Virginia School of Medicine, Charlottesville, Virginia 22908

BARTHOLOMEW M. SEFTON (12, 20), Molecular Biology and Virology Labortory, The Salk Institute, San Diego, California 92138

ALISTAIR T. R. SIM (40), Neurosciences Group, Faculty of Medicine, University of Newcastle, Newcastle, NSW 2308, Australia

GRETCHEN SNYDER (23), Laboratory of Molecular and Cellular Neuroscience, The Rockefeller University, New York, New York 10021

TATSUYA TAMAOKI (28), Kyowa Hakko Kogyo Co., Ltd., Tokyo Research Laboratories, Tokyo 194, Japan

NICHOLAS K. TONKS (37, 38), Cold Spring Harbor Laboratory, Cold Spring Harbor, New York 11724

YOSHIMASA UEHARA (31), Department of Antibiotics, National Institute of Health, Tokyo 141, Japan

KAZUO UMEZAWA (32), Department of Applied Chemistry, Faculty of Science and Technology, Keio University, Yokohama 223, Japan

VICTOR D. VACQUIER (22), Marine Biology Research Division, Scripps Institution of Oceanography, University of California, San Diego, La Jolla, California 92093

PETER VAN DER GEER (11), Molecular Biology and Virology Laboratory, The Salk Institute, San Diego, California 92138

DONAL A. WALSH (25), *Department of Biological Chemistry, School of Medicine, University of California, Davis, Davis, California 95616*

JEAN Y. J. WANG (6), *Department of Biology, University of California, San Diego, La Jolla, California 92093*

YUHUAN WANG (16), *Department of Biochemistry and Molecular Biology, Indiana University School of Medicine, Indianapolis, Indiana 46223*

MASATO WATANABE (27), *Department of Pharmacology, Nagoya University School of Medicine, Showa-ku, Nagoya 466, Japan*

MICHAEL J. WEBER (4), *Department of Microbiology, University of Virginia School of Medicine, Charlottesville, Virginia 22908*

RICHARD E. H. WETTENHALL (15), *Department of Biochemistry, University of Melbourne, Parkville, Victoria 3052, Australia*

MORRIS F. WHITE (7), *Research Division, Joslin Diabetes Center, Boston, Massachusetts 02215*

YEHIEL ZICK (5), *Department of Chemical Immunology, The Weizmann Institute of Science, Rehovot 76100, Israel*

Preface

The field of protein phosphorylation has grown and changed considerably since it was covered in 1983 in Volume 99 on Protein Kinases in the *Methods in Enzymology* series. At that time fewer than five protein kinase amino acid sequences were known. The number of identified protein kinases and the number of processes known to be regulated by protein phosphorylation have both increased enormously since then, and the end is not yet in sight. Fundamental to this proliferation has been the ability to isolate novel genes encoding protein kinases using the techniques of molecular biology. Equally important is the fact that the similarity in amino acid sequence of the catalytic domains of the protein kinases allows the instantaneous realization that a molecular clone isolated on the basis of biological function or partial amino acid sequence encodes a protein kinase. It is now clear that many, and perhaps most, aspects of growth regulation are controlled by a complex network of protein kinases and phosphatases.

The techniques that have already defined the unexpectedly large size and degree of complexity of the protein kinase gene family, and will continue to do so, are described in Volumes 200 and 201. These two volumes were consciously entitled Protein Phosphorylation, rather than Protein Kinases. This decision had two origins. One was the emerging realization that the protein phosphatases may prove to be of as much regulatory significance as the protein kinases. The other was that the study of protein kinases is sterile in the absence of the identification and characterization of both upstream regulators and downstream polypeptide substrates, many of which will not be protein kinases.

Of necessity, the first protein kinases identified and studied were those whose activity was prominent in tissues that could be obtained in large quantities. Most of the protein kinases that are important in growth control, however, are present at extremely low levels in cells. The development of sensitive techniques to study nonabundant proteins was, therefore, imperative. Considerable attention is given in these volumes to the use of recombinant DNA techniques for the preparation of large quantities of protein kinases, to means by which to detect trace quantities of specific polypeptides in complex mixtures of proteins, and to techniques with which to perform protein chemistry on vanishingly small quantities of phosphoproteins.

What does the future of this field hold? A major "watershed" will be the determination of the three-dimensional structure of a protein kinase. Techniques useful for the crystallization of cyclic AMP-dependent protein kinase are presented in Volume 200, but solution of the structure of the enzyme at atomic resolution has not yet been achieved. Knowledge of the structure of one or more protein kinases will almost certainly alter the study of these enzymes very significantly.

To date, with few exceptions, the study of protein phosphorylation has involved the study of the phosphorylation of proteins on serine, threonine, or tyrosine. The lack of attention paid to protein kinases generating acid-labile phosphoamino acids reflects not a lack of biological importance of these enzymes, since they clearly play a central role in bacterial chemotaxis, but rather the fact that methods for their study are few and poorly developed. Unanticipated and important roles for protein kinases may well become apparent if simple and reliable means with which to detect and study proteins containing labile phosphorylated amino acids are devised. No doubt the future will also hold other surprises, but we can only hope that four volumes are not needed the next time protein phosphorylation is covered in this series!

These volumes would never have seen "the light of day" without the diligence of Karen Lane. We thank her for her cheerful and tireless help.

BARTHOLOMEW M. SEFTON
TONY HUNTER

METHODS IN ENZYMOLOGY

VOLUME 68. Recombinant DNA
Edited by RAY WU

VOLUME 69. Photosynthesis and Nitrogen Fixation (Part C)
Edited by ANTHONY SAN PIETRO

VOLUME 70. Immunochemical Techniques (Part A)
Edited by HELEN VAN VUNAKIS AND JOHN J. LANGONE

VOLUME 71. Lipids (Part C)
Edited by JOHN M. LOWENSTEIN

VOLUME 72. Lipids (Part D)
Edited by JOHN M. LOWENSTEIN

VOLUME 73. Immunochemical Techniques (Part B)
Edited by JOHN J. LANGONE AND HELEN VAN VUNAKIS

VOLUME 74. Immunochemical Techniques (Part C)
Edited by JOHN J. LANGONE AND HELEN VAN VUNAKIS

VOLUME 75. Cumulative Subject Index Volumes XXXI, XXXII, XXXIV–LX
Edited by EDWARD A. DENNIS AND MARTHA G. DENNIS

VOLUME 76. Hemoglobins
Edited by ERALDO ANTONINI, LUIGI ROSSI-BERNARDI, AND EMILIA CHIANCONE

VOLUME 77. Detoxication and Drug Metabolism
Edited by WILLIAM B. JAKOBY

VOLUME 78. Interferons (Part A)
Edited by SIDNEY PESTKA

VOLUME 79. Interferons (Part B)
Edited by SIDNEY PESTKA

VOLUME 80. Proteolytic Enzymes (Part C)
Edited by LASZLO LORAND

VOLUME 81. Biomembranes (Part H: Visual Pigments and Purple Membranes, I)
Edited by LESTER PACKER

VOLUME 82. Structural and Contractile Proteins (Part A: Extracellular Matrix)
Edited by LEON W. CUNNINGHAM AND DIXIE W. FREDERIKSEN

VOLUME 83. Complex Carbohydrates (Part D)
Edited by VICTOR GINSBURG

VOLUME 84. Immunochemical Techniques (Part D: Selected Immunoassays)
Edited by JOHN J. LANGONE AND HELEN VAN VUNAKIS

VOLUME 85. Structural and Contractile Proteins (Part B: The Contractile Apparatus and the Cytoskeleton)
Edited by DIXIE W. FREDERIKSEN AND LEON W. CUNNINGHAM

VOLUME 86. Prostaglandins and Arachidonate Metabolites
Edited by WILLIAM E. M. LANDS AND WILLIAM L. SMITH

VOLUME 87. Enzyme Kinetics and Mechanism (Part C: Intermediates, Stereochemistry, and Rate Studies)
Edited by DANIEL L. PURICH

VOLUME 88. Biomembranes (Part I: Visual Pigments and Purple Membranes, II)
Edited by LESTER PACKER

VOLUME 89. Carbohydrate Metabolism (Part D)
Edited by WILLIS A. WOOD

VOLUME 90. Carbohydrate Metabolism (Part E)
Edited by WILLIS A. WOOD

VOLUME 91. Enzyme Structure (Part I)
Edited by C. H. W. HIRS AND SERGE N. TIMASHEFF

VOLUME 92. Immunochemical Techniques (Part E: Monoclonal Antibodies and General Immunoassay Methods)
Edited by JOHN J. LANGONE AND HELEN VAN VUNAKIS

VOLUME 194. Guide to Yeast Genetics and Molecular Biology
Edited by CHRISTINE GUTHRIE AND GERALD R. FINK

VOLUME 195. Adenylyl Cyclase, G Proteins, and Guanylyl Cyclase
Edited by ROGER A. JOHNSON AND JACKIE D. CORBIN

VOLUME 196. Molecular Motors and the Cytoskeleton
Edited by RICHARD B. VALLEE

VOLUME 197. Phospholipases
Edited by EDWARD A. DENNIS

VOLUME 198. Peptide Growth Factors (Part C)
Edited by DAVID BARNES, J. P. MATHER, AND GORDON H. SATO

VOLUME 199. Cumulative Subject Index Volumes 168–174, 176–194 (in preparation)

VOLUME 200. Protein Phosphorylation (Part A: Protein Kinases: Assays, Purification, Antibodies, Functional Analysis, Cloning, and Expression)
Edited by TONY HUNTER AND BARTHOLOMEW M. SEFTON

VOLUME 201. Protein Phosphorylation (Part B: Analysis of Protein Phosphorylation, Protein Kinase Inhibitors, and Protein Phosphatases)
Edited by TONY HUNTER AND BARTHOLOMEW M. SEFTON

VOLUME 202. Molecular Design and Modeling: Concepts and Applications (Part A: Proteins, Peptides, and Enzymes) (in preparation)
Edited by JOHN J. LANGONE

VOLUME 203. Molecular Design and Modeling: Concepts and Applications (Part B: Antibodies and Antigens, Nucleic Acids, Polysaccharides, and Drugs) (in preparation)
Edited by JOHN J. LANGONE

VOLUME 204. Bacterial Genetic Systems (in preparation)
Edited by JEFFREY H. MILLER

VOLUME 205. Metallobiochemistry (Part B: Metallothionein and Related Molecules) (in preparation)
Edited by JAMES F. RIORDAN AND BERT L. VALLEE

Section I

Analysis of Protein Phosphorylation

A. Phosphoamino Acid Determination
Articles 1 through 4

B. Antibodies to Phosphoamino Acids: Immunoprecipitation and Immunoblotting
Articles 5 through 10

C. Phosphopeptide Mapping and Phosphorylation Site Determination
Articles 11 through 17

D. Phosphopeptide Synthesis
Articles 18 and 19

E. Determination of Phosphorylation Stoichiometry
Articles 20 through 23

[1] Separation of Phosphotyrosine, Phosphoserine, and Phosphothreonine by High-Performance Liquid Chromatography

By DAVID P. RINGER

A fast, sensitive, high-performance liquid chromatography (HPLC) method for analyzing the products of protein kinase-catalyzed phosphorylations of the three hydroxyamino acids, tyrosine, serine, and threonine, has been developed. It incorporates the use of precolumn derivatization of amino acids with the fluorescent reagent, 9-fluorenylmethyl chloroformate (FMOC),[1] and their subsequent separation on an anion-exchange column under conditions of isocratic elution.[2]

The major steps in the use of this method for the analysis of proteins for phosphotyrosine, phosphoserine, and phosphothreonine contents are outlined in Scheme I.

Acid Hydrolysis of Protein

Because the phosphorylated esters of the hydroxyamino acids show reasonable chemical stability over a large pH range,[3] limited acid hydrolysis of phosphoproteins has been used for their analysis.[4,5] The complexity of the protein sample to be analyzed can vary from extractions of total cellular protein to individual proteins such as bands from sodium dodecyl sulfate (SDS)-polyacrylamide gels or products of immunoprecipitations.[6] Typically 10- to 500-μg samples of protein are lyophilized in glass tubes, redissolved in 1.0 ml of 6 N HCl (analytical grade), and placed in evacuated, sealed hydrolysis tubes for heating at 110°.[7] The length of hydrolysis time must be somewhat optimized for the specific protein sample. Optimal time for the hydrolytic release of all three phosphoamino acids in a variety of protein samples is 3 hr at 110° in 6 N HCl.[2] Acid hydrolysis times in excess of 3 hr result in a significant decomposition of all phosphoamino acids to yield free amino acid and inorganic orthophosphate. Following

[1] S. Einarsson, B. Josefsson, and S. Lagerkvist, *J. Chromatogr.* **282,** 609 (1983).
[2] J. S. Niedbalski and D. P. Ringer, *Anal. Biochem.* **158,** 138 (1986).
[3] R. H. A. Plimmer, *Biochem. J.* **35,** 461 (1941).
[4] L. Cohen-Solal, J. B. Lian, D. Kossiva, and M. J. Glimcher, *Biochem. J.* **177,** 81 (1979).
[5] L. J. Pike, D. F. Bowen-Pope, R. Ross, and E. G. Krebs, *J. Biol. Chem.* **258,** 9383 (1983).
[6] J. A. Cooper, B. M. Sefton, and T. Hunter, this series, Vol. 99, p. 387.
[7] D. B. Bylund and T. S. Huang, *Anal. Biochem.* **73,** 477 (1976).

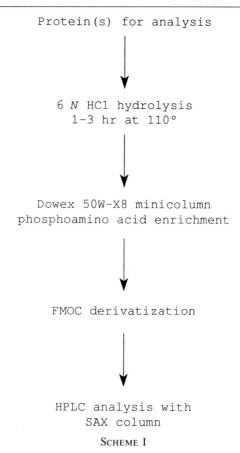

Protein(s) for analysis

6 N HCl hydrolysis
1–3 hr at 110°

Dowex 50W–X8 minicolumn
phosphoamino acid enrichment

FMOC derivatization

HPLC analysis with
SAX column

SCHEME I

acid hydrolysis samples are lyophilized and stored at −80° until needed for analysis.

For the measurement of phosphotyrosine, shorter hydrolysis times of 1 to 1.5 hr are recommended because it has a less acid-stable phosphate ester linkage than do phosphoserine and phosphothreonine.[5] An alternative method for the study of phosphotyrosine uses base hydrolysis of phosphoproteins (5 M KOH at 155° for 1 hr), yielding near-quantitative release of phosphotyrosine, but causing complete hydrolysis of phosphoserine and phosphothreonine to free amino acid and inorganic orthophosphate.[8] Caution is advised in the use of base hydrolysis if the sample contains proteins with phosphoramidate bonds, e.g., N-phospholysine or

[8] T. M. Martensen and R. L. Levine, this series, Vol. 99, p. 402.

N-phosphohistidine, because they will also be present in the hydrolysate and will complicate the HPLC profile during analysis.[9] The acid hydrolysis conditions routinely hydrolyze the phosphoramidate bonds, eliminating them from subsequent analysis steps.

Enrichment of Protein Hydrolysate for Phosphoamino Acids

Since phosphoprotein samples often have very low amounts of phosphoamino acids, i.e., less than 10 pmol/mg protein, it is generally desirable to enrich selectively the protein hydrolysate for phosphoamino acid content prior to analysis. This is especially true when one is unable to take advantage of phosphoamino acid detection by radioisotope labeling. A fast, convenient method described by Martensen[10] employs elution of phosphoamino acid from Dowex 50W-X8(H^+) with 0.1 N formic acid. A minicolumn version of this procedure routinely requires 30 min and consists of the following steps:

1. Minicolumn: Cut the bottom off a microfuge tube and insert a 1-ml disposable plastic pipette tip through the bottom. Pack the end of the pipette tip with a small amount of glass wool and transfer 0.35 ml of Dowex 50W-X8 as a 1 : 1 slurry in water into the pipettor tip.
2. Rinse the minicolumn with 1.0 ml of 0.1 N formic acid [1.07 ml formic acid (88%) in 250 ml distilled water].
3. Dissolve the lyophilized protein hydrolysate in 0.3 ml of 0.1 N formic acid.
4. Place a test tube under the minicolumn, apply the protein hydrolysate sample, and elute to the top of the Dowex surface.
5. Rinse the minicolumn twice with 0.8 ml of 0.1 N formic acid.
6. Discard the minicolumn and lyophilize the contents of the test tube.

Under these conditions, 96–98% of the unphosphorylated amino acids are retained on the minicolumn, while the phosphoamino acids are recovered in the following percentages: 80% for phosphotyrosine, and 100% for phosphothreonine and phosphoserine.[2]

FMOC Derivatization of Amino Acids

Prior to HPLC analysis, protein hydrolysates and phosphoamino acid standards are derivatized with FMOC[1] in the following manner.

---

[9] A. W. Steiner, E. R. Helander, J. M. Fujitaki, L. S. Smith, and R. A. Smith, *J. Chromatogr.* **202,** 263 (1980).
[10] T. M. Martensen, this series, Vol. 107, Part B, p. 1.

Reagents

FMOC reagent: Prepare just before use as a 15 mM stock by dissolving 77.5 mg FMOC in 20 ml of acetone

Borate buffer: Adjust a 1 M boric acid solution to pH 6.2 with NaOH

Procedure

1. Dissolve protein hydrolysate in 0.4 ml water and add 0.1 ml of borate buffer.

2. Add 0.5 ml FMOC reagent, vortex briefly, and allow the reaction to proceed for 40–60 sec to give complete derivatization.

3. In a chemical fume hood, add 2 ml pentane to the reaction mixture and vortex until the phases are well mixed. After allowing the phases to separate, aspirate and discard the pentane from the top of the tube.

4. Extract two more times with pentane, leaving the last extraction of pentane in the tube. Use a 1-ml tuberculin syringe equipped with a blunt-ended 26-gauge needle to withdraw the bottom aqueous layer. The sample is then transferred to a 3-ml syringe fitted with a 0.2-μm nylon syringe filter for filtering. The sample is now ready for HPLC.

HPLC Analysis of Phosphotyrosine, Phosphoserine,
    and Phosphothreonine

Preparation of Standards

Stock solutions of 1 mM phosphotyrosine, phosphoserine, and phosphothreonine are prepared and concentrations confirmed by the ninhydrin reaction. Stock solutions of each standard are aliquoted, 0.1 ml/tube, frozen at −20°, and used as needed. Typically a standard mix containing 4000 pmol of each phosphoamino acid is FMOC derivatized. An aliquot of the standards containing 25–50 pmol of each phosphoamino acid is injected at the beginning and end of each HPLC session for purposes of calibrating column retention times and quantitation of experimental samples.

Apparatus

HPLC pump (model 510; Waters, Milford, MA) equipped with a U6-K injector

HPLC fluorescence detector (FL748; McPherson, Acton, MA) equipped with a mercury lamp and a 24-μl flow cell. Filters are selected to allow excitation by 265 nm and fluorescence emission detection at 310 nm

Data module (745; Waters) used to integrate peak areas

Partisil-SAX column (250 × 4.6 mm; Alltech Assoc., Deerfield, IL) with 10-μm particle size

Chromatography

Chromatographic analysis of FMOC-derivatized phosphoamino acid standards on a strong anion-exchange column with isocratic elution by a buffer containing 55% (v/v) methanol, 1% (v/v) tetrahydrofuran, and 10 m*M* potassium phosphate, pH 3.9, is shown in Fig. 1. Calibration plots consisting of changes in log peak area as a function of phosphoamino acid concentration give straight-line graphs for levels between 5 and 1000 pmol

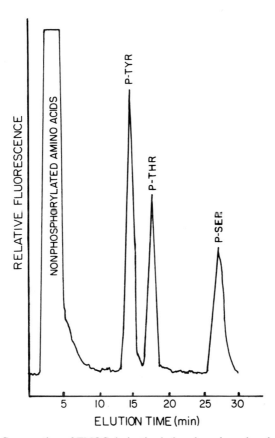

FIG. 1. HPLC separation of FMOC-derivatized phosphoserine, phosphothreonine, and phosphotyrosine (80 pmol each) on a Partisil 10 SAX column at 1.5 ml/min with a buffer containing 55% methanol, 1% tetrahydrofuran, and 10 m*M* potassium phosphate, pH 3.9. (From Niedbalski and Ringer.[2])

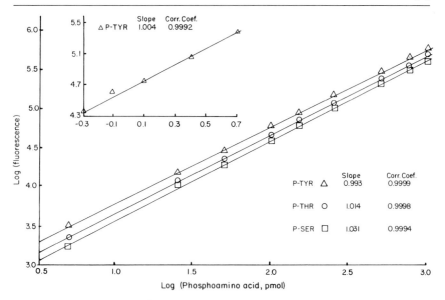

FIG. 2. Characterization of the quantitative relationship between the fluorescence detection of FMOC-derivatized phosphoamino acids and amount of phosphoamino acid injected during HPLC analysis. The inset demonstrates the ability to increase detectability of phosphotyrosine by using a higher fluorometer sensitivity setting. (From Niedbalski and Ringer.[2])

of phosphoamino acid (Fig. 2). Linear regressions of lines indicate essentially identical detection responses for all three phosphoamino acids. The range of detectability for all three phosphoamino acids could be extended to 0.5 pmol in a linear fashion, as shown for the case of phosphotyrosine (inset, Fig. 2).

Although this method was developed principally for the analysis of nonradioisotopically labeled samples, it may also be used for the analysis of labeled phosphoamino acids. In the analysis of ^{32}P-labeled phosphoamino acids, it should be noted that contaminating ortho[^{32}P]phosphate will coelute with phosphoserine. This difficulty can be overcome by adjusting the pH of the elution buffer to 3.7, which results in the elution of orthophosphate after phosphoserine, but at a cost of decreased resolution between phosphotyrosine and phosphothreonine.

The retention times of numerous common chromatographic phospho contaminants of phosphoamino acids have been determined and found not to interfere with the phosphoamino acid profile.[2] A less common contaminant which can cause problems is serine O-sulfate, which will coelute with phosphotyrosine under the standard elution conditions. Serine O-sulfate has been shown to be generated during the acid hydrolysis of proteins in the presence of sulfate.[11] It is advisable to check for the presence of serine *O*-sulfate contaminant in all work where phosphotyrosine quantitation is performed, especially if the protein sample has come in contact with sodium dodecyl sulfate during preparation. Resolution of serine *O*-sulfate from phosphotyrosine is possible by elution with 50% methanol, 1.0% tetrahydrofuran, 20 m*M* potassium phosphate adjusted to a pH of 5.2. While these conditions provide baseline resolution for all three phosphoamino acids and serine *O*-sulfate, elution time is doubled.

Concluding Remarks

The separation and quantitative analysis of FMOC-derivatized phosphorylated hydroxyamino acids by HPLC allow rapid, sensitive detection without the need for ^{32}P-labeling. This is an especially important feature when working with a sample from an *in vivo* source where radioisotopic labeling may not be possible or feasible. Some other inherent advantages of this approach for phosphoamino acid analysis are the ability to sample total phosphoamino acid content instead of just labeled forms, and the ability to distinguish new phosphorylations and phosphate turnover at pre-existing sites. Other HPLC procedures reported for the analysis of phosphoamino acids suffer from a number of disadvantages. These procedures, employing cation-exchange,[12] anion-exchange,[13] and reversed-phase[14,15] columns, have encountered such difficulties as lengthy separation times, incomplete resolution between phosphothreonine and phosphoserine, or coelution of aspartic and glutamic acids with phosphoamino acids. Others which overcame lengthy separation times[16] relied on ultraviolet absorption, allowing only micromolar detection of phosphoserine and phosphothreonine and nanomolar detection of phosphotyrosine. The use of postcolumn fluorescence derivatization of phosphoamino acids with compounds such as *o*-phthalaldehyde has met with greater success,[9,17] but requires a more complex in-line derivatizing reaction and does not allow removal of unreacted reagent as do the pentane extractions following FMOC derivatization. The anion-exchange HPLC of FMOC-fluorescent derivatives of phosphoamino acids, described here, is highly reproducible

[11] K. Murray, *Biochem. J.* **110,** 155 (1968).
[12] J.-P. Capony and J. G. Demaille, *Anal. Biochem.* **128,** 206 (1983).
[13] D. W. McCourt, J. F. Leykam, and B. D. Schwartz, *J. Chromatogr.* **327,** 9 (1985).
[14] L. Carlomagno, V. D. Huebner, and H. R. Matthews, *Anal. Biochem.* **149,** 344 (1985).
[15] N. Morrice and A. Aitken, *Anal. Biochem.* **148,** 207 (1985).
[16] J. C. Robert, A. Soumarmon, and M. J. M. Lewin, *J. Chromatogr.* **338,** 315 (1985).
[17] J. C. Yang, J. M. Fujitaki, and R. A. Smith, *Anal. Biochem.* **122,** 360 (1982).

and applicable for the analysis of phosphoproteins from both *in vivo* and *in vitro* sources.

Acknowledgments

This work was supported by The Samuel Roberts Noble Foundation, Inc., Ardmore, Oklahoma.

[2] Chemical Properties and Separation of Phosphoamino Acids by Thin-Layer Chromatography and/or Electrophoresis

By BERTRAND DUCLOS, SYLVIE MARCANDIER, and ALAIN J. COZZONE

One important criterion in demonstrating the occurrence of protein phosphorylation is showing that phosphoryl groups are covalently bound to certain amino acids of protein substrates and characterizing the nature of the linkage involved.

Phosphorylated amino acid residues in proteins are commonly classified into three main groups: (1) *O*-phosphates or *O*-phosphomonoesters are formed by phosphorylation of the hydroxyamino acids serine, threonine, and tyrosine; phosphohydroxyproline has not yet been found in proteins; (2) *N*-phosphates or phosphoramidates are produced by phosphorylation of the basic amino acids arginine, histidine, and lysine; (3) acyl phosphates or phosphate anhydrides are generated by phosphorylation of the acidic amino acids aspartic acid and glutamic acid. A fourth group, relatively less abundant, includes *S*-phosphates or thioesters of the sulfhydryl-containing amino acid cysteine.

Two types of intracellular phosphoproteins are known: enzymes that are intermediately phosphorylated, usually at their active sites, and proteins that are phosphorylated by protein kinases. It is generally assumed[1,2] that phosphorylated intermediates in enzymatic mechanisms are either phosphoramidates or acyl phosphates, but in some cases, such as phosphoglucomutase and alkaline phosphatase,[3] the active site of the enzyme is modified at a serine residue. On the other hand, phosphorylation by protein

[1] E. G. Krebs and J. A. Beavo, *Annu. Rev. Biochem.* **48,** 923 (1979).

[2] J. R. Knowles, *Annu. Rev. Biochem.* **49,** 877 (1980).

[3] L. Engström, P. Ekman, E. Humble, U. Ragnarsson, and O. Zetterqvist, this series, Vol. 107, p. 130.

TABLE I

CHEMICAL STABILITY OF PHOSPHORYLATED AMINO ACIDS[a]

Nature of phosphoamino acid	Stability in			
	Acid	Alkali	Hydroxylamine	Pyridine
O-Phosphates				
Phosphoserine	+	−	+	+
Phosphothreonine	+	±	+	+
Phosphotyrosine	+	+	+	+
N-Phosphates				
Phosphoarginine	−	−	−	−
Phosphohistidine	−	+	−	−
Phospholysine	−	+	−	−
Acyl phosphates				
Phosphoaspartate	−	−	−	−
Phosphoglutamate	−	−	−	−
S-Phosphates				
Phosphocysteine	+	+	+	+

[a] + indicates that the phosphoamino acid is stable, and − means that it is labile. For detailed information concerning the differential stability of the phosphoamino acids within a given group, see the text.

kinases is considered to take place essentially on hydroxyamino acids,[4] but the existence of enzymes specific for other amino acids cannot be excluded.[5]

Chemical Properties of Phosphoamino Acids

To demonstrate unequivocally that a certain amino acid of a protein is phosphorylated, it is usually necessary to isolate and identify the corresponding phosphoamino acid. However, there are several diagnostic tests that can be initially performed to obtain information on the nature of the phosphorylated amino acid. Such preliminary investigation is also useful for choosing the most appropriate technical procedure for preparing and analyzing the phosphorylated residues. Most of these tests are based on the chemical stability of the phosphomolecules to extremes of pH and/or to hydroxylamine and pyridine (Table I). Some other tests involve checking the reactivity with a variety of chemicals.

Of interest is the fact that the more commonly employed assay procedures and isolation methods in the investigations of both prokaryotic and

[4] C. S. Rubin and O. M. Rosen, *Annu. Rev. Biochem.* **44,** 831 (1975).
[5] J. M. Fujitaki and R. A. Smith, this series, Vol. 107, p. 23.

eukaryotic phosphoproteins involve the use of acid.[6] As described below, acid conditions are precluded in investigations of proteins containing N-phosphates or acyl phosphates. Since acid treatment of phosphoproteins during purification and determination of phosphoamino acids is routine, O-phosphomonoesters have been studied more often, and the existence of acid-labile phosphates has been largely overlooked. The latter class of phosphates may well be more widespread than is generally believed, even though the O-phosphates remain the major class in terms of quantity and extent of distribution in proteins.[6]

O-Phosphates

Phosphoserine (P-Ser), phosphothreonine (P-Thr), and phosphotyrosine (P-Tyr) resist acid treatment.[7] No significant loss of phosphoryl groups is observed after precipitation of O-phosphorylated proteins by perchloric acid or trichloroacetic acid. Incubation of phosphoproteins in 0.1 N HCl at 55° for 2 hr releases less than 10% of the phosphorylated moiety.[8] Treatment with 10% trichloroacetic acid at 90° for 20 min[9] leads to 80–90% recovery of the phosphoprotein in the precipitate. The stability of free O-phosphates to acid is in the order P-Thr > P-Ser > P-Tyr. In 1 N HCl at 100° the half-life of P-Thr and P-Ser is about 18 hr and that of P-Tyr about 5 hr.[7,10] Another measurement made in more concentrated HCl (6 N) at 105° indicates that the half-life of P-Ser is 8 hr while that of P-Thr is over 25 h.[8] From a general point of view, the detection of O-phosphates in acid hydrolysates of proteins is possible because the hydrolysis of the phosphomonoester bonds is considerably slower in acid than the hydrolysis of peptide bonds. But, because of the differential stability and recovery of O-phosphates, one must select an optimal time for phosphoprotein hydrolysis in order to allow complete polypeptide hydrolysis with only minimal loss of the phosphoamino acids. Usually this time is 2 hr in 6 N HCl at 110°, even though a shorter time (1 hr) would increase the yield of P-Tyr, and a longer time (4 hr) would favor the recovery of P-Thr.

Phosphoserine and phosphothreonine are labile in alkali whereas phosphotyrosine is not, which provides a means to differentiate between these molecules. Treatment with 1 N NaOH at 37° for 18–20 hr normally results

[6] T. M. Martensen, this series, Vol. 107, p. 3.
[7] R. H. Plimmer, *Biochem. J.* **35,** 461 (1941).
[8] L. Bitte and D. Kabat, this series, Vol. 30, p. 563.
[9] W. C. Schneider, *J. Biol. Chem.* **161,** 293 (1945).
[10] J. A. Cooper, B. M. Sefton, and T. Hunter, this series, Vol. 99, p. 387.

in the quantitative dephosphorylation of both P-Ser and P-Thr.[11,12] The lability of the phosphoryl groups of these two phosphoamino acids is, however, influenced by neighboring amino acid residues and, in some instances, P-Thr can become more resistant to alkali with a half-life of 7 hr in 3 N NaOH at 50°.[13,14] In general, the phosphoryl groups of P-Ser and P-Thr in peptide linkage are more labile to base than those in free phosphoamino acids.[7,15] By contrast, the stability of P-Tyr in proteins is the same as that of free P-Tyr. Treatment of various proteins containing P-Tyr with 5 N KOH at 155° for 30 min results in 80% recovery of free P-Tyr.[16,17] Similarly only 1% of free P-Tyr is hydrolyzed after treatment with 1 N NaOH at 100° for 5 hr.[17] Besides its specific resistance to alkali, P-Tyr can also be detected by reaction with fluorescamine or o-phthalaldehyde.[18,19]

All O-phosphates are stable in 1 M neutral hydroxylamine. They also resist aminolysis at pH 5.5, contrary to acyl phosphates and phosphoramidates.

Other Phosphoamino Acids

Phosphoramidates are extremely acid labile but relatively base stable, except for phosphoarginine (e.g., in 3 N KOH at 120° for 3 hr).[20,21] The corresponding hydrolysis curve starts high at low pH and sweeps low at high pH.[22] The phosphorylation of histidine is inhibited by diethyl pyrocarbonate and other histidine-specific reagents.[23] All N-phosphates are unstable in hydroxylamine. Pyridine also catalyzes their hydrolysis.[24]

Acyl phosphates are labile at both extremes of pH. This fact imposes a U shape on the hydrolysis curve, which is characteristic of this class of

[11] H. G. Nimmo, C. G. Proud, and P. Cohen, *Eur. J. Biochem.* **68,** 21 (1976).
[12] B. Ames, this series, Vol. 8, p. 115.
[13] P. J. Parker, F. B. Candwell, N. Embi, and P. Cohen, *Eur. J. Biochem.* **124,** 47 (1982).
[14] B. E. Kemp, *FEBS Lett.* **110,** 308 (1980).
[15] E. B. Kalan and M. Telka, *Arch. Biochem. Biophys.* **85,** 273 (1959).
[16] T. M. Martensen, *J. Biol. Chem.* **257,** 9648 (1982).
[17] T. M. Martensen and R. L. Levine, this series, Vol. 99, p. 402.
[18] J. C. Wang, J. M. Fujitaki, and R. A. Smith, *Anal. Biochem.* **122,** 360 (1982).
[19] G. Swarup, S. Cohen, and D. L. Garbers, *J. Biol. Chem.* **256,** 8197 (1981).
[20] R. A. Smith, R. M. Halpern, B. B. Bruegger, A. K. Dunlap, and O. Fricke, *Methods Cell Biol.* **19,** 153 (1978).
[21] W. P. Jencks and M. Gilchrist, *J. Am. Chem. Soc.* **87,** 3199 (1965).
[22] P. D. Boyer, M. Deluga, K. E. Ebner, D. E. Hultquist, and J. B. Peter, *J. Biol. Chem.* **237,** PC3306 (1962).
[23] E. W. Miles, this series, Vol. 47, p. 431.
[24] G. DiSabato and W. P. Jencks, *J. Am. Chem. Soc.* **83,** 4393 (1961).

phosphoamino acids even though the detailed shape of the U varies from one acyl phosphate to another.[25] These phosphate anhydrides are labile to hydroxylamine and pyridine,[24] and are destroyed by reductive cleavage.[26]

Thioesters of cysteine are cleaved in weak acid but are stable at very high or very low pH.[27,28] The hydrolytic behavior of these molecules is therefore expressed as an inverted U-shaped curve.[29] The phosphorylation reaction is blocked in the presence of heavy metals and chemical reagents that affect sulfhydryl groups.[30] The phosphate linkage is hydrolyzed by low concentrations of iodine (less than 2 mM)[29] and mercuric salts (5 mM $HgCl_2$ at pH 7.0),[31] but is relatively resistant to hydroxylamine or pyridine treatment.[27]

Techniques for Determination of Phosphoamino Acids

Phosphoamino acids may be characterized by various techniques including phosphorus-31 nuclear magnetic resonance (^{31}P NMR), conventional ion-exchange column chromatography, high-performance liquid chromatography (HPLC), and thin-layer or paper electrophoresis and/or chromatography. Also, immunoreactivity with monoclonal or polyclonal anti-phosphoamino acid antibodies affords an efficient means of investigation, as emphasized in a number of chapters in this volume.

^{31}P NMR presents several important practical advantages.[5,32] First, it is a nondestructive spectroscopic method which allows the recovery of intact samples after analysis. Therefore it eliminates the possible phosphoryl transfer reactions during the harsh conditions of either low or high pH often employed in the characterization and isolation of the phosphoamino acid moiety, and it does not require any nonphysiological or radioactive probes. In addition, this technique can provide a range of information about not only the nature of the residues phosphorylated in a protein, but also the interactions of the phosphoryl groups with other groups on this protein and the overall mobility of each phosphoryl group. On the other hand, caution must, however, be exercised in interpreting and quantitating NMR data because such factors as metal ion chelation, steric strain, and chemical anisotropy may affect the shape and position

[25] R. S. Anthony and L. B. Spector, *J. Biol. Chem.* **247**, 2120 (1972).
[26] C. Degani and P. D. Boyer, *J. Biol. Chem.* **248**, 8222 (1973).
[27] V. Pigiet and R. R. Conley, *J. Biol. Chem.* **253**, 1910 (1978).
[28] H. H. Pas and G. T. Robillard, *Biochemistry* **27**, 5835 (1988).
[29] S. Akerfeldt, *Acta Chem. Scand.* **14**, 1980 (1960).
[30] H. Neumann and R. A. Smith, *Arch. Biochem. Biophys.* **122**, 354 (1967).
[31] S. Akerfeldt, *Acta Chem. Scand.* **13**, 1479 (1959).
[32] M. Brauer and B. D. Sykes, this series, Vol. 107, p. 36.

of phosphorus resonances.[5] Moreover, a major disadvantage of ^{31}P NMR lies in its relative insensitivity, which requires high concentrations of phosphorylated sample (in the millimolar range) for adequate data acquisition within a reasonable period of time.

Aside from ^{31}P NMR, the other techniques utilized for determining the phosphoamino acids of a protein generally necessitate the hydrolysis of this protein.[6] The release of individual phosphoamino acids can be achieved by partial acid, enzymatic, or alkaline hydrolysis. Enzymatic degradation of proteins rarely goes to completion, probably because the peptide bonds adjacent to phosphoamino acids are poor substrates.[10] More frequently the protein sample is treated in either hot acid or hot base, depending on the chemical stability of the phosphoamino acids analyzed, as indicated above. Protein is usually obtained from a column eluate or extracted from a polyacrylamide gel fragment. In the latter case the losses incurred during sample preparation may be relatively important. To reduce them, a new technique has been developed which consists of transferring the protein to a membrane of poly(vinylidene difluoride) (Immobilon; Millipore, Bedford, MA) and subjecting it to hydrolysis while still bound to the membrane.[33]

In the following, we will only concern ourselves with the analysis of phosphohydroxyamino acids by thin-layer chromatography and/or electrophoresis.

Separation of O-Phosphates by One-Dimensional Thin-Layer Systems

When phosphate, phosphopeptides, and phosphoamino acids are the only compounds present, satisfactory separation can be obtained by using one-dimensional thin-layer electrophoresis. This applies principally to protein hydrolysates prepared from *in vitro* kinase reactions where radioactive ATP (or GTP) is used as phosphoryl donor. At pH 3.5, P-Ser, P-Thr, and P-Tyr can be separated, in that order, in a buffer containing acetic acid–pyridine–H$_2$O in the proportions $50:5:945$ (v/v/v).[34] Under these conditions, P-Ser and P-Thr migrate, close to each other, significantly faster than P-Tyr, probably due to their lower molecular mass and the absence of a hydrophobic benzene ring in their structure. By contrast, at pH 1.9, in a buffer made of 88% formic acid–acetic acid–H$_2$O $(50:156:1794,$ v/v/v) P-Tyr cannot be separated from P-Thr, while at the same pH P-Thr and P-Ser migrate quite distinctly from each other.[35]

[33] M. P. Kamps and B. M. Sefton, *Anal. Biochem.* **176,** 22 (1989).
[34] T. Hunter and B. M. Sefton, *Proc. Natl. Acad. Sci. U.S.A.* **77,** 1311 (1980).
[35] W. Eckhart, M. A. Hutchinson, and T. Hunter, *Cell (Cambridge, Mass.)* **18,** 925 (1979).

The three phosphohydroxyamino acids can be separated also by thin-layer chromatography on cellulose plates in an ascending solvent containing isobutyric acid–0.5 M NH$_4$OH (5 : 3, v/v).[10] An adequate separation can be achieved as well by using a solvent made of propionic acid–1 M NH$_4$OH–2-propanol (45 : 17.5 : 17.5, v/v/v).[36] In either case, P-Tyr migrates ahead, followed by P-Thr then P-Ser.

Separation of O-Phosphates by Two-Dimensional Thin-Layer Systems

Principles

The analysis of protein hydrolysates can be rendered erroneous by the possible presence in precipitates of contaminating phosphorylated compounds, namely when proteins are extracted from cells radioactively labeled *in vivo* with ortho[^{32}P]phosphate. Before hydrolyzing proteins, contaminating nucleic acids and polyphosphates can be removed by heating in 10% trichloroacetic acid at 90° for 15–20 min,[9] although this treatment may result in a limited loss of protein.[37] Phospholipids can be removed by extracting the acid precipitates sequentially with various solvents:[38] once with acetone, ethanol, and chloroform; twice with ethanol–ethyl ether (3 : 1, v/v), and then once with ethyl ether. Proteins precipitated with 10% trichloroacetic acid may also contain MgATP complexes. These can be eliminated by dissolving the precipitates in cold 0.1 M NaOH, then reprecipitating proteins with acid.[39]

Despite such preliminary treatments, some contaminating molecules may still be present in protein hydrolysates and comigrate with phospho-amino acids in one-dimensional systems. This concerns, in particular, nucleoside monophosphates arising from RNA degradation and ribose phosphates generated by depurination of nucleotides during acid hydrolysis.[40] Two-dimensional analytical systems are then required.

Reagents

O-Phospho-DL-serine (Sigma, St. Louis, MO)
O-Phospho-DL-threonine (Sigma)
O-Phospho-DL-tyrosine (Sigma)
5′-Adenosine monophosphate disodium salt (Serva, Westbury, NY)
5′-Cytidine monophosphate disodium salt (Serva)

[36] E. Neufeld, H. J. Goren, and D. Boland, *Anal. Biochem.* **177,** 138 (1989).
[37] D. Kennell, this series, Vol. 12, p. 636.
[38] J. N. Davidson, S. C. Fraser, and W. C. Hutchinson, *Biochem. J.* **49,** 311 (1951).
[39] P. J. Greenaway, *Biochem. Biophys. Res. Commun.* **47,** 639 (1972).
[40] M. Manaï and A. J. Cozzone, *Anal. Biochem.* **124,** 12 (1982).

5'-Guanosine monophosphate disodium salt (Serva)
5'-Thymidine monophosphate disodium salt (Serva)
5'-Uridine monophosphate disodium salt (Serva)
3'-Adenosine monophosphoric acid (Sigma)
3'-Cytidine monophosphoric acid (Sigma)
3'-Uridine monophosphoric acid (Boehringer, Mannheim, Germany)
3'-Guanosine monophosphate disodium salt (Boehringer)
Ortho[^{32}P]phosphoric acid, carrier free
[^{14}C]Uridine 5'-monophosphate
5'-Ribose phosphate disodium salt (Sigma)
3'-Ribose[^{32}P]phosphate

Solvents and Buffers

Solvent A: 5 vol isobutyric acid–3 vol 0.5 M NH$_4$OH
Solvent B: 7 vol 2-propanol–1.5 vol HCl–1.5 vol H$_2$O
Buffer at pH 1.9: 7.8% acetic acid–2.5% formic acid
Buffer at pH 3.5: 5% acetic acid–0.5% pyridine

Conditions of Migration

Ascending chromatography is carried out at room temperature in solvent A for 10–12 hr or in solvent B for 7–9 hr. Electrophoresis is performed at pH 1.9 for 50–60 min at 1600 V or at pH 3.5 for 40–50 min at 1100 V under cooling. Thin-layer cellulose plates (20 × 20 cm; 0.1-mm layer thickness) from two different commercial sources are used: either precoated plastic sheets (Polygram CEL 300; Macherey-Nagel Co., Düren, Germany), or TLC plastic sheet cellulose without fluorescent indicator from Merck Company (Darmstadt, Germany). The average amount of each unlabeled nucleotide analyzed is 4–10 nmol, and that of each phosphoamino acid is 10–15 nmol. For radioactive compounds the amount used corresponds to approximately 3000 cpm each.

Detection of Phosphorylated Molecules

Phosphoamino acids are visualized by staining with a solution of ninhydrin prepared by mixing 1 vol of 0.33% ninhydrin in *tert*-butanol with 1 vol of acetic acid–pyridine–H$_2$O (1 : 5 : 5, v/v/v). Unlabeled nucleoside monophosphates are detected by absorbance under shortwave ultraviolet light and radioactive molecules are revealed by autoradiography. Ribose monophosphates are detected by silver staining:[41] plates are treated first with a 1% solution of sodium metaperiodate in 50% aqueous acetone, then

---

[41] T. Yamada, M. Hisamatsu, and M. Taki, *J. Chromatogr.* **103**, 390 (1975).

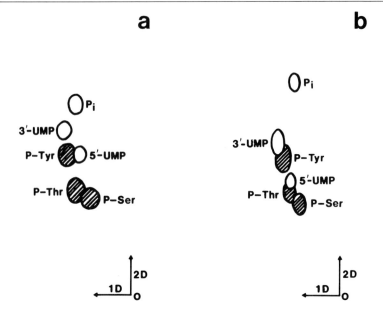

FIG. 1. Separation of phosphoamino acids and uridine monophosphate by double chromatography. Samples were chromatographed in solvent A in the first dimension (1D) and in solvent B in the second dimension (2D), then revealed by staining or autoradiography. The origin (O) is on the right, at the intersection of the two arrows. Analysis was performed by using cellulose plates from either Macherey-Nagel (a) or Merck (b). Shaded spots indicate the position of phosphoamino acids. Average data from three to five experiments are presented.

with a silver nitrate solution prepared by pouring 1 ml of saturated aqueous silver nitrate solution into 100 ml of 95% aqueous acetone, and finally with 1% potassium hydroxide.

Double Chromatography

In this system,[42] samples are chromatographed in solvent A in the first dimension, and in solvent B in the second dimension. Figure 1 shows that, in these conditions, the phosphoamino acids are rather well separated from one another even though P-Ser and P-Thr partially overlap. However, the separation of the phosphoamino acids from the nucleotide UMP, phosphorylated in either the 3′ or the 5′ position, is insufficient. Indeed, on the Macherey-Nagel plates (Fig. 1a), 5′-UMP contaminates P-Tyr and, on the Merck plates (Fig. 1b), 3′-UMP and 5′-UMP are not completely separated

[42] S. Nishimura, *Prog. Nucleic Acid Res. Mol. Biol.* **12,** 49 (1972).

from P-Tyr and P-Thr, respectively. It should be noted that the migration of 3'-UMP and 5'-UMP relative to that of phosphoamino acids varies with the commercial source of cellulose plates. In all cases, inorganic phosphate comigrates with several other compounds in the first dimension, but is individualized after migration in the second dimension.

Electrophoresis and Chromatography

Electrophoresis is carried out at pH 1.9 in the first dimension followed by ascending chromatography in solvent A in the second dimension.[34] The scheme of Fig. 2 shows that, after electrophoresis, 3'-UMP and 5'-UMP are well separated from the phosphoamino acids, and that P-Ser has a significantly higher mobility than P-Thr and P-Tyr. But the separation of the latter two phosphoamino acids is achieved only after chromatography in the second dimension. The nucleotides 3'-UMP and 5'-UMP comigrate during electrophoresis and hardly separate from each other after chromatography, as already observed in the previous two-dimensional system. Inorganic phosphate practically coincides with 5'-UMP.

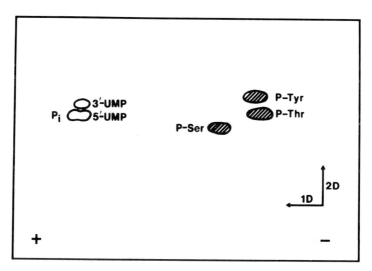

Fig. 2. Separation of phosphoamino acids and uridine monophosphate by electrophoresis and chromatography. Samples were subjected to electrophoresis at pH 1.9 in the first dimension and to ascending chromatography in solvent A in the second dimension. Analysis was performed by using cellulose plates from Merck. Other conditions are as described in Fig. 1. Average data from three to five experiments are presented.

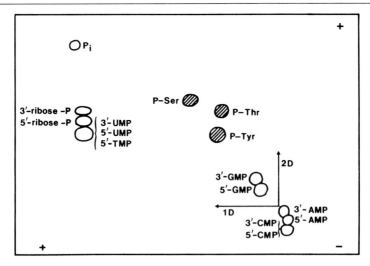

Fig. 3. Separation of phosphoamino acids and nucleoside monophosphates by double electrophoresis. Samples were subjected to electrophoresis at pH 1.9 in the first dimension and to electrophoresis at pH 3.5 in the second dimension. Analysis was performed by using cellulose plates from Merck. For other conditions see Fig. 1. Average data from five to eight experiments are presented.

Double Electrophoresis

Electrophoresis is performed at pH 1.9 in the first dimension and at pH 3.5 in the second dimension.[43,44] In principle, it would be possible to perform the two successive migrations in the other order, but it is better to do the pH 1.9 dimension first because the corresponding buffer has a greater capacity.[10]

The results summarized in Fig. 3 show that, at pH 1.9, the nine nucleotides studied separate well from the three phosphoamino acids:[40] TMP comigrates with UMP relatively far from the origin, GMP migrates much less than P-Thr and P-Tyr, and AMP and CMP migrate in the opposite direction, i.e., toward the cathode. At this pH value, P-Ser separates from both P-Thr and P-Tyr, which remain unresolved, as mentioned above.

During electrophoresis at pH 3.5, P-Thr separates from P-Thr, GMP migrates slightly toward the anode, while AMP and CMP migrate toward the cathode with a higher mobility for CMP. In addition, inorganic phos-

[43] B. M. Sefton, T. Hunter, K. Beemon, and W. Eckhart, *Cell (Cambridge, Mass.)* **20**, 807 (1980).
[44] M. L. Tripp, R. Pinon, J. Meisenhelder, and T. Hunter, *Proc. Natl. Acad. Sci. U.S.A.* **83**, 5973 (1986).

phate migrates much further than UMP and TMP, which still overlap each other. When comparing the electrophoretic behavior of the 3'-nucleotides to that of the corresponding 5'-nucleotides, no difference is observed for the pyrimidine derivatives (3'-CMP vs 5'-CMP, and 3'-UMP vs 5'-UMP). By contrast, in the case of purines, 3'-GMP has a higher mobility than 5'-GMP whereas 3'-AMP migrates more slowly than 5'-AMP.

In the conditions of protein hydrolysis (6 N HCl at 110° for 2 hr) used routinely for preparing phosphoamino acids, the only nucleotides arising from RNA degradation that remain stable are 3'-UMP and 3'-CMP. The purine nucleotides 3'-AMP and 3'-GMP are then depurinated and produce 3'-ribose monophosphate. In the double-electrophoresis system, the latter compound, as well as 5'-ribose monophosphate, migrates distinctly from the phosphoamino acids, close to UMP and TMP.

In conclusion, this system allows (1) complete separation of P-Ser, P-Thr, and P-Tyr from one another, (2) complete separation of these three phosphoamino acids from the pyrimidic and puric nucleoside monophosphates, and from ribose monophosphates, (3) complete separation of inorganic phosphate from all phosphoamino acids and nucleotides. Moreover it has proved suitable for separating a number of phosphopeptides, as well as some unidentified phosphorylated compounds which are produced on hydrolysis of both prokaryotic[40] and eukaryotic[10] cell proteins.

[3] Determination of Phosphoamino Acid Composition by Acid Hydrolysis of Protein Blotted to Immobilon

By Mark P. Kamps

Introduction

The identification of phosphorylated amino acids in a protein labeled biosynthetically with $[^{32}P]P_i$ is accomplished by chemical or enzymatic hydrolysis of amide bonds and resolution of the liberated phosphoamino acids by chromatographic techniques.[1] Theoretically, complete enzymatic hydrolysis represents the best approach for obtaining accurate determinations of the relative abundance and identity of phosphoamino acids within a protein, because it avoids the problem of hydrolysis of the phosphoester bond, which occurs during acid-catalyzed hydrolysis. The rates of acid-catalyzed hydrolysis of both peptide and phosphoester bonds are affected

[1] T. Hunter and B. Sefton, *Proc. Natl. Acad. Sci. U.S.A.* **77**, 1311 (1978).

by the nature of adjacent bonds and, consequently, the relative proportions of free phosphoamino acids after a partial hydrolysis are not necessarily an accurate reflection of their native abundance in the original protein. Because adjacent bonds are never completely resistant to hydrolysis by 5.7 N HCl at 110° for 1 hr, phosphoproteins will always release phosphoserine, phosphothreonine, or phosphotyrosine during acid hydrolysis. Consequently, if the purpose of phosphoamino acid analysis is simply to determine whether serine, threonine, or tyrosine is phosphorylated, then acid hydrolysis represents a fast, simple, and reproducible technique to achieve this goal.

Historically, the preparation of a protein for phosphoamino acid analysis required excision of the ^{32}P-containing phosphoprotein from a polyacrylamide gel, emulsification of the gel slice in buffer, extraction of the protein from the polyacrylamide slurry, separation of the protein from contaminating polyacrylamide, and precipitation of the protein with trichloroacetic acid (TCA).[2] This technique is time consuming, and cannot readily be used to analyze alkali-resistant phosphoamino acids from proteins contained in polyacrylamide gels that have been subjected to treatment with 1.0 N KOH. Partial alkaline hydrolysis of ^{32}P-labeled phosphoproteins in polyacrylamide gels is a useful method for enriching the relative abundance of radioactivity in phosphotyrosine.[3] However, after hydrolysis in the polyacrylamide gel, the protein has been cleaved into peptides that precipitate inefficiently with TCA. Large amounts of hydrolyzed acrylamide are also extracted along with the peptides. This soluble acrylamide precipitates with TCA and interferes with subsequent analysis of phosphoamino acids by electrophoresis.[3] A method has been devised for recovering phosphoamino acids after hydrolysis of the whole alkali-treated gel piece using ion-exchange chromatography, but this procedure is labor intensive.[2] In an effort to avoid the time-consuming steps of direct acid hydrolysis and the technical problems that interfere with the identification of phosphoamino acids in proteins from alkali-treated polyacrylamide gels, a simplified procedure for phosphoamino acid analysis has been developed, which takes advantage of the acid- and base-resistant properties of Immobilon (Millipore, Bedford, MA).[4] Immobilon is composed of poly-(vinylidene difluoride) and binds denatured proteins transferred electrophoretically from polyacrylamide gels. Immobilon exhibited the useful properties of releasing free phosphoamino acids during partial acid hydrolysis with 5.7 N HCl, retaining bound proteins during partial alkaline

[2] J. A. Cooper, B. M. Sefton, and T. Hunter, this series, Vol. 99, p. 387.
[3] J. A. Cooper and T. Hunter, *Mol. Cell. Biol.* **1**, 165 (1981).
[4] M. P. Kamps and B. M. Sefton, *Anal. Biochem.* **176**, 22 (1989).

hydrolysis with 1.0 N KOH, and maintaining structural integrity during sequential hydrolysis in 1.0 N KOH and 5.7 N HCl.

Acid Hydrolysis of Proteins Transferred to Immobilon

Transfer of Gel-Fractionated Proteins to Immobilon

Wet the Immobilon membrane in methanol for 10 min, and then soak it in transfer buffer [193 mM glycine, 25 mM Tris base, 0.1% sodium dodecyl sulfate (SDS) (w/v), 20% methanol] for an additional 30 min. Soak polyacrylamide gels (1.0 mm thick, 10% in acrylamide) containing proteins to be transferred to Immobilon in transfer buffer for 30 min. Transfer proteins electrophoretically from the gel to Immobilon at 50 V for 3 hr.[5]

Staining Immobilon with India Ink

If desired, proteins bound to Immobilon can be visualized by staining with India ink. The staining procedure does not interfere with the subsequent steps of either acid or base hydrolysis. The filter must be wetted before it is stained. Consequently, if the filter has dried after completing the transfer, it should be rewet for 1 min in methanol, then soaked in TN (50 mM Tris, pH 6.5, 150 mM NaCl) for 5 min with continuous agitation. If the filter is to be stained immediately after transfer, it can be placed directly into the solution containing India ink. Stain by incubating the wetted filter for 1.5 hr in TN containing 0.2% Tween 20 (polyoxyethylene sorbitan monolaurate, Sigma, St. Louis, MO) and 1 μl/ml Pelican brand India ink. Rinse the filter in 1 liter of TN for 5 min to remove the unbound ink. Seven percent of the proteins bound to Immobilon is released during this procedure. A 16.5-hr incubation in this staining solution releases 13.3% of the protein, and increasing the pH to 8.5 during the 16.5-hr incubation results in a loss of 25% of the protein. Consequently, both the shorter staining time (1.5 hr) and the slightly acidic pH (6.5) are important factors. Although eliminating Tween 20 from the staining solution prevents loss of protein from the filter, it also allows the India ink to bind nonspecifically to Immobilon. Tween 20 is therefore essential, and the 7% loss of protein must be accepted.

Acid Hydrolysis of Protein Bound to Immobilon

Before hydrolysis, it is important to remove from the filter the residual salt and detergent that remain from the staining or transfer procedure. Therefore, the wetted membrane should be soaked three times for 2 min

[5] H. Towbin, T. Staehelin, and J. Gordon, *Proc. Natl. Acad. Sci. U.S.A.* **76,** 4350 (1979).

each in 1 liter of water with continuous agitation. The rinsed membrane is then dried, marked with radioactive ink, and subjected to autoradiography. Using the exposed film as a template, cut out a narrow strip of Immobilon containing the ^{32}P-labeled phosphoprotein of interest. Rewet the strip by incubation for 0.5 min in methanol and 0.5 min in water. Remove the excess water by holding the strip vertically and touching its base to absorbent paper. Place the strip in a screwcap Eppendorf tube containing 200 μl; of 5.7 N HCl, making sure it is completely immersed in the acid. If the Immobilon strip is too long, cut it into smaller segments. Purge the tubes with nitrogen, secure their caps, and incubate them at 110° for 1 hr. After hydrolysis, spin the tubes at 10,000 to 14,000 rpm in a microfuge for 1 min. This sediments what appear to be carbon particles from the India ink. A majority of the India ink remains bound to the Immobilon. Taking care to avoid the small pellet at the bottom of the tube, transfer the supernatant to a new tube, and lyophilize in a Speed Vac concentrator (Savant, Hicksville, NY). Separate phosphoamino acids by electrophoresis at pH 1.9 and 3.5 on cellulose thin-layer plates.[1,2]

Evaluation

A comparison of the abundance of phosphoserine and phosphotyrosine in six cellular proteins has been made using either the procedure outlined above or that of extracting the proteins from polyacrylamide gels and performing an acid hydrolysis on the precipitated proteins.[4] Both techniques yielded the same relative abundance of phosphoamino acids, indicating that Immobilon does not preferentially bind some phosphoamino acids. Consequently, the qualitative results of phosphoprotein hydrolysis on Immobilon are consistent with previous determinations contained in the literature.

The quantitative recovery of total ^{32}P-labeled protein is also approximately the same for both techniques. For the conditions described above, the efficiency of electrophoretic transfer of total cellular phosphoproteins from the gel to Immobilon is 70%, and 88 to 96% of the radioactivity contained in the same six cellular proteins was released in the form of phosphopeptides, phosphoamino acids, and inorganic phosphate by acid hydrolysis on Immobilon. Therefore, the final recovery was approximately 61 to 67%. When extracted from the gel and precipitated with TCA, the recoveries of these same proteins were 54 to 78%.

Approximately one in eight proteins either did not bind efficiently to Immobilon or were not transferred efficiently from the polyacrylamide gel; consequently, when analyzing a single protein of interest, it is important to demonstrate that it undergoes successful transfer to Immobilon. In summary, the substantial advantage of using Immobilon is that it reduces the time required to analyze multiple samples fourfold or greater,[4] and eliminates tedium encountered in the use of glass homogenizers and in the extraction and elution steps required to purify proteins from polyacrylamide gels.

Base Hydrolysis of Proteins Transferred to Immobilon, and Subsequent
    Analysis of Phosphoamino Acids by Acid Hydrolysis

Hydrolysis of gel-fractionated phosphoproteins with 1.0 N KOH has been used to enrich selectively for phosphotyrosine, by reducing the content of phosphoserine, and, to a lesser extent, phosphothreonine.[3] The hydroxide anion catalyzes a substitution reaction that hydrolyzes the phosphoester bond in all three phosphoamino acids; however, it can also abstract a hydrogen atom (in the form of a proton) from the carbon atom at the β position (two carbons away) relative to the phosphate group of both phosphoserine and phosphothreonine, which catalyzes the formation of a double bond between the α- and β-carbon atoms and eliminates the phosphate group. Because it is part of an aromatic ring in phosphotyrosine, the hydrogen on the carbon atom in the β position to the phosphate is not acidic, and therefore it does not participate in this reaction. Consequently, phosphate in phosphotyrosine is much more stable to base hydrolysis than is that in phosphoserine or phosphothreonine. The analysis of alkali-stable phosphoamino acids in protein is complicated by two major factors. First, some sites of tyrosine phosphorylation occur between alkali-sensitive peptide bonds. After incubation in 1.0 N KOH, 100% of these phosphate-containing peptides diffuse from the polyacrylamide gel and, consequently, although the protein contains phosphotyrosine, it appears to contain none. An example of a protein that exhibits this behavior is p50, a cellular protein that is phosphorylated on tyrosine and binds to p60^{v-src} in chicken embryo fibroblasts, transformed by Rous sarcoma virus.[1] Second, after treatment of a polyacrylamide gel with 1.0 N KOH, analysis of the remaining phosphoamino acids is extremely difficult because much of the acrylamide has been hydrolyzed, and the protein, now reduced to shorter peptides, is difficult to precipitate with TCA.[3] When base hydrolysis is performed on proteins bound to Immobilon, both of these difficulties are substantially eliminated. Immobilon retains 90% of transferred proteins after a 2-hr hydrolysis in 1.0 N KOH at 55°, and during subsequent treatment with 5.7 N HCl it does not release hydrolysis products in quantities sufficient to interfere, substantially, with phosphoamino acid analysis.

Base Hydrolysis of Proteins Bound to Immobilon.

Either stained or unstained membranes can be used in this procedure. The alkaline treatment described below does not alter the pattern of India ink-stained proteins that are bound to Immobilon.[4] Wet the dried Immobilon filter for 1 min in methanol, transfer to TN for 5 min, and agitate continuously. Transfer the filter to 1.0 liter of 1.0 N KOH. Incubate at 55° for 2 hr. The filter turns brown during this procedure. Neutralize the filter by rinsing once in 500 ml TN for 5 min, once in 500 ml 1.0 M Tris, pH 7.0 for 5 min, and twice in 500 ml water for 5 min. The filter lightens during neutralization. Dry the filter under a heat lamp for 5 min, wrap in Saran, and expose to film.

Acid Hydrolysis of Immobilon That Was Treated Previously with 1.0 N KOH

After autoradiography, excise a strip of Immobilon containing protein of interest. Cut out as small a piece of Immobilon as possible. Proceed with the standard acid hydrolysis protocol outlined above. The combined procedures of alkaline and acid hydrolysis cause some of the Immobilon or some component of the Immobilon to become solubilized. If one analyzes a single sample equivalent to the total material derived from 20 mm^2 of Immobilon, the phosphoamino acid spots resolve well, but are more diffuse than those from smaller samples. Consequently, analysis of a portion of the sample derived from 5 mm^2, or less, is optimal.

Evaluation

Ninety percent of total protein is retained by Immobilon during the conditions used in base hydrolysis, whereas approximately 45% is retained by a polyacrylamide gel subjected to these same conditions.[4] The losses incurred by specific proteins also vary substantially during alkali treatment of phosphoproteins in polyacrylamide gels.[3,6] Quantitatively, therefore, Immobilon is superior in its retention of total protein. Qualitatively, the phosphate label in specific proteins was affected identically by base hydrolysis in acrylamide gels or on Immobilon.[4] Alkali treatment changed the distribution of total phosphate in proteins from BALB/c 3T3 cells transformed by p60^{v-src} and incubated in sodium orthovanadate from 90% phosphoserine, 4% phosphothreonine, and 6% phosphotyrosine to 45.5% phosphoserine, 11.5% phosphothreonine, and 43% phosphotyrosine.[4] The autoradiographic images after alkaline treatment of ^{32}P-containing proteins

[6] C. Bourassa, A. Chapdelaine, K. D. Roberts, and S. Chevalier, *Anal. Biochem.* **169**, 356 (1988).

on Immobilon were also much sharper than those images of identical proteins that were treated with base in a polyacrylamide gel.[4] The distinct India ink-staining pattern is also retained by Immobilon, whereas a majority of the Coomassie Blue stain is lost from a protein gel as a consequence of the standard base hydrolysis treatment. Whereas treatment of an acrylamide gel with base causes it to expand approximately 30%, the dimensions of Immobilon are not affected. This property facilitates the correlation of specific proteins before and after base hydrolysis. Qualitatively, therefore, alkaline hydrolysis of phosphoproteins on Immobilon is superior to that of hydrolysis within polyacrylamide gels. Finally, in subjecting base-treated Immobilon to acid hydrolysis, 92–93% of ^{32}P in phosphoproteins is released, and this material can be analyzed for phosphoamino acid content.[4]

Perspective

The usefulness of acid or base hydrolysis of proteins bound to Immobilon hinges on their ability to transfer efficiently from polyacrylamide gels and bind efficiently to Immobilon. Although approximately 85% of cellular phosphoproteins transfer with an efficiency of 70%, some proteins transfer at much reduced efficiencies. Consequently most, but not all, proteins are suitable for analysis using these techniques. A number of similar techniques have been described. Proteins can be transferred to nylon membranes and subjected to hydrolysis in 1.0 N NaOH.[7] However, unlike Immobilon, the nylon membrane dissolves in 5.7 N HCl. Consequently, it is not a suitable matrix for subsequent phosphoamino acid analysis. In preparation for phosphoamino acid analysis, proteins have also been transferred to and eluted from nitrocellulose,[8] but since nitrocellulose dissolves in both 5.7 N HCl and in 1.0 N KOH, it is an unsuitable membrane on which to perform phosphoprotein hydrolysis. The properties of Immobilon were previously shown to permit the purification of proteins by electrophoretic transfer from polyacrylamide gels and subsequent elution with SDS and Triton X-100.[9] Therefore, Immobilon is an ideal membrane on which to perform either the purification of phosphoproteins or the analysis of their phosphoamino acid content by acid or base hydrolysis.

[7] L. Contor, F. Lamy, and R. E. Lecocy, *Anal. Biochem.* **160,** 414 (1987).
[8] B. S. Parekh, H. B. Mehta, M. D. West, and R. C. Montelaro, *Anal. Biochem.* **148,** 87 (1985).
[9] B. Szewczyk and D. F. Summers, *Anal. Biochem.* **168,** 48 (1988).

[4] Comparison of Three Methods for Detecting Tyrosine-Phosphorylated Proteins

By LYNN M. KOZMA, ANTHONY J. ROSSOMANDO, and
MICHAEL J. WEBER

The three methods generally used for detection of phosphotyrosine-containing proteins are (1) direct phosphoamino acid analysis following acid hydrolysis,[1] (2) partial alkali hydrolysis, which preferentially removes phosphate from serine and threonine, thus enriching for phosphotyrosine,[1] and (3) Western immunoblotting with anti-phosphotyrosine antibodies.[2,3] In the course of our investigation of tyrosine-phosphorylated glycoproteins in *src*-transformed cells[4] we observed dramatically different apparent patterns of protein tyrosyl phosphorylation depending on which analytical methodology was employed. Each technique revealed a different protein as being the predominant tyrosine-phosphorylated protein. In addition, a number of low-molecular-weight tyrosine-phosphorylated proteins were detected poorly or not at all by the immunoblotting procedure. Although it is widely appreciated that alkali hydrolysis provides only a first approximation of tyrosine phosphorylation, since alkali-labile tyrosine phosphorylations and alkali-stable serine and threonine phosphorylations have been reported,[1,5] the large discrepancy between the results obtained using chemical versus immunological methodologies was more surprising. What follows is a summary of our comparison of the three methodologies and a brief discussion of the implications and possible causes for the observed differences.

Methodology

General Procedures

Antibodies against phosphotyrosine are prepared essentially as described by Kamps and Sefton (Ref. 3 and [10] in this volume). Phosphoamino acid analysis is as described by Cooper *et al.*[1] as modified by Kamps and Sefton,[3] except that separations of phosphoamino acids were

[1] J. A. Cooper, B. M. Sefton, and T. Hunter, this series, Vol. 99, p. 387.
[2] A. R. Frackelton, A. H. Ross, and H. N. Eisen, *Mol. Cell. Biol.* **3**, 1343 (1983).
[3] M. P. Kamps and B. M. Sefton, *Oncogene* **2**, 305 (1988).
[4] L. M. Kozma, A. Reynolds, and M. J. Weber, *Mol. Cell. Biol.* **10**, 837 (1990).
[5] J. A. Cooper and T. Hunter, *Mol. Cell. Biol.* **1**, 165 (1981).

performed by two-dimensional paper electrophoresis. Alkali hydrolysis is as described by Kamps and Sefton.[6]

Cell Culture and Mitogenic Stimulation

Growth and treatment of chicken embryo fibroblast cell cultures is as previously described.[4] Normal Swiss mouse 3T3 fibroblasts are grown to confluence (1×10^7 cells) in 100-mm culture dishes and confluent Swiss 3T3 cells expressing the temperature-sensitive *src* mutant LA29 are grown at the permissive temperature for at least 24 hr. Prior to mitogenic stimulation two plates of normal 3T3 cells are washed several times at room temperature with HEPES-buffered saline (5 ml/plate) and then incubated for 1.5 hr in the same buffer at 37°. After this incubation, one plate is incubated for an additional 10 min in the presence of 200 ng/ml epidermal growth factor (EGF). Following stimulation, the cells from both plates are lysed with hot Laemmli electrophoresis sample buffer containing 10 mM dithiothreitol (DTT), passed through a 22-gauge needle several times, and stored at $-70°$ until needed. 3T3 cells transformed by *src* are washed several times at room temperature (5 ml) with HEPES-buffered saline, lysed immediately in the same manner as the normal fibroblasts, and stored at $-70°$.

In Vivo Labeling of Cells

For labeling cells with ortho[^{32}P]phosphate, subconfluent cells are incubated with 3 mCi/ml ortho[^{32}P]phosphate (Du Pont, NEN, Boston, MA) in phosphate-free Dulbecco's modified Eagle's medium (DMEM) containing 10% conditioned medium for 8–12 hr. This is sufficient for equilibrium of inorganic phosphate with the ATP pool, which is essential when comparisons are made between immunological and radioisotopic methods for detecting phosphotyrosine. After the labeling period, cells are lysed as discussed below.

For [^{35}S]methionine labeling, subconfluent cells are labeled with DMEM containing 10% of the normal amount of methionine, 200 μCi/ml [^{35}S]methionine (Du Pont, NEN), and 10% conditioned medium for 12 hr at 35°. Cells are lysed in RIPA buffer [50 mM Tris-HCl, pH 7.2 containing 150 mM NaCl, 0.1% (w/v) sodium dodecyl sulfate (SDS), 1% (v/v) Nonidet P-40 (NP-40), and 1% (v/v) sodium deoxycholate], spun for 30 min at 100,000 g, and the proteins in the supernatant separated by SDS–polyacrylamide gel electrophoresis.

[6] M. P. Kamps and B. M. Sefton, *Anal. Biochem.* **176,** 22 (1989).

Anti-Phosphotyrosine Immunoblotting

One hundred microliters of each cell lysate (1×10^6 cells) is electrophoresed on a 10% polyacrylamide gel, transferred to nitrocellulose or Immobilon (Millipore, Bedford, MA), and immunoblotted with the anti-phosphotyrosine antisera. The filters are blocked at 37° for 2.0 hr in blocking solution [3% bovine serum albumin (BSA), 50 mM Tris-HCl (pH 7.5), 150 mM NaCl, and 0.05% Tween 20]. Following this incubation, the blocking solution is removed and 15 ml of fresh solution containing a 1 : 600 final antibody dilution is added. This is then incubated for an additional 1.5 hr at 37° with gentle rocking. Each of the filters is then washed several times for 5–10 min with 100 ml of blocking solution at the same temperature with vigorous agitation. Antibody detection is accomplished using [^{125}I]protein A (Amersham, Arlington Heights, IL) in 15 ml of the same solution at a concentration of 1 μCi/ml, again with gentle rocking and incubation at 37°. After a 1-hr incubation, the filters are washed several times as before at room temperature in 100 ml of blocking solution and a final wash is done using 100 ml of Tris-buffered saline solution (50 mM Tris-HCl, pH 7.5, 150 mM NaCl). Each filter is air dried and exposed to film overnight at room temperature.

When immunoblotting is performed in the presence of NP-40, the following procedure is utilized: All samples and transfers to nitrocellulose are identical to the ones above. Each filter is blocked in the same blocking solution (3% BSA, 50 mM Tris, pH 7.5, 150 mM NaCl), except no detergents are added during the blocking, and the incubation is done overnight at room temperature with gentle rocking. Detergents (0.1% NP-40, 0.05% Tween 20; v/v) are added to all other solutions except the final wash with Tris-buffered saline. All other conditions are comparable except each incubation is at room temperature and the [^{125}I]protein A binding continues for 2 hr. After the final washes each filter is dried and exposed to film at −70°.

Wheat Germ Agglutinin Chromatography

Whichever methods are ultimately used to detect tyrosine phosphorylation, it is essential that samples be prepared in a manner which preserves their *in vivo* phosphorylation state. Direct lysis of cells in hot sodium dodecyl sulfate, which is effective at inactivating phosphatases and proteases, is not suitable when native proteins or subcellular fractionation is desired. We have found the following procedures and inhibitors satisfactory for isolation of glycoproteins, with retention of 85% of the original total phosphotyrosine content. Note, however, that protein tyrosyl phosphorylation may not be lost uniformly from all the proteins in a sample.

For each sample, two 100-mm tissue culture dishes of nearly confluent cells are rinsed with phosphate-buffered saline (PBS) and lysed in 1.0 ml/dish RIPA containing 0.2 mM phenylmethylsulfonylfluoride (freshly prepared), 5 mM benzamidine, 0.1 mg/ml leupeptin, 10 μM pepstatin, 1 mM phenanthroline, 1 mM ethylenediaminetetraacetic acid (EDTA), 1 : 100 aprotinin, 1 mM pyrophosphate, and 10 mM sodium vanadate. Lysis is performed on ice. Samples are centrifuged at 100,000 g for 30 min at 4°. Wheat germ agglutinin–agarose (Vector Laboratories, Burlingame, CA) is packed in polypropylene columns (Evergreen, Los Angeles, CA) at 0.4 ml packed volume and washed once with 5 ml water and once with 5 ml 50 mM HEPES, pH 7.6 containing 0.1% NP-40, 10 mM Mg$_2$SO$_4$, 100 mM NaCl, 0.02% NaN$_3$, and the inhibitors listed above (wash I). The sample supernatants are then loaded onto the columns after fivefold dilution in wash I and allowed to incubate on ice for 30 min to maximize binding. The columns are then centrifuged at low speed in a Sorvall (Newtown, CT) RT6000B tabletop centrifuge for 32 min. The columns are washed with 5 ml wash I, centrifuged as before, then washed with wash II [same as wash I, except 0.1% (v/v) β-octylglucoside replaces 0.1% NP-40] and centrifuged. The columns are then closed off and 0.4 ml elution buffer (50 mM Tris-HCl, pH 7.6 containing 0.1% β-octylglucoside, 10 mM Mg$_2$SO$_4$, 100 mM NaCl, 0.02% NaN$_3$, 0.3 M N-acetylglucosamine, and the inhibitors listed above) is added and allowed to incubate for 30 min on ice to allow for complete elution. The columns were then centrifuged for 1 min to collect the eluates. Protein determinations are performed using a Lowry assay as modified by Markwell *et al* [7] Equal amounts of protein are concentrated in Amicon-30 concentration units (Amicon, Danvers, MA) and 2× Laemmli electrophoresis sample buffer is added, yielding a total volume of 70 μl, which could be applied directly to a 7.5% polyacrylamide gel.

Immunoprecipitation with Anti-Phosphotyrosine Antibodies

Wheat germ agglutinin (WGA) eluates from ^{32}P-labeled Rous sarcoma virus (RSV)-transformed and normal cells are incubated 1 hr on ice with a 1 : 25 dilution of unfractionated anti-phosphotyrosine antiserum, then 30 min on ice with Pansorbin (Calbiochem, Los Angeles, CA). Pellets are washed with 1 ml 50 mM Tris-HCl, pH 7.6 containing 1% NP-40, 0.5% sodium deoxycholate, 1 mM EDTA, and 1 M NaCl (high salt HO buffer). Pellets are then washed twice with 1 ml HO buffer containing 150 mM NaCl and once with 1 ml PBS. Immune complexes are dissociated by boiling the samples for 5 min in electrophoresis sample buffer containing

---

[7] M. A. Markwell, S. M. Haas, L. L. Bieber, and N. E. Tolbert, *Anal. Biochem.* **87**, 206 (1978).

5% 2-mercaptoethanol, the proteins are electrophoretically separated on a 7.5% polyacrylamide SDS gel, and then transferred to Immobilon membranes for autoradiography prior to phosphoamino acid analysis.

Comparison of Methods

^{32}P-Labeled Glycoproteins

Ortho[^{32}P]phosphate-labeled glycoproteins, isolated by WGA chromatography from normal and RSV-transformed cells and separated by SDS–PAGE, are shown in Fig. 1. The phosphoamino acid content (determined following partial acid hydrolysis) of ^{32}P-labeled glycoproteins extracted by WGA chromatography from normal cells (B) and RSV-transformed cells (A) is depicted in Fig. 2. It can be seen that, with few exceptions, the relative amount of phosphoserine on any given phosphoprotein was not greatly affected by transformation. However, *src* expression did result in a reduction in the relative phosphothreonine content of some < 68K proteins, and a marked increase in phosphotyrosine content on proteins of virtually every molecular weight. The protein with the highest amount of tyrosine phosphorylation, as detected by direct phosphoamino acid analysis, appeared to be a protein of M_r 190–195K with additional peaks around 135K and 95K.

Figure 3 shows the pattern of tyrosine-phosphorylated glycoproteins from normal and RSV-transformed cells as detected by alkali treatment, either after transfer to Immobilon (A) or directly in the fixed gel (B). This is compared to the pattern of tyrosine-phosphorylated glycoproteins from sister cultures of cells as determined by immunoblotting with anti-phosphotyrosine antibodies (C). Samples were run on the same gel, with pairs of lanes separated by prestained molecular weight marker proteins. One pair of lanes was cut from the gel and subjected to alkali hydrolysis. The remaining lanes were transferred to Immobilon and either subjected to alkali treatment or immunoblotting with anti-phosphotyrosine antibodies. As can be seen in Fig. 3, similar profiles were obtained when alkali treatment was performed in a fixed gel or after transfer to Immobilon membranes (lanes A vs B). In both cases the major alkali-stable glycoprotein phosphorylation in transformed cells had a molecular weight of 95K, and normal cell extracts contained low levels of alkali-stable phosphorylations. In contrast, the major tyrosine-phosphorylated glycoprotein in transformed cells as determined by immunoblotting with anti-phosphotyrosine antibodies, had a molecular weight of 135K. In addition, immunoblotting detected a 38K phosphotyrosine which was not evident in the ^{32}P-labeled samples, although most other low-molecular-weight tyrosine phosphorylations were detected less well or not at all with the antibody. Also, the

T N

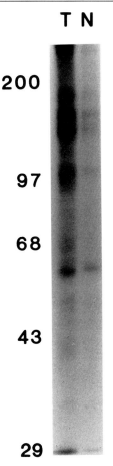

200

97

68

43

29

Fig. 1. Ortho[^{32}P]phosphate-labeled glycoproteins. Normal (N) and RSV-transformed (T) cells were labeled *in vivo* with ortho[^{32}P]phosphate as described in Methodology. WGA (50 μg) eluates were separated on a 7.5% polyacrylamide–SDS gel. The proteins were then transferred to an Immobilon membrane, dried, and visualized by autoradiography for 15 min using Kodak X-RP film.

Western blots did not detect tyrosine-phosphorylated glycoproteins in normal cells.

Quantitative Differences between Alkali Treatment in Fixed Gels or on Immobilon Membranes

As seen in Fig. 3, the pattern of alkali-stable protein phosphorylations was qualitatively the same whether alkali treatment was performed on a fixed gel or on proteins transferred to Immobilon. However, greater

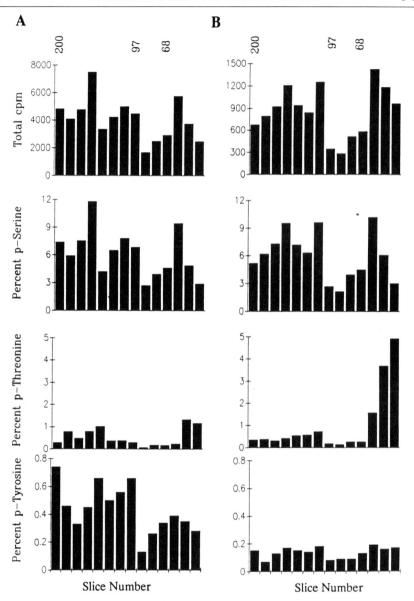

FIG. 2. Phosphoamino acid profiles of glycoproteins from normal cells (B) and RSV-transformed cells (A). The radioactive bands from the Immobilon membrane shown in Fig. 1 were excised, rehydrated in methanol followed by water, then subjected to phosphoamino acid analysis as described in Methodology. The results are expressed as a percentage of the total phosphoamino acid (cpm) in each sample. Molecular weight markers are designated at the top of each profile.

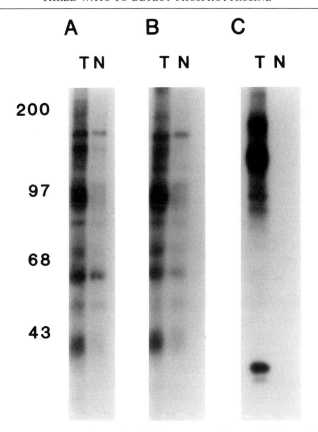

Fig. 3. Phosphoprotein profiles using alkali treatment or immunoblotting with anti-phosphotyrosine antibodies. Ortho[^{32}P]phosphate-labeled cells or unlabeled sister cultures of cells were subjected to WGA chromatography. WGA eluates (50 μg) of these normal (N) or RSV-transformed (T) cells were separated on a 7.5% gel. Alkali treatment was performed directly on the fixed gel for one set of ^{32}P-labeled samples (B). The remaining lanes were transferred to Immobilon and either alkali treated (A) or immunoblotted using anti-phosphotyrosine antibodies (C). For better comparison, the alkali-treated gel was allowed to remain in neutralizing solution 6 hr longer than originally described by Cooper et al.[1] to shrink the gel as close to its original size as possible. Phosphoproteins were visualized by autoradiography using Kodak X-RP film and intensifying screens for 1 hr (A), 6 hr (B), and 4 hr (C). The different exposure times required to obtain equivalent signal intensities in the two alkali-treated panels (A and B) reflect not only the twofold difference in retention of proteins but probably the absence of self-absorption of radioactive emission in the Immobilon-blotted samples.

phosphoprotein signals were obtained with alkali-treated Immobilon than with alkali-treated gels, since a much longer exposure was required for the gels to display equivalent signal intensities, raising the possibility that alkali treatment removed more proteins from fixed gels than from Immobilon. To test this possibility, lysates from cells labeled with [^{35}S]methionine were subjected to alkali treatment in a fixed gel or after transfer to Immobilon membranes, and the loss of protein was monitored both by determining the radioactivity released into the KOH solution during hydrolysis and by densitometric scans of films of equal exposure times both before and after hydrolysis (data not shown). The results demonstrated that proteins were indeed removed from both fixed gels and from Immobilon membranes during the alkali treatment since [^{35}S]methionine-labeled material was detectable in the 1 M KOH following hydrolysis. However, proteins were retained by the Immobilon much better than by the fixed gels (85 vs 44%), based on densitometric analysis.

Specificity of Anti-Phosphotyrosine Antibodies

To examine the hapten specificity of the phosphotyrosine antibodies in the immunoblotting procedure, various potential blocking reagents were added to the blotting buffer during the antibody-binding step. We first determined that antibody binding in the immunoblotting was completely blocked by 10 μM phosphotyrosine, but only partially by 1 μM phosphotyrosine (data not shown). Since some potential blocking reagents might be recognized by the antibody with a lower affinity, a 100 μM concentration of each reagent was used in our analysis. Figure 4, lane a, shows a representative profile using normal immunoblotting procedures with no potential competitor added. The ability to block antibody binding was then tested for tyrosine (d), phosphotyrosine (e), phosphoserine (f), phosphothreonine (g), phosphoarginine (h), phosphohistidine[8] (i), and tyrosine sulfate[9,10] (j). The results shown in Fig. 4 indicate that only exogenously added phosphotyrosine and phosphohistidine could effectively block the signal obtained by the immunoblotting procedure.

Since phosphohistidine is an acid-labile phosphoamino acid, whereas phosphotyrosine is not, we tested our glycoprotein samples for acid-labile phosphorylations. Prior to immunoblotting, samples bound to Immobilon were treated for 15 min at 55° in either 50 mM Tris, pH 7.6 and 150 mM NaCl (lane b), or in 0.2 N HCl (lane c) to remove acid-labile phosphates.

[8] S. J. Pilkis, M. Walderhaerg, K. Murray, A. Beth, S. D. Vankataramu, J. Pilkis, and M. R. El-Maghrbi, *J. Biol. Chem.* **258,** 6135 (1983).
[9] E. M. Danielson, *EMBO J.* **6,** 2891 (1987).
[10] W. B. Huttner, *TIBS* **12,** 361 (1987).

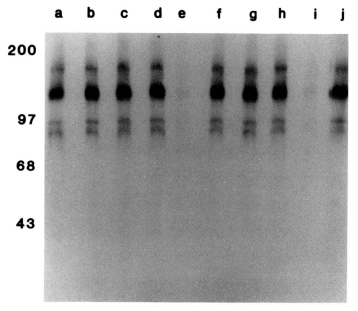

FIG. 4. Anti-phosphotyrosine antibody specificity. Fifty micrograms of WGA eluates from RSV-transformed cells was loaded per lane and separated on a 7.5% polyacryl-amide–SDS gel, then transferred to Immobilon. Immunoblotting with anti-phosphotyrosine antibodies was then performed either with (lanes b and c) or without (lanes a and d through j) pretreatment or in the presence (lanes d through j) or absence (lanes a, b, and c) of various exogenously added soluble reagents (100 μM each during the antibody-binding step). Lane a, untreated in the absence of exogenous reagents; lane b, 15-min pretreatment in 50 mM Tris-HCl, pH 7.6, 150 mM NaCl at 55°; lane c, 15-min pretreatment in 0.2 N HCl at 55°; lane d, tyrosine; e, phosphotyrosine; f, phosphoserine; g, phosphothreonine; h, phosphoarginine; i, phosphohistidine; j, tyrosine sulfate.

Acid treatment had no effect (compare lanes b and c), indicating that the signals observed are not likely to be due to acid-labile phosphoamino acids. We also determined that the anti-phosphotyrosine antibodies were unable to react with the bacterial sugar transport protein HPr[11] phosphory-lated on histidine-15 (data not shown). Thus, even though binding of these antibodies can be blocked by phosphohistidine, we do not believe that protein histidinyl phosphorylation contributes significantly to the signals generated in these immunoblots.

To directly determine whether the proteins detected by the anti-phos-photyrosine antibodies were indeed tyrosine phosphorylated, cells were

[11] J. Reizer, S. L. Sutrina, M. H. Saier, G. C. Stewart, A. Peterkofsky, and P. Reddy, *EMBO J.* **8**, 2111 (1989).

TABLE I
PHOSPHOAMINO ACID ANALYSIS OF ANTI-PHOSPHOTYROSINE-
IMMUNOPRECIPITATED PROTEINS

Band (M_r)	P-Ser (%)	P-Thr (%)	P-Tyr (%)
165	60.6	19.4	20.0
135	66.7	11.5	21.8
95	67.0	11.4	21.6
46	71.0	21.7	7.4
40	71.5	17.1	11.4

labeled with ortho[^{32}P]phosphate and the glycoprotein fraction was isolated by WGA chromatography. Tyrosine-phosphorylated glycoproteins were immunoprecipitated with unfractionated anti-phosphotyrosine antiserum and separated by SDS-gel electrophoresis. The profile of immunoprecipitated proteins was found to be identical to the profile obtained by immunoblotting with anti-phosphotyrosine antibodies (data not shown). This is an atypical finding, as the more usual experience is that immunoblotting detects more proteins than can be immunoprecipitated, presumably because the blotting procedure concentrates the proteins, facilitating low-affinity interactions. In addition, we chose to perform the immunoprecipitation with unfractionated serum because we suspect that higher affinity antibodies are sometimes lost during affinity purification on phosphotyrosine columns.[3] The phosphoproteins were then subjected to phosphoamino acid analysis, and all were found to contain phosphotyrosine (Table I). Interestingly, all the tyrosine-phosphorylated proteins are phosphorylated on threonine and serine as well.

Quantitative Differences between Different Antisera

Anti-phosphotyrosine antisera prepared from each of six different rabbits gave results similar to those shown in Fig. 3C when used to detect tyrosine-phosphorylated glycoproteins (data not shown). However, in the course of studies on the tyrosine phosphorylation of pp42/MAP kinase,[12] a 42-kDa tyrosine-phosphorylated protein, we observed that only two of the six sera gave a strong reaction with this protein. Therefore, we examined in more detail the possibility that different anti-phosphotyrosine antisera might detect different tyrosine-phosphorylated proteins.

Immunoblots comparing the specificity of four anti-phosphotyrosine antisera are shown in Fig. 5. SDS lysates from 3T3 fibroblasts which were unstimulated (−), EGF-stimulated (+), or transformed by the *src* oncogene (T) were prepared and analyzed with each of the four antisera. It can be

[12] A. J. Rossomando, D. M. Payne, M. J. Weber, and T. W. Sturgill, *Proc. Natl. Acad. Sci. U.S.A.* **86**, 6940 (1989).

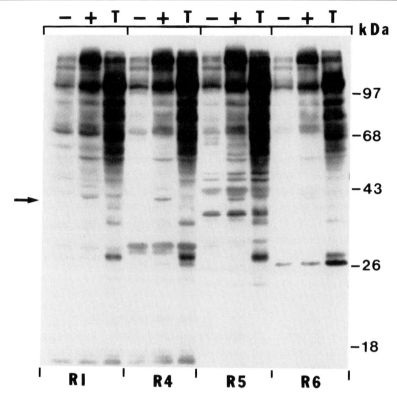

Fig. 5. Protein specificity of anti-phosphotyrosine antisera in the presence of NP-40. Immunoblots of whole-cell lysates from confluent 3T3 cells which had been left untreated (−), treated for 10 min with 200 ng/ml EGF (+), or transformed by *src* (T). R1–R6 are anti-phosphotyrosine antisera from different rabbits.

seen that the pp42/MAP kinase tyrosine phosphorylation which occurs in mitogen-stimulated cells (indicated with the arrow) was detected best with the R4 antiserum and was hardly detected at all with the R6 antiserum. On the other hand, a protein of approximately 26 kDa was detected best with R6, and hardly at all with the other sera. The four sera detected similar patterns of tyrosine phosphorylation in *src*-transformed cells; however, R5 detected proteins around 36 kDa better than the other sera, and R4 gave the best signal around 29 kDa. Thus each of these sera showed differences in the ability to recognize particular subsets of tyrosine-phosphorylated proteins.

The immunoblotting procedure utilized to generate the data for Fig. 5 included buffers with 0.1% NP-40, which was added to reduce nonspecific binding and lower the background. To determine whether the specificity differences seen in these immunoblots could be accounted for by a detergent-induced loss of low-affinity binding, we repeated the immunoblots at lower "stringency," omitting NP-40 from the buffers (Fig. 6). As anticipated, the blots became more similar. For example, R4, R5, and R6 could all now recognize the pp42/MAP kinase tyrosine phosphorylation, although this reaction was still weak with R1. Nevertheless, significant serum-specific differences were still evident. For example, a doublet around 18 kDa was best detected by the R1 antiserum; a 21-kDa protein was best detected with R4; and R5 gave the best signal with a 24-kDa protein.

The differences in antibody specificity described above could not be accounted for by immunoreactivity to proteins not phosphorylated on tyrosine, since neither preimmune serum nor serum preincubated with phosphotryosine showed significant reactivity with these protein bands (data not shown). We conclude that these different antisera contain different sets of anti-phosphotyrosine antibodies of varying affinities, and consequently they preferentially react with overlapping but nonidentical sets of tyrosine-phosphorylated proteins.

Discussion

The methodological comparison outlined above demonstrates large quantitative and even qualitative differences between the three standard methods for detecting phosphotyrosine. KOH treatment of ^{32}P-labeled proteins failed to distinguish alkali-stable serine and threonine phosphorylations from tyrosine phosphorylations, whereas immunoblotting with anti-phosphotyrosine antisera preferentially detected high-molecular-weight proteins and failed to detect many lower molecular weight tyrosine phosphorylations. In addition, different polyclonal anti-phosphotyrosine antisera displayed differences in reactivity with individual tyrosine-phosphorylated proteins.

Although much of the work presented here was performed with the glycoprotein fraction from cultured cells, comparably divergent results have been obtained when total cell lysates were analyzed for phosphotyrosine-containing proteins by the different methodologies: 30 or more tyrosine-phosphorylated proteins are evident in *src*-transformed cells, but the high-molecular-weight tyrosine-phosphorylated proteins are much more prominent if phosphotyrosine antibodies are used for detection[3,4] than if the phosphorylations are analyzed by direct phosphoamino acid analysis of proteins eluted from gel slices.[13] This discrepancy in the apparent pattern of tyrosine phosphorylated proteins appears to be due not only to

[13] R. Martinez, K. D. Nakamura, and M. J. Weber, *Mol. Cell. Biol.* **2,** 653 (1982).

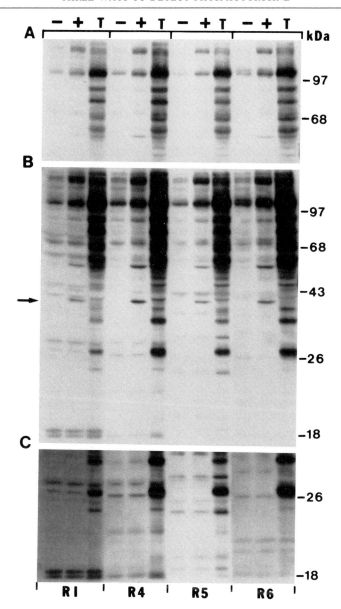

FIG. 6. Protein specificity of anti-phosphotyrosine antisera in the absence of NP-40. Conditions as for Fig. 5, but NP-40 was omitted from all buffers. (A) 7-hr exposure of high M_r proteins; (B) 18-hr exposure of entire immunoblot; (C) 48-hr exposure of low M_r proteins.

probable inefficiencies in eluting high-molecular-weight proteins from a gel prior to amino acid analysis, but also to the bias of the phosphotyrosine antibodies for high-molecular-weight proteins. Note that differences in elution from a gel or transfer to a membrane played no part in the comparisons reported here for the tyrosine-phosphorylated glycoproteins, since all the samples analyzed had been transferred to Immobilon or nitrocellulose filters regardless of the methodology employed for detecting phosphotyrosine.

The preferential detection of high-molecular-weight phosphorylations in the immunoblots could reflect a loss of low-molecular-weight proteins from the membrane, increased reactivity with proteins containing multiple tyrosine phosphorylations (such as some receptors) or both.

The different results obtained with the three methods may be explained in part by differences in the ability of these methods to detect tyrosine phosphorylations which occur in differing amino acid environments. For example, a phosphotyrosine residue may be contained within a peptide which is resistant to hydrolysis by 6 N HCl,[14] in which case it would not be detected by direct phosphoamino acid analysis. Alkali hydrolysis has been shown to remove phosphate from some phosphotyrosine residues[5] while some phosphoserine and phosphothreonine residues are relatively resistant to alkali hydrolysis.[1,5,6] Additionally, antibodies may be restricted from binding to a particular phosphotyrosine residue by surrounding amino acids, either due to charge or steric considerations. For example, with NP-40 in the blotting buffer, tyrosine phosphorylation of pp42/MAP kinase[14] could be detected only with two out of six antisera tested, which had been raised by the Kamps and Sefton method,[3] and with only one out of three monoclonal antibodies tested[2] (our unpublished data). We now know that the site of tyrosine phosphorylation in pp42/MAP kinase is very close to a phosphothreonine residue and to a glutamate.[15] It is possible that only a few antibody types will bind to this highly acidic environment with reasonable affinity. Thus, we suspect that the environment in which a particular phosphotyrosine residue occurs may determine the method(s) by which it can most readily be detected. Obviously, this cannot generally be known in advance, and therefore more than one detection method will have to be employed.

The following points should be considered when choosing which method to use for detecting tyrosine-phosphorylated proteins. Direct phosphoamino acid analysis is appropriate for confirming that a given protein is indeed tyrosine phosphorylated. In addition, this procedure can be used to compare a small number of samples with the advantage that the method

[14] T. M. Martensen, this series, Vol. 107, p. 3.
[15] D. M. Payne, A. J. Rossomando, P. Martino, A. K. Erickson, J. H. Her, J. Shabanowitz, D. F. Hunt, M. J. Weber, and T. W. Sturgill, *EMBO J.* **10,** 885 (1991).

can be internally controlled for protein concentration and labeling variations. However, it is time consuming and yields only low recovery of initial material. Additionally, it is chemically harsh and may destroy some tyrosyl phosphorylations. Alkali hydrolysis is rapid and requires few manipulations, making it suitable for comparing a small-to-moderate number of samples quickly. It is, however, chemically harsh and variable amounts of protein may be lost during the procedure, particularly if samples are not blotted onto Immobilon. Additionally, it is known that some phosphotyrosine residues may be lost while some phosphoserine and phosphothreonine residues remain rather resistant to hydrolysis. Immunoblotting with phosphotyrosine antibodies is specific, gentle, and rapid, making it ideal for screening a large number of samples simultaneously. Additionally, it can be used to study samples which cannot be labeled *in vivo,* such as biopsies. However, although all proteins detected with these antisera contain phosphotyrosine, not all tyrosine-phosphorylated proteins are detected by these antisera. In particular, many lower molecular weight protein-tyrosine phosphorylations are not detected, whether because of protein loss from the filters, or because of biases in the specificity or affinity of the antibodies. Thus, anti-phosphotyrosine antisera cannot be used to determine quantitatively the pattern of tyrosine phosphorylations in a sample.

It therefore seems that in choosing the method for detection of tyrosine-phosphorylated proteins, quick screening could be done most easily using alkali hydrolysis or immunoblotting, whichever method best detects the protein(s) of interest, and direct phosphoamino acid analysis should be used to confirm the presence of phosphotyrosine on that particular protein. Also, direct phosphoamino acid analysis may be the method of choice for detecting phosphorylation of lower molecular weight proteins. It also appears that since all methods are subject to artifacts of one form or another, that more than one method should be used to obtain a clear picture of all the tyrosine-phosphorylated proteins present in a sample.

Acknowledgments

We thank D. Michael Payne for his help in generating and characterizing the anti-phosphotyrosine antisera. John Fessler provided tyrosine sulfate and Fred Hess and Mel Simon the phosphoramidate used in synthesizing phosphohistidine. Jonathan Reizer provided the bacterial HPr protein and its kinase. This work was supported by grants from the USPHS National Cancer Institute: CA47815, CA40042, and CA39076. L.M.K. was supported as a USPHS predoctoral trainee by CA9109 and A.J.R. by GM08136.

[5] Generation and Use of Antibodies to Phosphothreonine

By Daphna Heffetz, Mati Fridkin, and Yehiel Zick

Introduction

Protein phosphorylation is a universal device exercised in various biochemical pathways to alter protein structure and function.[1,2] In spite of the importance of this regulatory mechanism, there are no simple and practical procedures to monitor changes in phosphate content of specific amino acids, as they occur under physiological conditions in intact tissues.

The most common procedure is to label intact cells or small tissue fragments with ^{32}P and subsequently to isolate ^{32}P-labeled proteins by conventional biochemical methods. In order to identify the specific amino acids that undergo phosphorylation, additional procedures for phospho-amino acid analysis are required.[3]

The whole method is long and tedious and suffers major drawbacks.[3] Immunoblotting of cellular proteins with antibodies directed against phosphoamino acids could provide an alternative procedure. This method is advantageous as it does not involve ^{32}P labeling, and can therefore be employed to monitor alterations in phosphorylation of specific proteins as they occur in intact organs or even whole animals. Indeed, mono-[4] and polyclonal[5] antibodies directed against phosphotyrosine residues were already generated and found useful in assessing tyrosine phosphorylation. We have previously reported on the generation of antibodies which specifically react with phosphothreonine residues (anti-P-Thr antibodies[6]). In this chapter we detail methods for the production characterization and use of these antibodies.

Preparation of Immunogen

Step 1. Production of Bromoacetic Acid N-Hydroxysuccinimide Ester (Ester I). Fifty millimoles of *N*-hydroxysuccinimide is dissolved in 100 ml of dry tetrahydrofuran. BrCH$_2$COOH (50 mmol) is added and the reaction mixture is cooled to 4°. A solution of *N,N'*-dicyclohexylcarbodiimide (50

[1] T. Hunter, *Cell* (*Cambridge, Mass.*) **50,** 823 (1987).
[2] A. M. Edelman, D. K. Blumenthal, and E. G. Krebs, *Annu. Rev. Biochem.* **56,** 567 (1987).
[3] J. A. Cooper, B. M. Sefton, and T. Hunter, this series, Vol. 99, p. 387.
[4] A. R. Frackelton, *Cancer Cells* **3,** 339 (1985).
[5] A. H. Ross, D. Baltimore, and H. N. Eisen, *Nature* (*London*) **294,** 654 (1981).
[6] D. Heffetz, M. Fridkin, and Y. Zick, *Eur. J. Biochem.* **182,** 343 (1989).

mmol) in tetrahydrofuran (25 ml) is added and the reaction mixture is stirred for 3 hr at 4°. The mixture is filtered on sintered glass to remove N,N'-dicyclohexylurea precipitates. The filtrate is evaporated at 22° under reduced pressure. Dry ethyl ether (100 ml) is added to the residual oily solution and precipitation is allowed to proceed for 24 hr at 4°. The white ester (65–75% yield; mp 100–103°; molecular mass 231, with $NaOCH_3$ in methanol–benzene[7]) (1 : 4, v/v) is collected by filtration on sintered glass, washed with dry ethyl ether, and dried *in vacuo*. It is kept dry under vacuum at −20° until further use.

Step 2. Coupling of P-Thr to Ester I. One millimole of P-Thr and 4 mmol of $KHCO_3$ are dissolved in 1 ml of H_2O. One millimole of ester I (dissolved in 1 ml of dimethylformamide) is added and the mixture is stirred for 2 hr at 22° to generate the activated hapten.

Step 3. Cross-Linking of P-Thr-Containing Hapten to Keyhole Limpet Hemocyanin (KLH) or Bovine Serum Albumin (BSA). The hapten generated in step 2 (solution of about 2 ml) is added immediately, following preparation, in small lots to 27.5 mg of KLH, dissolved in 2.5 ml of phosphate-buffered saline (PBS). The pH is maintained while stirring during the addition and for 96 hr thereafter at 9–9.5 by adding dropwise 1 M LiOH. The solution is kept at 4° during this procedure. A turbid mixture should be formed. After 96 hr the mixture is dialyzed against 1 liter of PBS, with six additional changes of PBS (0.5 liter each) over a 48-hr period. The dialyzed mixture is kept at −20°. To cross-link the P-Thr-containing hapten to BSA, the above procedure is carried out with 40 mg BSA.

Step 4. Cross-Linking of P-Tyr- or P-Ser-Containing Hapten to KLH or BSA. P-Tyr or P-Ser conjugated to BSA or KLH are generated as described above, save for the fact that P-Thr is replaced with either P-Tyr or P-Ser.

Antibody Induction and Purification

Immunization and blood collection follow common procedures. An equivalent of 1 mg KLH conjugate in 1 ml PBS is mixed with 1.5 ml of Freund's complete adjuvant. The mixture is emulsified by repeated uptake and expulsion with a small syringe. Portions of 2.5 ml are injected into New Zealand White rabbits at multiple intradermal sites along the spinal cord of the rabbit. Three weeks after the initial injection, each rabbit is given a booster injection of 2.5 ml of the same preparation. Additional booster injection is given 2 weeks later, containing 0.5 mg conjugate in

[7] M. Wilchek, M. Fridkin, and A. Patchornik, *Anal. Biochem.* **42**, 275 (1970).

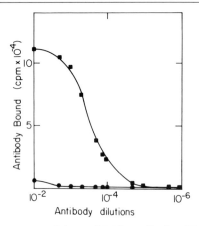

FIG. 1. Binding of serum containing anti-P-Thr antibodies (■) to P-Thr conjugated to BSA, as determined in a solid-phase radioimmunoassay (RIA). (●), NR.

1 ml PBS mixed with 1.5 ml Freund's incomplete adjuvant, and later in the course of immunization if the antibody titer becomes low. Blood is collected every 2 weeks starting on day 40, 10 days after the second booster. Blood is allowed to clot for 1–2 hr at room temperature, followed by overnight storage in the cold. Serum is removed by centrifugation at 4° for 10 min at 12,000 g. Serum for each rabbit and bleeding is kept separate and stored frozen.

Solid-Phase Radioimmunoassay (RIA). This method is used to evaluate individual antisera qualitatively. The procedure is essentially that described by Tsu and Herzenberg.[8] In brief, 50 ng conjugate of BSA–P-Thr, BSA–P-Tyr, or BSA–P-Ser in 50 μl is plated on polyvinyl 96-well microtiter plates (Falcon, Oxnard, CA) for 2 hr at 37°. After five washings with buffer I [1% (w/v) BSA, 130 mM NaCl, 20 mM Tris-HCl, pH 7.6] the wells are filled with buffer I and kept overnight at 4°. Next day, the buffer is removed and 50 μl of anti-P-Thr antibodies or normal serum is added at serial dilutions. The incubation is allowed to proceed for 2 hr at 37°. The plates are then washed five times with buffer I. Goat anti-rabbit ^{125}I-labeled antibodies (1 × 10^5 cpm in 50 μl buffer I) are added to each well and incubation continued for 2 hr at 37°. The plates are washed five times in buffer I and the wells are dried, cut, and counted in a γ counter. Figure 1 shows a typical RIA of a positive serum. Half-maximal and maximal

[8] T. T. Tsu and L. A. Herzenberg, *in* "Selected Methods in Cellular Immunology" (B. B. Mishel and S. M. Shiigi, eds.), p. 373. Freeman, San Francisco, California, 1980.

binding of the antibodies to BSA–P-Thr occurs at serum dilutions of 1 : 4000 and 1 : 1000, respectively. Binding of preimmune serum (NR) to BSA–P-Thr, or binding of anti-P-Thr antibodies to BSA alone, is negligible under these conditions. Individual serum samples are first screened in order to select several preparations with a high titer of the desired antibody for further purification and use.

Coupling of BSA–P-Thr Conjugate to Sepharose 4B. BSA–P-Thr conjugate (4 mg) is coupled to 1 g of Affi-Gel 15 (Pharmacia, Piscataway, NJ) according to instructions provided by the manufacturer.

Generation of Affinity-Purified Anti-P-Thr Antibodies. Affinity purification of anti-P-Thr antibodies is needed for two reasons: (1) excess of proteins such as albumin and IgG interfere in immunoprecipitation and immunoblotting, elevating the background signals, while (2) contaminating serum phosphatases and proteases can degrade or dephosphorylate the proteins under study. Ten to 20 ml of antiserum is gently mixed with an equal volume of a saturated solution of ammonium sulfate, pH 7.5. Stirring is continued for 120 min at 4° and the precipitate is recovered by centrifugation [Sorvall (Newtown, CT) SS34 rotor, 10,000 rpm for 20 min]. The precipitate is resuspended in 50% (v/v) ammonium sulfate, pH 7.5, up to the original serum volume and recovered by repeating the centrifugation. The pellet is dissolved in H_2O (to the original serum volume) and dialyzed at 4° against 1 liter of phosphate-buffered saline (pH 7.4) with six exchanges over a 72-hr period. Finally, the antibodies are dialyzed against 1 liter of buffer II (130 mM NaCl, 20 mM Tris-HCl, pH 7.6). The partially purified antibody solution is then applied to a column of BSA–P-Thr coupled to Sepharose 4B (4.5 × 1.5 cm) equilibrated with buffer II. The column is sealed and the antibodies are allowed to bind to the matrix for 16 hr at 4°. The column is then washed with 100 ml buffer II until $A_{280} > 0.02$. The column is further washed with 100 ml of buffer II containing 0.5 M NaCl, so as to elute additional proteins that bind nonspecifically. Finally, the column is washed with 100 ml of buffer II to remove excess NaCl. The column is sealed and 3 ml of 0.15 M P-Thr in buffer II is added. The agarose is mixed with the solution and left for 16 hr at 4°. The agarose is allowed to repack and the buffer is eluted. One-milliliter fractions are collected. An additional 10 ml of 0.15 M P-Thr in buffer II is added to the column and 1-ml fractions are collected. Residual antibodies (and additional proteins) that remain bound to the column are then eluted with 0.2 M HCl–glycine buffer, pH 2.7. Fractions of 0.7 ml are collected into tubes containing 0.3 ml of 1 M K_2HPO_4 so that the final pH of the solutions is maintained at 7.6. The protein concentration of each fraction is monitored by absorbance at 280 nm. A typical elution profile of affinity-purified

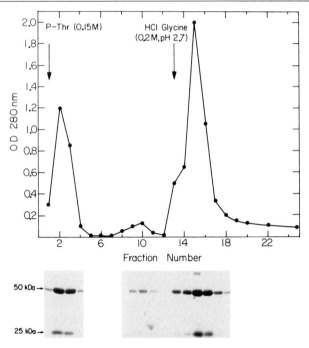

Fig. 2. Elution profile of anti-P-Thr antibodies from a column of BSA–P-Thr coupled to Affi-Gel 15. *Top:* Absorbance of the eluted fractions at 280 nm; *bottom:* Coomassie stain of individual fractions resolved by 7.5% SDS-PAGE.

anti-P-Thr antibodies is presented in Fig. 2. Eighty microliters from each fraction are mixed with 20 μl of 5× Laemmli sample buffer[9] and resolved by 7.5% SDS-PAGE. The proteins are stained with Coomassie Blue. Each affinity purification step yields about 5 mg of purified IgG (based on the assumption that OD 1.4 at 280 nm represents 1 mg/ml IgG). About 30% is eluted with P-Thr (peak I, Fig. 2) while 70% is eluted with acid (peak II, Fig. 2). Inspection of Fig. 2 reveals that the majority of proteins eluted either in peak I or peak II constitute IgG. Each peak fraction is pooled and dialyzed against 1 liter of buffer II with six exchanges over a 48-hr period. The dialyzed affinity-purified antibodies are aliquoted and stored at $-70°$.

Characterization of Affinity-Purified Anti-P-Thr Antibodies

The specificity of the antibody binding is determined using P-Thr, P-Ser, or P-Tyr conjugated to BSA as antigens. A typical RIA is presented

[9] U. K. Laemmli, *Nature (London)* **227,** 680 (1970).

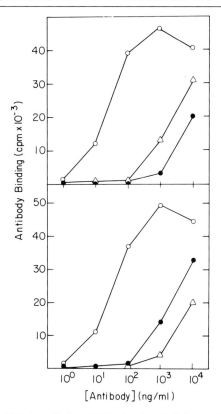

FIG. 3. Binding of affinity-purified peak I (top) and peak II (bottom) anti-P-Thr antibodies to P-Thr (○), P-Tyr (△), or P-Ser (●) conjugated to BSA, as determined by RIA.

in Fig. 3. Significant binding of anti-P-Thr antibodies to polyvinyl plates coated with 50 ng of BSA–P-Thr occurs at 10 ng/ml antibodies, while half-maximal binding is achieved at 30 ng/ml. BSA–P-Ser is 200-fold less potent in retaining the antibodies bound to the plates, while plates coated with BSA–P-Tyr retain significant amounts of anti-P-Thr antibodies only when the latter are added at 10 μg/ml. Binding of anti-P-Tyr antibodies to plates coated with BSA–P-Thr or BSA–P-Ser is negligible.[6]

The specificity of the anti-P-Thr antibodies can be further assessed by measuring the extent of inhibition exerted by P-Tyr, P-Ser, and P-Thr on binding of affinity-purified antibodies to BSA–P-Thr. As seen in Fig. 4, 5 mM P-Thr completely inhibits antibodies binding, with half-maximal effect at 40 μM. P-Ser is over 200-fold less effective than P-Thr, while P-Tyr is essentially ineffective. A variety of other phosphate-containing compounds, including nucleotides (ATP, CTP, AMP), sugars (glucose 1-

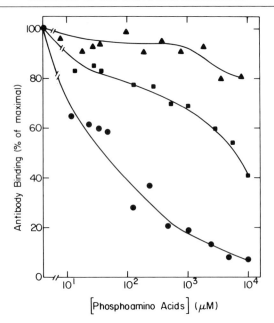

FIG. 4. Capability of P-Thr (●), P-Ser (■), or P-Tyr (▲) to inhibit binding of affinity-purified anti-P-Thr antibodies to BSA–P-Thr, as determined by RIA.

phosphate), phospholipids [phosphatidylserine (PtdSer), phosphatidylino-sitol (PtdIns), phosphatidylinositol phosphate (PtdInsP)], and nucleic acids (RNA, DNA) were found ineffective in inhibiting anti-P-Thr binding. Simi-larly, threonine itself, even when added at 1 mM, fails to inhibit antibody binding.[6]

Specific Binding of Anti-P-Thr Antibodies to Intracellular Proteins. The main purpose for the development of anti-P-Thr antibodies is their potential use as monitors of changes that occur in protein-threonine phos-phorylation in unlabeled cells or tissue fragments. We have already demon-strated[6] the utility of these antibodies in the detection by immunoblotting of enhanced threonine phosphorylation of the epidermal growth factor (EGF) receptor in intact A431 cells that were stimulated with EGF. This procedure is applicable to other systems as well. In the example depicted below we demonstrate the capability of anti-P-Thr antibodies to blot to several proteins present in extracts of the rat hepatoma Fao cells.[10]

Protein Extraction from Fao Cells. Rat hepatoma Fao cells are grown in 150-mm plates (Nunc, Delta, Denmark) in RPMI 1640 medium con-

[10] J. Deschatrette and M. C. Weiss, *Biochimie* **56,** 1603 (1974).

taining 10% fetal calf serum. Confluent monolayers are washed twice in PBS and frozen immediately on liquid nitrogen. Extraction of the cells is done by adding 1.0 ml buffer III (0.2 M sucrose, 50 mM HEPES, 80 mM β-glycerophosphate, 2 mM EGTA, 2 mM EDTA, 2 mM sodium orthovanadate, 100 mM NaF, 10 mM sodium pyrophosphate, 10 μg/ml aprotinin, 5 μg/ml leupeptin, 10 μg/ml soybean trypsin inhibitor, and 1 mM phenylmethylsulfonyl fluoride, pH 7.5). Cells are disrupted by repeated (three times) thawing and freezing on liquid nitrogen. The crude homogenate and cell debris are scraped from the plates and transferred into Eppendorf tubes. The tubes are centrifuged at 12,000 g for 15 min at 4°. The supernatant (~7 mg/ml protein) is collected. Samples (100 μl) are mixed with 25 μl of 5× Laemmli "sample buffer,"[9] run on 7.5% sodium dodecyl sulfate gels,[9] and transferred to nitrocellulose papers for Western blot analysis.

Western Blotting. Electrophoretic transfer of phosphoproteins from SDS gels to nitrocellulose papers is performed essentially as described by Burnette.[11] The transfer is carried out for 3 hr at a constant current of 200 mA in a buffer containing 50 mM glycine and 50 mM Tris-HCl, pH 8.8. The nitrocellulose papers are incubated for 2 hr at 22° in buffer IV [10 mM Tris-HCl, 140 mM NaCl, and 0.05% (v/v) Tween 20, pH 7.5] plus 1% (w/v) BSA. The papers are then incubated with 1 μg/ml affinity-purified anti-P-Thr antibodies in buffer IV (plus 0.1% BSA). To determine the specificity of antibody binding, duplicate papers are incubated with P-Thr, P-Ser, or P-Tyr, at concentrations ranging from 1 μM to 10 mM that are added together with the antibody solution. Following a 16-hr incubation at 4°, the papers are extensively washed (five times) in buffer IV, and the presence of anti-P-Thr antibodies is determined by adding 3 × 10⁵ cpm/ml of goat anti-rabbit ^{125}I-labeled antibodies in buffer IV/0.1% BSA. Binding of the labeled agent is carried out for 2 hr at 22°. This is followed by intensive washings in buffer IV. The papers are then dried and radiographed.

Specificity of Anti-P-Thr Antibody Binding. As shown in Fig. 5, anti-P-Thr antibodies blotted to several proteins present in Fao cells extracts. These range in molecular mass from 30,000 to 250,000 Da. Binding of the antibodies is restricted to P-Thr-containing proteins. Such a conclusion is based on the fact that P-Thr present at 10 μM completely competes with anti-P-Thr antibody binding. P-Ser at 0.1 mM is ineffective, but at 1 mM it is as effective as 10 μM P-Thr.

Immunoprecipitation of [³²P]Thr-Containing Proteins by Anti-P-Thr Antibodies. In order to assess the capability of anti-P-Thr antibodies to immunoprecipitate [³²P]Thr-containing proteins, Fao cells are labeled with ³²P and the phosphorylated proteins are extracted in Triton-containing

[11] W. W. Burnette, *Anal. Biochem.* **112,** 195 (1981).

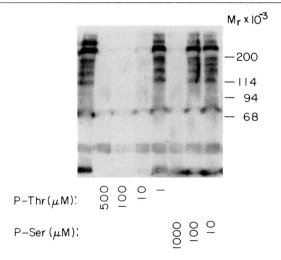

FIG. 5. Immunoblotting of Fao cell extracts with anti-P-Thr antibodies, carried out in the absence or presence of the indicated concentrations of P-Thr or P-Ser.

buffer.[6] We could demonstrate[6] that anti-P-Thr antibodies immunoprecipitate several ^{32}P-labeled phosphoproteins (pp38, pp55, pp85, pp100, pp155). Immunoprecipitation is specifically inhibited by 1 mM P-Thr but not by 1 mM P-Tyr.[6] ^{32}P-Labeled samples are incubated with anti-P-Thr antibodies and the immune complexes are separated from nonprecipitated proteins by adding immobilized protein A (Pansorbin, Calbiochem, San Diego, CA). Samples are then subjected to phosphoamino acid analysis. In the nonimmunoprecipitated fractions there is no significant reduction in [^{32}P]Thr content, suggesting that the anti-P-Thr antibodies immunoprecipitate a relatively small fraction out of the total phosphoproteins and therefore they are not useful yet for quantitative recoveries of [^{32}P]Thr-containing proteins. By contrast, phosphoamino acid analysis of total proteins precipitated by anti-P-Thr antibodies reveals that they contain 39% [^{32}P]Thr and 61% [^{32}P]Ser, with undetectable amounts of [^{32}P]Tyr (not shown). Since P-Thr residues constitute only 15% of the total phosphoamino acids,[3] these findings suggest that anti-P-Thr antibodies specifically precipitate and markedly enrich the immunoprecipitated samples with P-Thr-containing proteins. Since most of these proteins are likely to contain P-Ser residues as well, the presence of [^{32}P]Ser residues in these precipitates is expected.

Under our current assay conditions, a similar repertoire of P-Thr-containing proteins is either immunoprecipitated or immunoblotted by the antibodies. Nevertheless, we believe that the latter technique is advantageous for the following reasons: (1) the background signal is much lower in immunoblots; and (2) when immunoprecipitation of ^{32}P-labeled proteins is carried out with an antibody directed against a given phosphoamino acid, it is impossible to assess whether any change in ^{32}P content of a protein results due to an exclusive change in the content of that particular phosphoamino acid or whether the change actually reflects altered phosphorylation of different amino acids. This is not the case when employing the immunoblotting technique, where any change in the amount of bound antibody reflects, at least qualitatively, a change in phosphate content of the specific amino acid under study.

Conclusions

Our findings suggest that anti-P-Thr antibodies could be useful tools to probe proteins whose structure or function is mainly regulated through phosphorylation on threonine residues. By employing immunoblotting techniques it is now feasible to initiate *in vivo* studies in animal models in order to monitor the basal phosphothreonine contents of different proteins as well as their changes in response to various stimuli under physiological conditions.

[6] Generation and Use of Anti-Phosphotyrosine Antibodies Raised against Bacterially Expressed abl Protein

By JEAN Y. J. WANG

Protein-tyrosine kinases are important enzymes in the regulation of cell proliferation. Activation of protein-tyrosine phosphorylation is an essential step in the mitogenic signal transduction pathways of most growth factors. The studies of tyrosine kinases and their substrates have been facilitated by the isolation of antibodies specific for phosphotyrosine (P-Tyr). These antibodies have been used to identify and purify tyrosine-phosphorylated proteins. They can also be used to quantitate the content of P-Tyr in proteins. Antibodies for phosphotyrosine have been prepared using a variety of immunogens[1] (see also [7] through [10] in this volume). Most of the immunogens are synthesized chemically by linking phosphotyrosine or its analogs to carrier proteins such as bovine serum albumin or keyhole limpet hemocyanin. An unusual immunogen for the production

---

[1] J. Y. J. Wang, *Anal. Biochem.* **172,** 1 (1988).

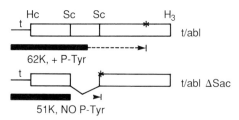

FIG. 1. *t/abl* fusion genes and their protein products from *E. coli*. The thin line represents a 240-bp coding sequence of SV40 t antigen beginning at the initiation codon. The open box represents the 3.6-kb v-*abl* coding sequence, the asterisk denotes the position of the termination codon. The functional tyrosine kinase domain spans the first 1.2 kb from the *Hinc*II site (Hc) to about 110 bp beyond the first *Sac*I site (Sc).[3] The nonkinase region of the v-abl protein is very unstable in *E. coli* but the tyrosine kinase domain is stable and can be accumulated to high levels. Deletion of the internal *Sac*I fragment truncates the protein by premature termination. The ΔSac protein lacks tyrosine kinase activity. H₃, *Hin*dIII.

of anti-phosphotyrosine (anti-P-Tyr) antibodies is the protein encoded by the v-*abl* oncogene in the Abelson murine leukemia virus.

The v-*abl* gene encodes a cytoplasmic tyrosine kinase which has the capacity to induce neoplastic transformation. The oncogene v-*abl* is a truncated form of a normal cellular gene c-*abl*. The truncation deletes an inhibitory domain of the c-abl protein. As a result, the tyrosine kinase activity of v-abl protein is deregulated.[2] The unregulated v-abl tyrosine kinase has an autokinase activity, thus the v-*abl*-encoded protein contains phosphotyrosine. The v-*abl* sequence encodes 1009 amino acids, of which the N-terminal 372 amino acids constitute a functional tyrosine kinase domain (Fig. 1). The v-abl protein can be overproduced in *Escherichia coli* when it is fused with the first 80 amino acids of the small t antigen from simian virus 40 (SV40) virus (Fig. 1). The expression of this *t/abl* fusion gene from the λ phage R_R promoter resulted in the accumulation of a large quantity of a truncated t/abl protein in *E. coli* cells.[3] The t/abl protein is truncated at the C-terminal end of v-abl, leaving a 60 to 62-kD N-terminal fragment which contains the SV40 t antigen and a functional tyrosine kinase domain. This overproduced t/abl protein fragment is stoichiometrically phosphorylated on tyrosine because phosphotyrosine residues are stable in bacterial cells. Deletion of an internal *Sac*I restriction fragment in v-*abl* results in the premature termination of translation and the inactivation of tyrosine kinase activity (Fig. 1). This t/ablΔSac protein is also overproduced, but it contains no phosphotyrosine.

[2] W. M. Franz, P. Berger, and J. Y. J. Wang, *EMBO J.* **8**, 137 (1989).
[3] J. Y. J. Wang and D. Baltimore, *J. Biol. Chem.* **260**, 64 (1985).

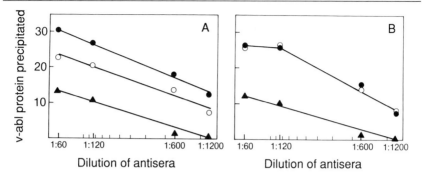

Fig. 2. Evidence that anti-phosphotyrosine antibodies are present in immune sera raised against t/abl protein. Immunoprecipitation of ^{35}S-labeled t/abl wild-type (A), or t/abl ΔSac (B) proteins, with dilutions of immune sera (●). The amount of ^{35}S-labeled protein precipitated was determined by densitometry of autoradiograms. Addition of t/abl protein (16 ng/μl) competed with the labeled antigen and inhibited immunoprecipitation of both proteins (▲). Addition of 40 mM phosphotyrosine partially inhibited the reaction with the wild-type protein but did not inhibit the immunoprecipitation of the mutant protein (○).

When immunized with the P-Tyr-containing t/abl protein fragment, rabbits produced antibodies to the SV40 small t antigen, the abl protein, and antibodies for phosphotyrosine. The presence of anti-phosphotyrosine antibodies in the immune sera could be demonstrated by the competition of immune reactivity with phosphotyrosine (Fig. 2). The immunoprecipitation of the wild-type t/abl fusion protein was partially inhibited by phosphotyrosine (Fig. 2A). This partial inhibition was not observed when the antisera were reacted with the ΔSac mutant protein, which lacked phosphotyrosine (Fig. 2B). The anti-P-Tyr antibodies could be purified using phosphotyrosine or phosphotyramine-Sepharose (see Purification of Anti-P-Tyr Antibodies, below). After purification, the anti-P-Tyr antibodies reacted only with the wild-type protein, which contained P-Tyr, but not with the mutant protein, which lacked P-Tyr. Moreover, the purified anti-phosphotyrosine antibodies reacted with a variety of other tyrosine-phosphorylated proteins.[4] These results establish that the t/abl fusion protein can induce the production of antibodies which recognize the phosphotyrosine determinant in many proteins.

Preparation of t/abl Fusion Protein from *Escherichia coli*

Expression of the t/abl fusion protein from the λ P_R promoter is heat inducible due to the presence of a temperature-sensitive λ repressor gene in the vector plasmid.[3] A 4-hr heat-induction period is sufficient to accumu-

[4] J. Y. J. Wang, *Mol. Cell. Biol.* **5**, 3640 (1985).

late a large amount of the t/abl protein fragment which is incorporated into the inclusion bodies as protein aggregates and can be readily separated from most of the bacterial proteins by a low-speed centrifugation.[3]

To obtain the highest possible level of protein production, it is important that freshly transformed *E. coli* be used. *Escherichia coli* strain HB101 is adequate for overproduction and the p*tabl*130 plasmid[3] is introduced into HB101 by the standard $CaCl_2$ transformation method. Ampicillin-resistant bacteria are selected on nutrient agar plates at 32°, the off temperature for the P_R promoter, to avoid selection against the expression plasmid.

Storing Freshly Transformed Bacteria

1. Inoculate 10 ml of L-broth plus ampicillin (50 µg/ml) with about 50 colonies from the agar plate and grow the culture with proper aeration at 32° to midlog phase (A_{650} 0.6–0.8). This should take 4 to 6 hr. Do not allow the culture to enter into stationary phase of growth.

2. Divide the culture into multiple 100-µl aliquots in 0.5-ml plastic Eppendorf tubes.

3. Add 8 µl of dimethyl sulfoxide (DMSO) to each aliquot, and quickly freeze tubes at −80°.

Preparing the t/abl Protein Aggregates

1. Inoculate 10 ml L-broth plus ampicillin with one tube of frozen bacteria (100 µl) and grow for 8 to 12 hr at 32° with proper aeration.

2. Dilute the 10-ml culture into 1 liter of fresh L-broth plus ampicillin and incubate at 32° for 2 to 4 hr until A_{650} reaches 0.4 to 0.6. At this time, shift the culture to a shaking water bath at 39° and incubate for 4 to 5 hr to induce the expression of t/abl protein.

3. Harvest bacteria by centrifugation at 5000 g [7000 rpm in a Sorvall (Newtown, CT) GS3 rotor] for 10 min at 4°. The bacteria pellet can be stored at −20° at this point.

4. Wash the pellet once with 250 ml of buffer A (10 mM Tris-HCl, pH 7.5, 1 mM EDTA, and 100 mM NaCl).

5. Resuspend the washed pellet in 10 ml of buffer A and disrupt the bacteria by sonication. Sonicate at 4° using full power for a total of 5 min, usually in 10 30-sec intervals with cooling periods in between. Spin the sonicate at 12,000 g (10,000 rpm in a Sorvall SS34 rotor) for 10 min. The t/abl protein aggregates are in the pellet fraction. The pellet should be whitish in color. Resuspend the pellet in 5 ml of buffer A and repeat the sonication step.

6. The final pellet obtained from the second sonicate contains mostly the t/abl fusion protein, which migrates as a series of five bands around 60 to 62 kDa on sodium dodecyl sulfate-polyacrylamide gel electrophoresis (SDS-PAGE).[3] Resuspend the pellet in 1 ml of buffer A and store the suspension at $-80°$. The concentration of the t/abl protein in the suspension can be estimated by SDS-PAGE and staining the Coomassie Blue.

Usually 10 mg of t/abl fusion protein can be obtained from 1 liter of bacterial culture.

Immunization and Analysis of Antisera

Renaturation of the t/abl protein aggregates is not necessary for the immunization of rabbits. All of the New Zealand White rabbits we have injected with the t/abl protein have produced anti-P-Tyr antibodies, although of varying titers and affinities. An initial injection into the inguinal lymph nodes (in the popliteal area of both hind legs[5]) gives better responses. If this route of injection is prohibited by the local regulations on animal use, subcutaneous injection does also induce anti-P-Tyr productions, albeit with lower titers.

Immunization Protocol

1. The primary injection is with 200 μg of t/abl protein/rabbit. Add SDS to the t/abl protein suspension at a final concentration of 0.4%. This will clarify the suspension but does not completely dissolve the aggregates. Prepare an emulsion with 50% (v/v) Freund's complete adjuvant. Use water to adjust the total volume so that 200 μg of the antigen is delivered to each rabbit in 1 ml of emulsion. Inject rabbits at multiple subcutaneous sites.

2. The second injection is given 4 weeks later, using 500 to 800 μg t/abl protein/rabbit. Prepare emulsion as described above but with incomplete adjuvant.

3. Rabbits can be bled 10 days later and the sera tested for anti-P-Tyr titers. Most rabbits respond quickly and produce the antibodies after the second injection. It may take four monthly injections to induce antibody production in some rabbits. The anti-P-Tyr titers usually maintain for 1 month after an injection.

[5] R. B. Goudie, C. H. W. Horne, and P. C. Wilkinson, *Lancet* **2**, 1224 (1966).

Preparation of P-Tyr–BSA

The anti-P-Tyr titers of the immune sera can be tested by competition experiments, as shown in Fig. 2. An easier was to test for anti-P-Tyr reactivity is to use P-Tyr–BSA. Phosphotyrosine can be coupled to bovine serum albumin using glutaraldehyde.

1. Dissolve 20 mg of BSA in 0.5 ml of 0.4 M phosphate buffer, pH 7.5.
2. Dissolve 5 mg of phosphotyrosine in 1.5 ml water and adjust pH to neutrality. Add the P-Tyr solution to the BSA solution.
3. Add 1 ml 20 mM glutaraldehyde dropwise while stirring over the course of 5 min. Continue stirring for 30 min.
4. Add one-tenth volume of 1 M glycine to block glutaraldehyde that has not reacted; continue stirring for 30 min.
5. Dialyze against phosphate-buffered saline (PBS) with four changes over 24 to 36 hr, and store at −80° in aliquots.

Purified anti-P-Tyr antibodies can detect as low as 0.5 ng of P-Tyr–BSA on immunoblots. Prepare strips of blots containing twofold serial dilutions of between 0.1 and 10 ng of P-Tyr–BSA. Use 1 : 500 to 1 : 1000 dilutions of the immune sera in immunoblotting. Either an enzyme-linked secondary antibody or ^{125}I-labeled protein A can be used to develop the blot. The anti-P-Tyr titer and affinity in an immune serum can be easily estimated by the sensitivity of its detection of P-Tyr–BSA.

Purification of anti-P-Tyr Antibodies

Anti-P-Tyr antibodies are routinely purified by affinity chromatography using Sepharose coupled with phosphotyramine or phosphotyrosine. Phosphotyramine is not yet commercially available and must be synthesized by the chemical phosphorylation of tyramine.[6] In our hands, phosphotyramine beads have a longer shelf-life and a higher binding capacity than phosphotyrosine beads. This is likely to be due to a more efficient coupling of phosphotyramine to the Sepharose. Since phosphotyrosine is commercially available, it would be easier to prepare phosphotyrosine–Sepharose. Use CNBr-activated Sepharose from Pharmacia (Piscataway, NJ). Between 15 and 20 mg of phosphotyrosine can be coupled to 1 g dry weight of beads following manufacturer's instructions. The procedure for the purification of anti-P-Tyr antibodies is as follows:

1. Ammonium sulfate precipitation of antibodies from sera: Add 22.6

[6] A. H. Ross, D. Baltimore, and H. N. Eisen, *Nature (London)* **294,** 654 (1981).

g of ammonium sulfate/100 ml of sera at 0° (40% saturation). Spin down protein pellet at 12,000 g for 20 min at 4°.

2. Dissolve the pellet in a minimal volume of PBS and dialyze against at least 20 vol of PBS for 4 hr to overnight at 4°.

3. Mix dialysate with phosphotyrosine–Sepharose in a plastic centrifuge tube and let the tube rotate overnight at 4°. Use 2 ml of packed beads to process antibodies from 100 ml of sera. If the titers are very high, more beads may be required. If the titer is low, reduce the amount of beads. The ratio of beads to sera can be optimized but it is not critical.

4. Pour the content of the tube into a disposable column [e.g., Bio-Rad (Richmond, CA) Econo column] to pack the beads. The flow-through fraction contains antibodies for SV40 t antigen (which also react with large T antigen) and antibodies for the abl protein (which react with c-abl proteins of human and mouse cells). Always monitor the flow-through fraction for anti-P-Tyr reactivity, for it is possible that the beads may have deteriorated.

5. Wash the beads extensively with PBS, about 50 column volumes, until A_{280} drops below 0.05.

6. Elute the bound anti-P-Tyr antibodies using a solution of 40 mM phenyl phosphate in PBS. Prepare this solution fresh each time. A majority of the antibodies will be eluted in the first two column volumes. Monitor the elution by reading A_{280}. The phenyl phosphate solution has an A_{280} of about 0.35. The peak fraction usually has an A_{280} reading of over 1.5. Antibodies of higher affinity may elute slower, so collect about four column volumes of eluant. Read the A_{280} of the combined eluant and determine the net A_{280} to estimate the antibody concentration. A net A_{280} of 0.5 is roughly equivalent to an antibody concentration of 0.4 mg/ml.

7. The pooled fractions must be dialyzed extensively to remove the bound phenyl phosphate from the antibodies. To stabilize the antibodies during dialysis, add BSA to the pooled fractions at a final concentration of 5 mg/ml. Dialyze against several changes of PBS plus 0.02% azide. At least 10^{-12} dilution of phenyl phosphate should be achieved in the dialysis step to free the antibodies from this bound ligand. The antibody solution can be concentrated by dialysis against 50% glycerol in PBS plus azide.

8. Store the dialyzed antibodies in aliquots at −20°. The antibodies can also be lyophilized and stored dry and reconstituted with the addition of water. There is not loss of activity with lyophilization.

9. The phosphotyrosine or phosphotyramine beads can be reused multiple times (at least 10 times). Regenerate the beads by washing with 1 column volume of an acid–urea solution (0.23 M glycine-hydrochloride, pH 2.6, 6 M urea), followed by washing with 10 column volumes of PBS. Store beads in PBS plus 0.02% azide at 4°.

Comments on Purification

Elution with phenyl phosphate recovers about half of the antibodies bound to the phosphotyramine beads. Another half of the bound antibodies can be stripped off the column using 50 mM diethylamine at pH 11. The high pH elution causes the denaturation of some antibodies because a white precipitate can be seen after neutralization and dialysis. Both the phenyl phosphate- and the alkaline-eluted antibodies detect the tyrosine-phosphorylated abl proteins. However, we have found that the alkaline preparation does not react with several other tyrosine-phosphorylated proteins as efficiently as the phenyl phosphate preparation. It is possible that some anti-P-Tyr antibodies in this polyclonal pool are more sensitive to alkaline denaturation. We have noticed that the titer and the affinity of anti-P-Tyr antibodies vary among different rabbits. The best rabbit we have had produced 100 μg of anti-P-Tyr antibodies/ml serum and the poorest produced about 10 μg/ml serum. On average, 30 to 60 μg of anti-P-Tyr antibodies can be obtained from 1 ml of immune serum. The anti-t or anti-abl antibodies, on the other hand, are present at the level of 1–2 mg/ml serum. Thus, the anti-P-Tyr antibodies represent a minority of total antibodies induced by the t/abl protein.

Use of Purified Anti-P-Tyr Antibodies

Immunoblotting

The polyclonal anti-P-Tyr antibodies generated by the above methods have been most useful in the detection of P-Tyr proteins on immunoblots. At a concentration of 1 to 3 μg/ml, these anti-P-Tyr antibodies can detect nanogram levels of tyrosine-phosphorylated proteins in lysates of 5×10^5 cells on immunoblots.[7]

Preparation of Whole-Cell Lysates. Attached cells are scraped off the culture dish in a small volume of cold PBS plus 5 mM EDTA and the cell suspension is transferred into an Eppendorf tube and spun for 10 sec. Add SDS sample buffer [66 mM Tris-HCl, pH 6.8, 2% (w/v) SDS, 10 mM EDTA, 10% (v/v) glycerol, and trace Bromphenol Blue] to solubilize the pelleted cells at a density of 1×10^7/ml and immediately place the cell lysate in a boiling water bath and heat for 10 to 15 min. The lysate will become nonviscous and can be used in SDS-PAGE. To determine the protein concentration of a whole-cell lysate, precipitate the solubilized proteins with trichloroacetic acid (50× volume of 5% TCA) and then use the Lowry method for protein determination.

---

[7] A. O. Morla and J. Y. J. Wang, *Proc. Natl. Acad. Sci. U.S.A.* **83,** 8191 (1986).

Immunoblotting is performed by the standard methods except for a few modifications:

1. Electrotransfer of proteins onto nitrocellulose paper or Immobilon membrane (Millipore, Bedford, MA) is performed using an SDS–methanol buffer (3 g Tris base, 14.4 g glycine, 1 g SDS, and 200 ml methanol/liter), at a constant current of 1000 to 1500 mA for 45 min to 1 hr. A water cooling system is used to reduce heat during transfer.

2. Use only fraction V BSA for blocking of the blots and for dilution of the antibody, the protein A, or the secondary antibody. Milk and some other preparations of BSA contain materials which inhibit the antibody reaction.

3. Establish specificity by competitions with phosphoamino acids. At 10 mM, phosphoserine and phosphothreonine do not compete with the antibodies whereas phosphotyrosine should completely abolish the antibody reaction. Phosphohistidine or tyrosine sulfate are not recognized by these anti-P-Tyr antibodies.

4. The anti-P-Tyr antibodies can be probed with ^{125}I-labeled protein A (use 1 μCi/ml of high specific activity ^{125}I-labeled protein A of 30–50 μCi/μg), or alkaline phosphatase-coupled secondary antibodies. The sensitivity of detection of the two methods is similar.

5. Diluted anti-P-Tyr antibodies are stable at 4° for at least 1 month and can be reused on blots. Diluted ^{125}I-labeled protein A can also be reused multiple times over a 2- to 4-week period.

Immunoprecipitation

Use 10 to 30 μg/ml anti-P-Tyr antibodies for the immunoprecipitation of P-Tyr proteins. Certain P-Tyr proteins which are readily detected by the antibodies on immunoblots cannot be efficiently immunoprecipitated. This could be due to the inaccessibility of the P-Tyr determinants in the native forms of some proteins. It is also possible that some P-Tyr sites are recognized with low affinity and can be detected only when the P-Tyr protein is immobilized on blots.

To improve the efficiency of immunoprecipitation, proteins can be first denatured in SDS and then diluted with a nonionic detergent before the addition of antibodies. The denaturation step also reduces the activity of phosphatases which can further decrease the antigen concentration. For example, prepare a whole-cell lysate at high cell density (1–2 × 10^7 cells/ml) in 2% SDS by the procedures described above. Then dilute that lysate 5- to 10-fold with a 1% Triton X-100 buffer containing protease inhibitor [2 mM phenylmethylsulfonyl fluoride (PMSF) and phosphatase inhibitors (2 mM vanadate, 10 mM EDTA, 5 mM NaF, and 50 mM pyrophosphate)].

The efficiency of immunoprecipitation can be increased for some but not all P-Tyr proteins by this denaturation protocol.

Anti-P-Tyr antibodies can be used to precipitate proteins labeled with radioactive amino acids or with [^{32}P]phosphate depending on the purpose of the experiment. P-Tyr proteins can be eluted from the immunoprecipitates by heating in SDS or by incubation with 40 mM phenyl phosphate. About 40 to 60% of the bound P-Tyr proteins can be eluted with 40 mM phenyl phosphate. This elution procedure can be used when it is necessary to retain the activity of a P-Tyr protein.

In Situ Staining of Phosphotyrosine Proteins in Cells or Tissues

Standard indirect immunofluorescence or immunoelectron microscopy methods have been used to stain cells or tissues with these purified anti-P-Tyr antibodies.[8,9] These methods have lower sensitivity than immunoblotting but are powerful in the localization of P-Tyr proteins in specimens of interest. Immunohistochemical methods have also be used to detect P-Tyr proteins in developing rat brains or in mouse testis (Dr. John G. Wood, 1988, and Dr. Debra Wolgemuth, 1987, personal communications).

Staining of bacterial colonies fixed on filter papers with anti-P-Tyr antibodies has been used to screen for tyrosine kinase mutants of v-*abl*.[10] This method has also been adapted to screen cDNA expression libraries for tyrosine kinase clones (see Volume 200, [47], in this series).[11] Colony hybridization with anti-P-Tyr antibodies may also have applications in screening for inhibitors or activators of tyrosine kinases.

Other Use of Anti-Phosphotyrosine Antibodies

Purification of tyrosine-phosphorylated proteins can be achieved using anti-P-Tyr antibodies coupled to beads. Because the yield of these polyclonal antibodies is low, they have not been used in preparative procedures. The hybridoma anti-P-Tyr antibodies described in Chapters [8] and [9] of this volume are much better reagents for preparative purposes.

Immunoblotting with anti-P-Tyr antibodies can also be used to measure the relative P-Tyr levels of proteins. Since all P-Tyr proteins identified thus far also contain phosphoserine and/or phosphothreonine, the incorporation of ^{32}P cannot be used as a measure of P-Tyr content. Phosphoamino

[8] E. B. Pasquale, P. A. Maher, J. Y. J. Wang, and S. J. Singer, *Proc. Natl. Acad. Sci. U.S.A.* **83**, 5507 (1986).
[9] K. Takata and S. J. Singer, *Blood* **71**, 818 (1988).
[10] E. T. Kipreos, J. G. Lee, and J. Y. J. Wang, *Proc. Natl. Acad. Sci. U.S.A.* **84**, 1345 (1987).
[11] E. B. Pasquale and S. J. Singer, *Proc. Natl. Acad. Sci. U.S.A.* **86**, 5449 (1989).

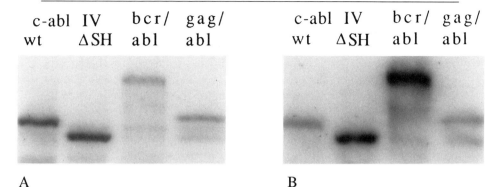

FIG. 3. Determination of relative phosphotyrosine content in proteins by immunoblotting with anti-P-Tyr antibodies. The four abl proteins were transiently expressed in COS cells using pSVL vector.[14] The wild-type (wt) abl protein was the murine type IV c-abl protein. The type IV cDNA was deleted internally from the first StuI site to the first HincII site in mutant ΔSH. This internal, in-frame deletion eliminated a regulatory domain and led to the activation of the c-abl autokinase activity.[2] The bcr/abl fusion protein was constructed from the human bcr and murine c-abl cDNAs.[15] This fusion protein is naturally found in human chronic myelogenous leukemia cells. The gag/abl fusion protein was that encoded by the genome of Abelson murine leukemia virus. (A) The anti-abl blot was probed with a monoclonal anti-abl antibody, 8E9. (B) The anti-P-Tyr blot was probed with affinity-purified anti-P-Tyr antibodies described here.

acid analysis is tedious and inaccurate because P-Tyr is lost during acid hydrolysis. Immunoblotting with anti-P-Tyr antibodies, on the other hand, is a simple way to measure the P-Tyr content of proteins. By immunoblotting serial 2-fold dilutions of P-Tyr–BSA between 1 ng and 1 μg, we have found that the signals on anti-P-Tyr immunoblots are linear over this 1000-fold range. Synthetic P-Tyr–BSA contains multiple P-Tyr (molar ratio during coupling is 64 mol P-Tyr to 1 mol BSA). If we assume that an average of 10 P-Tyr/BSA is present in our sample, then the signals on anti-P-Tyr immunoblots are linear in detecting between 0.04 and 40 ng of protein-bound phosphotyrosine. The P-Tyr content of proteins in most biological samples should fall into the linear range of detection by the anti-P-Tyr immunoblotting.

To determine the relative P-Tyr content, a protein of interest can be partially purified by immunoprecipitation and then probed with anti-P-Tyr antibodies and the anti-protein antibody on two separate immunoblots. This allows the determination of an P-Tyr/protein ratio from the signals on the blots. An example of P-Tyr/protein ratio determination is shown in Fig. 3. Four types of abl proteins were expressed in monkey COS cells and immunoprecipitated with a polyclonal anti-abl serum. The precipitated proteins were solubilized and applied to two gels. After transferring the proteins onto Immobilon membranes, immunoblotting was performed with either anti-abl monoclonal antibody (Fig. 3A) or anti-P-Tyr antibodies (Fig. 3B). The level of expression of the four proteins was not identical. Wild-type and ΔSH c-abl proteins accumulated to higher levels than the bcr/abl or the gag/abl oncoproteins (Fig. 3A). The level of P-Tyr in these four proteins was obviously different. The apparent P-Tyr/abl ratio of the bcr/abl protein was the highest. The gag/abl and the ΔSH proteins had comparable P-Tyr/abl ratio and the wild-type protein had the lowest P-Tyr/abl ratio. To increase the accuracy of determination of P-Tyr/abl ratio, an equal amount of each protein can be loaded onto SDS gels so that the anti-abl signals are the same for all four proteins. Immunoblotting of equal amount of these proteins with anti-P-Tyr antibodies will provide a direct determination of the relative P-Tyr level of these proteins.

The P-Tyr/protein ratio can also be used to determine the stoichiometry of tyrosine phosphorylation:

1. Prepare a standard of 1 mol/mol tyrosine-phosphorylated protein by immunoprecipitation with anti-P-Tyr antibodies.
2. The anti-P-Tyr precipitated sample is then divided between two immunoblots and probed with anti-protein and anti-P-Tyr antibodies. The P-Tyr/protein ratio of the 1 mol/mol P-Tyr protein can then be used as a standard to determine the stoichiometry of tyrosine phosphorylation of that protein in biological samples.[12] Again, loading similar amounts of proteins for anti-P-Tyr blotting will increase the accuracy of the determination.
3. This method cannot be used to determine the absolute stoichiometry if the protein of interest contains multiple P-Tyr sites, because it is not possible to obtain a 1 mol/mol tyrosine-phosphorylated standard by anti-P-Tyr immunoprecipitation. However, an estimation of the relative stoichiometry of phosphorylation can still be obtained by the described procedures.

Summary

Polyclonal antibodies for phosphotyrosine can be obtained by immunization with a t/abl fusion protein. These antibodies work well in immunoblotting and immunostaining experiments to identify P-Tyr proteins.

[12] A. O. Morla, G. Draetta, D. Beach, and J. Y. J. Wang, *Cell (Cambridge, Mass.)* **58,** 193 (1989).

For example, a 95-kDa thrombin-induced P-Tyr-protein in platelets[13] is recognized by these polyclonal antibodies but reacts poorly with other types of anti-P-Tyr antibodies (A. Golden and J. S. Brugge, 1988, personal communications). To obtain large quantities of anti-P-Tyr antibodies for preparative purposes, it is much easier to use hybridomas. An attempt to isolate hybridoma anti-P-Tyr antibodies by immunization with t/abl protein was unsuccessful, possibly because anti-P-Tyr antibodies were produced by a small percentage (~2%) of the immune cells responsive to the t/abl fusion protein. Monoclonal anti-P-Tyr antibodies, however, have been produced using other types of immunogens (see Chapters [8] and [9] in this volume). The different preparations of anti-P-Tyr antibodies appear to have distinct characteristics. Test several preparations to determine the optimal reagents for each experimental need.

[13] A. Golden and J. S. Brugge, *Proc. Natl. Acad. Sci. U.S.A.* **86**, 901 (1989).
[14] J. Y. J. Wang, *Oncog. Res.* **3**, 293 (1988).
[15] J. R. McWhister and J. Y. J. Wang, *Mol. Cell. Biol.* **11**, 1553 (1991).

[7] Preparation and Use of Anti-Phosphotyrosine Antibodies to Study Structure and Function of Insulin Receptor

By Morris F. White and Jonathan M. Backer

Introduction

Antibodies which bind phosphotyrosine [Tyr(P)] are extremely useful for the study of tyrosyl-specific protein kinases. The phosphotyrosine antibody (αPY) has remarkable practical value because phosphotyrosine-containing proteins are quantitatively rare in quiescent cells until a specific event, such as growth factor binding or oncogenic transformation, activates specific tyrosyl kinases. This chapter describes the preparation and use of a polyclonal αPY to elucidate the structure and function of the insulin receptor; this αPY has been used successfully in other systems as well.[1-3] The insulin receptor is a protein kinase that undergoes autophosphorylation on at least five tyrosyl residues during insulin binding, and

[1] D. R. Kaplan, M. Whitman, B. Schaffhausen, D. C. Pallas, M. F. White, L. Cantley, and T. M. Roberts, *Cell (Cambridge, Mass.)* **50**, 1021 (1987).
[2] C. E. Rudd, J. M. Trevillyan, J. D. Dasgupta, L. L. Wong, and S. F. Schlossman, *Proc. Natl. Acad. Sci. U.S.A.* **85**, 5190 (1988).
[3] D. K. Morrison, P. J. Browning, M. F. White, and T. M. Roberts, *Mol. Cell. Biol.* **8**, 176 (1988).

catalyzes tyrosyl phosphorylation of other proteins.[4] This αPY has been valuable during our efforts to elucidate the structure and function of the insulin receptor and its cellular substrates.

This chapter descibes the preparation of N-bromoacetyl-O- phosphotyramine, which is used to immunize rabbits for the generation of the αPY, and it describes the affinity purification of the αPY from rabbit serum. We also discuss several applications of the αPY, including (1) detection of phosphotyrosine-containing proteins in insulin-stimulated cells by immunoprecipitation and Western blotting, (2) determination of insulin receptor autophosphorylation stoichiometry *in vivo*, (3) determination of internalization of the phosphorylated insulin receptor,and (4) the identification of an insulin-stimulated phosphatidylinositol-3-phosphate kinase.

Preparation of Polyclonal Anti-Phosphotyrosine Antibody

Pang *et al.* first described the preparation of αPY by immunization of rabbits with O-phosphotyramine coupled to keyhole limpet hemocyanin (KLH).[5,6] The method involves the synthesis of phosphotyramine, addition of N-bromoacetyl bromide, and finally the coupling of N-bromoacetyl-O-phosphotyramine to KLH.

Synthesis of Phosphotyramine

1. Dissolve 1 g of tyramine (Aldrich, Milwaukee, WI) and 4.2 g of phosphorus pentoxide (Fisher, Fairlawn, NJ) in 4.1 ml of 85% phosphoric acid (Baker, Phillipsburg, NJ), and heat the mixture to 100° for 72 hr in a sealed tube.

2. Dilute the reaction to 50 ml with water and apply it onto a 20-ml Dowex 50 column (Sigma, St. Louis, MO) that was equilibrated with water. During elution of the resin with water, two peaks are observed at 268 nm. The first is relatively small and narrow, whereas the second is very broad and contains phosphotyramine. Over 1 liter of water may be necessary to completely elute pure phosphotyramine.

3. Reduce the volume of the second fraction to 100 ml on a rotary evaporator or by lyophilization, and crystallize the phosphotyramine from this solution by adding 100 ml of cold ethanol and incubating the solution at 4° for 24 hr. Collect the crystals by filtration and dry them *in vacuo*. By thin-layer chromatography, the product contains only phosphotyramine.

[4] C. R. Kahn and M. F. White, *J. Clin. Invest.* **82,** 1151 (1988).
[5] D. T. Pang, B. R. Sharma, and J. A. Shafer, *Arch. Biochem. Biophys.* **242,** 176 (1985).
[6] D. T. Pang, B. Sharma, J. A. Shafer, M. F. White, and C. R. Kahn, *J. Biol. Chem.* **260,** 7131 (1985).

Phosphotyramine can be separated from tyramine by thin-layer chromatography on silica gel developed with n-butanol : acetic acid : H_2O (7 : 2 : 5), and visualized with I_2. The R_f values of tyramine and phosphotyramine are 0.67 and 0.475, respectively.

Preparation of N-Bromoacetyl-O-phosphotyramine

1. Dissolve phosphotyramine (1 g) in 20 ml of 100 mM sodium borate and adjust the pH to 8.5 with 5 M LiOH.

2. Slowly add 8.8 g of bromoacetyl bromide (Aldrich) over a 15-min time interval. Stir the reaction with a Teflon-coated stirring bar. The pH falls rapidly during the addition of bromoacetyl bromide, and must be monitored continuously and maintained between 7.5 and 8.5 by careful addition of 5 M LiOH.

3. Dry the reaction solution by lyophilization, and macerate and extract the residue three times with ethyl ether, and then three times with ethanol. The product, n-bromoacetyl-O-phosphotyramine, migrates with an R_f of 0.42 on silica gel developed with n butanol : acetic acid : H_2O (7 : 2 : 5). The yield of this reaction can be estimated by the amount of O-phosphotyramine remaining in the mixture, and is generally >75%. Complete purification is not necessary as the starting materials are nonreactive in the next step.

Coupling N-Bromoacetyl-O-phosphotyramine to Keyhole Limpet Hemocyanin

1. Suspend 30 mg of $(NH_4)_2SO_4$-precipitated KLH (Calbiochem, San Diego, CA) in 5 ml of 100 mM sodium borate, pH 9.25, and exhaustively dialyze the KLH against 100 mM sodium borate.

2. Add 350 mg of N-bromoacetyl-O-phosphotyramine to the dialyzed KLH and incubate the mixture at 22° for 96 hr. Check the pH daily and maintain it above 9.0 by adding small portions of 5 M LiOH. After 96 hr, dialyze the solution against phosphate-buffered saline (PBS), pH 7.4, and concentrate the protein in the clear solution with an Amicon (Danvers, MA) ultrafiltration cell to 10 mg/ml.

Estimation of Yield of Coupled Phosphotyramine

The yield of the coupling reaction is estimated conveniently by UV spectrophotometry. The method is based on the fact that the phenol ring of phosphotyramine maximally absorbs at 268 nm and does not absorb at 293 nm.[5] The molar concentrations of the protein-linked hapten, [H], and the protein conjugate, [C], is estimated from the following equations:

$$[H] = \{Abs_{268}^{c} - [Abs_{290}^{c}(Abs_{268}^{p}/Abs_{290}^{p})]\}/\varepsilon_{268}^{h}$$
$$[C] = Abs_{290}^{c} (Abs_{280}^{p}/Abs_{290}^{p})/\varepsilon_{280}^{p}$$

Absorbance of KLH at each wavelength, Abs_{268}^{p}, Abs_{280}^{p}, and Abs_{290}^{p}, and absorbance at each wavelength of the KLH–phosphotyramine conjugate, Abs_{268}^{c}, Abs_{280}^{c}, and Abs_{290}^{c}, are measured for a 0.1 mg/ml solution of each protein. The values of ε_{268}^{h} and ε_{280}^{p} in 0.1% solutions are taken as 783 $cm^{-1}M^{-1}$ and 1.77 $cm^{-1}M^{-1}$, respectively. The incorporation of O-phosphotyramine residues/100 amino acyl residues of KLH is equal to $[H] \times 10^{4}/[C]$; our usual result is about 3 to 4 O-phosphotyramine residues/100 amino acyl residues.

Rabbit Immunization

Polyclonal αPY is raised in New Zealand White rabbits according to standard immunization protocols.[7] KLH–Tym(P) (1 mg) dissolved in 0.1 ml of PBS is emulsified in 0.9 ml of Freund's complete adjuvant (Difco, Detroit, MI). Portions (0.1 ml) of this suspension are injected into 10 different sites on the rabbit's back. After 21 days, similar inoculations are made with Freund's incomplete adjuvant containing KLH–Tym(P). The rabbits are bled after 14 days, reinoculated 7 days later, and followed by another series of bleedings and inoculations. The blood is allowed to clot on ice for 5–6 hr, and the serum is separated by centrifugation at 4° for 20 min at 2000 g and stored frozen at −20° until use.

Purification of αPY from Rabbit Serum by Affinity Chromatography on Immobilized Phosphotyramine

1. An affinity column of immobilized phosphotyramine is prepared by incubating 1 g of phosphotyramine with 5 ml of Affi-Gel (Bio-Rad, Richmond, CA) suspended in 10 ml of 50 mM NaHCO$_3$ for 1 hr at 22°. After the reaction, the Affi-Gel is washed with 50 mM NaHCO$_3$, incubated with 1 M ethanolamine, and washed extensively with 50 mM HEPES.

2. The αPY in 25 ml of serum is bound to 1 ml of phosphotyramine–Affi-Gel during a 10-hr batch incubation at 4°. The Affi-Gel is transferred to a column and washed successively with 50 ml of PBS (pH 7.4), 50 ml of 200 mM phosphate buffer (pH 7.0), and 50 mM HEPES containing 150 mM NaCl.

3. The αPY is eluted completely with three 5-ml portions of 200 mM p-nitrophenyl phosphate (Sigma) in 50 mM HEPES (pH 7.4). The Affi-Gel

[7] E. Harlow, and D. Lane, eds., "Antibodies: A Laboratory Manual," Cold Spring Harbor Lab., Cold Spring Harbor, New York, 1988.

is suspended in each 5-ml portion of eluate for 8 hr. The yield of αPY from 25 ml of serum is 7–8 mg assayed by the Bradford method (Bio-Rad) using IgG as the standard.

4. The *p*-nitrophenyl phosphate is completely removed from the eluate by exhaustive dialysis against 50 m*M* HEPES (pH 7.4) at 4°. Ordinarily, five changes of 2 liters each are sufficient to remove all of the *p*-nitrophenyl phosphate from 75 ml of eluate.

Identification of Phosphotyrosine-Containing Proteins with αPY

The αPY immunoprecipitates and immunoblots the insulin receptor β subunit and other proteins that undergo tyrosyl phosphorylation in cells during insulin stimulation. Protocols for both of these experiments are described below.

Immunoprecipitation of Radioactively Labeled Proteins

This protocol for labeling cells and immunoprecipitation of insulin-stimulated phosphotyrosine-containing proteins has been used in our laboratory for several years with many different cell types.[8] To illustrate the procedure, [^{32}P]phosphate labeling of a well-differentiated rat hepatoma cell line, Fao, is described; however, the method works successfully with [^{35}S]methionine labeling, surface ^{125}I iodination, and affinity labeling with [^{125}I]insulin by cross-linking as shown in Fig. 1.[9]

[^{32}P]Phosphate Labeling and Preparation of Cell Extracts

1. Incubate confluent Fao cells in 10-cm dishes for 2 hr with at least 5 ml of phosphate-free and serum-free RPMI 1640 medium (GIBCO, Grand Island, NY) containing 0.5 mCi/ml carrier-free ortho[^{32}P]phosphate (New England Nuclear, Boston, MA). Three milliliters of medium is sufficient to cover the monolayer if the dish is slowly rocked in the incubator with a Hoeffer "Red Rocker" (model No. PR 50); otherwise use 5 ml.

2. Add insulin (30 μl of the appropriate stock solution to 3 ml of medium) and continue the incubation at 37° for the desired time intervals. Stop phosphorylation and dephosphorylation quickly by removing the incubation medium and freezing the cell monolayers by pouring liquid nitrogen into the dishes.

3. Before the monolayer of cells thaws, add 1–2 ml of extraction buffer

---

[8] M. F. White, J. N. Livingston, J. M. Backer, V. Lauris, T. J. Dull, A. Ullrich, and C. R. Kahn, *Cell (Cambridge, Mass.)* **54**, 641 (1988).

[9] M. F. White, E. W. Stegmann, T. J. Dull, A. Ullrich, and C. R. Kahn, *J. Biol. Chem.* **262**, 9769 (1987).

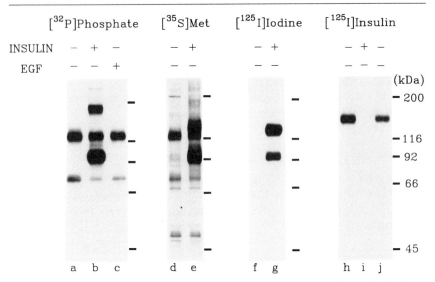

FIG. 1. Identification of the insulin receptor and the pp185 substrate in Fao cells. Confluent monolayers of Fao cells were labeled with ortho[^{32}P]phosphate (lanes a–c) as described in the text, or by labeling with [^{35}S]methionine (lanes d and e), [^{125}I]iodine (lanes f and g), or [^{125}I]insulin (lanes h–j) as described previously.[9] After incubation with 100 nM insulin (lanes b, e, and g) or epidermal growth factor (EGF) (lane c) for 1 min, or with 100 nM insulin for 3 hr during insulin binding (lane i), the cells were prepared for immunoprecipitation. The αPY was used to immunoprecipitate the phosphotyrosine-containing proteins from the whole-cell extracts. In one case (lane j), an anti-insulin receptor antibody was used to immunoprecipitate the insulin receptor previously precipitated with αPY and eluted with PNPP. The autoradiograms represent exposures of 6 hr (lanes a–c), 12 hr (lanes d and e), 12 hr (lanes f and g) and 4 days (lanes h–j).

containing Tris (50 mM, pH 8.2), NaCl (118 mM), NaF (100 mM), Na$_3$VO$_4$ (2 mM), Na$_3$P$_2$O$_7$ (10 mM), EDTA (4 mM), phenylmethylsulfonyl fluoride (PMSF) (1 mM), aprotinin (0.1 mg/ml), and Triton X-100 (1.0%, v/v). Scrape the frozen cells from the dishes and thaw the cell ice, but keep the samples ice cold. Vortex the extract and sediment the insoluble material by centrifugation at 50,000 rpm in a Beckman (Palo Alto, CA) 70.1 Ti rotor for 30 min. Separate and save the supernatant for immunoprecipitation.

Immunoprecipitation

1. Incubate the clarified cell extracts with 1 to 2 μg of affinity-purified αPY for at least 2 hr at 4°. An overnight incubation yields identical results.

2. Precipitate the antibody complex during a 1-hr incubation with 50 μl of 10% Pansorbin (CalBiochem) or protein A-Sepharose. We prefer Pansorbin because it yields a very tight pellet during centrifugation.

However, the Pansorbin must be washed exhaustively before use as previously described.[10]

3. Wash the Pansorbin three times by centrifugation and resuspension in 50 mM HEPES (pH 7.4) containing NaCl (118 mM), NaF (100 mM), Na$_3$VO$_4$ (2 mM), 0.1% (w/v) sodium dodecyl sulfate (SDS), and 1% (v/v) Triton X-100.

4. Elute the proteins from the washed precipitates with SDS–PAGE sample buffer.[10] Reduce the proteins with 100 mM dithiothreitol and separate them by SDS–PAGE on 6.0 to 7.5% gels. Our methods for SDS–PAGE have been described previously in this series,[10] but any established method will work. Alternatively, elute the phosphotyrosine-containing proteins by overnight incubation at 4° with 50 μl of 20 mM p-nitrophenyl phosphate in 50 mM HEPES (pH 7.4). Centrifuge and remove the supernatant and add 50 μl of 2X SDS–PAGE sample buffer. Use this method when problems are encountered with precipitation of nonspecific phosphoproteins; two sequential elutions are often necessary to completely elute the proteins.

5. Identify the phosphoproteins in the stained and dried gels by autoradiography at −70° using Kodak (Rochester, NY) X-Omat film and an intensifying screen. Quantify the radioactivity in the gel fragments by Cerenkov counting.

Immunoblotting with αPY

Immunoblot analysis with αPY has been used in many laboratories to identify phosphotyrosine-containing proteins and putative substrates for growth factor receptors, including the insulin receptor. Any standard immunoblotting procedure appears to work with this αPY. We briefly describe the protocol below. No systematic studies have been carried out to determine the optimum conditions so that the maximum number of proteins can be identified, so other protocols may be worth trying.

1. Prepare a cell extract from unlabeled cells as described above. Direct lysis of cells with SDS–PAGE sample buffer may be preferable in certain cases to prevent dephosphorylation. Separate the proteins in the extract by SDS–PAGE. Specific proteins can be purified from the Triton X-100 extract before the electrophoresis by affinity chromatography or immunoprecipitation, but this step is not required. We often concentrate the Tyr(P)-containing proteins before SDS–PAGE by immunoprecipitation with the αPY.

---

[10] M. Kasuga, M. F. White, and C. R. Kahn, this series, Vol. 109, p. 609.

2. Transfer the proteins to nitrocellulose by electroblotting, using standard techniques.[11] Use a standard transfer buffer composed of Tris (12 g), glycine (57.65 g), water (3.2 liters), and methanol (0.8 liters); addition of 0.1% SDS is helpful for transferring high-molecular-weight proteins. Carry out electroblotting for 2 hr at 100 V with appropriate cooling, or 15 hr at 10 V.

3. After the transfer, wash the nitrocellulose briefly with water, stain the nitrocellulose with 0.5% Ponceau S (Sigma) in 1% acetic acid, and destain with water to identify the standards. Mark the standards with pencil, as the Ponceau S staining will be lost.

4. Incubate the nitrocellulose with 1% bovine serum albumin/PBS (BSA/PBS) for 1 hr at 37° to block nonspecific binding. Wash for 1 hr at 22° with PBS.

5. Incubate the nitrocellulose with gentle rocking in a small container or sealed plastic bags for 1 hr at 22° with 20 ml of BSA/PBS containing 20–30 μg of αPY. After incubation remove and store the antibody solution, as it can be reused, and wash the nitrocellulose with PBS three times at 22° for 10 min each.

6. Incubate nitrocellulose with ^{125}I-labeled protein A (10 μCi/20 ml) in BSA/PBS for 1 hr at 22° with gentle rocking. After incubation remove and store the ^{125}I-labeled protein A solution, as it can be reused, and wash the nitrocellulose with PBS three times at 22° for 10 min each. Locate the phosphoproteins on the dried nitrocellulose by autoradiography at −70° using Kodak X-Omat film.

Determination of Stoichiometry of Tyrosyl Autophosphorylation of Insulin Receptor by Sequential Immunoprecipitation with αPY and Anti-Insulin Receptor Antibody

Immunoprecipitation or immunoblotting of phosphotyrosine-containing proteins does not provide an estimate of the stoichiometry of phosphorylation. However, combined with a protein-specific antibody, the stoichiometry of tyrosyl phosphorylation can be determined. The method described here was used with the insulin receptor in Fao cells, but has been applied successfully in other cell type. It involves the sequential immunoprecipitation of [^{35}S]methionine-labeled receptor from unstimulated or insulin-stimulated cells, first with the αPY and then with an anti-insulin receptor antibody. Other labeling techniques also work, including surface iodination and affinity labeling by cross-linking [^{125}I]insulin to

[11] L. G. Davis, M. D. Dibner, and J. F. Battey, "Basic Methods in Molecular Biology." Elsevier, New York, 1986.

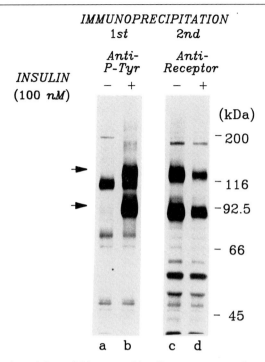

IMMUNOPRECIPITATION
1st   2nd

FIG. 2. Estimation of the stoichiometry of insulin receptor tyrosyl autophosphorylation in Fao hepatoma cells. The cells were labeled with [³⁵S]methionine, treated without (−) or with (+) insulin for 1 min, and immunoprecipitates were prepared as described in the text. The αPY immunoprecipitates are shown in lanes a and b, whereas the subsequent anti-insulin receptor antibody immunoprecipitates obtained from the αPY-depleted extracts are shown in lanes c and d.

the receptor with disuccinimidyl suberate. Immunoprecipitation of these labeled receptors with αPY provides an estimate of the fraction of receptors that contain Tyr(P), whereas ^{32}P labeling yields a different result owing to multiple phosphorylation sites and changing stoichiometries during the experiments. For the insulin receptor in [³⁵S]methionine-labeled Fao hepatoma cells, 70% of the labeled receptors is immunoprecipitated with αPY from insulin-stimulated cells (Fig. 2, lane c). Thus, before insulin stimulation, none of the receptors contain Tyr(P) (Fig. 2, lane a) and all of them are recovered with the anti-insulin receptor antibody (Fig. 2, lane b), whereas after stimulation only 30% of the receptors cannot be recognized by αPY and are immunoprecipitated only by the anti-insulin receptor antibody (Fig. 2, lane d).

Sequential Immunoprecipitation

1. Incubate the Fao cells in 10-cm dishes with 0.5 mCi of [^{35}S]methionine in 5 ml of methionine-free RPMI-1640 medium for 15 hr.[9] Stimulate the cells with 100 nM insulin and prepare a cell extract as described above for the [^{32}P]phosphate-labeled cells.

2. Immunoprecipitate the Tyr(P)-containing insulin receptors with αPY as described above. Use two sequential immunoprecipitations with the αPY to be certain that all of the Tyr(P)-containing receptors are removed from the extract. After immobilization on protein A and washing, the immunoprecipitates can be pooled.

3. Add the protein-specific anti-insulin receptor antibody to the αPY-depleted supernatant and immunoprecipitate the remaining receptors. Specific parameters for this step will depend on the characteristics of the antibody used. Again, two sequential immunoprecipitations are useful to ensure complete clearance of the protein, and the immunoprecipitations can be pooled after immobilization on protein A and washing.

4. Elute the immunoprecipitated proteins and separate by SDS–PAGE under reducing conditions. Detect the proteins by autoradiography and quantify the proteins by scintillation counting or densitometry.

Internalization of Tyrosyl-Phosphorylated Receptors

The αPY can be used to evaluate the tyrosyl phosphorylation of cell surface receptors during ligand-stimulated endocytosis. This method was designed to study the insulin-stimulated internalization of phosphorylated insulin receptors in rat hepatoma cells.[12] It combines the use of surface iodination, proteolytic removal of surface receptor, and immunoprecipitation of intact internalized receptors with the αPY. The method can be applied to other receptor and cell systems, but the choice of protease to determine internal receptors may vary. For the insulin receptor, the rate of internalization of Tyr(P)-containing receptors can be readily estimated from the appearance of trypsin-resistant α subunits (Fig. 3).

Surface Iodination of Fao Hepatoma Cells

1. Incubate confluent monolayers of Fao cells (100-mm dishes) overnight in serum-free medium (RPMI 1640, GIBCO). Before surface labeling, wash the cells two times in ice-cold PBS containing 2 mM CaCl$_2$ and 1 mM MgCl$_2$ (buffer A); leave cells on ice in buffer A.

2. Add 3 ml of ice-cold buffer A containing 10 mM β-D-glucose and 30

---

[12] J. M. Backer, C. R. Kahn, and M. F. White, *J. Biol. Chem.* **264**, 1694 (1988).

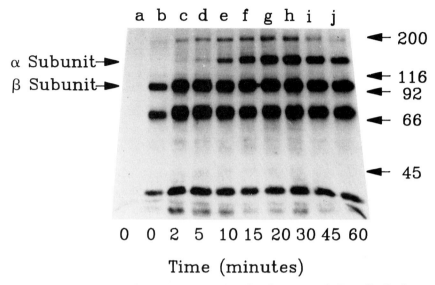

a b c d e f g h i j

α Subunit→
β Subunit→

← 200

← 116
← 92

← 66

← 45

0   0   2   5   10 15 20 30 45 60

Time (minutes)

FIG. 3. Internalization of tyrosyl-phosphorylated insulin receptors in Fao cells. Confluent monolayers of Fao cells were surface iodinated at 4° and stimulated with 100 nM insulin at 37°. After the indicated time intervals, the cells were solubilized (lane a), or trypsinized at 4°, solubilized, and immunoprecipitated with αPY. Labeled proteins were separated by SDS–PAGE under reducing conditions and visualized by autoradiography. The appearance of intact α subunit indicates the internalization of Tyr(P)-containing insulin receptor. Note that trypsinization without any insulin stimulation stimulates tyrosyl phosphorylation of the insulin receptor, a well-known phenomenon (lane b). However, this does not complicate the interpretation of the results, as all of this receptor is digested by trypsin, indicating that it is on the cell surface.

μl of 200 units/ml lactoperoxidase (Sigma) suspended in 42% $(NH_4)_2SO_4$. Place the ice tray on a rocking platform in a fume hood.

3. For each dish of cells, mix 0.5–1 mCi carrier-free [^{125}I]iodine (350–600 Ci/ml) (New England Nuclear) into 30 μl of commercially prepared glucose oxidase solution (100 units/ml, Sigma). Add this solution to the cells and rock them slowly for 30 min on ice.

4. Wash the cells five times with ice-cold buffer A to remove volatile iodine. Leave the last ice-cold wash on the cells, remove the cells from the hood, and keep them on ice until use.

Receptor Internalization

1. Remove the ice-cold buffer A from the iodinated cells immediately prior to beginning insulin stimulation. Add 10 ml RPMI 1640 medium (37°) containing 50 mM HEPES, pH 7.4, 0.1 mg/ml bovine serum albumin, and

the appropriate concentration of insulin. Incubate cells at 37° for desired length of time.

2. After incubation, rapidly remove the medium, place cells on ice, and add 10 ml of ice-cold buffer A containing 50 mM HEPES (pH 7.4) and 2 mM sodium orthovanadate (buffer B). The cells should be processed immediately to determine the total and intracellular phosphorylated receptors.

Determination of Total and Intracellular Tyrosyl-Phosphorylated Receptors

Total Tyr(P)-Containing Receptors. Remove the ice-cold buffer B and solubilize the cells directly in 1 ml of ice-cold 100 mM Tris (pH 8.0) containing 1% Triton X-100, 2 mM vanadate, 1 μg/ml leupeptin, 0.1 mg/ml aprotinin, and 0.35 mg/ml PMSF (diluted from a stock of 35 mg/ml in ethanol). Remove the insoluble material by centrifugation at 100,000 g for 30 min, and immunoprecipitate the Tyr(P)-containing receptors with the αPY as described above.

Internalized Tyr(P)-Containing Receptors

1. To determine internalized tyrosyl-phosphorylated receptors, remove buffer B and add 1 ml of 0.5 mg/ml N-tosyl-1-phenylalanine chloromethyl ketone (TPCK)-treated trypsin (Worthingon, Freehold, NJ) in ice-cold buffer B. Incubate cells with occasional rocking on ice for 30 min.

2. Stop trypsinization by adding 100 μl of 25 mg/ml soybean trypsin inhibitor. Mix well. Scrape and transfer the cells into ice-cold microfuge tubes. Wash the cells two times with ice-cold buffer B containing 2.5 mg/ml soybean trypsin inhibitor.

3. Solubilize the cell pellet in 1 ml of ice-cold 100 mM Tris (pH 8.0), containing 1% Triton X-100, 2 mM vanadate, 1 μg/ml leupeptin, 0.1 mg/ml aprotinin, and 0.35 mg/ml PMSF. Remove the insoluble material by centrifugation at 100,000 g for 30 min.

4. Immunoprecipitate and separate the proteins by SDS–PAGE as described above. Internalized receptors are those which are resistant to cell-surface trypsinization. In the case of the insulin receptors, internalization is quantified by measuring trypsin-resistant α subunits, since the β subunit is relatively trypsin insensitive in whole cells (Fig. 3).

Determination of Insulin-Stimulated Phosphatidylinositol 3-Kinase in αPY Immunoprecipitates

Insulin stimulation of CHO cells expressing transfected human insulin receptors leads to an activation of phosphatidylinositol 3-kinase (PI-3-

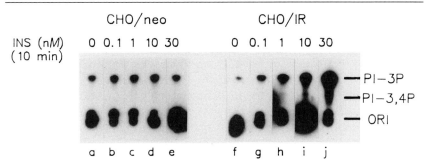

FIG. 4. The αPY immunoprecipitates PI-3-kinase activity from insulin-stimulated CHO cells. Subconfluent CHO cells transfected with the neomycin resistance gene (CHO/*neo*) and CHO cells transfected with the human insulin receptor (CHO/IR) were stimulated with the indicated concentrations of insulin for 10 min. The cells were lysed and incubated with the αPY. PI was added to the immunoprecipitate and the reaction was started by addition of [γ-^{32}P]ATP. The products were extracted with chloroform/methanol and separated by thin-layer chromatography (TLC). Migration of PI-3-P is indicated. The radioactivity at the origin (ORI) is due to [^{32}P]ATP and other water-soluble ^{32}P-labeled material that were not completely extracted from the chloroform layer. Mobile phase: chloroform–methanol–water–ammonium hydroxide (60 : 47 : 11.3 : 2, v : v).

kinase) activity in αPY immunoprecipitates.[13,14] While the presence of this activity in the αPY immunoprecipitates suggests that the PI-3-kinase is itself tyrosyl phosphorylated during insulin stimulation, this has not yet been clearly established. The assay of insulin-stimulated PI-3-kinase is similar to that described originally for platelet-derived growth factor (PDGF)-stimulated cells using this αPY, and is shown in Fig. 4.[1,15]

Preparation of Protein A-Sepharose, Phosphatidylinositol, and TLC Plates for Phosphatidylinositol 3-Kinase Assay

Protein A-Sepharose. Transfer 5 ml protein A–Sepharose into a 50-ml conical tube. Add 40 ml 10 mM Tris, pH 7.5, containing 1% (w/v) BSA, and incubate the suspension at 22° for 30 min on a rotating wheel. Finally, wash the Sepharose three times with water and suspend it (50%, v/v) in PBS containing 0.2% (w/v) azide. Store at 4°.

Aluminum Silica Gel-60 Thin-Layer Chromatography (TLC) Plates. Use aluminum silica gel-60 TLC sheets, 0.2 mm thick, without indicator

[13] G. Endemann, K. Yonezawa, and R. A. Roth, *J. Biol. Chem.* **265,** 396 (1990).
[14] N. Ruderman, R. Kapeller, M. F. White, and L. C. Cantley, *Proc. Natl. Acad. Sci. U.S.A.* **87,** 1411 (1990).
[15] K. R. Auger, L. A. Serunian, S. P. Soltoff, P. Libby, and L. C. Cantley, *Cell (Cambridge, Mass.)* **57,** 167 (1989).

(#5553-7; EM Science, Gibbstown, NJ). Place the sheets in a TLC tank, and slowly fill the tank with 1% (w/v) potassium oxalate. Drain the oxalate solution from tank slowly by siphon, and bake the plates at 100° overnight. The treated TLC plates can be stored at room temperature, but should be baked at 100° for 1 hr before use.

Phosphatidylinositol. Bovine phosphatidylinositol (PI) is obtained from Avanti (Birmingham, AL) in chloroform/methanol solution. Evaporate the solvent under a nitrogen stream in a polypropylene microfuge tube. Suspend the PI in 10 mM HEPES containing 1 mM EGTA (pH 7.5) at a concentration of 2 μg/μl. Sonicate the lipid in a Heat Systems Ultrasonics (Farmingdale, NY) or equivalent bath sonicator for 10 min; low-power bath sonicators are insufficient.

Immunoprecipitation of PI-3-Kinase from Insulin-Stimulated Chinese Hamster Ovary Cells with αPY

1. Chinese hamster ovary (CHO) cells expressing the human insulin receptor are arrested at 85% confluence by overnight incubation in Ham's F12 medium (GIBCO) containing 0.5% insulin-free BSA.[14] Stimulate the cells by adding insulin to this medium. At appropriate time intervals, rapidly discard the medium, place the cells on ice, and add 10 ml of ice-cold 20 mM Tris (pH 7.5), 137 mM NaCl, 1 mM MgCl$_2$, 1 mM CaCl$_2$, and 100 μM Na$_3$VO$_4$ (extraction buffer). Wash the cells two times in ice-cold extraction buffer.

2. Add 1 ml of extraction buffer containing 10% glycerol, 1% nonidet P-40 (NP-40; Sigma), 1 mM PMSF (diluted from a 35 mg/ml ethanol stock), 0.1 mg/ml aprotinin, and 1 μg/ml leupeptin. Rock the dishes at 4° for 20 min. Scrape and transfer the cells to cold polypropylene microfuge tubes, and remove the insoluble material by centrifugation in a microfuge at 10,000 g for 20 min at 4°.

3. Add 1–2 μg of αPY to the supernatant and incubate this mixture at 4° for 2 hr. Precipitate the αPY with 50 μl of 50% protein A–Sepharose for 1 hr at 4°. Wash the protein A–Sepharose pellet with the following solutions:

   a. Three times with PBS containing 1% NP-40 and 100 μM vanadate
   b. Three times with 100 mM Tris (pH 7.5) containing 500 mM LiCl$_2$ and 100 μM vanadate
   c. Two times in 10 mM Tris (pH 7.5) containing 100 mM NaCl, 1 mM EDTA, and 100 μM vanadate

4. Suspend the pellet in 50 μl of 10 mM Tris (pH 7.5) containing 100 mM NaCl, 1 mM EDTA, and 100 μM vanadate.

Assay of Phosphatidylinositol 3-Kinase

1. Add 10 μl of 100 mM MgCl$_2$ and 10 μl of sonicated lipid (20 μg/assay point). Initiate the phosphorylation reaction by adding 10 μl of 440 μM ATP containing 30 μCi of [^{32}P]ATP and 10 mM MgCl$_2$. Incubate the mixture for 10 min at 22° with occasional agitation. Stop the reaction by adding 20 μl of 8 N HCl.

2. Extract the PI from the reaction mixture with 160 μl of chloroform : methanol (1 : 1). Vortex the mixture and separate the phases by centrifugation for 2–3 min in a microfuge. Remove 50 μl of the lower organic layer and apply it to a TLC plate next to PI-4-P and PI-3,4-P$_2$ standards. Develop the plates in CHCl$_3$: CH$_3$OH : H$_2$O : NH$_4$OH (60 : 47 : 11.3 : 2). Adequate separation can be obtained with a 10-cm plate.

3. Allow the plates to dry and identify the lipid products by autoradiography using Kodak X-Omat film and an intensifying screen (Fig. 4). Determine the position of lipid standards by iodine staining. Radioactivity in the lipid products can be quantified by cutting or scraping the TLC plate and Cerenkov counting (Fig. 4).

Acknowledgments

This work has been supported in part by National Institutes of Health Grants DK38712 (M.F.W.) and a Diabetes and Endocrinology Research Center Grant DK36836. J.M.B. is a recipient of a National Research Service award, DK08126; M.F.W. is a scholar of the PEW Foundation, Philadelphia.

[8] Generation of Monoclonal Antibodies against Phosphotyrosine and Their Use for Affinity Purification of Phosphotyrosine-Containing Proteins

By A. Raymond Frackelton, Jr., M. Posner, B. Kannan, and F. Mermelstein

The first polyclonal and monoclonal antibodies reactive with phosphotyrosine (PY)-containing proteins were generated by immunizing animals with *p*-aminobenzylphosphonic acid (ABP) diazotized to carrier protein.[1,2] This analog of phosphotyrosine was chosen over phosphotyrosine itself as the haptenic group because the phosphonate group, in contrast to the

[1] A. H. Ross, D. Baltimore, and H. N. Eisen, *Nature (London)* **294**, 654 (1981).

[2] A. R. Frackelton, Jr., A. H. Ross, and H. N. Eisen, *Mol. Cell. Biol.* **3**, 1343 (1983).

phosphate group of PY, was resistant to enzymatic hydrolysis, and thus might be a more persistent and efficient immunogen. Polyclonal and monoclonal antibodies to ABP, however, frequently cross-reacted with a number of other phosphate-containing compounds (for example, nucleotides).[2,3] One ABP-induced monoclonal antibody, 2G8, had been extremely useful for characterizing and purifying a variety of phosphotyrosyl proteins.[2,4–8] It had, however, two limitations: (1) it had a relatively low binding constant for phosphotyrosyl proteins and therefore could be used effectively only when coupled at high density to an activated matrix such as CNBr–Sepharose; (2) it cross-reacted with mononucleotides and phosphohistidine.[2] Because of these limitations, a substantial and successful effort was made to produce a panel of monoclonal antibodies that has higher affinity and greater specificity for phosphotyrosyl proteins.

To accomplish this goal, we tested a variety of phosphotyrosine–analog conjugates, immunizing protocols, and screening procedures. The most successful immunogens were conjugates of phosphotyrosine or phosphotyramine (PYA) coupled via their free amino groups to keyhole limpet hemocyanin (KLH) using either the heterobifunctional cross-linking agent succinimidyl-4-(p-maleimidophenyl)butyrate (SMBP) or the homobifunctional cross-linker glutaraldehyde.

We describe here methods for preparing the PY–protein and PYA–protein conjugates, immunizing mice, and screening hybridoma culture fluid for specific antibodies. We compare and contrast the affinities and fine specificities of these antibodies with those of the earlier anti-ABP monoclonal antibody (2G8), and provide detailed methods for using the most avid and specific monoclonal antibody (1G2) for purifying PY-containing proteins from cells.

Preparation of Immunogens and Antigens for ELISA

Preliminary studies demonstrated that the immune system of mice frequently recognizes the immunizing PY or PYA haptenic group in the

[3] P. M. Comoglio, M. R. Di Renzo, G. Tarone, F. G. Giancotti, L. Naldini, and P. C. Marchisio, *EMBO J.* **3**, 483 (1984).

[4] A. R. Frackelton, Jr., P. M. Tremble, and L. T. Williams, *J. Biol. Chem.* **259**, 7909 (1984).

[5] B. Friedman, A. R. Frackelton, Jr., A. H. Ross, J. M. Conners, H. Fujiki, T. Sugimura, and M. R. Rosner, *Proc. Natl. Acad. Sci. U.S.A.* **81**, 3034 (1984).

[6] J. G. Foulkes, M. Chow, C. Gorka, A. R. Frackelton, Jr., and D. Baltimore, *J. Biol. Chem.* **260**, 8070 (1985).

[7] T. O. Daniel, P. M. Tremble, A. R. Frackelton, Jr., and L. T. Williams, *Proc. Natl. Acad. Sci. U.S.A.* **82**, 2684 (1985).

[8] Y. Yarden, J. A. Escobedo, W. J. Kuang, T. L. Yang-Feng, T. O. Daniel, P. M. Tremble, E. Y. Chen, M. E. Ando, R. N. Harkins, U. Francke, V. A. Fried, A. Ullrich, and L. T. Williams, *Nature (London)* **323**, 226 (1986).

context of the bifunctional cross-linking reagent.[9] To minimize clonal dominance by responses to PY (or PYA) cross-linker, we found it essential to use generically different chemical cross-linking agents for the original immunizations and the final antigenic challenge.

PY or PYA Conjugates with Keyhole Limpet Hemocyanin Using SMBP

SMBP is a heterobifunctional agent that reacts with sulfhydryl groups via a maleimido group and with amino groups via a succinimidyl group.[10] PY (or PYA) is first linked via its free amino group to SMBP. Meanwhile, disulfide bonds of KLH are reduced with sodium borohydride. The PY–maleimide conjugate is then reacted with the free sulfhydryl groups of KLH.

Materials for SMBP Conjugation

SMBP (Pierce, Rockford, IL), 20 mM (7.3 mg/ml) in tetrahydrofuran
Keyhole limpet hemocyanin (KLH) (Calbiochem, San Diego, CA), 50 μM (5 mg/ml, assuming 10^5 g/mol), thoroughly dialyzed against 0.1 M EDTA and 6 M urea
O-Phospho-L-tyrosine (Sigma, St. Louis, MO), 10 mM (2.61 mg/ml) in 0.05 M sodium phosphate buffer, pH 7.0
O-Phospho-DL-tyramine, prepared as described by Ross et al.,[1] 10 mM (2.17 mg/ml) in 0.05 M sodium phosphate buffer, pH 7.0
Dichloromethane (30 ml)
Sodium borohydride (20 mg)
Butanol (0.2 ml)
NaH_2PO_4 (1 ml), 0.1 M
Acetone (0.4 ml)
Nitrogen gas

Procedure

1. Preparation of PY (or PYA)–Maleimide Conjugate: Combine 1 ml of 20 mM SMBP in tetrahydrofuran with 2 ml of 10 mM PY (or PYA) in 0.05 M sodium phosphate buffer, pH 7.0. Incubate at 30° for 30 min with occasional stirring. Remove tetrahydrofuran by gently bubbling nitrogen through the solution, and then extract excess SMBP with three 10-ml aliquots of dichloromethane, saving the aqueous phase for subsequent reaction with reduced KLH.

2. Reduction of KLH Disulfide Bonds: To 10.2 mg of KLH in 2 ml of 6 M urea, 0.1 M EDTA, alternately add small portions of sodium borohydride and *n*-butanol, totaling 20 mg and 0.2 ml, respectively, taking care

[9] A. R. Frackelton, Jr., unpublished observations (1984).
[10] T. Kitagawa, T. Kawasaki, and H. Munechika, *J. Biochem.* (*Tokyo*) **92,** 585 (1982).

that effervescence does not cause the reaction to spill over the tube. Incubate the reaction for 30 min at 30° (during which time considerable aggregation of KLH will be observed). Excess sodium borohydride is then decomposed by the addition of 1 ml of 0.1 M NaH_2PO_4 and 0.4 ml of acetone.

3. Reaction of PY (or PYA)–Maleimide Conjugate with Reduced KLH: One-half of the aqueous phase containing PY (or PYA)–maleimide is combined with the reduced KLH and incubated at 25° for 2 hr. Protein aggregates are then dispersed by homogenization, and the whole reaction mixture is exhaustively dialyzed against phosphate-buffered saline (PBS). These two conjugates contained 8 PY and 20 PYA haptenic groups per KLH 100,000 daltons, respectively, determined by inorganic phosphate analysis.[11]

PY–Protein or PYA–Protein Conjugates Using Glutaraldehyde

Glutaraldehyde is a homobifunctional cross-linking agent reacting with amino groups. It is quite different structurally from the SMBP agent, and for this reason was used to produce KLH conjugates for the final immunogenic challenge, as well as to produce bovine serum albumin (BSA) conjugates for use in enzyme-linked immunosorbent assays (ELISA). The procedure detailed below is adapted from Reichlin[12] and Konopka *et al.*[13]

Materials for Glutaraldehyde Conjugation

Keyhole limpet hemocyanin, 0.15 mM (15 mg/ml in PBS, pH 7.2)

Bovine serum albumin [radioimmunoassay (RIA) grade; Sigma], 0.46 mM (30 mg/ml) in buffer A (0.15 M NaCl and 0.1 M sodium phosphate, pH 7.2), and dialyzed against this buffer

O-Phospho-L-tyrosine, 9 mM (2.35 mg/ml) in buffer A

O-Phospho-DL-tyramine, 9 mM (1.95 mg/ml) in buffer A

Glutaraldehyde (Sigma), 5 M in sealed ampoule. Prepare 20 mM glutaraldehyde in 0.15 M NaCl and 0.1 M sodium phosphate, pH 7.2

Procedure

1. Combine 15 mg of protein with 1–5 μmol of PY or PYA (0.1 ml to 0.5 ml of stock solution, above; the amount of hapten added will affect the haptenic valence of the conjugate, see below).

[11] J. E. Buss and J. T. Stull, this series, Vol. 99, p. 7.
[12] M. Reichlin, this series, Vol. 70, p. 159.
[13] J. B. Konopka, R. L. Davis, S. M. Watanabe, A. S. Ponticelli, L. Schiff-Maker, N. Rosenberg, and O. N. Witte, *J. Virol.* **51,** 223 (1984).

2. Add 1 ml of 20 mM glutaraldehyde and allow to react for 5 hr at room temperature.

3. Dialyze the reaction mixture against PBS containing 0.05% (w/v) sodium azide as a preservative. These haptenated proteins can be stored at 4° for >5 years.

By quantitative phosphate analysis, the PY–glutaraldehyde–KLH conjugate prepared with 5 μmol PY contained 1.5 PY groups/KLH (100,000 Da) whereas PYA–glutaraldehyde–BSA conjugates prepared with 1 and 5 μmol PYA contained 1.5 and 7 PYA groups, respectively, per molecule of BSA.

Immunization Protocol

Of several immunization protocols tested, the most successful involves immunizing mice (BALB/c, female) with an intraperitoneal injection of 50 μg of PY_8–SMBP–KLH, emulsified in Freund's complete adjuvant, challenging 2 months later with 50 μg of PYA_{20}–SMBP–KLH in Freund's complete adjuvant, and finally challenging another 2 months later with $PY_{1.5}$–glutaraldehyde–KLH, just 3 days before taking spleen cells (fusing as described previously with P3U1 myeloma-derived fusing partners[2,14]).

Serum from the immunized mice is screened for antibody activity in a solid-phase enzyme-linked immunosorbent assay (see method below) using PYA linked at low valency via glutaraldehyde to BSA (1.5 haptenic groups/molecule of BSA). The use of low-valency antigen and low antigen concentration in the ELISA favors the detection of only high-affinity antibodies.[15,16] All of the immunized mice had high titers of antibodies reactive with PYA in this assay. However, when these antibodies were screened using a much more demanding criterion—the ability to bind [^{32}P]phosphotyrosyl epidermal growth factor (EGF) receptor in a solid-phase assay (see method below and Fig. 1 legend)—antibodies from only those mice immunized by the protocol described above performed well. Fusion of the spleen cells from one of these mice with the P3U1 fusion partner resulted in approximately 3000 hybridomas, hundreds of which produced antibody to PY as judged by the ELISA. Of the 30 most reactive by ELISA, 4 hybridoma culture fluids showed strong reactivity in the EGF

[14] D. E. Yelton, B. A. Diamond, S. P. Kwan, and M. D. Scharff, in "Current Topics in Microbiology and Immunology" (F. Melchers, M. Potter, and N. L. Warner, eds.), p. 81. Springer-Verlag, New York, 1978.

[15] T. T. Tsu and L. A. Herzenberg, in "Selected Methods in Cellular Immunology" (B. B. Mishell and S. M. Shiigi, eds.), p. 373. Freeman, San Francisco, California, 1980.

[16] A. Nieto, A. Gaya, M. Jansa, C. Moreno, and J. Vives, Mol. Immunol. 21, 537 (1984).

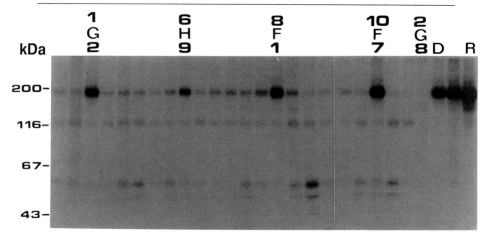

FIG. 1. Screening hybridoma fluid for ability of their antibodies to bind [^{32}P]Tyr-EGF receptor. Microtiter wells were coated with rabbit anti-mouse Ig (50 μl of 50 μg/ml) as described in the text. Hybridoma culture fluids (70 μl) were incubated in the wells for 4 hr, after which the wells were rinsed several times with PBS. Meanwhile, a partially purified preparation of phosphotyrosyl EGF receptors was made as described in the text. Aliquots of the crude EGF receptors were dispensed into each microtiter well. After 14 hr at 4°, the wells were washed extensively and eluted with hot Laemmli SDS gel sample buffer. The eluates were boiled, resolved by SDS–polyacrylamide gel electrophoresis, and ^{32}P-labeled proteins were visualized by overnight autoradiography. Lanes 1G2, 6H9, 8F1, and 10F7 contain the 170-kDa EGF receptor purified by antibodies from this fusion. Lane D and the next lane contain EGF receptor purified by 1 and 10 μg, respectively, of affinity-purified PY antibody 2D2 from a different fusion. Lane 2G8 shows the inability of the earlier ABP antibody (10 μg of affinity-pure 2G8) to purify EGF receptors. Lane R shows the EGF receptor purified by EGF receptor antibody.

receptor-binding assay: hybrids 1G2, 6H9, 10F7, and 8F1 (Fig. 1). In contrast the 2G8 monoclonal antibody that had been so useful in the past was completely ineffectual. Each of these four hybrids were passed as an ascites in mice, and the monoclonal antibodies were affinity purified from the ascites fluid by phosphotyramyl–Sepharose affinity chromatography, specifically eluting with the phosphotyrosine analog, phenyl phosphate, as described previously.[2]

ELISA and RIA Protocols of PY Antibodies

1. Dilute PYA$_{1.5}$–glutaraldehyde–BSA to 100 ng/50 μl with PBS. *Immediately* dispense 50 μl into each well of an Immulon (Dynatech, Alexandria, VA) or Corning (Corning, NY) 96-well microtiter plate. Incubate overnight at 4°.

2. Flick out the contents of the wells and wash once with PBS using a squirt bottle.

3. Completely fill the wells with 5% (w/v) Milkman nonfat milk in PBS–azide. Incubate for 1 hr at 37°.

4. Wash two times with PBST [PBS containing 0.1% (v/v) Tween 20].

5. Dispense antibody samples (or 10 μl of hybridoma fluid) into wells (in 50 μl of PBST). Incubate at 37° for 1–2 hr.

6. Flick out the solution and wash the wells three times with PBST.

7. RIA: For RIA, add sheep (Fab'$_2$) anti-mouse ^{125}I-labeled Ig (~10^5 cpm/well in 50 μl of PBST). Incubate 1–2 hr at 37°. Wash four times with PBST. Count wells.

7. ELISA: For ELISA, add 50 μl of a 2000-fold dilution of sheep anti-mouse Ig (peroxidase labeled; Jackson ImmunoResearch Labs, West Grove, PA) in PBST. *Note:* Be certain that no azide is used from here on as it inhibits peroxidase. Incubate 1–2 hr at 37°.

8. Wash four times with PBST.

9. Add 100 μl of substrate solution (10 ml of 0.1 M citrate buffer, pH 4.5, containing 4 μl of 30% (w/w) H$_2$O$_2$ and 10 mg of o-phenylenediamine. Prepare just before use and protect from light). Color develops in 2–15 min at room temperature.

Preparation of PY Proteins from A431 Cells Stimulated with EGF

1. Seed A431 cells 16–20 hr before experiment at 1–2 $\times$ 10^5 cells in 2 ml culture medium [with 10% fetal calf serum (FCS)] in 35-mm culture wells.

2. The next day, rinse wells two or three times with 2 ml of phosphate-free Dulbecco's minimal essential medium (DMEM), then add 1 ml of phosphate-free DMEM containing 1 mM sodium L-pyruvate, 2 mM L-glutamine, nonessential amino acids, 0.1% (w/v) BSA, and 1 mCi of ^{32}P. Incubate at 37° in 5% CO$_2$ for 3–4 hr.

3. Add 100 ng EGF/well. Mix and incubate at 37° for 5 min.

4. Transfer the plate quickly to ice, aspirate the culture fluid to radioactive waste, draining the fluid thoroughly.

5. Add phenylmethylsulfonyl fluoride (PMSF) (from 100 mM room-temperature stock) to an aliquot of ice-cold extraction buffer (see recipe below); mix it quickly and then dispense 0.4 ml onto A431 cell monolayer.

6. Dislodge monolayer with a cell scraper (or rubber policeman). Transfer to a 1.4-ml conical plastic centrifuge tube.

7. Rinse the well with another 0.4-ml aliquot of extraction buffer, transferring this rinse into the same tube.

8. Incubate at 0° for 10–20 min, with intermittent vigorous vortexing.

9. Centrifuge mixture at 8000 g for 15 min. Phosphotyrosyl proteins are purified from the supernatant as described below. (For use in screening hybridoma culture fluids for antibodies to PY proteins, we enrich this

extract for EGF receptors, taking advantage of their ability to bind to wheat germ agglutinin. The extract is passed over a small affinity matrix bearing wheat germ agglutinin (Pharmacia, Piscataway, NJ), and partially pure EGF receptors are eluted with 0.3 M N-acetylglucosamine.)

Antigenic Fine Specificity of Monoclonal Antibodies

In order to begin the process of selecting the best antibody for further use, one can determine first the apparent binding constants of each antibody for $PYA_{1.5}$–BSA and for a number of potentially cross-reacting molecules. To evaluate the hybrids described above, for example, microtiter plate wells are coated with $PYA_{1.5}$–BSA and then incubated sequentially with monoclonal antibody and sheep anti-mouse ^{125}I-labeled immunoglobulin (one can also use the ELISA assay described above). The concentration dependence of binding (Fig. 2) provides estimates of the apparent binding constant of each monoclonal antibody, ranked in order of decreasing affinity: 1G2 > 6H9 = 2G8 > 10F7 > 1H9 ≥ 2D2, with 1G2 having an apparent K_d of about 4×10^{-10} M, nearly 10-fold better than our original anti-ABP antibody (2G8).

The specificity of the most avid antibody, 1G2, was tested by determining the ability of various phosphate-containing compounds to inhibit its binding to PYA–BSA in the radioimmunoassays just described above.

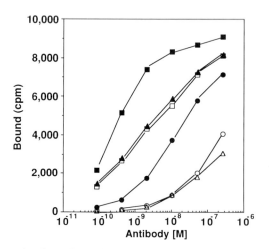

FIG. 2. Concentration dependence of monoclonal antibody binding to PYA–BSA. Microtiter plate wells were coated with 25 ng of PYA_1–glutaraldehyde–BSA, adsorptive sites were blocked with milk, and the wells were incubated sequentially with monoclonal antibody and sheep anti-mouse ^{125}I-labeled Ig as described in the text. ■, 1G2; ▲, 2G8; □, 6H9; ●, 10F7; ○, 1H9; △, 2D2.

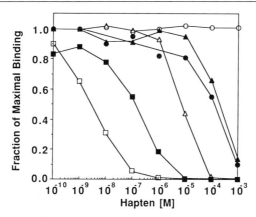

FIG. 3. Inhibition of ^{125}I-labeled 1G2 binding to PYA–BSA by phosphotyrosine and its analogs. Microtiter plate wells were coated with antigen and then blocked with milk as described in the text. The various phosphotyrosine analogs and other phosphate-containing molecules were diluted to twice the indicated molarity in 25 μl PBST and added to the wells. Then ^{125}I-labeled 1G2 (50,000 cpm in a total volume of 25 μl of PBST) was added to each well. After 3 hr at room temperature, the wells were rinsed quickly with PBST four times and radioactivity remaining bound was quantitated. □, PYA–BSA; ■, PY–BSA; ●, PYA; ▲, PY; △, phenyl phosphate; ○, others (5′-adenosine monophosphate, ATP, phosphoserine, phosphothreonine, ribose 5′-phosphate, ribose 1-phosphate, sodium pyrophosphate, and phosphoseryl-BSA).

The 1G2 antibody was inhibited only by phosphotyrosine analogs (Fig. 3), with an apparent K_i for PYA–BSA of 3×10^{-9} M, compared to 1×10^{-7} M for our original antibody (2G8, data not shown). Ribose phosphates, phosphoserine, phosphothreonine, phosphoseryl-BSA, and a variety of nucleotides (e.g., AMP, ATP) failed to inhibit. Similar analyses of the other monoclonal antibodies revealed that 1H9 and 2G8 cross-react with mononucleotides, while 2D2 cross-reacts with ATP and pyrophosphate. Antibodies 6H9 and 10F7, like 1G2, reacted only with close phosphotyrosine analogs (e.g., phenyl phosphate, p-nitrophenyl phosphate, and phosphotyramine).

Thus, the new monoclonal antibodies generally lack the cross-reactivities that 2G8 and other antibodies to ABP exhibited for mononucleotides and certain other phosphate-containing compounds.[2,3] Notice that of this panel of monoclonal antibodies, only 1H9 and 2D2 share this cross-reactivity.

It is interesting to contrast the fine specificity of the 1G2 antibody with another monoclonal antibody, PY20, produced by immunizing mice with conjugates of PY to KLH and ovalbumin, using 1-ethyl-3-(3-dimethylaminopropyl)-carbodiimide (EDAC) as a bifunctional cross-linking agent (see [9] in this volume).[17] The 1G2 antibody binds free phenyl phosphate (apparent K_d of 7×10^{-6} M), about 30-fold stronger than free PY, and about 15-fold stronger than free PYA, suggesting that both the free amino group and carboxyl group on PY interfere with antibody binding. Of course, the PY amino and carboxyl groups would be absent in a tyrosine-phosphorylated protein, being involved in peptide bonds. The PY20 antibody, on the other hand, binds free phenyl phosphate slightly less well than PY, with half-maximal inhibition in a similar ELISA occurring at about 5×10^{-4} M phenyl phosphate.[17]

Use of 1G2 MAb for Purifying PY Proteins

Phosphotyrosine-containing proteins can be purified on a small scale for analytical purposes, from about 1×10^6 cells, typically labeled *in vivo* with [^{32}P]P$_i$, or on a large scale for preparative purposes, using 1×10^9 cells or more. For both of these purposes we favor using the antibody as a solid-phase immunosorbent. This not only allows virtually unlimited reuse of the antibody, but also takes advantage of the "persistent hapten" effect, thereby increasing the apparent avidity of the antibody for any unusual PY protein that may bind weakly to the antibody.[2] For the purposes of illustration, we will describe the steps to purify PY proteins from EGF-stimulated cells on an analytical scale, and from chronic myelogenous leukemia cells on a preparative scale.

Microbatch Purification

Extraction Buffer

Triton X-100 (1%)
SDS (0.1%, w/v) (optional)
Tris base (10 mM)
EDTA (5 mM)
NaCl (50 mM)
Sodium pyrophosphate (30 mM)
Sodium fluoride (50 mM)
Sodium orthovanadate (100 μM)
BSA, crystallized (0.1%, w/v) (Sigma, optional)
Phenylmethylsulfonyl fluoride (PMSF; 1 mM)
Adjust pH to 7.6 at 4° with HCl

PMSF (from a 100 mM stock in 2-propanol or absolute ethanol) must be added to the buffer *immediately* before it is used to lyse cells. The extraction buffer (lacking PMSF) stores many weeks at 4°.

[17] J. R. Glenney, Jr., L. Zokas, and M. J. Kamps, *J. Immunol. Methods* **109**, 277 (1988).

Eluting Buffer

Eluting buffer: Same as the extraction buffer, except that the 0.1% BSA is omitted, and 1.0 mM phenyl phosphate is included as a competing hapten to displace PY proteins. Adjust pH to 7.6 with HCl. Store aliquots at −20°

Affinity Purification of PY Proteins

1. After the detergent extract (see above) has been clarified by centrifugation, the extract is transferred to a 1.5-ml tube containing 10 μl of anti-phosphotyrosine–Sepharose beads (15 mg of PY antibody covalently linked to CNBr-activated Sepharose 4B; Pharmacia). This suspension is mixed end over end for 1–3 hr at 0°.

2. After this incubation, the anti-phosphotyrosine beads are sedimented (microcentrifuge, about 7 sec), and the supernatant fluid is carefully aspirated (leaving about 10–20 μl of fluid above the beads).

3. The beads are washed by resuspending them in 1 ml of extraction buffer lacking BSA, sedimenting the beads, and aspirating the supernatant fluid. This wash is repeated three times, with the last allowed to mix for 10 min or more (this helps to remove proteins that are nonspecifically adsorbed to the beads).

4. Phosphotyrosyl proteins that have bound to the anti-phosphotyrosine beads are specifically and gently eluted with the PY analog, phenyl phosphate. Add 30 μl of eluting buffer to the washed beads. Incubate 5–10 min, occasionally resuspending the beads.

5. Prepare a hemostat with a 25 gauge needle protruding five- to six-hundredths of an inch from the side of the jaws. With another needle make a pinhole in the cap of the tube that contains the beads and eluting buffer. Wipe dry the bottom of the tube and puncture a single, straight hole in the bottom using the 25-gauge needle–hemostat. Place this tube piggyback onto an open 1.5-ml tube, and then insert them together into an appropriate carrier.

6. Centrifuge at 1500 g for 1 min. The eluate (45 μl) will collect in the lower tube, free of any Sepharose beads. The PY proteins can then be analyzed further by, e.g., immunoprecipitation with antibodies to specific proteins and SDS–PAGE.

Notes

1. When pipetting beads, cut off about 1/4-in. of pipette tip to afford a wider bore.

2. To regenerate used anti-phosphotyrosine beads: Wash the beads several times with a solution of one part extraction buffer and one part 2 M NaCl. Beads have been successfully regenerated and used through ~100 cycles.

3. To minimize artifactual dephosphorylation, do not exceed 2×10^6 cells/ml of extraction buffer unless 0.1% SDS is included in the buffer.

We have used the microbatch procedure to isolate PY proteins from retrovirally transformed cells, from cells stimulated with EGF, transforming growth factor α (TGF-α), PDGF, fibroblast growth factor (FGF), insulin, colony-stimulating factor 1 (CSF-1), and interleukin 3 (IL-3), and from many other normal and transformed cells, including human chronic myelogenous leukemia cells expressing the aberrant p210[bcr–abl] tyrosine kinase.[2,4–8,18–22]

Antibody specificity has been evaluated by phosphoamino acid analysis of purified proteins,[19,20] as well as by phenyl phosphate inhibition of binding to the antibody (Fig. 4). Binding of the *abl* p210 was inhibited >95% by as little as 0.04 mM phenyl phosphate, and 0.3 mM phenyl phosphate quantitatively eluted PY proteins.

Large-Scale Isolation of PY Proteins

Extraction Buffer

Extraction buffer: As in the microbatch procedure, but eliminating BSA

Eluting Buffer

Eluting buffer: As in the microbatch procedure, but using 40 mM phenyl phosphate. If ultrafiltration steps will be used to subsequently concentrate the PY proteins, substitute 0.7% (w/v) β-octylglucoside for Triton X-100.

Extraction of Nonadherent Cells

1. Nonadherent cells growing in spinner culture are harvested by centrifugation, and the cell pellets are treated with extraction buffer (all subsequent steps at 4°): 2×10^9 cells are resuspended in 20 ml of ice-cold PBS and vigorously vortexed while adding 1 liter of extraction buffer.

2. The extract is clarified by centrifugation at 1500 g for 30 min and subsequent filtration through fluted Whatman (Clifton, NJ) #1 filter paper.

[18] J. C. Bell, L. C. Mahadevan, W. H. Colledge, A. R. Frackelton, Jr., M. G. Sargent, and J. G. Foulkes, *Nature (London)* **325**, 552 (1987).
[19] R. D. Huhn, M. R. Posner, J. G. Foulkes, S. Rayter, and A. R. Frackelton, Jr., *Proc. Natl. Acad. Sci. U.S.A.* **84**, 4408 (1987).
[20] R. D. Huhn, M. E. Cicione, and A. R. Frackelton, Jr., *J. Cell. Biochem.* **39**, 129 (1989).
[21] R. Isfort, R. D. Huhn, A. R. Frackelton, Jr., and J. N. Ihle, *J. Biol. Chem.* **263**, 19203 (1988).
[22] A. Sengupta, W.-K. Liu, Y.-G. Yeung, D. Yeung, A. R. Frackelton, Jr., and E. R. Stanley, *Proc. Natl. Acad. Sci. U.S.A.* **83**, 8062 (1988).

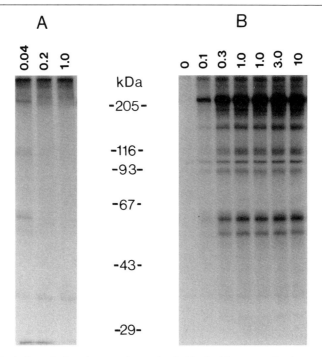

FIG. 4. Isolation of phosphotyrosyl proteins is blocked by competing hapten. Chronic myelogenous leukemia cells (RWLeu4), which express an aberrant, 210-kDa abl protein with constitutive tyrosine kinase activity, were labeled and extracted as indicated in the text. (A) Phenyl phosphate was added to the cell extracts at 0.04, 0.2, or 1.0 mM as indicated. (B) Phosphotyrosyl proteins were purified as usual, except that the indicated concentration (mM) of phenyl phosphate was used for specific elution.

3. Gently stir the clarified extract overnight with 6 ml of the 1G2–Sepharose 4B immunosorbent.

4. Collect the immunosorbent beads on a sintered glass funnel (coarse), and wash with 100 ml of extraction buffer.

5. Transfer the immunosorbent into a 50-ml polypropylene culture tube, and wash three times for 10 min each with 50 ml of filtered extraction buffer (filtered through 0.45-μm nitrocellulose to remove the ubiquitous keratins and any other trace contaminant proteins).

6. Transfer the beads into a small open-bed column, and wash with 12 ml of filtered extraction buffer.

7. Specifically elute PY proteins from the immunosorbent using 12 ml of the phenyl phosphate-containing elution buffer.

These proteins can be concentrated by ultrafiltration, and further purified by standard techniques including fast protein liquid chromatography (FPLC), glycerol gradient ultracentrifugation, and one- and two-dimensional SDS–PAGE. The eluted PY proteins are typically nondenatured, and thus can be analyzed for a variety of functional activities. These proteins can be stored at $-70°$ for >1 year with no obvious changes in their SDS electrophoretic properties.

We have used this procedure with good success for isolating PY proteins from chronic myelogenous leukemia cells and, with minor modifications, for isolating PY proteins from growth factor-stimulated adherent cells. Greater than 90% of the protein phosphotyrosine in EGF-stimulated cells bound to the 1G2 immunosorbent (data not shown). The 1G2 hybridoma cells can be obtained from the ATCC (Rockville, MD) Safe Deposit collection, and the antibody itself is commercially available.

Acknowledgment

This work was supported in part by Grants R01-CA39235 and 5P30-CA13943 from the National Cancer Institute.

[9] Isolation of Tyrosine-Phosphorylated Proteins and Generation of Monoclonal Antibodies

By John R. Glenney

Introduction

The mechanism by which tyrosine kinases exert their effects is not well understood. Although tyrosine-specific protein kinase activity was described more than 10 years ago, the relevant substrates that mediate the effects of tyrosine phosphorylation have been elusive. One approach to identification of tyrosine kinase substrates has been to analyze the phosphoamino acid content of proteins that are thought to play a role in the process affected by the tyrosine kinase. This approach assumes that we can deduce the identity of the relevant substrates and simply test them for the presence of phosphotyrosine. It is quite possible that some mediators of the tyrosine kinase signal are novel proteins that have not been previously studied. To gain insight into the function and regulation of such substrates we have taken the approach of (1) isolation of tyrosine-phosphorylated proteins using immunoaffinity chromatography with a high-affinity monoclonal anti-phosphotyrosine antibody and (2) generation of monoclonal antibodies to the polypeptide backbone of individual substrates. The antibodies to substrates can then be used to study the distribution, turnover, and phosphorylation of the protein in addition to cloning the cDNA encoding the protein from an expression library. In this chapter two protocols for the isolation of phosphotyrosine-containing proteins are described, one using harsh denaturing conditions [boiling sodium dodecyl sulfate (SDS)] for lysis and extraction, and the second using more gentle buffers. Also included are the generation and use of the anti-phosphotyrosine affinity column and convenient assays for monitoring the protein and phosphotyrosine content of the eluted fractions. Methods used for the monoclonal antibody production to generate antibodies to the polypeptide backbone of individual substrates[1] are described.

Purification of Phosphotyrosine-Containing Proteins from Tissue Culture Cells by Affinity Purification on Anti-Phosphotyrosine Monoclonal Antibody Columns

Isolation of Proteins under Nondenaturing Conditions

Tissue culture cells are grown on plastic 150-mm tissue culture dishes to approximately 90% confluence. If the cells contain a constitutively active tyrosine kinase such as v-src, the cells are lysed directly. In the case of growth factor receptors such as the epidermal growth factor (EGF) receptor and platelet-derived growth factor (PDGF) receptor, the kinase must be first activated by ligand. The growth factor-containing medium may be reused. Medium is removed and the plates placed in a cold room or refrigerator for 5 min prior to lysis of the cells. Six milliliters of ice-cold buffer A [10 mM imidazole, 0.5 M NaCl, 1% (w/v) Triton X-100, 0.2 mM sodium vanadate, 0.2 mM phenyl methyl sulfonyl fluoride (PMSF), 2 mM sodium azide, pH 7.3] is added to each plate. Generally 10 plates are lysed at a time and are then stacked on a shaker in the cold room and left to extract for 30 min with gentle swirling. Forty to 50 plates can be harvested conveniently within 10–15 min in this way. The viscous lysate is scraped into the tubes of a Beckman (Palo Alto, CA) 45TI rotor and centrifuged at 40,000 rpm for 1.5 hr at 4°. After centrifugation the supernatant is carefully removed and added to a 250-ml conical centrifuge bottle (Corning, Corning, NY) together with 1–2 ml of immobilized anti-phosphotyrosine antibody. [We use a monoclonal antibody, PY20,[2] coupled at a concentration of 10 mg antibody/ml Affi-Gel (Bio-Rad, Richmond, CA).] PY20

[1] J. R. Glenney, Jr. and L. Zokas, *J. Cell Biol.* **108,** 2401 (1989).
[2] J. R. Glenney, Jr., L. Zokas, and M. P. Kamps, *J. Immunol. Methods* **109,** 277 (1988).

Affi-Gel is available through ICN Biomedicals (Costa Mesa, CA; Cat. #152-369). The tube is capped and shaken vigorously for 3–6 hr at 4°. This is followed by centrifugation at 500 g for 5 min to collect the affinity matrix. The supernatant is carefully removed from the Affi-Gel pellet, and the pellet is resuspended in a small volume of buffer B (10 mM imidazole, 0.1 M NaCl, and 0.1% Triton X-100) and placed into a small chromatography column. The column is washed with a minimum of 20 column volumes of buffer B at 4° and the substrates then eluted with buffer B containing 5 mM phenyl phosphate. A low flow rate (1 ml/hr) during elution helps sharpen the peak of phosphotyrosine-containing protein. Fractions (0.5 ml) are monitored as described below.

Isolation of Phosphotyrosine-Containing Proteins under Denaturing Conditions

In order to ensure complete solubilization of the phosphotyrosine-containing proteins (since many are known to be in an insoluble cytoskeletal matrix), we use a method to isolate the proteins after solubilization in SDS. The SDS lysis also has the advantages of (1) inhibiting phosphatases that could lead to a reduced recovery of the phosphotyrosine-containing proteins on the affinity column and (2) inactivating tyrosine kinases that may phosphorylate substrates after cell lysis.

Cells are grown as described above and, after removal of the medium, SDS lysis buffer [3 ml of 1% (w/v) SDS, 10 mM Tris, pH 7.0, kept at 100°] is added and swirled quickly to lyse the cells. After adding lysis buffer to five plates, the plates are then put in a microwave oven and further heated on a setting of high for 5 sec. The plates are then restacked in a different order and reheated in the oven for an additional 5 sec. The plates are removed and the lysis buffer is then swirled on the bottom of the plate for 5–10 sec. The plates are allowed to stand on the bench at a slight angle to allow liquid to drain to one side of the plate. The viscous lysate is scraped into the tubes of a Beckman 45 TI rotor and centrifuged at 40,000 rpm for 2 hr at 10–15°. The supernatant is carefully separated from the translucent pellet and then used either directly for affinity chromatography or stored at −70°. We find it convenient when working up several hundred plates of cells to grow up and lyse 50 plates at a time and to store the lysate until all of the plates have been harvested in this way. The lysate from 50 plates is diluted fivefold to a final concentration of 0.2% SDS, adjusted to 1% Triton X-100, 50 mM NaCl, 10 mM imidazole, 0.2 mM sodium vanadate, and 1 mM EDTA, pH 7.3, and added to a 250-ml conical centrifuge tube, to which is then added 1–3 ml of PY20 Affi-Gel/tube and shaken as described above. In this case the lysate can be shaken with the Affi-Gel overnight. The affinity resin is collected by centrifugation and resuspended in buffer C (10 mM imidazole, 50 mM NaCl, 1 mM EDTA, 1% Triton X-100, 0.1% SDS) and loaded into a small chromatography column. The column is then washed with a minimum of 20 vol of buffer C and then 2–3 vol of buffer C (reducing the Triton X-100 to 0.1% and the SDS to 0.05%). The column is then eluted with the same low-detergent buffer with the addition of 5 mM phenyl phosphate. Fractions are collected (0.5 ml) and monitored as described below.

Assay of Affinity Column

The column fractions eluted from the anti-phosphotyrosine column can be assayed in a variety of ways. The appropriate way will be dictated by the amount of protein from the column and the sensitivity needed. Since the elution buffer contains high concentrations of phenyl phosphate or phosphotyrosine in addition to detergents, monitoring the absorbance at 280 nm is not practical. We find that in a standard preparation from transformed chick fibroblasts in which 100 plates of cells are used, the overall yield is approximately 0.5 to 1.0 mg of phosphotyrosine-containing protein. With this amount of protein one can monitor by standard protein assays. We find it convenient to monitor using the BCA protein assay (Pierce, Rockford, IL) since detergent and phenyl phosphate do not interfere. The assay is performed in 96-well test plates with comparison to known bovine serum albumin (BSA) concentrations that have been made up and diluted in the exact buffer used for the elution of the protein from the PY20 column. An aliquot (10–15 μl) of each fraction is placed in the wells of a 96-well test plate, to which 200 μl of the BCA protein assay mix is added. The plate is incubated at 37° for 30–60 min. The absorbance at 570 nm is then monitored using a microplate reader. The absorbance or protein concentration of each fraction is then plotted as a function of the fraction number.

As an alternative to this direct protein assay, fractions can be monitored by Western blots using an anti-phosphotyrosine antibody (either polyclonal or monoclonal) to assess the yield of individual phosphotyrosine-containing proteins. We have observed, for instance, that some proteins are eluted by phenyl phosphate from the PY20 column very early, whereas others trail off for some time. Some proteins, like calpactin I and map kinase, are poorly retained on the column. Thus, assaying by Western blotting would be important if one were interested in proteins with unusual properties. To assay by Western blotting, 15 μl of each fraction is mixed with 15 μl of 2 × SDS sample buffer and subjected to SDS–polyacrylamide gel electrophoresis. After electrophoresis, the proteins can be stained with silver or transferred to either nitrocellulose or Immobilon (Millipore, Bedford, MA) for Western blotting. After blocking nonspecific sites with 3% BSA the blots are treated with anti-phosphotyrosine antibodies followed by an iodinated second antibody (see above and [10] in this volume for methods). The peak fractions assessed using the protein determination, silver staining, or Western blotting are pooled and can be concentrated using an Amicon (Danvers, MA) Centricon concentrator to approximately 0.5–1 mg protein/ml.

Immunization of Mice and Production of Hybridomas Directed Against Individual Phosphotyrosine-Containing Proteins

Immunization

Phosphotyrosine-containing protein (0.1–0.5 mg), isolated as described above and to be used to immunize BALB/c mice, is emulsified with Freund's complete adjuvant. We use multiple subcutaneous, intraperitoneal, and intramuscular immunizations. Second and third immunizations are performed at 1-month intervals using one-half the amount of antigen used in the first immunization and substituting Freund's incomplete adjuvant while maintaining the same routes of immunization. One week after the third immunization the mouse is immunized by injection of soluble phosphotyrosine-containing proteins in the tail vein. This is followed by another intravenous immunization 2–3 days later. For the final intravenous immunization we use the phosphotyrosine-containing proteins dialyzed for at least 4–5 days against phosphate-buffered saline (PBS) to reduce the detergent concentration. Intravenous (tail vein) immunizations are performed using a 27-gauge needle and 100–200 μl total volume containing 100–200 μg protein. The fusion is performed 2 days after the last immunization. In some cases the final series of intravenous immunizations with soluble antigen will be lethal to the mouse, due to excessive detergent in the antigen preparation or anaphylactic shock. For this reason the parental myloma cell line is readied for immediate hybridoma production in case the mouse dies unexpectedly.

Fusion

One to 2 days following the last immunization, the mouse is sacrificed and the spleen removed. A small amount of blood is routinely collected by aspiration after severing the jugular vein. This serum can be very useful in developing an assay for screening the hybridomas. The spleen cells are dissociated by first mincing the spleen with a razor blade and scissors and then passing it through a sterile 5-ml syringe with an 18-gauge needle 10–20 times. The cell suspension is placed in a sterile 15-ml centrifuge tube and allowed to settle for 2–3 min. The suspension of cells is carefully removed from the top, leaving the large clumps of tissue and cells at the bottom. These are discarded. Both the spleen cell suspension and the parental myeloma cells are then centrifuged. We routinely use the PAI myeloma cell line as fusion partner. For each fusion we use the cells from five or six 150-mm plates of subconfluent myeloma cells growing exponentially and subcultured daily. Within this range the exact number of either spleen or myeloma cells does not appear to be critical. The cells are collected by centrifugation at 1000 g for 5 min. The myeloma cells are mixed with the spleen cells in a 50-ml conical centrifuge tube and brought up to 50 ml with serum-free Dulbecco's modified Eagle's medium (DMEM) and centrifuged as above. The medium is carefully removed and 2–3 ml of fusion solution added. Fusion solution consists of 5 g Kodak (Rochester, NY) polyethylene glycol 1450 (PEG), 0.5 ml dimethyl sulfoxide (DMSO), and 5 ml phosphate-buffered saline adjusted to a pH of 8.0 and sterile filtered. The prewarmed fusion solution is added over 45 sec with tapping of the side of the tube every few seconds to resuspend the cells. The cell suspension is then diluted with 50 ml of prewarmed DMEM over a period of 90 sec and allowed to sit for a further 8 min at room temperature, followed by 2–3 min at 37°. The cells are collected by centrifugation and resuspended in DMEM containing 10% horse serum supplemented with HAT (hypoxanthine, aminopterin, thymidine), purchased as a 50× stock solution (Sigma, St. Louis, MO). The cell suspension is directly plated into 96-well tissue culture plates. In most cases it is convenient to use 10 plates, to which 20% of the total cell suspension is dispersed, and another 10 plates for the dispersal of 80% of the cell suspension. The plates that have one to three growing hybridomas/well are used for assay. The plates are then placed back in a 37° incubator and allowed to remain undisturbed for 5–6 days.

Assay of Hybridomas

First screen: ELISA

Optimally one would want to test plates that have one to three growing hybridomas per well. Generally we assay for antibodies at 9–11 days following the fusion. The fastest growing hybridomas in most cases do not yield antibodies. Do not be impatient if some of the wells turn yellow due to overgrowth. It is important that the enzyme-linked immunosorbent assay (ELISA) be standardized several times using a single batch of phosphotyrosine-containing protein antigen prior to screening hybridomas, since there are only a few days to test and select cells for further studies.

Both the serum obtained at the time of fusion (see above) and the immunizing antigen used to coat the plates are titered to develop the ELISA. As a general guide, the serum can be diluted 1/1000 and the amount of antigen on the plate varied from 1 ng to 1 μg/well. Immulon 4 plates (Dynatech, Baton Rouge, LA) are used and 75 μl of the antigen dilution in H_2O is added to each well. The protein is dried on the bottom of the well, using a hair dryer pointed at the plate for ~1 hr, or simply by incubating the plates overnight (not stacked) under a fume hood that has good air flow. After the plates are dry the nonspecific protein-binding sites are blocked by the addition of 200 μl 5% (w/v) BSA in 10 mM Tris-HCl, pH 7.2 to 7.4, 100 mM NaCl to each well. The blocking is allowed to proceed at least 1 hr at 25°, but the plates can be conveniently stored in blocking buffer at 4° for several weeks. After blocking, the plates are washed in ELISA wash buffer [10 mM Tris-HCl, pH 7.2, 100 mM NaCl, 0.05% (v/v) Tween 20] several times. Washes are accomplished by flicking the liquid from the plate into a sink and refilling using a 500-ml squirt bottle filled with ELISA wash buffer. Cultural supernatant (75 μl) is then added to each well using a multichannel pipette and sterile pipette tips under a laminar flow hood. Make sure that the designations of the culture plate and test plate are the same, including the orientation, to prevent later uncertainty in aligning positives from the ELISA with cultures in the tissue culture plate. The test plates are incubated in first antibody for 2 hr, washed five to six times as above and then in second antibody (goat anti-mouse peroxidase-conjugated IgG diluted 1/1000 in ELISA wash buffer containing 3% BSA) for 2 hr. After a final five or six washes the assay is developed with the addition of 150 μl 50 mM sodium citrate (pH 5.5) containing 0.4 mg o-phenylenediamine (available in preweighed tablet form from Sigma) and 0.5 μl 30% H_2O_2/ml. The plates are shielded from sunlight and checked every 5 min. The reaction is stopped before the background becomes excessive (the background can vary from plate to plate) by the addition of 50 μl 4 N H_2SO_4/well. The color reaction can be quantified by an ELISA reader; however, judgments made by eye give the same results.

To select positives from the culture plates, the test plate and culture plate are placed side by side and the wells that were positive in the test plate are indicated by marks on both the top and bottom of the culture plate. False positives are very common in the side and corner wells, so if there is an abundance of positives these wells could be avoided. Even if false positives are selected in this first round of screening they will be identified as such in the second round (see below) before cloning. One should make sure that there is enough antigen on hand to assay 10 plates at least two or three times. It is not unusual when testing the hybidomas to have all of the wells give a positive reaction in the first assay. This could be due to residual spleen cells that survived for the first 4–5 days and continue to produce IgG directed against substrates, or excessive sensitivity of the assay. In any case, if all wells are positive, feed the cells (75 μl medium) and repeat the assay the very next day. After 10–11 days in the well of a 96-well plate, cells will very shortly overgrow and die, so that decisions as to whether they are positive or negative need to be made as soon as possible after the assay. We use the ELISA to initially screen but not to make final decisions on whether to clone and pursue these antibodies.

Second Screen: Western Blot

Select as many hybridomas as possible that show a positive reaction by ELISA and rescreen by Western blotting or immunoprecipitation. Many antibodies will not recognize a protein in a Western blot or immunoprecipitation, even though they react strongly by ELISA. Valuable time can be wasted in trying to determine the nature of the antigen in these instances. For this reason we use Western blotting as a routine part of our initial screening. Cells from 96-well plates are *gently* resuspended and transferred to 24-well plates where they are allowed to grow again in HAT medium for a few more days. Each plate is examined microscopically every day and culture supernatants are removed when the cells are just coming to confluence. One milliliter is removed and replaced with 1 ml of fresh medium. This culture medium is tested in a Western blot by one of two methods, depending on the number to be tested:

1. If 1–20 hybridomas are to be tested per day it is usually easier to test "strips" of the antigen displayed on a Western blot. A single well gel is constructed and the phosphotyrosine-containing protein resolved by SDS–PAGE and transferred to nitrocellulose or Immobilon. The nitrocellulose can then be blocked and cut into strips for use in a standard blotting assay. From a standard size gel, 30–50 strips can be made.

2. Alternatively, if many culture supernatants (20–100) are to be tested in a single day, we have found that a device in which 28 can be tested at a time is particularly useful (Immunetics, Cambridge, MA). This involves running the phosphotyrosine-containing protein on a minigel containing only one wide sample slot, transferring to nitrocellulose and, after blocking with a high-protein solution, placing the entire blot in a frame in which individual "chambers" are formed on the surface of the nitrocellulose. Each device holds 2 blots, with a total capacity of 56 antibody solutions

(only 50 μl of antibody solution is needed). The blot can be washed while in the frame, then removed from the frame and treated in secondary antibody. If multiple blots are used, these can all be treated in a common solution of secondary antibody. If using either strips or a frame, a positive antibody control should be included. Either a monoclonal antibody to a protein known to be phosphorylated on tyrosine in the system, or the serum from the mouse used for fusion (diluted 1/100–1/1000 as determined previously) can be used as a control. A negative control may also be required. In our experience, using proteins isolated in native form from v-src-transformed chick fibroblasts, we were able to find approximately 200–300 antibodies which reacted with substrates by ELISA, and of these 50 reacted with substrates by Western blotting. Many of these reacted with the same protein and thus, by using Western blotting in the second stage assay, it is possible to focus on a small number of antibodies to each protein. Based on these assays a small number of hybridomas (<30) are cloned by limiting dilution directly from the 24-well plate. Rescreening is performed by ELISA as described above. We always clone the hybridomas at least two times to ensure that a stable hybridoma has been isolated. If the frequency of positives in the second cloning is less than 100%, then an additional round of cloning is performed.

Summary

The methods used above allow one to generate the tools to study individual phosphotyrosine-containing proteins. While all of the substrates we have identified in this way contain phosphotyrosine, this cannot be assumed. Some polypeptides may bind to the anti-phosphotyrosine column because of association with other phosphotyrosine-containing proteins or by nonspecific binding to the anti-phosphotyrosine affinity column. The antibodies generated to these substrates allow one to assess potential questions directly. At a minimum, phosphoamino acid analysis must be performed on the candidate substrate labeled in intact cells with $^{32}PO_4$ and precipitated with the antibody to the substrate. Methods for this can be found in Cooper and Hunter.[3]

[3] J. A. Cooper and T. Hunter, *Mol. Cell. Biol.* **1**, 394 (1981).

[10] Generation and Use of Anti-Phosphotyrosine Antibodies for Immunoblotting

By MARK P. KAMPS

Introduction

Since first reported in 1981 by Ross *et al.*,[1] antibodies reactive with phosphotyrosine have become an invaluable tool for identifying substrates of the protein-tyrosine kinases. Whereas conventional techniques that identify phosphotyrosine in a protein require two steps—first, isolation of the protein by specific antisera or by two-dimensional gel electrophoresis and, second, phosphoamino acid analysis[2]—anti-phosphotyrosine immunoblot analysis (AIA) provides a single assay that detects phosphotyrosine in specific polypeptides among total cellular proteins. This chapter will discuss (1) the uses of anti-phosphotyrosine (anti-PY) antibodies, (2) factors important in the preparation of antibodies to phosphotyrosine, (3) a concise protocol for making, purifying, and using high-affinity, polyclonal rabbit anti-PY antibodies for AIA, and (4) an assessment of the strengths and weaknesses of these antibodies.

A common use of anti-PY antibodies has been the cataloguing of substrates of the protein-tyrosine kinases that stimulate cell division, in an effort to determine whether the consistent phosphorylation of certain proteins correlates with the positive regulation of growth. To this end, AIA was used to identify numerous substrates in cells transformed by the oncogenic viral protein-tyrosine kinases,[3–5] and in cells treated with epidermal growth factor (EGF), platelet-derived growth factor (PDGF), and insulin[6–8] (see [6–8] in this volume). AIA also revealed that tyrosine protein phosphorylation is highly elevated in cells subjected to heat shock,[9]

[1] A. H. Ross, D. Baltimore, and H. Eisen, *Nature (London)* **294,** 654 (1981).
[2] J. A. Cooper and T. Hunter, *Mol. Cell. Biol.* **1,** 165 (1981).
[3] M. P. Kamps and B. M. Sefton, *Oncogene* **2,** 305 (1988).
[4] M. P. Kamps and B. M. Sefton, *Oncog. Res.* **3,** 105 (1988).
[5] M. Hamaguchi, C. Grandori, and H. Hanafusa, *Mol. Cell. Biol.* **8,** 3035 (1988).
[6] T. Kadowaki, S. Koyasu, E. Nishida, K. Tobe, T. Izumi, F. Takaku, H. Sakai, I. Yahara, and M. Kasuga, *J. Biol. Chem.* **262,** 7342 (1987).
[7] M. F. White, R. Maron, and C. R. Kahn, *Nature (London)* **318,** 183 (1985).
[8] A. R. Frackelton, Jr., P. M. Tremble, and L. T. Williams, *J. Biol. Chem.* **259,** 7909 (1984).
[9] P. A. Maher and E. B. Pasquale, *J. Cell Biol.* **108,** 2029 (1989).

hyperosmotic shock,[10] and treatment with the phosphatase inhibitor, orthovanadate,[11] and that the same normal cellular proteins whose phosphorylation was stimulated by vanadate were also the substrates of the src oncoprotein.[4] Together, these studies indicated normal cellular tyrosine protein kinases and viral tyrosine protein kinases have numerous similar polypeptide substrates. More recent studies have focused on developing monoclonal antibodies to each substrate by immunizing mice with total populations of phosphotyrosine-containing proteins from transformed chicken fibroblasts, isolated by immunoaffinity to anti-PY monoclonal antibodies[12] (see [9] in this volume).

More specific questions have also been addressed through the use of AIA. Morla et al.[13] found that interleukin 3 (IL-3) stimulates tyrosine phosphorylation of a set of proteins in IL-3-dependent murine cell lines. Cartwright et al.[14] and Giordano et al.[15] identified new tyrosine-phosphorylated proteins in human cancers, and Lindberg et al.,[16] Letwin et al.,[17] and Kornbluth et al.[18] cloned novel protein-tyrosine kinases from prokaryotic cDNA expression libraries by screening filters using AIA (see [47] in volume 200 of this series).

The uses of antisera to phosphotyrosine are not limited to the simple identification of phosphotyrosine in proteins. The majority of bands that are visualized by AIA can be isolated by immunoprecipitation with some polyclonal rabbit antisera[5] or, on a larger scale, by affinity purification with monoclonal antibodies to phosphotyrosine[4,12,19] (see [8] and [9] in this volume). The ability to isolate pure populations of tyrosine-phosphorylated proteins using nondenaturing conditions permits investigation of the effects of tyrosine phosphorylation on the activity of proteins in vitro. Affinity purification has already been exploited, in a crude sense, to recover phospholipase activity in extracts of A-431 cells, treated with EGF.[20] This discovery led to the direct identification of phosphotyrosine in phos-

[10] C. R. King, I. Borrello, L. Porter, P. Comoglio, and J. Schlessinger, Oncogene 4, 13 (1989).
[11] J. K. Klarlund, Cell 41, 707 (1985).
[12] J. R. Glenney, Jr. and L. Zokas, J. Cell Biol. 108, 2401 (1989).
[13] A. O. Morla, J. Schreurs, A. Miyajima, and J. Y. J. Wang, Mol. Cell. Biol. 8, 2214 (1988).
[14] C. A. Cartwright, M. P. Kamps, A. I. Meisler, J. M. Pipas, and W. Eckhart, J. Clin. Invest. 83, 2025 (1989).
[15] S. Giordano, M. F. DiRenzo, D. Cirillo, L. Naldini, L. Chiado Piat, and P. M. Comoglio, Int. J. Cancer 39, 482 (1987).
[16] R. A. Lindberg, D. P. Thompson, and T. Hunter, Oncogene 3, 629 (1988).
[17] K. Letwin, S.-P. Yee, and T. Pawson, Oncogene 3, 621 (1988).
[18] S. Kornbluth, K. E. Paulson, and H. Hanafusa, Mol. Cell. Biol. 8, 5541 (1988).
[19] J. R. Glenney, L. Zokas, and M. P. Kamps, J. Immun. Meth. 109, 277 (1988).
[20] M. I. Wahl, T. O. Daniel, and G. Carpenter, Science 241, 968 (1988).

pholipase C-γ.[21,22] The assay of the activities of phosphorylated populations of proteins should represent an important second generation application for anti-PY antibodies.

Considerations Involved in Making Antisera

Antigen

We have found that the profile of cellular proteins to which anti-PY
antibodies bind is not affected substantially by the exact nature of the
immunogen to which the antibodies were raised.[3] Antibodies against the
phosphorylated tyrosine residues in the Abelson tyrosine kinase (see [6]
in this volume) as well as to (1) phosphotyrosine, (2) polymerized mixtures
of phosphotyrosine, alanine, and glycine, or (3) phosphotyrosine, alanine,
and threonine [each coupled to keyhole limpet hemocyanin (KLH)], all
bind to essentially the same proteins when used for AIA. In addition,
antibodies raised to phosphotyramine, p-aminobenzyl phosphonic acid
(ABPA), and arsanylic acid, each a structural homolog of phosphotyrosine, bind essentially the same proteins.[3] Therefore, the choice of antigen
can be based on factors such as immunogenicity and ease of synthesis.

Antigenicity

The use of glutaraldehyde to cross-link phosphotyrosine to KLH results in the formation of an antigen that is approximately one-tenth as
immunogenic as that prepared using the cross-linking reagent, 1-ethyl-3-
(3-dimethylaminopropyl) carbodiimide (EDAC). Diazotization of ABPA
and subsequent coupling to KLH produce an antigen as immunogenic as
phosphotyrosine coupled to KLH with EDAC.[3] Diazotization requires a
primary aromatic amino group; therefore, phosphotyrosine itself cannot
be cross-linked by this method. Because phosphotyrosine potentially
could be hydrolyzed by serum phosphatases, ABPA was initially used in
its stead.[1] Since the phosphorus atom in ABPA is attached to the aromatic
ring by a –CH$_2$– bridge, rather than an –O– bridge, it cannot be hydrolyzed
by phosphatases. However, our results indicate that if enzymatic dephosphorylation of phosphotyrosine in injected conjugates occurs, its rate is
insufficient to affect, substantially, the immune response of the rabbit.

[21] J. Meisenhelder, P.-G. Suh, S. G. Rhee, and T. Hunter, *Cell* **57**, 1109 (1989).
[22] B. Margolis, S. G. Rhee, S. Felder, M. Mervic, R. Lyall, A. Levitzki, A. Ullrich, A.
Zilberstein, and J. Schlessinger, *Cell* **57**, 1101 (1989).

Consequently, there is no inherent advantage to the use of ABPA instead of phosphotyrosine and, since phosphotyrosine is the true determinant, its use is suggested.

Polymerization of mixtures of three amino acids, one of them phosphotyrosine, to KLH using EDAC consistently generated an immunogen that gave a high-titer response in rabbits. These conjugates contained 45–90 phosphotyrosine residues per molecule of KLH and, because EDAC catalyzes a condensation reaction between amino and carboxyl groups, the phosphotyrosine residues are presented within the context of a polypeptide backbone and are separated from the carrier by a variable spacer arm. The procedure is also very simple, and large quantities of antigen are synthesized in a relatively short time. Three rabbits immunized with conjugates produced by EDAC-mediated coupling of phosphotyrosine to KLH all produced high circulating levels of anti-PY antibodies, ranging from 600 to 100 μg/ml.[3] The protocol for synthesis of this antigen is fast and reproducible, and has been adopted as our standard method for preparing immunogen.

Preparation and Use of Anti-PY Antibodies

Synthesis of Antigen

The synthesis of antigen containing phosphotyrosine, alanine, and glycine is described below. In a 15-ml Corex tube, dissolve 0.64 nmol phosphotyrosine (160 mg) in 3.0 ml water by slowly adding small volumes of 2.0 N NaOH. Do not use more NaOH than is necessary. When dissolved, add 0.64 nmol alanine (56 mg) and 0.65 nmol glycine (47 mg). Stir until dissolved, and adjust the pH to 5.7. Add water to a final volume to 4.0 ml. Add 10 mg KLH (dissolved in 50% glycerol; Calbiochem, San Diego, CA). Position the tube above a magnetic stirring apparatus and stir the solution vigorously at room temperature. Add 2.56 nmol EDAC (472 mg) in 80-mg aliquots over the course of 1.0 hr, and maintain the pH at 6.0 by the addition of 1.0 M HCl. The pH becomes basic as the EDAC is added. The first aliquot of EDAC will increase the pH much more than successive additions. Maintain the pH at 6.0 during a second hour of stirring. Continue stirring overnight. Dialyze at 4° for 48 hr against 1.0 liter of 50 mM Tris, pH 7.2; 150 mM NaCl. Change dialysis buffer twice during this period. During dialysis some of the conjugate precipitates. Collect the conjugate, including the precipitated material, and store at −80°. It is very important to avoid the use of any buffers or reagents containing nucleophilic amino groups, such as glycine methyl esters or ammonium cations, since they will react with the coupling reagent.

Immunization of Rabbits and Screening Sera

For immunization, use the unfractionated, dialyzed conjugate emulsified in complete Freund's adjuvant. We have not compared directly a variety of injection protocols. Routinely, primary immunizations with 1.5 mg of conjugate were administered intradermally in the back. Approximately one-fourth of all attempted intradermal injections were, in fact, subcutaneous. Sera were tested for reactivity with phosphotyrosine-containing proteins at weeks 8 and 13. Sera can be analyzed by AIA at a 1 : 200 dilution, as described below (using p60^{v-src}-transformed and untransformed control cell extracts), or by enzyme-linked immunosorbent assay (ELISA) or Western blot analysis using phosphotyrosine conjugates. If analysis by ELISA is desired, make a conjugate with either bovine serum albumin, (BSA) or chicken ovalbumin (OVA), remove the precipitate that occurs after dialysis by centrifugation, and use the soluble conjugate as primary antigen to coat microtiter wells. Western blot analysis is possible as well, because a substantial amount of the soluble antigen, using BSA or OVA as carrier, can be resolved by electrophoresis through 10% polyacrylamide gels containing sodium dodecyl sulfate (SDS). This material can be transferred to nitrocellulose or Immobilon (Millipore, Bedford, MA) and probed with a 1 : 200 dilution of sera.

Affinity Purification

Purification of anti-PY antibodies is achieved by virtue of the affinity of the antibodies for phosphotyramine. Because phosphotyramine is not available commercially, investigators have substituted phosphotyrosine in this protocol, which apparently does the job as well as phosphotyramine. Incubate sera at 58° for 30 min, centrifuge at 8000 rpm at 4° for 30 min in a Sorvall (Newtown, CT) HB4 rotor to remove insoluble material, and isolate immunoglobulins by precipitation with ammonium sulfate. Dissolve the precipitated proteins in TN (0.1 M NaCl, 0.05 M Tris-HCl, pH 7.2) and dialyze at 4° for 24 hr against 2 liters of TN. To make the affinity resin, incubate Affi-Gel 10 or 15 (Bio-Rad, Richmond, CA) in a 10-fold excess volume of 75 mM phosphotyramine, pH 7.5 for 15 hr at 4° with end-over-end rotation. These conditions provide a substantial molar excess of phosphotyramine over activated ester. Equilibrate columns containing 4 ml phosphotyramine resin with TN. Perform all affinity purification steps at 23°. Pass purified immunoglobulins, dissolved in TN, over columns twice. As the anti-PY antibodies bind, the translucent resin turns white. Consequently, the volume of serum that can be processed by a single column can be determined simply by observing the progressive saturation of the resin. Usually, 4 ml of resin will bind at least 30 mg of anti-PY antibodies (30 ml of high-titer serum). Wash column with 40 ml TN, followed by a wash of 40 ml 5 mM sodium phosphate, pH 7.2, 5 mM phosphoserine (Sigma, St. Louis, MO), 5 mM phosphothreonine (Sigma), 100 mM NaCl. This wash removes antibodies that cross-react with phosphoserine and phosphothreonine and other phosphorylated compounds. Elute the specific antibodies with 12 ml of 40 mM phenyl phosphate (Sigma), 90 mM NaCl, 30 mM Tris-HCl, pH 7.2, collecting 1.0-ml fractions. Measure the A_{280}, and pool those fractions containing 90% of the protein. Dialyze the antibodies for 5 days against 1 liter of TN, containing 0.01% (w/v) sodium azide. changing the dialysis buffer daily.

Preparation of Samples for AIA

By comparison with cells lysed in hot SDS, proteins extracted in the immunoprecipitation buffers, NP-40 buffer or RIPA buffer, lose up to 95% of phosphotyrosine as a consequence of the activity of cellular phosphatases.[3] The majority of tyrosine dephosphorylation can be prevented by inclusion of 100 μM sodium orthovanadate (Sigma), an excellent inhibitor of phosphotyrosine phosphatases, and by preparation of cellular extracts at a concentration of 5 × 10⁶ cells/ml or less.[3] Very little tyrosine phosphorylation occurs after cell lysis if EDTA (2 mM) is included in the buffer. In preparation for AIA, cellular extracts are mixed with an equal volume of 2× protein sample buffer (2× PSB) and boiled. PSB (1 ×̇) contains 5 mM sodium phosphate, pH 6.8, 2% (w/v) SDS, 0.1 M dithiothreitol (DTT), 5% (v/v) 2-mercaptoethanol, 10% (v/v) glycerol, 0.4% (w/v) Bromphenol Blue.

To prepare samples directly from cells grown in tissue culture, wash the cells twice with 10 ml of 50 mM Tris, pH 7.4, 140 mM NaCl, 3.3 mM KCl, and aspirate the buffer completely. Add 1× PSB at 100°. Use approximately 1 ml/5 × 10⁶ cells. The PSB does not need to contain orthovanadate. In the case of adherent cells, swirl the plate at a 45° angle until the viscous cell lysate is completely free of the tissue culture dish. Pour the lysate into a 5-ml polypropylene tube and boil 5 min. Shear the sample 10 times through a 22-gauge needle, 10 times through a 27-gauge needle, and store at −70° prior to use. Prolonged storage at −20° will result in protein degradation.

AIA Protocol

Western blotting is performed essentially according to Towbin et al.[23] Assess the relative concentration of proteins in each sample by polyacrylamide gel electrophoresis followed by staining with Coomassie Brilliant Blue R (Sigma). Load as much protein as possible on a 1.0-mm thick gel, but avoid obvious overloading (for a well that is 5 mm wide, this amount

[23] H. Towbin, T. Staehelin, and J. Gordon, Proc. Natl. Acad. Sci. U.S.A. **76,** 4350 (1979).

is approximately equivalent to 4×10^4 fibroblastic cells). Transfer the gel-fractionated proteins to a nitrocellulose filter by electrophoresis at 40 V for 1.5 hr in transfer buffer (57.8 g glycine, 12.0 g Tris base, 3.0 g SDS, 800 ml methanol; water to 4 liters). Block nonspecific protein-binding sites by incubation of the filter for 16 hr in blocking buffer [5% BSA (essentially fatty acid and immunoglobulin free; Sigma) and 1% OVA dissolved in TNA (10 mM Tris-HCl, pH 7.2, 0.9% NaCl, 0.01% sodium azide)]. Incubate the filters for 2 hr at room temperature with 2 μg/ml anti-PY antibodies in blocking buffer, and then rinse twice for 10 min in TNA, once for 10 min in TNA supplemented with 0.05% NP-40, and twice more for 10 min in TNA. To detect rabbit immunoglobulins, add 40 ml blocking buffer containing 0.5 μCi/ml of ^{125}I-labeled protein A (low specific activity; New England Nuclear, Boston, MA), incubate for 30 min, and rinse the filter following the same procedure used after the antibody incubation, as indicated above. These conditions produce the highest levels of antibody binding to antigen and the lowest levels of nonspecific binding of either antibody or protein A. The use of Tween 20 or sodium deoxycholate in buffers for blocking or rinsing the filter diminishes the amount of specific signal. In our experience nonfat dry milk cannot be used to block sites that bind protein nonspecifically because anti-PY antibodies react strongly with a component in nonfat dry milk (see below).

Characterization of Polyclonal Rabbit Anti-PY Antibodies

Do All Proteins That React with Anti-PY Antibodies Contain PY?

A majority of proteins that have been identified by AIA have been purified subsequently by affinity chromatography with anti-phosphotyrosine antibodies and shown to contain phosphotyrosine.[3,5] However, we have observed two examples of possible nonspecificity in AIA analysis. When prestained markers are run on the gel, the antibodies bind to a number of the markers, approximately to the same extent as they would to a weak band in extracts from normal cells. The suggestion that anti-PY antibodies can exhibit minor cross-reactivity indicates that the presence of phosphotyrosine should be confirmed by phosphoamino acid analysis after affinity chromatography. The appearance of a new immunoreactive band after transformation of cells or treatment of normal cells with a growth factor is usually a good indication that the protein contains phosphotyrosine.

A more intriguing observation is that anti-PY antibodies bind very strongly to components in nonfat dry milk. We have analyzed nonfat dry milk proteins themselves by AIA, and find a number of anti-PY-reactive

proteins. Do these milk proteins contain phosphotyrosine? Possibly. However, it is also possible that the large abundance of phosphoserine in casein
causes these proteins to bind the antibody. We have not yet resolved this
issue.

What Fraction of Phosphotyrosine-Containing Proteins Are Recognized by AIA?

It is likely that proteins containing 50 to 90% of total cellular phosphotyrosine are recognized by anti-PY antibodies. Using affinity purification
with mouse monoclonals or rabbit polyclonal anti-PY antibodies, approximately 50%[19] or 94%,[5] respectively, of total phosphotyrosine can be removed from cellular extracts. The mouse and rabbit antibodies used in
these studies recognize almost identical proteins when analyzed by AIA,
and the bands visualized by AIA correlate well with the phosphorylation
pattern of proteins that are affinity purified from the same cells. In addition,
each oncogenic tyrosine protein kinase that we have examined—those
encoded by the v-*yes*, v-*fgr*, v-*abl*, v-*fps*, v-*fes*, *bcr/abl*, and *neu*
genes—react with anti-PY antibodies in AIA. Likewise, the receptors for
EGF, PDGF, and insulin, each of which is phosphorylated on tyrosine in
response to binding ligand, bind anti-PY antibodies in AIA. Consequently,
the preponderant evidence suggests that a majority of proteins containing
phosphotyrosine will appear as bands on anti-PY immunoblots. It should
be kept in mind, however, that a small fraction of the tyrosine-phosphorylated proteins may contain the majority of the phosphotyrosine in total
cellular protein. If this is the case, then some nonabundant substrates may
not be visualized by AIA because of sensitivity limitations.

Do Anti-PY Antisera Vary in Their Affinities for Phosphorylation Sites?

In general, the AIA pattern appears to be at least a fair representation
of the relative abundance of phosphotyrosine in individual proteins. However, there are exceptions. We have studied the ability of p56^{lck} to react
with anti-PY antibodies on immunoblots. p56^{lck} that is phosphorylated on
tyrosine exclusively at its autophosphorylation site reacts at least 10 times
better with anti-PY antibodies than does p56^{lck} that contains the same
amount of phosphotyrosine distributed at other sites. Apparently, there
are differences in the affinity of anti-phosphotyrosine antibodies for different sites of tyrosine phosphorylation. Consequently, one cannot assume
that the intensity of staining of a band in AIA reflects accurately its content
of phosphotyrosine.

Are Mouse Monoclonal Antibodies as Good as Rabbit Polyclonal Antibodies for AIA?

Monoclonal antibody PY-20[19] binds the same proteins from cells transformed by p60$^{v\text{-}src}$ as do polyclonal rabbit antibodies.[4] When using PY-20, it is important not to extend the rinsing periods, because one begins to lose the specific signal. This may be because PY-20 has a lower binding constant than do polyclonal rabbit antibodies. However, if one adheres to the protocol described above, using ^{125}I-labeled goat anti-mouse antibodies rather than ^{125}I-labeled protein A to identify the bound monoclonal antibody, the specific signal remains strong. When examining less abundant substrates, such as those in human carcinoma cell lines,[14] we have observed that rabbit antibodies give a better ratio of signal to background. Consequently, we suggest employing polyclonal rabbit antibodies for AIA and monoclonal antibodies for affinity purification.

Are Rabbit Antibodies as Good as Monoclonal Mouse Antibodies for Affinity Purification?

In our experience, all proteins that bind monoclonal antibody PY-20 during affinity purification contain phosphotyrosine, whereas a number of proteins that bind columns containing polyclonal rabbit antibodies do not. Consequently, we suggest that only monoclonal antibodies be used for large-scale affinity purification of phosphotyrosine-containing proteins.

Can Each Band Visualized by AIA Be Purified by Affinity Chromatography?

A majority of the substrates of p60$^{v\text{-}src}$ that appear as bands by AIA can be affinity purified on PY-20 monoclonal anti-phosphotyrosine antibody columns[4] (see [9] in this volume). However, two putative phosphotyrosine-containing proteins, identified by AIA in cell lines from human carcinomas,[14] bound very poorly to PY-20 antibody columns and could not be purified by this technique. Therefore, our current data suggest that most, but not all, of the proteins represented as bands on an AIA can be purified by affinity to anti-phosphotyrosine resins.

Summary

The techniques of detecting phosphotyrosine-containing proteins by AIA and purifying these proteins by affinity chromatography with anti-phosphotyrosine monoclonal antibodies are both reliable and consistent. Although not every tyrosine-phosphorylated protein can be detected and purified by using these techniques, a majority can. No doubt future studies will employ these approaches both for analyzing the specific role that tyrosine protein phosphorylation plays in regulating cell division and for investigating the properties of heretofore uncharacterized proteins that contain phosphotyrosine.

[11] Phosphopeptide Mapping and Phosphoamino Acid Analysis by Two-Dimensional Separation on Thin-Layer Cellulose Plates

By William J. Boyle, Peter van der Geer, and Tony Hunter

Introduction

Peptide mapping is a powerful technique used to help determine peptide structure and composition of proteins. Peptide maps or fingerprints of proteolyzed proteins are usually obtained by resolution on either one-dimensional sodium dodecyl sulfate-polyacrylamide gel electrophoresis (SDS–PAGE),[1] reversed-phase high-performance liquid chromatography (HPLC),[2] or by two-dimensional separation on thin-layer cellulose (TLC) plates. Perhaps the most common applications of peptide mapping are (1) to compare proteins suspected to be encoded by the same or related genes, (2) to prepare individual peptides to determine amino acid composition and sequence, and (3) to determine the precise location of amino acid residues that are posttranslationally modified by either fatty acid acylation, glycosylation, methylation, acetylation, or phosphorylation. Because of the fact that often only vanishingly small amounts of a protein are available for analysis, it is often difficult or impossible to carry out peptide mapping without labeling proteins with radioactive precursors.

Two-dimensional separation of proteolytic digests by electrophoresis and chromatography on TLC plates is a technique that is well suited to analysis of labeled protein samples. First, it is an extremely sensitive technique that requires only a few disintegrations per minute (dpm) of metabolically labeled product. Second, because digests are resolved in two dimensions, a variety of information is derived that often yields subtle but important clues about a given peptide that may help reveal its composi-

[1] D. W. Cleveland, S. G. Fischer, M. W. Kirschner, and U. K. Laemmli, *J. Biol. Chem.* **252**, 1102 (1977).
[2] H. Juhl and T. R. Soderling, this series, Vol. 99, p. 37.

tion. Finally, because cellulose is an inert substance, the peptide material can be recovered for secondary analysis, such as determining amino acid composition and sequence, or determining the presence and position of phosphoamino acid residues.

Peptide mapping experiments are often frustrated by problems that arise during sample preparation and that lead to poor resolution and inconclusive results. The following method for two-dimensional peptide analysis on TLC plates has been used in our laboratory with repeated success. This technique focuses on the use of the Hunter thin-layer electrophoresis system (HTLE-7000; CBS Scientific, Del Mar, CA) to resolve peptides in the first dimension, and has been used successfully to map phosphorylation sites in several different proteins in our laboratory.[3–9] It features a unique clamping system and an inflatable nylon air bag to remove excess liquid from the surface of the plate and to prevent buffer from siphoning from the electrode compartments onto the plate. This system provides uniform cooling because of close and even contact of the plate with the built-in cooling surface. Since good results require careful sample preparation, we describe here in detail our procedures for sample preparation and analysis. This protocol may be used to prepare phosphopeptides to be analyzed by a variety of techniques, including resolution by HPLC. In addition, several techniques that might be useful for actual identification of phosphotryptic peptides and the assignment of phosphorylation sites in a known sequence are described.

Sample Preparation and Proteolytic Digestion

This method is useful for mapping ^{14}C-,^{35}S-, or ^{32}P-labeled proteins. For ^{32}P-labeled samples a small portion can be used for phosphoamino acid analysis while the remainder can be used for phosphopeptide mapping. The only difference between ^{14}C- or ^{35}S-labeled samples and ^{32}P-labeled samples is that one can monitor the recovery of the ^{32}P label at each step by Cerenkov counting. For ^{14}C- or ^{35}S-labeled samples one must proceed blindly until the final transfer step. The entire protocol will usually last for 3–4 days, but can be accelerated if necessary. For many steps the timing

[3] T. Hunter, N. Ling, and J. A. Cooper, *Nature (London)* **311**, 480 (1984).
[4] J. A. Cooper, K. L. Gould, C. A. Cartwright, and T. Hunter, *Science* **231**, 1431 (1986).
[5] K. L. Gould, J. R. Woodgett, J. A. Cooper, J. E. Burs, D. Shalloway, and T. Hunter, *Cell* **42**, 849 (1985).
[6] C. M. Isacke, I. S. Trowbridge, and T. Hunter, *Mol. Cell. Biol.* **6**, 2745 (1986).
[7] D. W. Meek and W. Eckhart, *Mol. Cell. Biol.* **8**, 461 (1988).
[8] G. A. Weinmaster, D. S. Middlemas, and T. Hunter, *J. Virol.* **62**, 2016 (1988).
[9] P. van der Geer and T. Hunter, *Mol. Cell. Biol.* **10**, 2991 (1990).

TABLE I

PHOSPHOPEPTIDE MAPPING AND PHOSPHOAMINO ACID ANALYSIS

Day	Protocol
1	Identify protein bands of interest by lining up the exposure with the gel. Cut out the bands, remove the paper backing, grind up the gel, and elute in 50 mM NH$_4$HCO$_3$, 0.1% SDS, and 0.5% 2-ME in two steps with a combined volume of 1.2 ml. Elute 2–3 hr at room temperature each time
2	Clear the combined eluate by spinning in a microcentrifuge (10 min at 10,000 rpm), add carrier protein (10–20 μg/ml) and TCA to 20% (250 μl), and precipitate 1 hr on ice. Spin down the precipitates (10 min, 10,000 rpm, 4°), wash once with 95% ethanol, air dry
	For PAAs: Hydrolyze precipitate in 50 μl 6 N HCl for 1 hr at 110°, lyophilize, dissolve pellets in 6 μl pH 1.9 buffer containing ~60 μg/ml cold phosphoamino acids. Spot up to 5 μl and run the electrophoresis in two dimensions. Stain, after drying the plates, with ninhydrin (0.25% in acetone) to visualize marker phosphoamino acids, and expose to film at −70° with an intensifier screen
	For peptide maps: Dissolve pellet in 50 μl cold performic acid (nine parts 98% formic acid and one part 30% hydrogen peroxide, incubated 30 min at room temperature), incubate 60 min on ice, add 400 μl H$_2$O, freeze, and lyophilize
3	Resuspend pellet in 50 μl NH$_4$HCO$_3$, pH 8.0, add 10 μl trypsin (1 mg/ml), incubate for 2–3 hr at 37°, repeat once. Add 400 μl H$_2$O, vortex lyophilize, repeat once
4	Dissolve pellet in 400 μl buffer (pH 1.9, pH 4.72, or H$_2$O depending on which electrophoresis system will be used), spin 10 min at 10,000 rpm, transfer supernatant, and lyophilize
5	Resuspend pellet in 6 μl electrophoresis buffer (pH 1.9 or 4.72) or H$_2$O, spin 5 min at 10,000 rpm, spot up to 5 μl, and run the electrophoresis dimension. Dry plates and run the chromatography dimension. Dry, mark, and expose the plates at −70° with an intensifier screen

is flexible, and there are convenient points to stop overnight. The protocols given below are summarized in Table I.

Elution from Polyacrylamide Gel

The most common way to prepare samples is to localize either protein "bands" from one-dimensional (Laemmli) gels, or "spots" from two-dimensional gels. This can be accomplished by either staining the gels with Coomassie Blue or by autoradiography. Treating gels with fluors, such as diphenyloxazole (PPO) or sodium salicylate, to help detect weak signals is not recommended because this can impede extraction and digestion procedures. However, PPO-treated gels can be soaked in glacial acetic acid for ~30 min and washed several times in water to remove PPO, and then used for elution of proteins. For optimal recovery we prefer to dry untreated gels onto Whatman (Clifton, NJ) 3MM filter paper.

After drying the gel, make marks around the sides of the gel on the paper backing with ^{35}S-labeled radioactive India ink (10 μCi/ml) to provide alignment marks, and expose to film. After developing the film, align the ink marks on the dried gel with their images on the film over a light box. Once completely aligned, staple the film to the paper. Place this sandwich film-side down on the light box and trace the band using a pencil on the back of the paper. For Coomassie Blue-stained gels, where the band of interest may be easily visualized, this may not be necessary. Remove the staples and film from the dried gel and place it on a hard flat surface such as a thick glass plate. Carefully excise the band using a new single-edge razor blade. Alternatively, if the film is not needed, one can cut through the film into the gel on either side of the band of interest while they are still stapled together. Next, peel the paper backing from the gel slice while immobilizing it with forceps, and remove any paper fibers by scraping the back of the slice with a razor. Try to remove as much of the paper as possible without shaving down the gel slice. Gloves should be worn to avoid contaminating the gel piece with "finger" proteases.

Place the gel slice in a Kontes disposable tissue grinder tube (#K 749520-0000; Kontes, Vineland, NJ) and add about 500 μl of freshly prepared 50 mM ammonium bicarbonate (400 mg/100 ml), pH 7.3–7.6. The pH of freshly prepared ammonium bicarbonate is usually within the range of 7.3–7.6. It will drift to a more alkaline pH (~8.3) overnight and can be used during the enzymatic cleavage steps (see below). If the sample is labeled with ^{32}P, then determine how many counts per minute (cpm) are present by counting Cerenkov radiation before proceeding.

Let the gel piece rehydrate for about 5 min at room temperature (RT). Using the disposable plastic pestles which accompany the tubes, grind the gel piece until it is fine enough to pass through a yellow 200-μl pipette tip. For multiple samples the pestles may be inserted into a hand-held, variable-speed electric drill to homogenize each gel piece, discarding or washing the pestles thoroughly between samples. Transfer the gel suspension to a screw-cap microcentrifuge tube (1.5 ml, #72.692 or 2.0 ml, #72.693; Sarstedt, Princeton, NJ). Rinse the tissue grinder and tube twice with 250 μl of ammonium bicarbonate to transfer residual gel bits quantitatively (total volume of ammonium bicarbonate solution used is 1.0 ml). Add 50 μl of 2-mercaptoethanol (2-ME) and 10 μl of 10% (w/v) SDS. Tightly screw on the lid, vortex briefly to mix (but try to avoid foaming), and then boil the sample for 3–5 min.

Remove the tube and incubate at 37° or room temperature for at least 90 min with shaking (we sometimes incubate tubes overnight). Next, the suspension is vortexed and then briefly centrifuged for 5–10 min at low speed (3000 rpm in a swinging bucket table-top centrifuge, or 2 min in a microfuge) to remove the gel bits. The microfuge compresses the gel bits better and is what we use most often. Carefully transfer the supernatant, using a disposable plastic transfer pipette (avoiding any debris), to a new tube and measure its volume. The disposable transfer pipettes from Research Products International Corporation (#147500; Mount Prospect, IL) are particularly good for this purpose because they have a very narrow and drawn-out tip. It is important to transfer the supernatant immediately, because otherwise the gel pellet reswells, and if the sample was centrifuged in a microfuge, the surface of the pellet may "fall over," making it difficult to aspirate all the supernatant. A second volume of ammonium bicarbonate plus SDS and 2-mercaptoethanol is added to wash the gel bits and to elute any residual protein. Since the volume of total eluate collected should be roughly 1.2 ml, the second volume of buffer to be added should equal 1.2 ml minus the volume of the first supernatant. After adding the second volume of elution buffer, replace the cap and vortex the tube briefly. The suspension may be reboiled for another 3–5 min (optional) before incubating again at 37° for 30–60 min with or without shaking.

The total elution period, steps 1 and 2 combined, should be at least 1.5 hr and may last 12 hr. After the second elution step the gel bits are removed again by centrifugation, and the second supernatant is removed and combined with the first (should be a final volume of about 1.2 ml). Clarify the total eluate by spinning in a microfuge for 5–10 min to pellet any residual debris, then carefully transfer the supernatant to another new tube. It is essential that no particulate debris remain in the clarified supernatant. Place the tube on ice and chill (samples can be frozen and stored at this stage if desired).

Instead of eluting protein from a gel piece, elution from SDS-polyacrylamide gels can also be accomplished by the electrophoretic transfer of proteins to nitrocellulose, nylon, or Immobilon (Millipore, Bedford, MA) membranes.[10,11] Protein eluted in this fashion can either be used to generate a tryptic digest or to carry out partial acid hydrolysis for phosphoamino acid determination (acid hydrolysis can be carried out only with Immobilon)(see under Two-Dimensional Phosphoamino Acid Analysis). To generate a tryptic digest, the membrane fragment containing the labeled protein or peptide is incubated directly with trypsin and the tryptic peptides are released in a soluble form. Labeled proteins transferred to membranes can

[10] R. H. Aebersold, J. Leavitt, R. A. Saavedra, L. E. Hood, and S. B. H. Kent, *Proc. Natl. Acad. Sci. U.S.A.* **84**, 6970 (1987).
[11] K. Luo, T. R. Hurley, and B. M. Sefton, *Oncogene* **5**, 921 (1990).

be visualized and excised using the same method outlined above. This technique is described in detail.[12] Briefly, the piece of membrane corresponding to the protein band is soaked in 0.5% polyvinylpyrrolidone (PVP-360; Sigma, St. Louis, MO) in 100 mM acetic acid at 37° for 30 min to block nonspecific absorption of trypsin, washed briefly in water, and then with fresh 50 mM ammonium bicarbonate. Digestion is achieved by incubating the piece of membrane in 200 μl of 50 mM ammonium bicarbonate with 10 μg trypsin for 2 hr at 37°, followed by addition of another 10 μg of trypsin and a second 2-hr incubation at 37°. Tryptic peptides are released from the membrane in this process, and in general >90% of the ^{32}P radioactivity in a protein is released as phosphopeptides, although very large hydrophobic peptides may be retained. The digest is lyophilized and washed and then, if necessary, the peptides are oxidized in performic acid as described below.

Precipitation

If the protein has been eluted from a gel piece, the next step is to concentrate the labeled protein and to remove any SDS by precipitating the proteins with trichloroacetic acid (TCA), followed by washing with a volatile organic solvent to remove TCA. The proteins are then oxidized in the presence of performic acid, which will fully oxidize all methionine and cysteine residues present to a single oxidation state (methionine sulfone and cysteic acid, respectively). This will prevent artifactual separation of oxidation isomers during chromatography.

To save time, while the eluate from the previous step is chilling, performic acid is prepared by adding 900 μl of formic acid (~98%) and 100 μl of hydrogen peroxide (30%) to a 1.5-ml microcentrifuge tube. Performic acid, formed during incubation at room temperature for 60 min, is then chilled by placing on ice.

The eluted protein is precipitated in the presence of carrier protein and 15–20% TCA. We usually use boiled pancreatic RNase A because it is both a good carrier and degrades contaminating RNA. However, other proteins, such as IgG and bovine serum albumin (BSA), can be used as well. Add 20 μg of carrier (RNase A) to the chilled eluate, followed by 250 μl of ice-cold TCA (100%). Mix well, then leave on ice for at least 60 min. Collect the TCA precipitate by centrifugation at >8000 g in a microfuge for 10 min at 4°. Carefully remove the supernatant, leaving the final 50–100 μl behind (the pellet may not be visible at this stage). Recentrifuge briefly to bring the liquid on the walls of the tube to the bottom, and

[12] K. Luo, T. R. Hurley, and B. M. Sefton, this volume [12].

remove the remaining supernatant with a disposable microtransfer pipette, combining it with the first (TCA supernatant). Wash the precipitate with 500 μl of either absolute ethanol or acetone (ice cold), and centrifuge again for 5 min at 4°. A small white pellet at the bottom of the tube should be obvious at this stage, although occasionally the carrier protein precipitates along the side of the tube. Carefully remove the supernatant as before, and air dry the pellet (do not lyophilize).

If the samples are ^{32}P labeled, then determine Cerenkov counts remaining and determine the percentage recovery. The yield should be about 70% of the starting radioactivity unless the protein was unusually large or a very high concentration acrylamide gel was used during the purification. If the recovery was low, check both the TCA supernatant and ethanol/ acetone wash for residual counts. If the majority of your counts remained in the TCA supernatant, replace it on ice for an additional 1–4 hr and repeat the centrifugation and washing steps.

Oxidation and Proteolytic Digestion

Usually only trypsin[13] or chymotrypsin[13] are capable of digesting TCA-precipitated protein to completion, although a variety of proteases, such as *Staphylococcus aureus* V8 protease,[14] thermolysin,[15] *Pseudomonas fragi* protease,[16] and proline-specific endopeptidase,[17] will yield reproducible fingerprints (see Table V for sources). The example treatment used below will illustrate the use of trypsin, but any of the enzymes listed above could be used instead, using the same enzyme concentration and sample treatment. Prepare stocks of N-tosyl-L-phenylalanine chloromethyl ketone (TPCK)-treated trypsin (Worthington Biochemical Corp., Freehold, NJ) and other enzymes in 0.1 mM HCl at a concentration of 1.0 mg/ml and store in small aliquots in liquid nitrogen. The choice of brand or batch of microcentrifuge tubes for the precipitation and subsequent oxidation may be important. We have had occasional problems with "spot doubling" in the chromatographic dimension; this may be due to the mold-release compound used during the manufacturing of the tubes causing alkylation of peptides during oxidation. If there are problems, it might be worth

[13] D. G. Smyth, this series, Vol. 11, p. 214.
[14] G. R. Drapeau, this series, Vol. 47, p. 189.
[15] R. L. Heinrikson, this series, Vol. 47, p. 175.
[16] G. R. Drapeau, *J. Biol. Chem.* **255,** 839 (1980).
[17] T. Yoshimoto, R. Walter, and D. Tseru, *J. Biol. Chem.* **255,** 4786 (1980).
[18] J. E. Folk, *in* "The Enzymes," Vol. 3 (P. D. Boyer, ed.), p. 57. Academic Press, New York, 1971.

testing different brands of microcentrifuge tubes to find out which gives the best result.

Beginning with the washed and dried TCA precipitate one has several options: (1) proceed directly with trypsinization, (2) use the entire sample to determine phosphoamino acid content (see below), or (3) use portions for both tryptic peptide analysis and for phosphoamino acid analysis. For either direct trypsinization or for combined tryptic and phosphoamino acid analysis, dissolve the precipitate in 50–100 μl of ice-cold performic acid (made as described above) by vortexing thoroughly, and then let the sample oxidize at 0° for 60 min. Be careful not to warm the sample. This will result in unwanted side reactions. Remove a small portion of the dissolved sample at the end of this period if phosphoamino acid analysis is needed (see below). After complete oxidation, add 400 μl of deionized water to dilute the performic acid, vortex, and freeze the sample on dry ice. The oxidation step can be omitted if none of the peptides of interest contains methionine or cysteine, although oxidized proteins are in general more efficiently digested with trypsin.

Since performic acid can cleave certain peptide bonds during extended incubation at higher temperatures it is important to both dilute the sample in water and freeze as soon as the oxidation step is finished. The frozen sample is then lyophilized to completion. This usually takes 3–4 hr, but can be left overnight for convenience. A Speed-Vac from Savant (Hicksville, NY) is very good for this process because the sample is concentrated in the bottom of the tube. The sample should appear as white and stringy material like a miniature cottonball, not as a hard pellet.

Resuspend the oxidized protein pellet in about 50 μl of ammonium bicarbonate, pH 8.0–8.3. The ammonium bicarbonate solution prepared on the previous day should have the correct pH. Vortex well and spin the tube briefly to bring all the liquid to the bottom of the tube. For standard trypsinization, add 10 μl of a stock 1.0 mg/ml solution of TPCK–trypsin (10 μg), close the lid tightly, then incubate for 3–5 hr (or overnight) at 37°. At this time vortex the digest well (at least 30–40 sec), centrifuge to condense the sample as before, then add an additional 10 μl of stock trypsin and incubate again for 3–5 hr at 37°.

Incubation periods lasting 2–3 hr each should be sufficient, and one can use less trypsin (1–2 μg total). Short, sequential additions (repeated five times) of 0.1 μg of trypsin for 30–60 min have been used successfully. This may be valuable when trying to obtain complete tryptic digests while eliminating any contaminating chymotryptic activity. However, we follow the original protocol because it is very reproducible and seems to solubilize the maximum number of counts from the precipitate.

Washing and Lyophilization

After digestion is complete, the next step is to remove the ammonium bicarbonate by repeated lyophilization. Undigested material is then removed, and soluble peptides are dissolved in running buffer and then applied to a TLC plate. If any salt or debris contaminate the sample it may cause peptides to streak or smear in the first dimension (see below). This part of the procedure is crucial for preparing samples for two-dimensional analysis, and must be done carefully to obtain "pretty" maps. Hydrophobic peptides may be lost by irreversible adsorption to the walls of the tube if one lyophilizes the sample to dryness. This loss can be avoided by lyophilizing down to ~10 μl at each step rather than to dryness.[19] Because variable losses of hydrophobic peptides may occur, this practice, while somewhat inconvenient for large numbers of digests, might be worth adopting in special cases.

To the digest (~70 μl of ammonium bicarbonate solution) add 400 μl of deionized water, freeze, and lyophilize (or just lyophilize if using a Speed-Vac). Add another 300 μl of water, vortex well (at least 60 sec), then relyophilize. Check to make sure that no salt is left, i.e., no fine white powder should be visible on the sides of the tube. Next, resuspend the sample in 300 μl of either pH 1.9 buffer, pH 4.72 buffer, or water (see below for explanation of choices) and vortex well again. Centrifuge the sample for at least 2 min in the microfuge at RT, then transfer the supernatant with a microtransfer pipette, carefully avoiding any debris, to a new tube and relyophilize. If any particulate material remains, resuspend the sample in 100 μl of buffer, vortex, and centrifuge again. Carefully remove the supernatant and check to see if it is debris free.

When preparing a [14]C- or ^{35}S-labeled sample, remove 3 μl, spot on a glass fiber filter, and count by liquid scintillation counting. Lyophilize the rest of the sample until it is completely dry, then resuspend in 5–10 μl of the same buffer. For ^{32}P-labeled samples estimate total yield by Cerenkov counting. The samples are now ready for application onto the TLC plates and can be stored at $-20°$ until used.

First Dimension: Thin-Layer Electrophoresis

Rationale behind Electrophoretic Separation

The first dimension of separation is the electrophoresis of the sample on a TLC plate. The optimal conditions are often determined empirically by using various volatile buffers and varying the voltage and time of

---

[19] M. J. Hubbard and P. Cohen, *Eur. J. Biochem.* **180,** 457 (1989).

TABLE II
COMPOSITION OF BUFFERS

Electrophoresis buffers[a]		Chromatography buffers[a]	
Component	Volume (ml)	Component	Volume (ml)
pH 1.9 buffer		Regular chromatography buffer	
Formic acid (88%)	50	n-Butanol	785
Glacial acetic acid	156	Pyridine	607
Deionized water	1794	Glacial acetic acid	122
pH 3.5 buffer		Deionized water	486
Glacial acetic acid	100	Phospho chromatography buffer	
Pyridine	10	n-Butanol	750
Deionized water	1890	Pyridine	500
pH 4.72 buffer		Glacial acetic acid	150
n-Butanol	100	Deionized water	600
Pyridine	50	Isobutyric acid buffer	
Glacial acetic acid	50	Isobutyric acid	1250
Deionized water	1800	n-Butanol	38
pH 8.9 buffer		Pyridine	96
Ammonium carbonate (in g)	20.0	Glacial acetic acid	58
Deionized water	2000	Deionized water	558

[a] For 2 liters.

electrophoresis. We use three volatile buffer systems for electrophoresis for 20–30 min at 1.0 kV in the first dimension: pH 1.9 buffer, pH 4.72 buffer, and pH 8.9 buffer (named for final pH; see Table II). In general, more basic peptides are resolved in pH 1.9 buffer, and more acidic peptides in pH 8.9 buffer. At pH 4.72 peptides with free side-chain carboxyl groups such as glutamate are resolved well because the pK values of the acidic side chain are near 4.7. Included also is pH 3.5 buffer, which is usually used for second dimension electrophoretic separation of phosphoamino acids. We sometimes use this buffer in the first dimensional separation of tryptic peptides and when plate-purifying peptides phosphorylated *in vitro*. At this pH [γ-^{32}P]ATP is well resolved from phosphopeptides. pH 6.5 (10% pyridine, 0.4% acetic acid) is another volatile buffer, which is foul smelling, but can be useful for separating peptides that contain histidine from neutral peptides, and for analysis of phosphohistidine, which is unstable at pH values below 6.

There are no general rules that help predict which electrophoretic system to choose. pH 1.9 buffer is often used as a starting point because most peptides are soluble in it, and it resolves peptides well. If peptides of known sequence are to be resolved their predicted mobilities (see below) may dictate the buffer to be used. The choice of buffer system will determine which buffer is used to resuspend the tryptic digest during the final wash stage, except for the pH 8.9 buffer, which uses deionized water for the final wash.

Preparation of Electrophoresis Buffers

Table II lists the composition of the most commonly used buffers. Reagent-quality solvents and deionized water should be used at all times. Pyridine is particularly unstable. It should be stored under nitrogen and should not be used if it is yellow in color. Buffers can be prepared well in advance of electrophoresis if stored in glass containers with air-tight lids or stoppers. Check the pH of solutions after they are prepared. The actual pH of the buffer may be slightly different depending on the grade of solvents or the type of deionized water used. For example, the pH 1.9 buffer we use is sometimes actually pH 1.8. If the pH of the solution varies by more than 0.2 pH unit, check that the dilution of each added component is correct. Do not adjust the pH of the buffers by using concentrated acid or base. Record the pH of the buffers as they are prepared.

Preparation of Thin-Layer Cellulose Plates

Select a TLC plate. We use 100-μm 20 × 20 cm microcrystalline cellulose glass-backed plates without fluorescent indicators (#5716; E. M. Science, Gibbstown, NJ), but other types of cellulose plates can be used, and plastic-backed plates work for most purposes. Plates are poured in one direction, and it is best to orient the plate so that electrophoresis proceeds in the direction of pouring, i.e., the direction with gaps on the edges of the plate. First, make sure that the cellulose layer is smooth, with an even surface that does not have any gouges or nicks. Place the plate on a light box and illuminate from beneath to see if the thickness of cellulose is even across the area to be used for electrophoresis and chromatography. Based on these criteria most plates are suitable, although we often come across boxes that contain several plates with an unusable cellulose surface. The sides of the plate where the cellulose forms a beaded edge should be at the top and bottom. The smooth edges on which the cellulose flows up to the end of the plate should be on the left- and right-hand sides and will make contact with the buffer wicks during electrophoresis.

Make several orientation marks on the plate. This is best done by placing an actual size outline, as shown in Fig. 1, on a light box, then putting the plate on top of it, cellulose side up. Using a very soft-leaded blunt pencil, make a small cross mark (+) where the sample and the marker

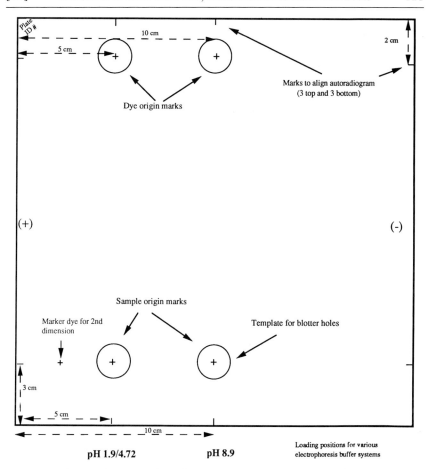

FIG. 1. Diagram for a stencil outline to mark the 20 × 20 cm TLC plates for two-dimensional separation of tryptic phosphopeptide digests by electrophoresis (horizontal plane) and chromatography (vertical plane). The stencil can be used both to mark the TLC plate, using an extra-soft lead pencil, and to serve as a template to make the blotter used to wet the plate as described in the text. The positions of the sample origins and marker dye origins depend on the electrophoresis buffer system employed (bottom).

dye origins will be without gouging the cellulose surface (see Fig. 1). Note that the position of the origin and dye spots will vary depending on the electrophoresis buffer used, e.g., to the left-hand side for pH 1.9 and 4.72 buffers and in the middle for pH 8.9 buffer. Make short marks on the sides of the plate adjacent to the sample and dye origin marks (Fig. 1). In the upper left-hand corner of the plate put an identification number which refers to a specific sample. Place the plate in a rack (#05-718-35; Fisher Scientific, Pittsburgh, PA) while additional plates are prepared.

Sample and Marker Dye Application

To apply the sample to the TLC plate the following are required: (1) an area on the lab bench with a flat surface, (2) a microfuge to centrifuge the samples, (3) an air hose (equipped with a Nalgene 0.2-μm nitrocellulose filter unit to trap aerosols and particulate matter) or a cold air blower to dry the samples as they are applied, and (4) Oxford ultramicropipettes (1–5 μl, #8501-8505; Oxford, St. Louis, MO) with disposable plastic capillary tips (#809; Oxford) to apply the samples. A marker dye, composed of a mixture of 5 mg/ml ε-dinitrophenyl (DNP)-lysine (yellow) and 1 mg/ml xylene cyanol FF (blue) solubilized in water containing 30–50% (v/v) of pH 4.72 buffer is also needed. The dye mixture, which appears green in solution, will separate during electrophoresis into its yellow and blue components. The separation distance depends on the pH of the run, and can be used to ascertain that the electrophoresis lasted the correct time. The exact position of the ε-DNP-lysine marker relative to the origin is useful in determining the position of an electrically neutral zone (see below). The same marker dye is also used to determine the extent of migration during the chromatography steps (see below).

To apply the samples and dye, place a marked TLC plate on a flat surface. Remove 1–2 μl of dye mix using the Oxford micropipette and "spot" a single drop (~0.5 μl) on the mark for the dye. Dry the spot using an air stream from a filtered air line fitted with a 1-ml plastic syringe at the end to focus the flow. Never touch the plate with either the pipette tip or the air nozzle. Before applying the sample, centrifuge it for 30–60 sec to pellet any particulate material. Sample volumes are in the 5 to 10-μl range, and we try to load about 3–7 μl of the sample. Usually the number of counts per minute needed will determine the amount of each sample loaded. For either ^{14}C or ^{35}S load at least 1000 cpm and possibly up to 20,000 cpm, depending on the number of labeled peptides. For ^{32}P load at least 100–200 cpm and up to 1000 cpm for proteins that are multiply phosphorylated. Use the pipette to remove only the top two-thirds of sample, making sure not to pick up any debris from the bottom of the tube.

Spot single 0.5-μl drops on the sample mark by carefully touching the surface of the drop to the cellulose, and dry with the air stream. The air line and pipette can be held in different hands so one can load and dry sequentially. A 5-μl sample is usually applied in about 10 controlled spottings. Try to keep the area wetted during each application to a minimum, i.e., between 1 and 2 mm in diameter, and try to superimpose the position exactly as each drop is added. Usually a faint brown ring will form around the circumference of the spot and the cross mark should be directly in the center. Once samples are loaded, return the plate to the plate holder. The samples are quite stable on the plates and can be spotted the day before electrophoresis.

Preparing Apparatus for Electrophoresis

Once all the plates have been loaded, begin to prepare the electrophoresis apparatus. We use the Hunter thin-layer electrophoresis unit (HTLE 7000; CBS Scientific, Inc., Del Mar, CA). The apparatus should be installed on a flat surface and connected to an electrophoresis supply capable of delivering at least 1.0 kV at 100 mA, running water (at ~16°), and a constant-pressure air line with a regulator valve (see manufacturer's instructions for complete installation). Connect the electrodes to the buffer tanks and the power supply. Remove the pins that fasten the restraining lid of the machine and lift the lid up together with the attached nylon air bag. Remove the thick black Neoprene pad and the Teflon sheet, and set them aside. Fill each electrode tank with about 500 ml of the appropriate buffer. Cut two sheets of thin (0.004 in.) polyethylene plastic sheeting large enough to cover the surface of the electrophoresis bed (about 25 × 45 cm). Make sure that there are no large creases or folds in the sheeting that will keep the plate from laying completely flat while under pressure. Gloves should be worn for the next step and all subsequent steps. Place one sheet so it covers the Teflon cover of the cooling plate and tuck it down on either side between the cooling plate and the buffer tanks.

Next, prepare two electrophoresis wicks from pieces of Whatman 3MM filter paper cut to 20 × 28 cm. Fold the paper in half lengthwise to give a wick of 14 × 20 cm, and sharpen the crease using the back of the hand. Wet the wicks by placing them in a tray containing electrophoresis buffer. Slide the wicks into each buffer tank with the creased edge up through the slot that is formed next to the cooling plate when the lid is closed, and fold the wick over the polyethylene sheet covering the electrophoresis bed. Place the second polyethylene sheet over the wicks, so that it spreads out over the top of the buffer tanks, followed by the top Teflon sheet and the Neoprene pad. Close the lid of the apparatus and replace the pins so that they are securely in place. Turn the pressure to 10 lb/in.^{2} to inflate the air bag. This will squeeze out excess liquid from the wick and completely flatten its edges. When ready, shut off the pressure and disassemble the apparatus as before; remove the pins, lift up the lid, remove the Neoprene pad and top Teflon sheet, and place aside. Carefully lift up the top sheet of polyethylene by one corner and slowly peel it away, and then wipe up any excess buffer from the top and bottom sheets of polyethylene using a Kimwipe. The apparatus is now ready to run when the plate is ready. It is best to keep the wicks under pressure until electrophoresis is started so the wicks do not soak up excess buffer.

The plate needs to be wetted with the same running buffer before it is ready for electrophoresis. We use a special blotter that wets the plate while also concentrating the sample. It is composed of two sheets of Whatman 3MM paper sewn together and has two holes about 1.5–2.0 cm in diameter bored through with a cork borer at positions corresponding to the sample and dye marks (see Fig. 1). Soak the blotter in buffer in a large tray. Let the blotter drain to remove excess liquid; then, using gloved hands, blot it against dry paper until it is only slightly damp. Blotters can be reused until they literally fall apart. Used blotters should be labeled according to buffer used, and rewetted only with the same buffer. Place the TLC plate on a flat surface, cellulose side up, then carefully place the damp blotter down on the plate. Concentrate first on wetting the areas around the sample and then the dye marks. Do this by pressing around the edges of all the holes to direct the wetting flow toward the center of the circle. If done properly all the liquid converges on the center of the circle and acts to concentrate the spotted sample, which should form a tight brownish spot. If the sample does not wet evenly and quickly, it probably means that the digest contains some foreign material and in this case it is unlikely that the peptides will resolve well. Pat the blotter over the rest of the area to make sure the entire surface of the plate is wet. Extra liquid can be added at this stage by dipping gloved fingers in the buffer and applying it directly to the back of the blotter. Remove the blotter and the plate is ready for electrophoresis. It requires only a minimal amount of liquid for the peptides to migrate in. The plate should look dull gray and not shiny because of too much liquid on the plate. Areas where the buffer has puddled should be blotted carefully with a Kimwipe.

Lift up the wicks from the disassembled apparatus, wipe away any liquid on the plastic sheet, and place the plate flat on the surface of the bed. The plate should be oriented so that the dye and sample marks are on a vertical axis from top to bottom, i.e., perpendicular to the electric field. In our system the anode (+) is on the left, and the cathode (−) on the right. Most of the peptides will migrate toward the cathode or to the right during electrophoresis if using either pH 1.9 or 4.72 buffer. Next, place the wicks so they overlap onto the top of the plate against the cellulose by about 1 cm. Cover the plate and wicks with the top polyethylene sheet, followed by the Teflon sheet and the Neoprene pad. Close the lid and again secure the pins. Next, turn the air pressure up to 10 lb/in.^{2} as before and make sure that the water supply is on for cooling the device during the run (there is a safety interlock on the HTLE-FS model which ensures that the power cannot be applied without the water flowing). Close the safety cover of the apparatus, then set the voltager timer and turn the power supply on. For a typical run using pH 1.9 buffer electrophorese for 25 min at 1.0 kV (~21 mA). After the run is complete, turn the power supply off, then release the air pressure. Disassemble the apparatus and carefully remove each layer and then the plate. Air dry the plate for about 30 min using an electric fan. Never oven dry the plates, because you may bake the peptides onto the cellulose. The apparatus can be used to electrophorese plates sequentially. We have reused the same buffer up to 10 times without any obvious change in electrophoresis patterns. Simply wet the next plate in line and place it on the apparatus when the previous plate is finished.

Preparative Electrophoresis

An alternate method can be used for spotting samples with large volumes (10–30 μl) or which contain large amounts of protein (>50 μg). Spot the sample on a vertical line emanating about 1 cm upward from the sample origin mark. Spot 0.5 μl at 1-mm increments, drying after each drop, until the entire line has been spotted once. Repeat this again until the entire sample is loaded. The plate is wetted using a modified blotter that has a hole cut in the shape of the sample origin (a long rectangle). The sample is electrophoresed and dried as described above.

Second Dimension: Thin-Layer Chromatography

Rationale behind Chromatography

The second dimension is thin-layer chromatography using organic solvent buffer to separate peptides based on their ability to partition between the mobile (buffer) and stationary (cellulose) phases. We use large glass tanks (Shandon 500, tank size 57 × 23 × 57 cm) that were traditionally used for descending paper chromatography with 46 × 57 cm paper sheets. One can also use the smaller glass "brick" tanks instead, if they are equilibrated with buffer using Whatman 3MM paper to line three sides of the tank up to the top. The chromatography tanks should be kept in an quiet area of the laboratory or in a separate room to avoid vibrations that disturb chromatography. Buffers should be made and poured into the tank in advance of the chromatography step to allow complete equilibration with the gas phase. Using silicone or vacuum grease, make sure that the lid forms a tight seal with the tank.

Choice and Preparation of Chromatography Buffers

We use three different chromatography buffer systems and choose the appropriate buffer depending on the type of separation required (Table II).

> Regular chromatography buffer: This buffer is used for maps of proteins containing peptides with solubilities ranging from very hydrophilic to very hydrophobic. When performing total protein fingerprints, such as for [35S]methionine-labeled or 125I-iodinated proteins, regular chromatography buffer is probably the best choice.
>
> Phospho chromatography buffer: This buffer contains more aqueous solvents, and as its name implies it is good for separating relatively hydrophilic peptides such as phosphopeptides.
>
> Isobutyric acid buffer: This buffer, which is malodorous, is the most aqueous system of the three and is used in special cases to resolve extremely hydrophilic peptides.[20]

Unfortunately, there are no strict rules that can be applied to help predict which buffer to choose. Migration patterns will probably have to be tested using each buffer system to analyze a given protein. It is probably best to begin using regular chromatography buffer for 14C-, 35S-, or 125I-labeled proteins and phospho chromatography buffer for phosphopeptide analysis. The mobilities of some amino acids differ significantly in isobutyric acid buffer and this can be useful in determining the identity of peptides (see below).

Preparing Chromatography Tank and Chromatography

Prepare the chromatography buffer(s) and pour enough buffer into a tank so that it is about 2 cm deep. Make sure that the tank is level, then measure the depth. There should be enough liquid to wet the bottom of the cellulose side of the plate while it is on a slant, 1–2 cm from the bottom. Do this the day before chromatography so the buffer completely equilibrates with the gas phase of the system. To accelerate the equilibration step place Whatman 3MM paper along the inside of the tank and allow the buffer to climb up the walls. The tank should be kept undisturbed with the lid on for several hours prior to each run. The buffer will last for several months in the tanks or in glass storage bottles if tightly capped or stoppered. It is a good idea to keep a log of how many plates are run in each buffer tank and how long since the last buffer change. Because each buffer contains pyridine it is best to use amber glass bottles for storage.

---

[20] K.-H. Scheidtmann, B. Echle, and G. Walter, *J. Virol.* **44,** 116 (1982).

Before chromatography apply a drop of the same green dye marker (0.5 μl) to a spot at the same level as the sample origin but over in the left-hand margin (but not within the compressed area left by the electrophoresis wicks) (see Fig. 1). This will act as a colored marker for the second dimension and as a reference for calculating the relative mobilities of peptides. Note that the dye may partially resolve into individual yellow and blue components. To begin chromatography, carefully remove the lid of the tank and place the marked, thoroughly dried plates into the tank so that each plate is at right angles to the direction of electrophoresis with the bottom corresponding to the origin end of the plate. They should be placed at roughly an 80° angle, with the top of the plate resting on the walls of the tank. Up to eight plates can be loaded per tank if glass bottles are used to form a center support. Once the plates are in position quickly replace the lid and make sure an air-tight seal is formed. Chromatography is usually carried out for 6–8 hr until the buffer front has reached about 3 cm from the top of the plate. The period needed to chromatograph plates will vary depending on the buffer used, the separation desired, and the batch of plates. Do not disturb a run or open the tank while a run is in progress. When complete, remove the lid and carefully lift each plate from the tank, making sure not to drip any buffer on plates remaining inside. Air dry the plates in a fume hood until you can no longer smell any residual buffer (this usually takes about 60 min). If peptides are to be extracted from the plate for further analysis, then we do not recommend drying the plates by baking in an oven; otherwise they can be. Once dry the peptide maps are completed and ready for fluorography (^{14}C or ^{35}S) or autoradiography (^{32}P or ^{125}I).

Two-Dimensional Phosphoamino Acid Analysis

Partial Acid Hydrolysis

The method for two-dimensional phosphoamino acid analysis (PAA) has been previously described in detail.[21] The protocol has been updated below, with particular emphasis on using the HTLE-7000 system. One can use either TCA-precipitated proteins taken to the ethanol wash stage or proteins after oxidation in performic acid and lyophilization. Resuspend the dried sample using 50–100 μl of 6 N HCl (or constant boiling 5.7 M HCl), vortex, then centrifuge briefly, and incubate at 110° for 60 min. If using snap-cap microfuge tubes a second test tube rack is taped firmly on top of the rack holding the tubes so that the lids of the tubes remain closed.

---

[21] J. A Cooper, B. M. Sefton, and T. Hunter, this series, Vol. 99, p. 387.

One can also use Sarstedt screw-top tubes, which remain closed without taping. It is essential to note that this is a partial amino acid hydrolysis and that incubation times significantly longer that 60 min will result in increasing liberation of free phosphate from phosphoamino acid residues.

After hydrolysis, dry the sample by lyophilization in a Speed-Vac (equipped with an NaOH trap to collect acid) and determine Cerenkov counts remaining. Resuspend the hydrolysate in 5–10 μl of pH 1.9 buffer (same buffer used for peptide mapping) which contains 15 parts buffer to 1 part cold phosphoamino acid standards (1.0 mg/ml of each phosphoserine (P-Ser), phosphothreonine (P-Thr), and phosphotyrosine (P-Tyr) in deionized water). As described by Cooper et al.,[21] four samples can be analyzed on a single plate (Fig. 2).

Loading Samples

Using the diagram in Fig. 2 as a guide, the 100-μm TLC plates should be marked with a soft-leaded pencil as described above for peptide mapping. To load the samples, follow the same procedure used for loading tryptic digest samples: (1) centrifuge the sample prior to loading to remove particulate debris, (2) load a spot of marker at the top of the plate, (3) load the samples by adding small, ~0.5-μl drops, using an Oxford micropipette and air drying between each drop. Between 10 and 100 cpm is needed for each PAA. We have successfully determined the phosphoamino acid composition of samples where no counts per minute could be detected using a scintillation counter! The plate is now ready for electrophoresis. For larger sample volumes or loads the sample can also be spotted on a line using the protocol for preparative peptide maps.

Two-Dimensional Electrophoresis

Set up the electrophoresis apparatus as above, using pH 1.9 buffer. The plates are to be wetted using a blotter with five holes in it (corresponding to the four sample origins and one for the dye marker; Fig. 2) that has been dampened with pH 1.9 buffer. Wet the plate by overlaying the blotter, and squeeze out the buffer from the blotter around each of the five holes as quickly as possible, so that all samples focus as small spots on their respective origins. Remove the blotter and place the plate in the apparatus. Electrophoresis is carried out for 20 min at 1.5 kV. Then remove the plate when electrophoresis is completed and let it air dry, using a fan, for at least 30 min. Be sure that the plate does not smell of acetic acid.

Change the buffer in the apparatus to pH 3.5 buffer by washing out the pH 1.9 from each buffer tank, adding pH 3.5 buffer, and replacing the old wicks with new ones. We routinely add EDTA to the pH 3.5 buffer used

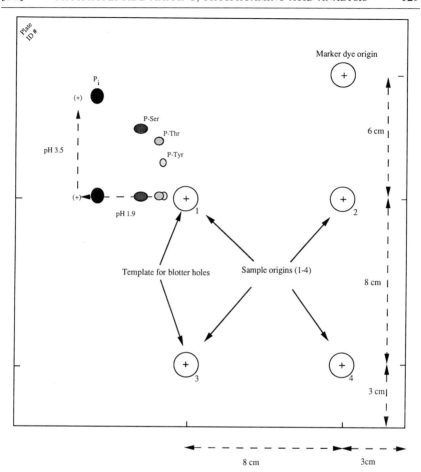

FIG. 2. Diagram for a stencil outline to mark the 20 × 20 cm TLC plates for phosphoamino acid analysis using pH 1.9 buffer (horizontal plane) and 3.5 buffer (vertical plane) as described in the text. The stencil outline can be used for both marking the TLC plate and to serve as a blotter template. The sample origins 1–4 are for applying acid hydrolysates of phosphoprotein or isolated phosphopeptides, while the origin in the top right corner is used to apply the marker dye. The upper left quadrant depicts the resolution of the individual ^{32}P-labeled products obtained after acid hydrolysis, in the first and second dimensions.

for wetting the blotters (0.5 m*M* final) to prevent streaking in the second dimension. When the plate is completely dry it is blotted with pH 3.5 buffer for the second dimension. To do this, strips of single-thickness Whatman 3MM paper that are custom designed for wetting each part of the plate (3 cm wide for the bottom, 6.5 cm wide for the middle, 10 cm wide for the top) are used. Thoroughly wet the three blotters in pH 3.5

buffer plus EDTA. Place these on the plate in rows parallel to the first dimension of separation about 1 cm from the line of origin (see Fig. 3). Press along the edges to squeeze buffer out of the wicks to wet the plate so that the buffer from two adjacent edges meets along the line of the origin to concentrate the sample for the second dimension. A sharp brown line should appear where the buffer fronts meet, which should be as straight as possible. An additional piece of dampened blotting paper or gloved fingers dipped in buffer can be used to dampen any dry areas to make sure that the plate is completely wet. Remove the blotters and soak up any puddled buffer with a Kimwipe. Place the plate in the apparatus making sure to rotate the plate 90° counterclockwise, and reassemble the apparatus as described above. Electrophorese the plate for 16 min at 1.3 kV. Disassemble the apparatus, remove the plate, and let it air dry, using a fan for 30 min as before, or bake in the oven at 65° for 10 min.

Detection of Phosphoamino Acids

To visualize the cold phosphoamino acid standards spray the plate with 0.25% (w/v) ninhydrin in acetone. Return the plate to the 65° oven for at least 15 min to develop the stain. Distinct purple spots will appear that correspond to the positions of P-Ser, P-Thr, and P-Tyr (see Fig. 3). The PAA is complete and is now ready for autoradiography.

Acid Hydrolysis of Immobilized Phosphoproteins

Phosphoamino acids can also be recovered by acid or alkaline hydrolysis of ^{32}P-labeled proteins bound to polyvinylidene difluoride (PVDF) membranes (Immobilon; Millipore, Bedford, MA) after electrophoretic transfer from SDS-polyacrylamide gels.[22] This method is described in detail by Kamps.[23] Briefly, the piece of Immobilon membrane containing the protein band is first washed in a large volume of water to remove glycine and SDS from the membrane, dried, if necessary rewetted briefly in methanol and rinsed in water. The proteins are then hydrolyzed by incubation in 200 μl 6 N HCl at 110° for 60 min in a microcentrifuge tube. After the hydrolysis the tube is centrifuged and the acid is removed to a new tube and lyophilized. The dried sample is treated as described above. *Note:* Nitrocellulose membranes cannot be treated with concentrated acid because this will dissolve the membrane. Instead, one can first solubilize the protein from the membrane by proteolysis and perform PAA on the eluted peptides. Alternatively, protein can be eluted directly from nitrocellulose with 50%

[22] M. P. Kamps and B. M. Sefton, *Anal. Biochem.* **176,** 22 (1989).
[23] M. P. Kamps, this volume [3].

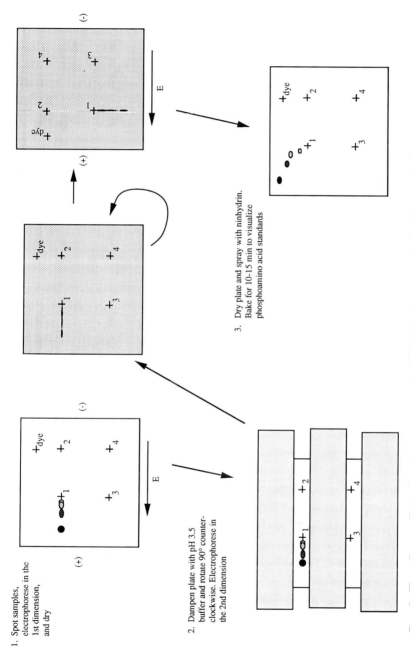

FIG. 3. Flow diagram illustrating how to prepare the TLC plate for separation in the second dimension of two-dimensional phosphoamino acid analysis. Detailed explanation of steps 1–3 is provided in the text. The direction of the electric field (E) during electrophoresis is indicated by an arrow under the TLC plate in steps 1 and 2.

pyridine in 100 mM ammonium acetate and then subjected to acid hydrolysis.[24]

Autoradiography and Fluorography

Before autoradiography or fluorography we mark the edges of the plates with radioactive India ink. Several landmarks are chosen to help orient the film; usually the marks are placed on the sides of the plates to line up with the sample and dye origins. Often a small piece of tape or sticker with the plate identification number written in radioactive ink is placed in the upper left corner as a reference.

For ^{32}P-labeled peptides or amino acids we perform autoradiography using presensitized Kodak (Rochester, NY) XAR film and an intensifying screen at $-70°$. Place the presensitized (flashed) side of the film up and lay it on top of the plate followed by an intensifying screen. Expose to film for 1 to 7 days, depending on the number of counts loaded. Exposure to presensitized film for longer than 7 days is not recommended because the background signal becomes excessive.

For either ^{14}C- or ^{35}S-labeled proteins it is usually best to use a fluorescent enhancer to detect signals. Plates can either be dipped in molten 2-methylnaphthalene and 0.4% PPO or sprayed with commercial fluors such as En^{3}Hance (New England Nuclear). Allow the plates to dry before exposing to film ($\sim$ 20 min), but do not dry too long or the fluor will sublime. Lay a piece of presensitized film, flashed side down, on top of the plate and expose at $-70°$ for 1–15 days, depending on the number of counts loaded. When not using an intensifying screen one can expose the film for considerably longer periods of time (up to years).

Analysis of Peptide Maps

Predicting Phosphopeptide Mobility Based on Amino Acid Composition

Tryptic phosphopeptide mapping can be used as a means of studying the phosphorylation status of a protein of interest. The protein of interest is isolated from ^{32}P-labeled cells, usually by immunoprecipitation, and tryptic peptide maps are prepared. To determine whether treatments with specific stimuli affect phosphorylation, peptide maps from control and treated cells can be compared and examined for the appearance of new phosphopeptides. In addition phosphopeptide mapping can be used to identify individual phosphorylation sites. Most proteins generate a large

---

[24] L. Contor, F. Lamy, and R. E. Lecocq, *Anal. Biochem.* **160**, 414 (1987).

number of tryptic peptides, and it can be difficult to identify individual spots on a map by microchemical analysis, especially with femtomole amounts of peptide. However, for a protein of known sequence it is possible to deduce the identity of a specific phosphopeptide indirectly, using a combination of techniques including secondary digestion with endopeptidases of defined specificity and Edman degradation. To facilitate this procedure the number of possible candidate peptides is narrowed down by making use of peptide mobility predictions. The mobility of a given peptide in both the electrophoretic and chromatographic dimensions can be predicted as described below and then compared with the experimentally determined mobility of the phosphopeptide in question. This information is used to eliminate theoretical peptides with obviously dissimilar mobilities.

To begin, make a list of the predicted tryptic peptides of the protein of interest assuming a limit digest of trypsin, which cleaves after lysine and arginine. Select as the initial candidate pool only peptides which contain a phosphorylatable residue (i.e., serine, threonine, or tyrosine). Keep in mind that partial digestion products may be generated when multiple Arg and Lys residues appear in tandem. In addition, trypsin does not cleave either Arg–Pro or Lys–Pro and does not efficiently cleave the sequences Arg–Asp/Glu and Lys–Asp/Glu. These sequences will lead to multiple partial digestion products, which should be cataloged as potential individual tryptic digestion products. Trypsin also does not cleave efficiently at Arg or Lys in the sequence Arg–X–P-Ser/P-Thr, while working quite efficiently to cleave Arg–P-Ser/P-Thr bonds. This preference can be of use in determining positions of phosphorylation sites.

The mobility in the electrophoresis dimension is determined by the electrical charge and the mass of a peptide, and is most accurately described by $m_r = keM^{-2/3}$, where m_r is the relative mobility, M is the mass, and e the electrical charge of the peptide in the buffer used.[25] To simplify calculations we generally use $m_r = eM^{-1}$. The net charge on an amino acid at a particular pH can be calculated from the experimentally determined pK_a values of the α-NH$_2$ and α-COOH groups, and of any ionizable side-chain groups. The contributions of the α-NH$_2$ and α-COOH groups of the amino acid can be ignored when it is part of a peptide, although each peptide contains an N-terminal α-NH$_2$ group and a C-terminal α-COOH group that must be taken into account. The net charge on a peptide can be calculated by summing the charges of the constituent amino acid side chains together with the charges on the α-NH$_2$ and α-COOH groups, which are all listed for each pH in Table III. One must include the predicted

[25] R. E. Offord, *Nature (London)* **211**, 591 (1966).

TABLE III

EXPECTED CHARGES ON AMINO ACIDS IN PEPTIDES AT DIFFERENT pH VALUES

Amino acid	pH				
	1.9	3.5	4.7	6.5	8.9
N-Terminal NH$_2$ group	+1	+1	+1	+1	~+0.5
C-Terminal COOH group	Neutral	~−0.5	−1	−1	−1
Arg	+1	+1	+1	+1	+1
Asp	Neutral	Neutral	~−0.7	−1	−1
Cys-SO$_3$H[a]	~−1	−1	−1	−1	−1
His	+1	+1	+1	~+0.5	Neutral
Glu	Neutral	Neutral	~−0.5	−1	−1
Lys	+1	+1	+1	+1	+1
P-Ser	−1	−1	−1	~−1.3	−2
P-Thr	−1	−1	−1	~−1.3	−2
P-Tyr	−1	−1	−1	~−1.3	−2

[a] Cys-SO$_3$H is the product of performic acid oxidation, which we routinely carry out prior to tryptic digestion. Cysteine itself has no charge.

value for the charge for either P-Ser, P-Thr, or P-Tyr for peptides that are predicted to contain these phosphoamino acids. The total charge is then used to determine the charge-to-mass ratio of a given peptide. Because there is a considerable endoosmotic effect during electrophoresis on TLC plates (i.e., neutral molecules move up to several centimeters toward the cathode, with the exact distance depending on the pH), electrophoretic mobilities are calculated with reference to an electrically neutral marker, i.e., for each amino acid and peptide the distance migrated must be plotted from the experimentally determined position of a neutral marker rather than from the origin. Between pH 3 and pH 8 ε–DNP-lysine is a good neutral colored marker. For pH 1.9 phosphotyramine is a neutral marker detectable by staining with ninhydrin. For pH 8.9 proline or γ-amino-butyric acid (GABA) are stainable neutral markers. From the known pK_a values lysine has, by definition, a net charge of +1 between pH 3 and pH 9, and is a useful +1 marker. At pH 1.9 arginine, lysine, and histidine have charges of about +1.5, but one can use tyramine as a stainable +1 marker between pH 1.9 and 8. Histamine is a good, stainable +1 marker between pH 7 and 9, and a +2 marker between pH 1.9 and 4. The most convenient negative marker is the blue dye, xylene cyanol FF, which has a net charge of −1 between pH 1.9 and 9.

The mobility of a peptide in the chromatographic dimension depends on the hydrophobicity of the peptide. One can obtain a rough estimate of the hydrophobicity of a peptide by summing the hydrophobicities of all the amino acids in the peptide. For this purpose we have experimentally

TABLE IV
MOBILITIES OF AMINO ACIDS DURING CHROMATOGRAPHY

Amino acid	One-letter code	Molecular weight	R_f		
			Regular buffer[b]	Phospho chromatography buffer[c]	Isobutyric buffer[d]
Ala	A	89	0.35	0.41	0.61
Arg	R	174	0.20	0.31	0.69
Asn	N	132	0.15	0.21	0.44
Asp	D	133	0.14	0.22	0.46
Cys-SO$_3$H[a]	C	121	0.08	0.19	0.23
Glu	E	147	0.22	0.31	0.53
Gln	Q	146	0.20	0.29	0.51
Gly	G	75	0.23	0.30	0.50
His	H	155	0.19	0.29	0.67
Ile	I	131	0.76	0.77	0.92
Leu	L	131	0.81	0.81	0.95
Lys	K	146	0.14	0.26	0.58
Met-sulfone[a]	M	149	0.26	0.45	0.59
Phe	F	165	0.77	0.78	0.93
Pro	P	115	0.40	0.47	0.73
Ser	S	105	0.25	0.33	0.48
P-Ser		185	ND[e]	0.20	0.35
Thr	T	119	0.34	0.41	0.57
P-Thr		199	ND	0.23	0.43
Trp	W	204	0.69	0.69	0.72
Tyr	Y	181	0.64	0.66	0.72
P-Tyr		261	ND	0.28	0.53
Val	V	117	0.56	0.62	0.82

[a] Cys-SO$_3$H and Met-sulfone are the oxidation states obtained after performic acid oxidation.
[b] n-Butanol : pyridine : acetic acid : H$_2$O (97 : 75 : 15 : 60) is the chromatography buffer we use for ^{35}S-, ^{125}I-, or ^{14}C-labeled tryptic peptides.
[c] n-Butanol : pyridine : acetic acid : H$_2$O (75 : 50 : 15 : 60) is a buffer we use for phosphopeptide maps.
[d] Isobutyric acid : n-butanol : pyridine : acetic acid : H$_2$O (65 : 2 : 5 : 3 : 29) is a buffer we use for phosphopeptide maps.
[e] ND, Not determined.

determined the mobilities of each amino acid and the three phosphoamino acids in all three chromatography buffers relative to the ε–DNP-lysine. These R_f values are listed in Table IV. To calculate the mobility of a peptide the R_f values of all the constituent amino acids are summed, and the total is divided by the number of amino acids in the peptide. This value is not a very accurate determination of mobility, because the amino acids in a peptide lack charges on their α-NH$_2$ and α-COOH groups which contributed significantly to the R_f values of the individual amino acids. As a result R_f values calculated for peptides tend to be lower and more bunched together than they really are. However, the relative mobility ranking is, in general, correct. The predicted mobility in two dimensions is obtained by simply combining the predicted mobilities for each dimension.

Predicted mobilities should be plotted in two dimensions relative to the neutral marker position and the position of the ε–DNP-lysine in the chromatography dimension. This map can then be compared with the actual tryptic phosphopeptide map. The relative positions of phosphopeptides compared with the predicted positions will often give a direct clue to the identity of the phosphopeptide in question. However, additional information can be obtained by secondary digestion with other proteases, by Edman degradation, and by double-labeling experiments with individual amino acids. These procedures are described in the following sections.

We have developed a program for use on a Digital VAX computer, in conjunction with the University of Wisconsin Genetics Computer Group software,[26] that will carry out these calculations and plot the predicted two-dimensional mobilities of tryptic peptides and phosphopeptides. Any protein sequence recovered from the database can be analyzed by this program. It is available from Tony Hunter or Lisa Caballero in Computer Services at the Salk Institute (La Jolla, CA).

Recovering Individual Phosphopeptides from Thin-Layer Cellulose Plates

Since cellulose is an inert substance, the phosphopeptide identified by autoradiography can be recovered for subsequent analysis, including phosphoamino analysis, secondary digestion with site-specific endopeptidases, and direct sequence analysis by manual or automated Edman degradation. To do this, align the autoradiogram with the radioactive ink marks, then tape the edges of the plate to the autoradiogram. The film should be placed against the cellulose side of the plate. Place this sandwich on a light box so that the film side is down. By shining the light through the sandwich, one can localize the position of the phosphopeptide of interest. Circle the spot on the back side of the plate with a marker, then remove the film. Place the plate, with the back side down and the cellulose side facing up, then trace the mark on the surface of the plate using a soft-leaded blunt pencil. This makes it possible to recover the cellulose area containing the

[26] J. Devereux, P. Haeberli, and O. Smithies, *Nucleic Acids Res.* **12**, 387 (1984).

phosphopeptide. Alternatively, one can trace the position of the peptide and the alignment marks from the autoradiogram onto tracing paper, and then align the tracing over the plate and retrace the spots so that an impression is made on the cellulose.

To elute the peptide, take a blue Eppendorf pipette tip and slice off the first few millimeters of the pointed end, and then push a 6.5-mm diameter porous polyethylene disk (#006651; Omnifit, Atlantic Beach, NY) into the large end until it is sealed firmly in place. Alternatively, commercially available TLC zone recovery pipettes can be employed (#06-641; Fisher Scientific, Pittsburgh, PA). Attach a vacuum line to the large end using flexible polyethylene tubing and turn on the vacuum line. Scrape the cellulose from the marked area on the plate, using the sliced end of the tip, so that the loose material is sucked in and becomes caught against the polyethylene disk. Carefully remove the vacuum line, and place the blue tip large end down into a 1.5-ml microcentrifuge tube. The phosphopeptide is eluted from the cellulose by two sequential washes with 100 μl of pH 1.9 buffer, followed by one wash with deionized water. This is achieved by placing the tubes in a microcentrifuge, and adding buffer through the top of the blue tip followed by brief centrifugation (<5 sec). The eluted material is collected in the Eppendorf tube. The efficiency of elution should be checked by Cerenkov counting of the tip and the eluate. If the peptide is very hydrophobic it may not elute efficiently, and then 50% pyridine in water may be tried. Small cellulose particles that come through the sinter are cleared from the eluate by centrifugation. The supernatant is transferred with a microtransfer pipette to a new microcentrifuge tube, and is then lyophilized as described above to reduce the volume and to remove organic buffers. In the dried state the peptide is stable and can be either stored or used immediately for subsequent analysis.

Phosphoamino Acid Analysis and Fingerprinting of Individual Phosphopeptides

To determine the phosphoamino acid content of individual phosphopeptides the purified phosphopeptide should be hydrolyzed in 6 N HCl for 60 min at 110° and processed further as described above. Fingerprints of phosphopeptides can be obtained by a more limited hydrolysis in HCl. Because partial acid hydrolysis proceeds at a different rate for each peptide bond, depending on the amino acids forming the bond, the map of partial hydrolysis products is characteristic for a given phosphopeptide and is referred to as a fingerprint.[7] Identical or related peptides will give rise to identical or very similar fingerprints. By using this method one can verify experimentally whether two phosphopeptides are related, possibly as a result of partial proteolysis or differential phosphorylation. Purified phosphopeptides isolated from cellulose plates are dissolved and hydrolyzed for 30 min at 110° in 6 N HCl as described for phosphoamino acid analysis. Then lyophilize the HCl, dissolve in 6 μl deionized water, and spot the hydrolysis products on a 100-μm TLC plate. The products are resolved by electrophoresis at pH 8.9 or 3.5 at 1 kV for 25 min in the first dimension and by chromatography using the phospho chromatography buffer in the second dimension. Alternatively, multiple samples can be run in parallel by electrophoresis at pH 3.5 for 30 min at 1.5 kV.

Edman Degradation

To obtain additional clues to the identity of a phosphopeptide of interest it may be useful to determine at which position in a peptide the phosphorylated residue(s) exists. This can be done by carrying out manual Edman degradation[3] on individual phosphopeptides isolated from a TLC plate. For optimal resolution one needs at least 50–100 cpm/cycle. For manual Edman degradation, the purified peptide is dissolved in 20 μl water in a microcentrifuge tube and an aliquot is taken as starting material. The volume is restored to 20 μl with water, and 20 μl 5% phenyl isothiocyanate in pyridine is added. The sample is incubated at 45° for 30 min, and then extracted twice with 200 μl of heptane : ethyl acetate (10 : 1) and twice with 200 μl heptane : ethyl acetate (2 : 1), vortexing for 15 sec each time and spinning briefly in a microcentrifuge to resolve the phases. The final aqueous phase is frozen and lyophilized, and then taken up in 50 μl trifluoroacetic acid and incubated for 10 min at 45°. The trifluoroacetic acid is removed *in vacuo*, and the residue is taken up in a volume of water equal to 20 μl minus the volume taken out as a sample of the starting material. An aliquot equal to that taken for the starting material is removed (cycle 1), the volume restored to 20 μl, and the whole procedure is repeated. The reaction products of each cycle can be analyzed by spotting them on a TLC plate (about 1 cm apart) followed by electrophoresis for 25 min at 1 kV at pH 1.9 or 3.5. Free [^{32}P]phosphate should be used as a marker. The release of free [^{32}P]phosphate at a particular cycle, which results from β elimination during cyclization, indicates the presence of a P-Ser or P-Thr residue. P-Tyr is stable to cyclization and is released as the anilinothiazolinone derivative of P-Tyr. This can be converted to the phenylthiohydantoin (PTH) of P-Tyr by incubation in 0.1 N HCl for 20 min at 80°. Marker PTH–P-Tyr is readily synthesized by reacting P-Tyr with phenyl isothiocyanate as described above. PTH–P-Tyr marker can be detected as a dark spot when the TLC plate is examined under a hand-held UV light.

In addition to the cycle at which free [^{32}P]phosphate or [^{32}P]-PTH–P-Tyr is released, information on the sequence of the peptide may be obtained by the electrophoretic mobility shifts detected at each cycle. Thus if a positively or negatively charged amino acid is removed at a cycle before the phosphoamino acid there will be a corresponding shift in the mobility of the peptide. Remember that if the peptide contains a C-terminal lysine this will react with phenyl isothiocyanate at cycle 1, which will cause the loss of a positive charge. It may be difficult to determine the position of a second more C-terminal phosphorylated residue, and in practical terms it is usually hard to carry out more than five cycles. Nevertheless this analysis should make it possible to determine the position of phosphorylated residues within the first five residues of a particular peptide. Comparing these data with the amino acid sequence of candidate peptides may lead to elimination of several candidates.

One can also carry out automated sequence analysis of phosphopeptides. In older spinning cup sequenators, 1–5% of the free [^{32}P]phosphate or [^{32}P]PTH–P-Tyr released at a given cycle is extracted into the butyl chloride phase, but if enough radioactivity is available this can identify phosphorylated residues. Little if any [^{32}P]phosphate or [^{32}P]PTH–P-Tyr is extracted in the gas-phase sequenator. However, two different strategies for the sequencing of ^{32}P–labeled phosphopeptides using a gas-phase sequenator have been developed. These methods are discussed in detail by Roach and Wang[27] and Wettenhall.[28]

Further Digestion with Proteases

Sensitivity for digestion with sequence-specific proteases depends on the presence of a recognition site for the protease used (Table V).[24] Determination of the sensitivity of phosphopeptides for digestion therefore gives information about the amino acid sequence of the peptide under investigation. Isolate the peptide from a TLC plate as described above. Dissolve in the appropriate buffer, usually 50 μl 50 mM ammonium bicarbonate, pH 8.0, and digest the peptide with 1–5 μg of protease at 37° for 60 min under appropriate conditions (Table V). Add 400 μl deionized water and lyophilize, and then repeat this procedure. Analyze by spotting samples on TLC plates and separation in two dimensions as described above for peptide mapping. As controls, undigested peptide, and a mix of

[27] P. J. Roach and Y. Wang, this volume [16].
[28] R. E. H. Wettenhall, R. H. Aebersold, L. E. Hood, and S. B. H. Kent, this volume [15].

TABLE V

PROTEASE SPECIFICITY AND DIGESTION CONDITIONS[a]

Protease	Source[b]	Specificity	Conditions	References
Endoproteinase Asp-N	Boehringer (P. fragi)	X-CSO$_3$H; X-D	50 mM NH$_4$HCO$_3$, pH 7.6, 37°	16
Endoproteinase Glu-C	ICN (V8)	E-X	50 mM NH$_4$HCO$_3$, pH 7.6, 37°	14
TPCK-Trypsin	Worthington	R-X; K-X	50 mM NH$_4$HCO$_3$, pH 8.0, 37°	13
α-Chymotrypsin	Worthington	F-X; W-X; Y-X	50 mM NH$_4$HCO$_3$, pH 8.0, 37°	13
Thermolysin[c]	Calbiochem	X-L; X-I; X-V	50 mM NH$_4$HCO$_3$, pH 8.0, 1 mM CaCl$_2$, 55°	15
Proline-specific endopeptidase	ICN	P-X	50 mM NH$_4$HCO$_3$, pH 7.6, 37°	17
Carboxypeptidase B	Sigma	X-R-COO$^-$; X-K-COO$^-$	50 mM NH$_4$HCO$_3$, pH 8.0, 37°	18

[a] Enzymes that are commonly used for further digestion of purified peptides are listed, with the conditions we use in our laboratory.

[b] Enzymes can be obtained from Boehringer Mannheim Biochemical (Indianapolis, IN), ICN Biomedicals (Lisle, IL), Worthington Biochemical Corporation (Freehold, NJ), Calbiochem (San Diego, CA), and Sigma (St. Louis, MO).

[c] The activity of thermolysin is not very specific; it will recognize most apolar residues to some extent.

digested and undigested material, should be analyzed in parallel. Before mixing the digested and undigested samples, it is important to inactivate the protease by making the sample 1% in 2-mercaptoethanol and boiling for 1 min. A minimum of 30–50 cpm should be loaded per plate. Up to three samples can be analyzed on one TLC plate, using origins 5 cm apart. If the phosphopeptide of interest proves to be sensitive to digestion with certain proteases, candidate peptides lacking recognition sites for those proteases can be eliminated. It is important to remember that most endopeptidases do not work well as exopeptidases, and therefore there may be very inefficient cleavage of peptide bonds adjacent to the N- or C-terminal residues. In addition it should be noted that chymotrypsin does not cleave after P-Tyr. Finally, manual Edman degradation of phosphopeptides generated by secondary digestion can be informative, since the hydroxyamino acid of interest may now be within the first five residues.

Double Labeling

To determine whether a specific phosphopeptide contains methionine or cysteine, proteins can be labeled *in vivo* with [^{35}S]methionine and ortho[^{32}P]phosphate, or with [^{35}S]cysteine and ortho[^{32}P]phosphate. Alternatively proteins can be labeled *in vivo* using [^{35}S]methionine or [^{35}S]cysteine and subsequently labeled *in vitro* by autophosphorylation or used as a substrate for a particular protein kinase by incubation with [γ-^{32}P]ATP and a divalent cation. To analyze double-labeled samples, the protein is purified and digested as described above. After separation of the digest in two dimensions, the plates are marked with ^{35}S-containing India ink and exposed first with four layers of aluminum foil between the plate and the film and intensifying screen (do not cover the ink marks). This allows high-energy β radiation emitted by the [^{32}P]phosphate to hit the film and screen but absorbs the low-energy β radiation emitted by the ^{35}S. To obtain the ^{35}S exposure, either let the ^{32}P decay after obtaining the ^{32}P exposures, or else start with about one-tenth the amount of ^{32}P compared to ^{35}S. Enhance by spraying with En^{3}Hance (New England Nuclear) and expose to obtain ^{35}S maps. Using the marks, one can line up the ^{32}P and ^{35}S exposures. This makes it possible to find out whether a particular phosphopeptide contains methionine or cysteine. This procedure can also be applied to proteins labeled with individual ^{3}H-labeled amino acids, but this is considerably more difficult because of the low efficiency of autoradiographic detection of ^{3}H.[29] Unfortunately the success of such an experiment depends heavily

[29] T. Patschinsky, T. Hunter, F. Esch, J. A. Cooper, and B. M. Sefton, *Proc. Natl. Acad. Sci. U.S.A.* **79**, 973 (1982).

on the stoichiometry of phosphorylation of the peptide under investigation. It is important to realize that the phosphorylated peptide has a different mobility than the unphosphorylated peptide. Thus, if only 1% of a peptide is phosphorylated, all the ^{32}P will comigrate with only 1% of the ^{35}S, while the remaining 99% of the ^{35}S will run at a different position, and it is very unlikely that 1% of an ^{35}S-labeled peptide will be detected as a separate peptide. It is important to keep these considerations in mind when performing a double-labeling analysis. The ratio of radioactivity in the phosphorylated and unphosphorylated forms of a peptide has been used as a method of determining phosphorylation stoichiometry at a particular site.[30]

Using a combination of these approaches one can narrow down the number of possible candidates and often guess the identity of phosphopeptides on the actual map. We have found the best way to do this is to set up a table of all the relevant peptides listing the predicted mobilities, sensitivity to secondary proteases, positions of Ser, Thr, and Tyr, and methionine and cysteine content. Comparison with the actual properties of the phosphopeptides can then readily be made (for example, see van der Geer and Hunter[9]).

Phosphoisomers and Partial Tryptic Digestion Products

In the process of locating phosphorylation sites, another clue can be obtained from the pattern of peptides lying on a diagonal. These can be either phospho isomers or else partial trypsin digestion products due to multiple adjacent basic residues. Phospho isomers are usually generated by multiple phosphorylations of a single tryptic peptide containing more than one phospho acceptor site. Addition of a single phosphate to a given peptide will affect its mobility in both the electrophoretic and chromatographic dimensions. The negative charge of a phosphate group will alter its charge-to-mass ratio, rendering it more negatively charged and thereby changing its mobility in the electrophoretic dimension. The polar phosphate group will also render the peptide more hydrophilic and therefore decrease its mobility in the chromatographic dimension. Phosphopeptides with varying degrees of phosphorylation (phospho isomers) are resolved as a group of peptides that lie on a diagonal sloping toward the anode. Shown in Figs. 4 and 5 are two examples of phosphopeptide maps that contain phosphopeptides known to be phospho isomers present in tryptic digests of the v-myb product and the CSF-1 receptor. Each spot represents the addition of an extra phosphate group, so that one can usually count the maximal number of phosphates, assuming that the peptide migrating

[30] B. M. Sefton, T. Patschinsky, C. Berdot, T. Hunter, and T. Elliot, *J. Virol.* **41**, 813 (1982).

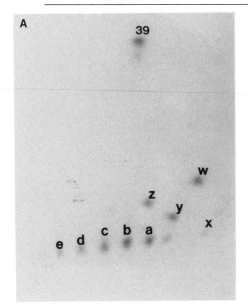
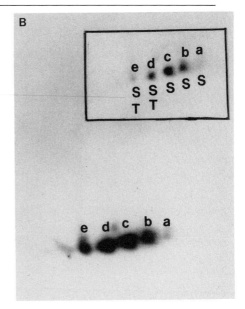

FIG. 4. Two-dimensional peptide map analysis of [^{35}S]cysteine and ^{32}P-labeled p48^{v-myb} isolated from metabolically labeled avian myeloblastosis virus (AMV)-transformed chicken myeloblast BM2/C3 cells. (A) ^{32}S-Labeled tryptic digest and (B) ^{32}P-labeled tryptic digest. Samples were resolved in parallel by electrophoresis at pH 1.9 followed by ascending chromatography in phospho chromatography buffer. (B, insert) Result of phosphoamino acid analysis of the individual peptides a–e as indicated by S (P · Ser) and T (P · Thr). Peptides a–e, isolated by the black outline, are phospho isomers of the v-myb tryptic peptide containing residues 267–303. Peptides w, x, y, and z are unidentified peptides, while peptide 39 corresponds to residues 304–317.

slowest toward the anode contains only a single phosphate. Thus the v-myb peptide 267–303 can be phosphorylated at up to five positions. The CSF-1 receptor peptide contains at least five phosphorylatable residues. Some of our maps resolve as many as seven different phospho isomers. Peptide maps of proteins that yield similar diagonal migration patterns should alert the investigator to look for tryptic peptides that contain several phosphorylatable residues, and may predict candidate peptides without further experimentation. Phospho isomers can also be resolved by isoelectric focusing, and this technique also allows one to count the number of phosphate residues per peptide.[31]

We have also observed examples where two monophosphorylated

[31] C. J. Fiol, A. M. Mahrenholz, Y. Wang, R. W. Roeske, and P. J. Roach, J. Biol. Chem. **262,** 14042 (1987).

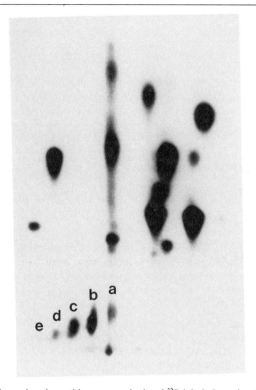

FIG. 5. Two-dimensional peptide map analysis of ^{32}P-labeled murine CSF-1 receptors. CSF-1 receptors were isolated by immunoprecipitation from SV40-immortalized murine macrophage BAC1.2F5 cells. Receptors were labeled *in vitro* by incubation of the immune complex with [γ-^{32}P]ATP and Mn^{2+} for 30 min at 37°. Tryptic digests were resolved in two dimensions by electrophoresis at pH 1.9 followed by ascending chromatography in isobutyric acid buffer. Peptides a–e represent a series of phospho isomers, which contain P-Tyr and P-Ser.

forms of the same peptide resolve during chromatography, but not during electrophoresis. This can result from phosphorylation of two different positions within a peptide, which affect its hydrophobicity to different degrees due to context. This was particularly apparent in the case of a peptide that was phosphorylated on either a serine or a tyrosine.[6]

Peptides that are related by the addition of single basic residues also fall on a diagonal. However, this diagonal runs in the opposite direction to that observed for phospho isomers, because each additional basic residue adds a single positive charge at the same time as increasing the hydrophilicity of the peptide.[3,7] This information can also be useful in

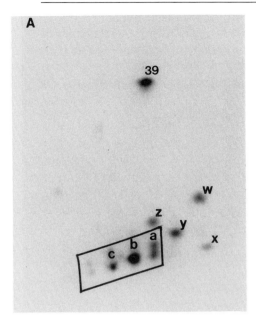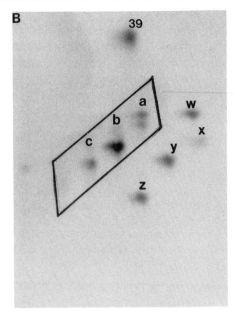

FIG. 6. Differential chromatography of [³⁵S]cysteine-labeled p48$^{v\text{-}myb}$ isolated from metabolically labeled AMV-transformed chicken myeloblast BM2/C3 cells. Equal amounts of digests were resolved in parallel by electrophoresis at pH 1.9, followed by ascending chromatography in either phospho chromatography buffer (A) or isobutyric acid buffer (B). The designation of peptides is the same as in Fig. 4.

identifying phosphorylation sites, since the sites will lie in regions where there are adjacent basic residues.

Differential Chromatography

Since the relative mobility of a given peptide in the chromatographic dimension is dependent on the relative mobilities of its constituent amino acids, it is possible to make use of conditions that exploit differences in peptide composition to predict amino acid content. From the data in Table IV, one can see that the relative mobilities for some amino acids are considerably greater during chromatography in isobutyric acid buffer than in phospho chromatography buffer. Of particular interest is the fact that Pro and His show 155 and 238% relatively greater mobility, respectively. This suggests that peptides enriched in either Pro and/or His will have a greater percentage increase in chromatographic mobility in isobutyric acid buffer than other peptides. A graphic example of this is seen in the analysis of a tryptic digest of [³⁵S]cysteine-labeled v-*myb* protein (Fig. 6). Phosphopeptides a–c (Fig. 6), which are a set of phospho isomers, have a greater percentage increase in mobility than all the other cysteine-containing peptides, and correspond to residues 267–303, which contain six Pro and four His (~27% Pro + His). Thus, differential chromatography of unknown phosphopeptides might reveal peptides with similar behavior, and this may facilitate their identification.

Confirming Identification of Phosphorylation Sites

Ultimately, the identity of the phosphorylation site must be confirmed by another method. This can involve synthesis of a peptide of the same sequence, which can then be phosphorylated *in vitro* and tested for comigration, or site-directed mutagenesis of the phosphoacceptor amino acid to a nonphosphorylatable residue. If large enough amounts of protein are available, direct sequencing can be conducted. Finally, antibodies directed against a specific peptide sequence can be used. These methods are described below.

Phosphorylation of Chemically Synthesized Peptides

To confirm the identity of a particular phosphopeptide such a peptide can be synthesized and phosphorylated *in vitro,* by incubation with a purified protein kinase. Phosphorylation reactions should be carried out in the appropriate protein kinase buffer in presence of [γ-^{32}P]ATP. It is possible that the peptide of interest will prove to be a poor substrate for the protein kinase used, possibly because the peptide is not long enough or because residues needed for recognition by the protein kinase lie outside the peptide of interest. This can be circumvented by synthesizing a larger peptide that can be trimmed with the appropriate proteases after phosphorylation. As an alternative, peptides can be phosphorylated chemically by incubation with [γ-^{32}P]ATP and Mn^{2+}.[32] After phosphorylation the phosphopeptide should be purified. This can be done by HPLC or by separation in one or two dimensions on TLC plates as described above. When separation on TLC plates is used the phosphopeptide can be isolated from the TLC plate as described above. The purified phosphopeptide can than be tested for comigration with the phosphopeptide phosphorylated *in vivo* and isolated from a phosphopeptide map as described above.[3–5,8,9,33] Methods for automated synthesis of phosphopeptides are improving, and as an alternative to synthesizing peptides followed by phosphorylation it may be possible to synthesize the appropriate phosphopeptide for comigra-

[32] G. Schieven and S. Martin, *J. Biol. Chem.* **263,** 15590 (1988).
[33] K. L. Gould, J. R. Woodgett, C. M. Isacke, and T. Hunter, *Mol. Cell. Biol.* **6,** 2738 (1986).

tion studies. Phosphopeptide synthesis strategies are described in detail by Perich.[34,35]

In Vitro Mutagenesis

After identification of a phosphorylation site, and if a cDNA clone is available, the site can be mutated into a residue that cannot function as a phosphate acceptor. Phosphopeptide maps of the mutant protein should now lack the phosphopeptide if the site was correctly identified. This evidence is regarded as absolute proof of the identity of a particular phosphopeptide. Mutants must be expressed in the appropriate cell lines and can be tested for absence of the phosphorylation site under investigation by ^{32}P labeling and tryptic phosphopeptide mapping.[4,8,9] This will also allow the investigator to look for phenotypes of the mutated protein. In addition to mutation to residues that resemble the unphosphorylated acceptor (Ala for Ser or Thr, and Phe for Tyr) one can change the acceptor site to an amino acid that resembles the phosphorylated amino acid (for instance, Asp for P-Ser and P-Thr), whose presence may mimic the effect of phosphorylation of the wild-type protein.

Identification of Phosphorylation Sites by Automated Sequencing

Large quantities of bacteria- or baculovirus-expressed protein can be used for identification of phosphorylation sites by microchemical analysis. When the protein of interest is a protein kinase and the phosphorylation sites of interest are the autophosphorylation sites the protein can be phosphorylated in the presence of [γ-^{32}P]ATP and millimolar concentrations of cold ATP or it can be labeled in the intact bacterial or insect cells with ^{32}P. When the protein is a substrate it can be phosphorylated by incubation with the appropriate protein kinase and 10–100 μM concentrations unlabeled ATP containing [γ-^{32}P]ATP or it can be labeled *in vivo* with ^{32}P. After gel purification and phosphopeptide mapping, the ^{32}P-labeled phosphopeptide can be identified and isolated by two-dimensional separation on a TLC plate or preferably by microbore HPLC separation. Purified peptide can now be subjected to automated peptide sequencing. If the peptide contains P-Ser or P-Thr, then there will be no identifiable PTH derivative at this position.[27,28] We have successfully used this method to identify the autophosphorylation site in a novel protein kinase expressed in *Escherichia coli* (R. Lindberg, manuscript in preparation). It is also possible to recover P-Tyr-containing peptides for direct sequencing from

[34] J. W. Perich, this volume [18].
[35] J. W. Perich, this volume [19].

a total tryptic digest by immunoaffinity absorption with anti-P-Tyr antibodies, followed by microbore HPLC separation.[36] One should bear in mind that 10- to 100-pmol quantities of peptide are needed for an accurate determination of the amino acid sequence.

Immunoprecipitation of Tryptic Peptides with Peptide-Specific Antibodies

In rare cases where several different anti-peptide antibodies against a particular protein are available, phosphorylated tryptic peptides can be recognized by anti-peptide antibodies. Peptides can be immunoprecipitated from a complete tryptic digest as follows. An immunoabsorbent is made out of formalin-fixed *Staphylococcus aureus* bacteria by incubation with anti-peptide antibody followed by washing. The immunoabsorbent is added to the ^{32}P-labeled tryptic digest in 10 μl of 50 mM ammonium bicarbonate, pH 7.6, and incubated for 60 min at 0°. The bacteria are collected by centrifugation in a microfuge, and the supernatant is removed and lyophilized. The bacterial pellet is washed twice with 50 μl of ammonium bicarbonate, and the bound peptides are eluted with two applications of 20 μl of pH 1.9 electrophoresis buffer. The two eluates are pooled and lyophilized. Unbound and bound peptides and a total digest are separated by two-dimensional electrophoresis and chromatography. If the anti-peptide antibody is directed against the phosphorylation epitope (and its binding is not inhibited by phosphorylation), then if the phosphopeptide has been correctly identified the peptide in question should be missing from the unbound map, and be present in the bound map.[37]

Conclusion

We have reviewed a method for identification of phosphorylation sites in proteins based on separation of proteolytic degradation products in two dimensions on TLC plates. This method has been successfully applied in our laboratory for identification of phosphorylation sites in several different proteins. The major advantage of separation on TLC plates compared with the use of HPLC and microchemical analysis is that only radiochemical amounts of protein (i.e., attomoles or femtomoles) are required to obtain good results.

[36] H. E. Tornqvist, M. W. Pierce, A. R. Frackelton, R. A. Nemenoff, and J. Avruch, *J. Biol. Chem.* **262**, 10212 (1987).
[37] K.-H. Scheidtmann, A. Kaizer, A. Carbone, and G. Walter, *J. Virol.* **38**, 59 (1981).

Acknowledgments

The protocols and discussions outlined above are the cumulative effort of many workers at the Salk Institute who have refined the method originally employed by Wade Gibson.[38] These include Jon Cooper, Suzanne Simon, Jill Meisenhelder, Kathy Gould, Clare Isacke, Jim Woodgett, Ellen Freed, Gerry Weinmaster, David Meek, David Middlemas, and Martin Broome.

[38] W. Gibson, *Virology* **62**, 319 (1974).

[12] Cyanogen Bromide Cleavage and Proteolytic Peptide Mapping of Proteins Immobilized to Membranes

By Kunxin Luo, Tamara R. Hurley, and Bartholomew M. Sefton

Peptide mapping is widely used for the study of sites of protein phosphorylation. In the case of proteins that must be purified by gel electrophoresis, the samples are eluted from a preparative sodium dodecyl sulfate (SDS)-polyacrylamide gel and then precipitated with trichloroacetic acid (TCA) in the presence of carrier proteins[1,2] (see [11] in this volume). This procedure of sample preparation can, however, be laborious and tedious and, on occasion, result in a considerable loss of the radioactive material. Previously, Aebersold *et al.*[3] demonstrated that proteins immobilized on a nitrocellulose membrane can be subjected to preparative tryptic digestion for high-performance liquid chromatography (HPLC) analysis. We have found that both cyanogen bromide (CNBr) cleavage and proteolytic digestion of phosphoproteins bound to membranes yield samples suitable for one- and two-dimensional peptide analyses. The digestion of membrane-bound proteins is rapid and minimizes losses.

CNBr Cleavage of Immobilized Proteins

1. ^{32}P-Labeled proteins should be fractionated by SDS-polyacrylamide gel electrophoresis, using whatever procedure resolves the protein of interest well.

[1] W. Gibson, *Virology* **62**, 319 (1974).
[2] K. Beemon and T. Hunter, *J. Virol.* **28**, 551 (1978).
[3] R. H. Aebersold, J. Leavitt, R. A. Saavedra, L. E. Hood, and S. B. H. Kent, *Proc. Natl. Acad. Sci. U.S.A.* **84**, 6970 (1987).

2. Transfer the gel-fractionated proteins electrophoretically to nitrocellulose (Schleicher & Schuell, Keene, NH) under conditions that are best for the protein under study. Neither Immobilon-P (Millipore, Bedford, MA) nor nylon (GeneScreen, New England Nuclear, Boston, MA) can substitute. Nylon dissolves in formic acid, and no radioactivity is recovered after CNBr digestion of proteins bound to Immobilon-P. Prior to transfer, the nitrocellulose membrane should be wetted in water and then soaked in transfer buffer [193 mM glycine, 25 mM Tris base, 0.1% (w/v) SDS, 20% (v/v) methanol].[4] We usually transfer at 60 V for 1 hr for a 60K protein using a Bio-Rad (Richmond, CA) transblot apparatus.[4] It is probably advisable to transfer proteins with greater molecular weights for longer times.

3. After transfer, rinse the membrane with water, wrap it in Saran Wrap to keep it moist, apply radioactive or fluorescent alignment markers, and expose it to X-ray film to localize the protein of interest. If the membrane dries out, it can be rewetted with water.

4. Excise the band containing the protein of interest, and incubate the piece of membrane with 50–100 mg/ml CNBr[4a] (Sigma, St. Louis, MO) in 70% (v/v) formic acid for 1 to 1.5 hr at room temperature. Increased time time will not increase yield. The volume of the reaction can vary depending on the size of the piece of nitrocellulose, but it should be enough to cover the piece of membrane completely. A larger volume increases the recovery of the peptides released from the filter during the reaction. We usually use a volume of 200 μl for a 3 $\times$ 8 mm piece of nitrocellulose.

5. At the end of the digestion, centrifuge the samples in a microfuge for 5 min, and then transfer the liquid to a new microfuge tube. Eighty-five to 95% of the counts should be present in the liquid.

6. Dry the supernatant on a Speed-Vac (Savant, Hicksville, NY). It takes about 30 min to dry a sample containing 200 μl.

7. Redissolve the residue in 30–40 μl of H_2O and dry again. This will remove the last traces of formic acid.

8. Dissolve the dried peptides in standard SDS–sample buffer. The dye should remain blue. If the Bromphenol Blue turns yellow, residual formic acid is present in the sample. Tris-HCl (pH 9) can be added to raise the pH. Otherwise, the fragments will not run well on the gel.

9. Analyze the fragments by electrophoresis on a 24% (w/v) acrylamide/0.054% (w/v) bisacrylamide gel. A Tricine cathode buffer [0.1 M N-

[4] M. P. Kamps and B. M. Sefton, *Oncogene* **2**, 305 (1988).
[4a] We usually make a stock solution of CNBr at approximately 300 mg/ml by adding 70% formic acid directly to the container in which the CNBr is supplied. This solution of CNBr can be stored stably at −70°.

tris(hydroxymethyl)methylglycine (Tricine), 0.1% SDS, 0.1 M Tris base, pH 8.25] can help resolve low-molecular-weight peptides.[5]

10. Dry the gel and expose with an intensifying screen (Lightning Plus, Dupont, Wilmington, DE).

Tryptic Peptide Mapping of Immobilized Proteins

1. ^{32}P-labeled proteins are fractionated by SDS-polyacrylamide gel electrophoresis and transferred electrophoretically to a membrane as described in the CNBr cleavage procedures. Nitrocellulose, Immobilon-P, and nylon membranes can all be used, but we prefer nitrocellulose membranes. Prior to transfer, an Immobilon-P membrane should be wetted in methanol for 1 min, washed with water for 2 min, and then soaked in transfer buffer. Nylon and nitrocellulose membrane should be wetted in water and then soaked in transfer buffer.

2. Rinse the blot with water after transfer. Wrap it in Saran Wrap to keep the membrane moist, apply alignment markers, and detect the band containing the protein of interest by autoradiography.

3. Excise the membrane piece of interest. If the membrane dries out, you can rewet the nitrocellulose and nylon membranes with water and the Immobilon-P membrane with methanol followed by water.

4. Soak the pieces of membrane immediately in 0.5% PVP-360 (polyvinylpyrrolidone, M_r 360,000; Sigma) in 100 mM acetic acid for 30 min at 37°. The exact molecular weight of polyvinylpyrrolidone may not be important.[3] This step increases the efficiency of elution of the peptides from the membrane.

5. Aspirate the liquid. Wash the membrane with H_2O extensively (5 times, 1 ml each) and then with freshly made 0.05 M NH$_4$HCO$_3$ once or twice.

6. The procedure for digestion of the immobilized protein with trypsin is essentially the same as the procedure used for the digestion of an eluted protein described in [11] in this volume. Briefly, the piece of membrane is incubated with 10 μg tosylphenylalanine chloromethyl ketone (TPCK)-treated trypsin (Worthington, Freehold, NJ) for 2 hr in 150–200 μl freshly made 0.05 M NH$_4$HCO$_3$ at 37°, and then for another 2 hr with an additional 10 μg of fresh enzyme. The volume of the reaction should be enough to cover the piece of membrane completely.

7. At the end of the digestion, add 300 μl (or more) H_2O to the sample, and centrifuge in a microfuge for 5 min. Transfer the liquid to a new

[5] H. Schägger and G. von Jagow, *Anal. Biochem.* **166,** 368 (1987).

microfuge tube. Approximately 90% of the radioactivity should be present in the liquid.

8. Dry the liquid on a Speed-Vac. It takes about 3–4 hr to dry.

9. If necessary, the dried tryptic peptides can now be oxidized in 50 μl performic acid as described in [11] in this volume. At the end of oxidation, add 1 ml of H_2O and dry the sample on a Speed Vac.

10. Analyze the sample by electrophoresis and chromatography as described in [11] in this volume.

Comments

In contrast to the protocol for digestion of eluted proteins, it is very important that the oxidation be done after the tryptic digestion of immobilized proteins. No peptides can be eluted from the Immobilon-P membrane by trypsin if the Oxidation is carried out first. It is likely that strong acid makes the binding of the protein to the Immobilon-P membrane irreversible. In addition, nylon membranes are dissolved in the strong acid (our unpublished observation). Chymotryptic peptide mapping of proteins immobilized to membranes can also be carried out using this procedure.[6]

For the analysis of tryptic phosphopeptides on 0.1-mm cellulose thin-layer plates, we usually load 30% or less of the total digest. Since tryptic digestion of immobilized proteins eliminated the need for carrier protein, it may be possible to use less than 20 μg of trypsin and still achieve complete digestion, and to load more than 30% of a given sample on the cellulose plate. We have not, however, tested this.

It has been reported that the recovery of some hydrophobic tryptic peptides is low after tryptic digestion of protein bound to nitrocellulose.[3] Therefore, some caution in the use of this procedure may be warranted. The hydrophilicity provided by the phosphate moieties of phosphorylated peptides may, however, render them particularly suitable for analysis by this technique. This procedure has been used successfully with p56^{lck},[7] p60^{c-src} (H. T. Adler, personal communication), p53,[7] simian virus 40 (SV40) large T antigen,[7] and proteins encoded by c-*abl*, v-*abl* (E. T. Kipreos, personal communication), and *bcr-abl* (A. M. Pendergast, personal communication).

[6] K. Luo and B. M. Sefton, *Oncogene* **5**, 803 (1990).
[7] K. Luo, T. R. Hurley, and B. M. Sefton, *Oncogene* **5**, 921 (1990).

[13] Analysis of the *in Vivo* Phosphorylation States of Proteins by Fast Atom Bombardment Mass Spectrometry and Other Techniques

By Philip Cohen, Bradford W. Gibson, and
Charles F. B. Holmes

The most popular method for analyzing the *in vivo* phosphorylation state of a protein is to radiolabel it by incubating cells and tissues with ^{32}P-labeled inorganic phosphate.[1] The cells/tissues are then lysed and the phosphoprotein of interest is separated and identified by peptide mapping or immunoprecipitation with specific antibodies. This procedure is sensitive and extremely useful, but a number of problems are associated with the methodology. First, a change in ^{32}P-labeling of a protein taking place in response to an extracellular signal can only be equated with a change in the phosphate content if equilibration has occurred with the added isotope. However, due to multiple pools of phosphate within cells (inorganic phosphate, phospholipids, RNA, DNA, etc.) a true steady state is rarely, if ever, achieved. Second, different phosphorylation sites on the same protein turn over at different rates, and it is virtually impossible to assess when, and whether, each site has reached isotopic equilibrium. Third, estimation of the stoichiometry of phosphorylation at a particular site assumes that isotopic equilibrium has been reached between intracellular ATP and that site, and that the specific radioactivity of the pool of intracellular ATP which labels that site can be measured. It is extremely difficult to test the validity of either of these assumptions, and where they have been tested they have been shown to be invalid.[2] Further problems arise in trying to identify the number of phosphorylation sites in a protein by these procedures. For example, if phosphoproteins are digested with proteinases and peptides separated either by high-performance liquid chromatography (HPLC) on a C_{18} column or sodium dodecyl sulfate-polyacrylamide gel electrophoresis (SDS–PAGE), the number of ^{32}P-labeled peptides observed cannot be equated with the number of phosphorylation sites, because partial proteolytic cleavage is a frequently encountered problem. Identification of the positions of phosphorylated residues within peptides by amino acid sequencing[3] can also be difficult if a peptide is phosphorylated at more than one site.

---

[1] J. C. Garrison, this series, Vol. 99, p. 20.

[2] S. E. Mayer and E. G. Krebs, *J. Biol. Chem.* **245,** 3153 (1970).

[3] Y. Wang, A. W. Bell, M. A. Hermodsen, and P. J. Roach, *J. Biol. Chem.* **261,** 16909 (1986).

An alternative procedure has been to determine the amount of inorganic phosphate in a protein or peptide directly, either by alkaline hydrolysis[4] or by ashing in the presence of magnesium nitrate.[5] It has been applied to tissues, such as skeletal muscle, which take up ^{32}P-labeled inorganic phosphate rather poorly, and to proteins that can be isolated in large amounts. However, this method lacks sensitivity, and contamination of reagents or glassware, whether by traces of inorganic phosphate or substances that interfere with colorimetric analysis, is a serious hazard. In addition, some phosphoserine and phosphothreonine residues are resistant to hydrolysis.[6] Furthermore, no information can be gained about the location of phosphorylation sites within the protein or peptide, and the method is inadequate in situations where a peptide contains more than one phosphorylated residue.

Over the past few years, we have adopted a new and powerful strategy for analyzing the *in vivo* phosphorylation states of proteins that is described below and which overcomes the potential disadvantages discussed above.

Principle

The presence of a phosphorylated residue increases the molecular mass of a peptide by 80 Da. Phosphorylated and dephosphorylated forms of a peptide can therefore be detected very easily by fast atom bombardment mass spectrometry (FABMS) or liquid secondary ion mass spectrometry (LSIMS). Cells/tissues are homogenized under conditions that prevent phosphorylation and dephosphorylation, and the protein of interest is purified in the presence of kinase and phosphatase inhibitors. Following proteolytic digestion, peptides are chromatographed on a C_{18} column to resolve the dephosphopeptide from phosphorylated derivatives, which are identified by FABMS. Phosphorylation stoichiometries are determined by quantitative amino acid analysis and by integration of the ultraviolet absorbance peaks corresponding to each derivative.[7-10] The positions of

[4] P. S. Guy, P. Cohen, and D. G. Hardie, *Eur. J. Biochem.* **114**, 399 (1981).
[5] J. E. Buss and J. T. Stull, this series, Vol. 99, p. 7.
[6] B. E. Kemp, *FEBS Lett.* **110**, 308 (1980).
[7] C. F. B. Holmes, N. K. Tonks, H. Major, and P. Cohen, *Biochim. Biophys. Acta* **929**, 208 (1987).
[8] L. Poulter, S. G. Ang, B. W. Gibson, D. H. Williams, C. F. B. Holmes, F. B. Caudwell, and P. Cohen, *Eur. J. Biochem.* **175**, 497 (1988).
[9] C. MacKintosh, D. G. Campbell, F. B. Caudwell, and P. Cohen, *FEBS Lett.* **234**, 189 (1988).
[10] P. Dent, D. G. Campbell, F. B. Caudwell, and P. Cohen, *FEBS Lett.* **259**, 281 (1990).

phosphoserine residues within a peptide are identified by sequence analysis after conversion to S-ethylcysteine.[11–13] The positions of phosphothreonine and phosphotyrosine residues are determined by FABMS or by tandem mass spectrometry of the molecular ion after collisional activation.[14,15]

Analytical Methods

High-Performance Liquid Chromatography

Peptides are purified by HPLC on a Vydac 218TP54 reversed-phase C_{18} column (Separations Group, Hesperia, CA) equilibrated in either 0.1% (by volume) trifluoroacetic acid, pH 1.9 (Rathburn Chemicals, Peebleshire, Scotland) or 10 mM ammonium acetate, pH 6.5. The columns are developed with linear water/acetonitrile gradients with an increase in acetonitrile concentration of 0.33%/min and analyzed online for ultraviolet absorbance and ^{32}P radioactivity. The flow rate is 1.0 ml/min and fractions of 0.5 ml are collected.

Conversion of Phosphoserine to S-Ethylcysteine

In order to convert phosphoserine residues to S-ethylcysteine, peptides (0.1–1.0 nmol) are dissolved in 50 μl of a reaction mixture consisting of ethanethiol (60 μl), water (200 μl), dimethyl sulfoxide (200 μl), ethanol (100 μl), and 5 N NaOH (65 μl) and incubated for 1 hr at 50° under nitrogen.[12] Incubations are terminated by cooling and adding 10 μl of glacial acetic acid. After dilution to 1 ml with water and freezing, reactions are dried, redissolved in 0.5 ml of water, redried, and this process is repeated twice more. Finally the sample is redissolved in 0.1 ml of 0.1% (by volume) trifluoroacetic acid containing 50% (by volume) acetonitrile and subjected to amino acid analysis and protein sequencing as described below.

Amino Acid Analysis

This is carried out using a Waters PICOTAG system (Millipore, Ltd., Watford, England) although other analyzers employing phenyl isothiocyanate (PITC) precolumn derivatization and of similar speed (22-min cycle,

[11] H. E. Meyer, E. Hoffmann-Posorske, H. Korte, and L. M. G. Heilmeyer, Jr., *FEBS Lett.* **204**, 61 (1986).
[12] C. F. B. Holmes, *FEBS Lett.* **215**, 21 (1987).
[13] H. E. Meyer, E. Hoffmann-Posorske, and L. M. G. Heilmeyer, Jr., this volume [14].
[14] K. Biemann and H. A. Scobie, *Science* **235**, 305 (1987).
[15] B. W. Gibson and P. Cohen, this series, Vol. 193, p. 480.

including reequilibration time) and sensitivity can be used. Peptides are hydrolyzed in the vapor phase *in vacuo* for 1 hr at 150° with 6 M HCl/2 mM phenol, dried, redried in ethanol/water/triethylamine (2 : 2 : 1 by volume), and derivatized with PITC/ethanol/water/triethylamine (1 : 7 : 1 : 1 by volume). The phenylthiocarbamyl (PTC) derivatives are then separated by chromatography on a C_{18} column. Serine, threonine, and tyrosine are corrected for 17, 13, and 18% destruction during hydrolysis, respectively.

Protein Sequencing

Edman sequencing is carried out using a gas-phase or pulsed liquid sequencer (Applied Biosystems, Foster City, CA) equipped with an one-line reversed-phase chromatography system.

FABMS, LSIMS, and Tandem Mass Spectrometry

FABMS is generally performed in the positive ion mode using a VG70-250SE mass spectrometry (VG Analytical, Manchester, England) or an MS50 mass spectrometer (Kratos Analytical, Manchester, England), although other mass spectrometers of equivalent specification can be used. An Ion-Tech source operating at 8 kV is employed to generate the fast xenon beam, or Cs^+ in the case of LSIMS. Peptides (2 μl, 10–1000 pmol) in 10% (by volume) acetic acid are concentrated to ~0.5 μl on a stainless steel probe tip and 0.5 μl of matrix [dithiothreitol : dithioerythritol (3 : 1) in 1% HCl] is added before insertion into the machine. Mass spectra are then obtained by scanning the magnet at a rate of 100 sec/decade, or faster, over the mass range of interest. In the FAB or LSIMS mass spectra of peptides, one usually observes an abundant protonated or deprotonated molecular ion $[MH^+$ or $(M-H)^-]$, depending on whether positive or negative ions are being selected. Fragment ions originating from bond cleavages along the peptide backbone are also observed, but to a lesser extent. From the observed molecular ions, molecular masses of peptides are deduced by either subtracting or adding the mass of a proton. To sequence a peptide from its mass spectrum, mass differences between fragment ion types are correlated to the masses of amino acid side chains and used to deduce a partial or complete sequence for the phosphopeptide in question. Since there are three possible cleavage sites for each amino acid residue (at either side of the α carbon and at the amide linkage), with charge retention at either the N-terminus or C-terminus, six different fragment ion types can be encountered (a_n, b_n, and c_n for N-terminal ions and x_n, y_n and z_n, for C-terminal ions).[13,16] In the case of phosphoserine, phos-

---

[16] J. A. McCloskey, ed., this series, Vol. 193.

phothreonine, and phosphotyrosine, the mass differences between two ions of the same type (e.g. a_6 to a_7) are 80 Da higher relative to their unphosphorylated counterparts, i.e., 167 vs 87 Da [Ser(P) vs Ser], 181 vs 101 Da [Thr(P) vs Thr], or 243 vs 163 Da [Tyr(P) vs Tyr]. Thus, one must first assign the observed fragment ions to these various ion types to arrive at a correct interpretation of the spectrum. For a more complete description of the analytical process, readers are referred to another volume in this series.[16]

Tandem mass spectrometry can also be carried out on the phosphopeptide and has considerable advantages for sequence analysis. In this strategy, the peptide does not need to be homogeneous, since the molecular ion for the peptide in question can be selected from a mixture of peptides in the first mass spectrometer (MS-I), prior to analysis of sequence ions in the second mass spectrometer (MS-II). Tandem mass spectrometry can be carried out on a number of commercially available four-sector mass spectrometers (Kratos Concept II HH; VG 70SE or ZAB 4F, VG Analytical; JEOL HX110/HX110, JEOL USA, Peabody, MA). Either FAB or LSIMS is used to generate the peptide molecular ion as described above for a two-sector experiment. The isotopically pure ^{12}C component of the parent molecular ion is then selected in MS-I and collisionally activated. Typically, helium is used in a collision cell located between MS-I and MS-II at a pressure sufficient to attenuate the parent ion beam to $\sim$30% of its initial value. The fragment ions ("daughter ions") resulting from decomposition of the collisionally activated parent ions are then mass analyzed and detected in MS II. In general, much more extensive fragmentation is observed in these tandem spectra and a complete (or more complete) sequence determination can be made.[17]

Procedure

In order to preserve the *in vivo* phosphorylation states of proteins, tissues are homogenized in 3 vol of buffer containing 5 mM EDTA (to chelate Mg^{2+} and inhibit protein kinases), 50 mM NaF and 5 mM PP$_i$ (to inhibit protein-serine/threonine phosphatases), and 1 mM sodium orthovanadate (to inhibit protein-tyrosine phosphatases). The protein of interest is then highly purified in the presence of fluoride and vanadate to prevent dephosphorylation. In the case of an enzyme whose activity is altered by phosphorylation, the effectiveness of 50 mM NaF in preventing dephosphorylation can be established by demonstrating that its kinetic parameters

---

[17] B. W. Gibson, *in* "Biological Mass Spectrometry" (A. L. Burlingame and J. A. McCloskey, eds.), p. 315. Elsevier, Amsterdam, 1990.

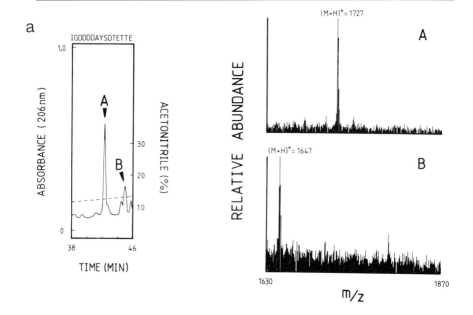

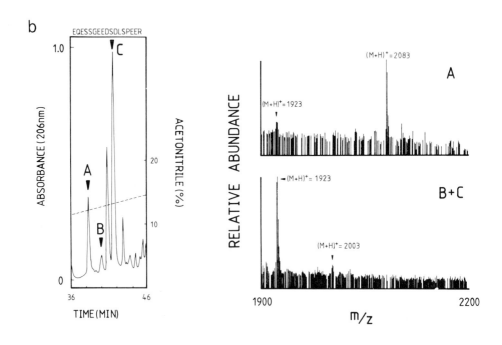

do not alter during purification.[18] It is not essential that the protein be homogenous and the procedures described below have been applied successfully to proteins that are only 50–70% pure.[7,9]

The protein is mixed with a trace (about 1% by weight) of ^{32}P-labeled protein that has been purified in the absence of phosphatase inhibitors and then labeled by incubation with all the protein kinases known to phosphorylate it in vitro. This trace of "marker" protein enables the purification of each peptide of interest to be followed and recoveries to be calculated, but is present in too low an amount to interfere with subsequent analyses.

The protein is then digested with a proteinase (usually trypsin) under conditions that generate the required peptides in high yield and of a convenient size (5–30 residues) for FABMS and sequence analysis. Because phosphorylation sites are located on exposed regions of proteins, it is usually possible to release phosphopeptides by brief (~10 min) tryptic attack of the native protein, under conditions that hardly digest the rest of the protein.[8–10,19,20] The optimal conditions can be determined very rapidly by pilot experiments using the ^{32}P-labeled marker). The release of trichloroacetic acid-soluble peptides from the native protein is extremely useful because it frequently allows peptides to be purified by a single chromatography on a C_{18} column at pH 1.9. If further purification is necessary, then the C_{18} column is preceded by gel filtration on a 150 × 1.2 cm column of Sephadex G-50 (Superfine Grade, Pharmacia, Piscataway,

[18] P. J. Parker, N. Embi, F. B. Caudwell and P. Cohen, Eur. J. Biochem. **124**, 47 (1982).
[19] P. Cohen, D. C. Watson, and G. H. Dixon, Eur. J. Biochem. **51**, 79 (1975).
[20] R. E. H. Wettenhall and P. Cohen, FEBS Lett. **140**, 263 (1982).

FIG. 1. Reversed-phase chromatography and FABMS analysis of peptides from rabbit skeletal muscle protein phosphatase inhibitor-2. In the upper left diagram the phosphorylated form of the peptide (IGDDDDAYSDTETTE), comprising residues 78–92 of inhibitor-2 (A), and the dephosphorylated form (B) were resolved on a Vydac C_{18} column. Fractions eluting at 42–43 min (A) and 44–46 min (B) were pooled and 1.5-nmol aliquots subjected to FABMS to identify the derivatives. In the lower left diagram, the diphosphorylated (A), monophosphorylated (B), and dephosphorylated (C) forms of the peptide EQESSGEEDSDLSPEER were resolved on the same column. Fractions eluting at 38–39 min (A) and 40–43 min (B + C) were pooled and 2-nmol aliquots analyzed by FABMS. The full line shows absorbance at 206 nm and the broken line the acetonitrile gradient. The 206-nm absorbance peak eluting between peaks B and C in the lower left diagram is a peptide corresponding to residues 33–40 of inhibitor-2. The proportions of the phosphorylated and dephosphorylated forms were quantitated by amino acid analysis. Following conversion of phosphoserine to S-ethylcysteine, sequence analysis established that the peptide 78–92 and the first two serines in the peptide 117–133 were the sites of phosphorylation. (Taken from Holmes et al.[7])

NJ), i.e., a procedure that does not resolve phosphorylated and dephosphorylated forms of a peptide.[10]

Following chromatography on a C_{18} column at pH 1.9, it is necessary to locate the phosphorylated and dephosphorylated forms of each peptide. The ^{32}P-labeled marker usually reveals the positions of monophosphorylated peptides. The dephosphorylated peptide, being more hydrophobic, elutes at a 0.5–1.0% higher acetonitrile concentration, while derivatives containing more than one phosphorylated residue elute at 0.5–1.0% lower acetonitrile concentrations than the monophosphorylated peptide. The different species are identified by subjecting aliquots of each fraction to FABMS and amino acid analysis. If the various derivatives are completely resolved from one another and from impurities (Fig. 1), then phosphorylation stoichiometries can be determined very accurately from the relative amounts of each species measured by both quantitative amino acid analysis and integration of the ultraviolet absorbance peaks.[7–10]

In some situations, phosphorylated and dephosphorylated derivatives are not resolved by chromatography on the C_{18} column at pH 1.9. This frequently happens when investigating peptides labeled by cyclic AMP-dependent protein kinase, which phosphorylates sequences of the type Arg–Arg–Xaa–Ser–. If the serine is not phosphorylated, trypsin usually cleaves the Arg–Xaa bond to yield a dephosphorylated peptide commencing Xaa–Ser–. Phosphorylation renders the Arg–Xaa bond resistant to trypsin, and only the Arg–Arg bond is cleaved to yield the phosphorylated derivative Arg–Xaa–Ser(P), which coelutes with the dephosphopeptide Xaa–Ser–.[9,10] Sometimes, trypsin will cleave the dephosphopeptide at either the Arg–Arg bond or the Arg–Xaa bond to yield two different derivatives (see Fig. 2).[10] In these situations, or if the phosphorylated and/or dephosphorylated derivatives are too impure to be quantitated accurately, the entire region containing all the derivatives is pooled and rechromatographed on the C_{18} column at pH 6.5. Resolution of phosphorylated and dephosphorylated peptides is greater at pH 6.5 because of the extra negative charge on the phosphate moiety, and phosphopeptides frequently elute at 2–3% lower concentrations of acetonitrile than dephosphopeptides (e.g., Fig. 2). Phosphorylation stoichiometries are then determined by amino acid analysis and integration of peak areas as described above.

Determination of Phosphorylation Stoichiometries by FABMS

In a few cases, it has proved impossible to adequately resolve phosphorylated and dephosphorylated derivatives. One example is a 12-residue tryptic peptide on glycogen synthase which is phosphorylated on three

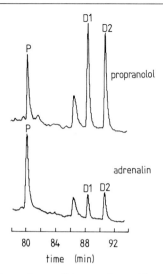

FIG. 2. Separation of different forms of the tryptic peptide containing a serine residue on the glycogen-binding subunit of protein phosphatase-1 which is phosphorylated by cyclic AMP-dependent protein kinase. The peptide was isolated from the skeletal muscle of rabbits injected with adrenalin (lower trace) or the adrenalin antagonist propranolol (upper trace) and resolved into monophosphorylated (P) and dephosphorylated derivatives (D1 and D2) by chromatography on a C_{18} column at pH 6.5 (detected by monitoring absorbance at 214 nm). Sequence and FABMS analysis showed that peptide D1 had the sequence VSFADNFGFNLVSVK, while D2 contained an additional arginine at the N terminus. Peptide P was identical to D2 except that serine-3 was phosphorylated. The other peptide absorbing at 214 nm is an impurity. (Taken from Dent *et al.*[10])

serine residues by glycogen synthase kinase 3 (Fig. 3). In this peptide, the dephosphorylated, monophosphorylated, diphosphorylated, and triphosphorylated peptides could not be separated by chromatography on a C_{18} column because, for unknown reasons, each derivative elutes as a broad peak which overlaps with another species.[8] A second example is a 27-residue tryptic peptide corresponding to the C terminus of glycogen synthase which contains one of the serines phosphorylated by cyclic AMP-dependent protein kinase. Due to its relatively large size the presence of a single phosphate does not alter the elution position of the peptide very much, while partial tryptic cleavage of the Arg–Asn bond at the C terminus of the protein introduced further heterogeneity that prevented adequate resolution of the phosphorylated and dephosphorylated forms.[8] In these situations, the stoichiometry of phosphorylation can be estimated by FABMS alone, by measuring the relative abundance of each molecular ion in the FAB mass spectrum (Fig. 3). However, for reasons discussed

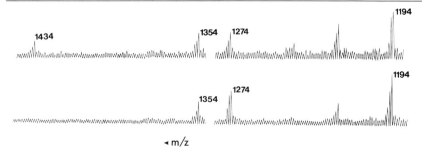

FIG. 3. Molecular ion regions from the positive ion FAB mass spectrum containing different derivatives of the glycogen synthase peptide PASVPPSPPSLSR. The peptide was isolated from the skeletal muscle of rabbits injected with adrenalin (upper spectrum) or propranolol (lower spectrum). The species at m/z 1194, 1274, 1354, and 1434 correspond to dephosphorylated, monophosphorylated, diphosphorylated, and triphosphorylated forms of the peptide, respectively. Other species are contaminants.

below, this method tends to underestimate the phosphorylation stoichiometry and should be used only if complete separation of the phosphorylated and dephosphorylated derivatives on a C_{18} column proves to be impossible.

FABMS is a surface-sampling technique and the abundance of molecular ions of peptides in the mass spectrum reflects to a large extent their surface activities.[21,22] In particular, it has been shown that hydrophilic peptides are suppressed in the presence of hydrophobic peptides, which occupy the surface of the matrix preferentially.[23] As a result, there is a danger that a signal from a phosphorylated peptide will be suppressed to some extent by the presence of the more hydrophobic dephosphorylated derivative. There are a number of ways to minimize this problem. First, it is necessary to integrate molecular ion abundancies over an extended period of time in case the signal from the phosphorylated derivative appears more slowly.[8,24] Second, one should try to generate a relatively large peptide (20–30 residues) since the dephosphorylated peptide will tend to suppress the phosphorylated derivative(s) to a lesser extent due to the smaller contribution of the phosphorylation site to the difference in surface activity. Third, peracetylation with acetic anhydride/pyridine (1 : 1, by volume) or esterification with various alcohols in the presence of HCl or 0.2 M acetyl chloride can be used to generate more hydrophobic deriva-

[21] W. V. Ligon and S. B. Dorn, *Int. J. Mass Spectrom. Ion Processes* **57,** 75 (1984).

[22] W. W. Ligon, *Anal. Chem.* **58,** 485 (1986).

[23] S. A. Naylor, A. F. Findeis, B. W. Gibson, and D. H. Williams, *J. Am. Chem. Soc.* **108,** 6359 (1986).

[24] L. Poulter, S. G. Ang, D. H. Williams, and P. Cohen, *Biochim. Biophys. Acta* **929,** 296 (1987).

tives whose relative ionization efficiencies are reduced or eliminated.[24] Fourth, the use of a more hydrophobic hydroxyl-containing matrix, such as 1,2,6-hexanetriol, or the addition of certain strong acids (e.g., $HClO_4$) to the matrix, have been shown to yield molecular ion abundancies that more accurately reflect the relative concentrations of phosphorylated peptides.[24]

Identification of Phosphorylation Sites within Peptides

Phosphoserine undergoes β elimination to dehydroalanine upon incubation with 1 M NaOH and can then be converted to S-ethylcysteine upon addition of ethanethiol (see Analytical Methods).[11-13] The location of phosphoserine residues in peptides is therefore revealed during amino acid sequencing by the presence of the phenylthiohydantoin (Pth) derivative of S-ethylcysteine at a particular cycle of Edman degradation, which is eluted at a characteristic position, just before diphenylthiourea. The positions of serine residues that are not phosphorylated is revealed by the presence of Pth-Ser and its dithiothreitol (DTT) adduct. A typical result is illustrated in Fig. 4.

This method is excellent and can also be used for the selective isolation of phosphoserine-containing peptides,[12] but the following pitfalls and limitations should be borne in mind:

1. If phosphoserine is the N-terminal or C-terminal residue, conversion to S-ethylcysteine does not take place on incubation with 1 M NaOH and ethanethiol.[25,26]

2. The presence of Pth-S-ethylcysteine at a particular cycle of Edman degradation does not prove that phosphoserine was present before modification. Conversion to S-ethylcysteine will also occur if the serine was esterified in some other way (e.g., acetylation or glycosylation). It is therefore essential that the presence of a phosphorylated residue be established by FABMS prior to conversion to S-ethylcysteine.

3. Phosphoseryl residues are very resistant to β elimination if followed by a proline residue.[6] In such situations a much longer incubation with NaOH and ethanethiol (5 hr or more) may be necessary.[10]

4. Ethanethiol has a strong and unpleasant smell and all reactions must be carried out in an extremely well-ventilated hood.

5. The method is not applicable to peptides containing phosphothreonine or phosphotyrosine residues.

[25] H. E. Meyer, K. Swiderek, E. Hoffman-Posorske, H. Korte, and L. M. G. Heilmeyer, Jr., *J. Chromatogr.* **296**, 129 (1987).

[26] A. J. Garton, D. G. Campbell, P. Cohen, and S. J. Yeaman, *FEBS Lett.* **229**, 68 (1988).

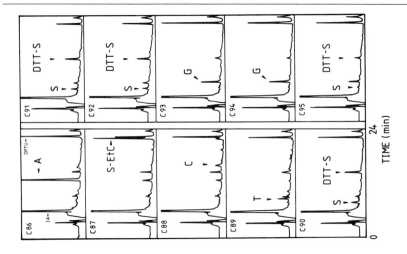

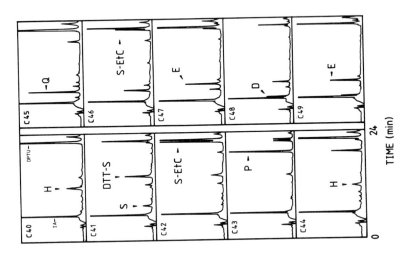

Fortunately, ~90% of phosphorylated residues are phosphoserine. If phosphothreonine or phosphotyrosine is suspected, however, or if conversion of phosphoserine to *S*-ethylcysteine cannot be used for one of the above reasons, then the positions of phosphorylated residues can be determined by FABMS. The sequences of many peptides can be determined by analyzing the fragment ions that appear in the FAB mass spectrum.[8,14] However, due to incomplete fragmentation this does not always succeed, and much larger amounts of material (several nanomoles) are required than are needed to simply determine the molecular mass of a peptide (10–1000 pmol). However, phosphopeptides can be analyzed by tandem mass spectrometry using either an FAB or LSIMS source. In this procedure, the isotopically pure ^{12}C component of the parent molecular ion is selected in MS-I and collisionally activated to produce fragment ions, which are then mass separated in MS-II to yield a tandem mass spectrum (Fig. 5). The much more extensive fragmentations that are observed in tandem spectra, as compared to two-sector spectra, make it possible to sequence routinely entire phosphopeptides (<2 kDa) using 50–500 pmol of peptide.[14] An alternative method is to perform manual Edman degradation on a peptide using the method of Tarr,[27] and to redetermine the molecular mass of the truncated peptide after each cycle of

[27] G. E. Tarr, *in* "Methods in Protein Sequence Analysis" (M. Elzinga, ed.), p. 223. Humana Press, Clifton, New Jersey, 1984.

FIG. 4. Gas-phase sequencer analyses of the *S*-ethylcysteinyl derivatives of two peptides derived from the C-terminus of glycogen synthase. The peptide comprising residues C-40–C-53 (HSSPHQSEDEEEPR) of the C-terminal cyanogen bromide fragment (CB–C) was isolated from the skeletal muscle of rabbits injected with propranolol, and resolved into diphosphorylated, monophosphorylated, and dephosphorylated derivatives by chromatography on a C_{18} column at pH 1.9.[8] The diphosphorylated derivative was incubated with 1 *M* NaOH and ethanethiol and a 100-pmol aliquot sequenced. The reversed-phase chromatograms on the left show Pth derivatives corresponding to residues C-40–C-49 (Edman cycles 1–10). The peptide comprising residues C-85–C-95 of CB–C (RASCTSSSGGSKR) was also incubated with NaOH and ethanethiol and the *S*-ethylcysteine-containing derivative was purified by chromatography on the C_{18} column at pH 1.9 and sequenced. The reversed-phase chromatograms on the right show Pth derivatives corresponding to residues C-86–C-95 (Edman cycles 2–11). The ultraviolet detector sensitivity was absorbance = 0.02 at full scale. Serine was detected by the appearance of two Pth derivatives, namely Pth-Ser itself (S) and its dithiothreitol adduct (DTT-S). Other abbreviations: IA, injection artifact; DPTU, diphenylthiourea; H, Pth-histidine; P, Pth-proline; Q, Pth-glutamine; E, Pth-glutamic acid; D, Pth-aspartic acid; A, Pth-alanine; G, Pth-glycine; T, Pth-threonine; C, Pth-cysteine, S-EtC, Pth-S-ethylcysteine. The experiments demonstrated that the second and third serines in the peptide C-40–C-53 and the first serine in C-85–C-96 were phosphorylated, while the other serines were not.

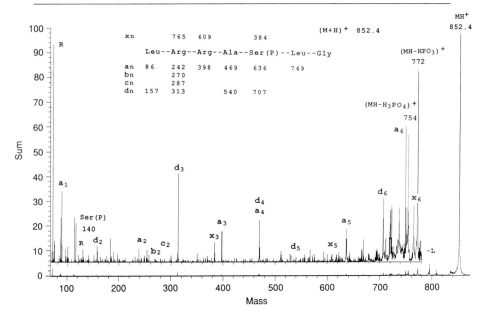

FIG. 5. Tandem collision-induced dissociation mass spectrum of the protonated molecular ion of phosphorylated kemptide, Leu–Arg–Arg–Ser(P)–Leu–Gly (MH$^+$ = 852.4). Note that the position of phosphoserine is defined by the assignments of a_4 and a_5 (m/z 439 and 636, Δm = 167 Da) and by the d_5 ion which corresponds to the loss of H_3PO_4 from a_5 (m/z 540 and 636, Δm = 96 Da).

Edman degradation. The presence of phosphoserine, phosphothreonine, or phosphotyrosine is indicated when the molecular mass of the peptide decreases by 167, 181, or 243 Da after a particular cycle of Edman degradation, instead of the 87, 101, or 163 Da expected for unmodified serine, threonine, or tyrosine residues.

Comments on Procedure

Advantages

The described methodology does not suffer from the disadvantages of ^{32}P-labeling techniques discussed in the introduction. Although not as sensitive as the ^{32}P labeling of cells, the procedures are nevertheless quite sensitive, and have been applied successfully to low-abundance proteins composing only 0.02–0.2% of the soluble protein in muscle.[7–10] Determination of the molecular mass of a peptide by FABMS usually requires only 10–100 pmol of material, and is therefore of comparable sensitivity to amino acid analysis and amino acid sequencing. Furthermore, phosphorylated and dephosphorylated forms of peptides can be readily identified by FABMS even when they are impure (e.g., Fig. 3). Advances, such as in tandem mass spectrometry, can in principle sequence (and hence identify phosphorylation sites in) peptides of 20–25 residues in a single experiment, and this technique was used to identify the labile phosphoaspartic acid residue in the bacterial chemotactic peptide Che Y after conversion to the stable homoserine analog.[28] Another exciting development is in electrospray MS and matrix-assisted laser desorption time of flight MS, which promise to increase the sensitivity of mass spectrometry (especially of hydrophilic peptides) by an order of magnitude or more[15] and to allow for the direct analysis of unfragmented proteins. In a recent experiment, a 46-residue phosphopeptide from protein phosphatase inhibitor-1 was accurately mass analyzed with 1 pmol of material using electrospray MS (C. F. B. Holmes, unpublished).

The procedures described above are also very rapid. The determination of a molecular mass by FABMS takes only a few minutes and it is possible to obtain molecular masses and amino acid compositions of every fraction from a C_{18} column within a day or two. Purification of the protein and peptide generation is usually the rate-limiting step.

A further advantage of the procedures described in this chapter is that no radioactive isotopes are needed, apart from the tiny amounts required to make the ^{32}P-labeled marker.

Assumptions and Limitations

There is one assumption specific to this procedure, namely that the dephosphorylated and phosphorylated forms of the peptide are recovered in equal yields from proteolytic digestion of the protein until separation of these derivatives by reversed-phase column chromatography. A further assumption implicit in all methods for determining the *in vivo* phosphorylation states of proteins is that the level of phosphorylation does not alter during extraction and homogenization of the tissue.

The procedures are best applied when the amino acid sequence of the protein of interest is known. While this was a serious limitation a few years ago, the explosive increase in cDNA cloning has ensured that the structures a great number regulatory proteins have now been determined.

The methodology is particularly effective for analyzing the phosphorylation states of residues from proteins which are already suspected from *in vitro* studies to undergo phosphorylation. In order to detect unknown phosphorylation sites without the benefit of ^{32}P-labeled marker peptides,

---

[28] D. A. Sanders, B. L. Gillece-Castro, A. L. Burlingame, and D. E. Koshland, *J. Biol. Chem.* **264**, 21770 (1989).

aliquots of each fraction from a C_{18} column are subjected to amino acid analysis after NaOH/ethanethiol incubation to convert phosphoserine to S-ethylcysteine. The PTC derivative of S-ethylcysteine elutes at a characteristic position between PTC-methionine and PTC-isoleucine.[12] However, detection of unknown phosphothreonine and phosphotyrosine residues requires detailed and time-consuming analysis of each peptide in the digest using amino acid and sequence analysis in combination with FABMS. Owing to the advent of online liquid chromatography (LC)-linked electrospray MS, these procedural limitations may be significantly reduced in the near future, since fractions from the C_{18} column are mass analyzed directly and correlated to the liquid chromatogram. This should enhance the ability to identify and quantify clusters of multiply phosphorylated peptides.

Sulfation of tyrosine residues is a posttranslational modification that also increases the molecular mass of a peptide by 80 Da.[15] However, there is little danger of confusing phosphorylation with sulfation, because the latter modification is acid labile. Desulfation occurs upon incubation for 5 min at 37° in the presence of 6 M HCl, conditions which do not hydrolyze phosphorylated residues.[29] In addition, desulfation is extensive during FABMS in the positive ion mode, whereas phosphorylated residues are stable under these conditions. FABMS must be carried out in the negative ion mode in order to detect sulfated tyrosines.[15]

Concluding Remarks

The emphasis in this chapter has been to describe in practical terms a new approach to the identification of *in vivo* phosphorylation sites, and the measurement of phosphorylation stoichiometries, that makes use of a range of modern techniques and advances in protein chemistry. Although the detailed procedures described in this chapter have so far only been applied to three proteins,[7–10,30] it is already evident that, together, they constitute a new and powerful approach. The determination of molecular masses by FABMS is simple and can be carried out without detailed knowledge of the underlying theory. For a more detailed discussion of the mass spectrometric aspects of this topic, readers are referred to another chapter in this series.[15]

[29] W. B. Huttner, this series, Vol. 107, p. 200.
[30] P. Clarke and D. G. Hardie, *EMBO J.* **9**, 2439 (1990).

[14] Determination and Location of Phosphoserine in Proteins and Peptides by Conversion to S-Ethylcysteine

By Helmut E. Meyer, Edeltraut Hoffmann-Posorske, and Ludwig M. G. Heilmeyer, Jr.

In proteins the most abundant phosphoamino acid is phosphoserine. Phosphothreonine and phosphotyrosine are found in decreasing order; nitrogen–phosphate linkages occur rarely. Treatment with alkali easily splits off phosphate from phosphoserine and phosphothreonine by β elimination; in contrast, phosphotyrosine is not affected.[1] β Elimination of phosphoserine produces dehydroalanine, which is stable when it is surrounded by peptide bonds. Nucleophiles such as amines (methylamine) or sulfite can be added to the double bond.[2] Upon hydrolysis of the peptide the newly formed amino acid can be detected. Thus, β-methylaminoalanine and cysteic acid have been proposed as stable derivatives for the determination of phosphoserine.[3] Some cysteine residues, however, can eliminate hydrogen sulfide, which yields dehydroalanine as well. Furthermore, problems arise in the quantitative determination of the newly formed amino acids. In principle, cysteic acid presents the same difficulties as phosphoserine. This highly negatively charged amino acid is retained in the reaction cartridge of the gas-phase sequencer due to ionic interactions with Polybrene and its insolubility in butyl chloride. Therefore, a positive identification is not possible and it is not easily detected during amino acid analysis. Depending on the kind of amino acid analysis performed, it may be difficult for β-methylaminoalanine to be resolved from other amino acids, especially in the highly sensitive PTC-amino acid[4] analysis. In our hands optimal results are obtained with ethanethiol, which yields S-ethylcysteine by nucleophilic addition to dehydroalanine. S-Ethylcysteine can easily be determined by PTC-amino acid analysis.[5] PTH-S-ethylcysteine is detected as a well-resolved species during sequence analysis.[6] β Elimina-

[1] T. M. Martensen, this series, Vol. 107, p. 3.
[2] T. R. Soderling and K. Walsh, *J. Chromatogr.* **253**, 243 (1982).
[3] W. D. Annan, W. Manson, and J. A. Nimmo, *Anal. Biochem.* **121**, 62 (1982).
[4] PTC, Phenylthiocarbamyl; PTH, phenylthiohydantoin; HPLC, high-performance liquid chromatography; PITC, phenyl isothiocyanate.
[5] H. E. Meyer, K. Swiderek, E. Hoffmann-Posorske, H. Korte, and L. M. G. Heilmeyer, Jr., *J. Chromatogr.* **397**, 113 (1987).
[6] H. E. Meyer, E. Hoffmann-Posorske, H. Korte, and L. M. G. Heilmeyer, Jr., *FEBS Lett.* **204**, 61 (1986).

tion of cysteine residues can be prevented by oxidation to cysteic acid, a derivative which is stable under the applied modification conditions. In the following we describe methods to determine phosphoserine as S-ethylcysteine in proteins and peptides, and during sequence analysis.

Phosphoserine Determination as S-Ethylcysteine

Reagents

All reagents must be freshly prepared.

Performic acid: Formed from 95 parts formic acid (sequence grade) and 5 parts hydrogen peroxide (30% by mass)

Modification mixture: Composed of 80 μl ethanol, 65 μl of 5 N NaOH, 60 μl ethanethiol, and 400 μl H_2O, which should be added as the last component.

Procedure

1. Add 50 to 1000 μl of performic acid to the dried phosphoprotein (0.05 to 1 nmol) in a 5 × 60 mm test tube (Pyrex glass).

2. Incubate for 1 hr at 4° in a refrigerator.

3. Dry down in a SpeedVac (Savant, Hicksville, NY) (takes about 10 min, vacuum 0.2 mbar).

4. Add 20 μl of the modification mixture; seal the tube with Parafilm and incubate the mixture at 50° for 1 hr in a heat block.

5. Add 5 μl of acetic acid after cooling down to room temperature.

6. Dry immediately, employing a vacuum of less than 0.01 mbar. The sample should be dry in less than 1 hr.

7. Hydrolyze the protein employing the gas-hydrolysis procedure and perform PTC-amino acid analysis as described below.

The phosphoprotein sample employed should be free of buffer substances such as tris(hydroxymethyl)aminomethane, N-(2-hydroxyethyl)-piperazine-N'-2-ethanesulfonic acid, and morpholinopropanesulfonic acid, as some of these substances produce a high number of components which cannot be identified in a chromatogram of the PTC-amino acid derivatives. Specifically, these substances interfere with S-ethylcysteine determination.

Performic acid oxidation is a necessary step to avoid false-positive results arising from cysteine residues which in some instances undergo β elimination (most probably depending on neighboring amino acids).

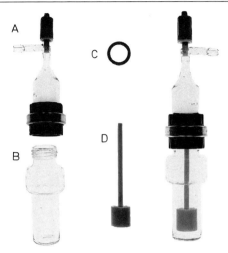

FIG. 1. Protein and peptide hydrolysis vessel. The optimized hydrolysis vessel consists of the following parts: head cap (A) with a temperature-resistant (200°) Rotaflo valve (HP 6/6) fused with a shortened adapter piece (No. 703-03) connected with a screw cap (No. 701-13). Vessel body (B) is made from an adapter piece (No. 703-03) by an in-house glass blower. All parts from (A) and (B) are purchased from QVF Corning (Wiesbaden, Germany). The seal (C) between the head cap and the vessel body is made from KAL-REZ (Du Pont, Newton, CT) and purchased from Applied Biosystems (No. 221010). The holder (D) for the sample test tubes is made from Teflon in house. The assembled vessel is shown on the right-hand side.

Carboxymethylation and N-vinylpyridylation are not useful, since both modifications enhance S-ethylcysteine formation.

Gas-phase hydrolysis of the S-ethylcysteine-modified protein is a crucial step; in the worst case, oxygen will destroy all of the formed S-ethylcysteine. Therefore, extreme caution is essential to remove all oxygen from the hydrolysis vessel. The vessel is evacuated repeatedly and carefully flushed with argon (99.999%). This procedure is carried out successfully in a vessel as shown in Fig. 1. Parallel processing of more than 10 samples at a time is not recommended; dirt will accumulate in the vacuum station and in the oil of the vacuum pump, negatively affecting the yield of S-ethylcysteine.

Figure 2A demonstrates the phosphoserine determination in troponin I isolated in a phosphate-containing form[7] following performic acid oxidation. Figure 2B shows a control omitting the S-ethylcysteine modification. It is remarkable that the amount of phosphate determined by ashing and

[7] K. Swiderek, K. Jaquet, H. E. Meyer, and L. M. G. Heilmeyer, Jr., *Eur. J. Biochem.* **671,** 1 (1988).

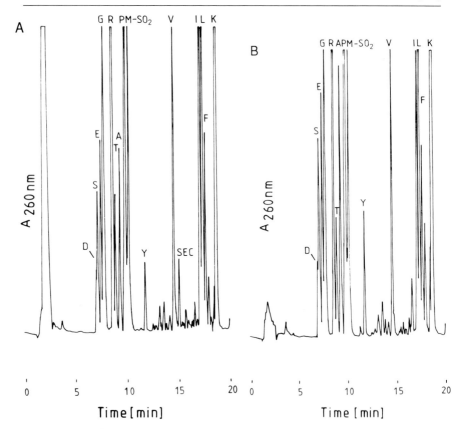

FIG. 2. Phosphoserine determination in cardiac troponin I. Cardiac troponin I (200 pmol) is peroxidized as described in the text. Gas-phase hydrolysis is carried out with (A) or without (B) prior *S*-ethylcysteine modification. Fifty percent of the S-ethylcysteine-modified sample (A) and of the unmodified sample (B) are analyzed by PTC-amino acid analysis as described in the text. Two moles of S-ethylcysteine is determined per mole of troponin I. Control (B) shows that no disturbing signal is present at the position where PTC-*S*-ethylcysteine elutes. Amino acids are designated with the single-letter code: A, alanine; D, aspartic acid; E, glutamic acid; F, phenylalanine; G, glycine; I, isoleucine; K, lysine; L, leucine; P, proline; R, arginine; S, serine; T, threonine; V, valine; Y, tyrosine. SEC, *S*-ethylcysteine; M-SO₂, methionine sulfone.

the amount of phosphoserine determined as *S*-ethylcysteine is identical within the standard error of the methods. This is true even without performic acid oxidation when the holoprotein is employed. Other examples are the α and the β subunits of phosphorylase kinase[8-10] and cardiac troponin. T.[11] However, the amount of *S*-ethylcysteine varies greatly with

[8] H. E. Meyer, G. F. Meyer, H. Dirks, and L. M. G. Heilmeyer, Jr., *Eur. J. Biochem.* **188**, 367 (1990).

and without performic acid oxidation when peptides derived by proteolytic digestion of the same proteins are employed, as will be shown in the following section.

Screening Procedure for Phosphoserine-Containing Peptides

Reagents

All reagents must be freshly prepared.

Performic acid (see above)
Buffer for digestion: Proteins or peptides are digested by endoproteinase Lys-C in 25 mM tris(hydroxymethyl)aminomethane hydrochloride, pH 8.6, containing 1.0 mM ethylenediaminetetraacetic acid and 0.025% sodium dodecyl sulfate (SDS). Alternatives are required according to the protease employed
Endoproteinase Lys-C: Five micrograms of sequence-grade endoproteinase Lys-C (Boehringer Mannheim, Germany) is solubilized in 100 μl H$_2$O. This solution can be employed for a maximum of 2 days if stored at 4°
Modification mixture (see above)

Procedure

1. Add 1 ml performic acid to the dried phosphoprotein (ca. 1 mg) in a 7 × 100 mm polypropylene test tube.
2. Incubate for 1 hr at 4° in a refrigerator.
3. Dry down in a SpeedVac (ca. 10 min, 0.2 mbar)
4. Dissolve the oxidized phosphoprotein in 2 ml digestion buffer. Add 5% acetonitrile to improve digestion with endoproteinase Lys-C.
5. Add 100 μl of the Lys-C solution. The ratio between the phosphoprotein and the protease is 1 : 200 (w/w). Incubate for 18 hr at 35° (optimize for protein and protease).
6. Separate the peptide mixture by reversed-phase high-performance liquid chromatography (HPLC). Collect peptide-containing fractions detected by UV absorbance at 214 nm.
7. Transfer an aliquot from each peptide-containing fraction (we take

[9] N. F. Zander, H. E. Meyer, E. Hoffmann-Posorske, J. W. Crabb, L. M. G. Heilmeyer, Jr., and M. W. Kilimann, *Proc. Natl. Acad. Sci. U.S.A.* **85,** 2929 (1988).
[10] M. W. Kilimann, N. F. Zander, C. C. Kuhn, J. W. Crabb, H. E. Meyer, and L. M. G. Heilmeyer, Jr., *Proc. Natl. Acad. Sci. U.S.A.* **85,** 9381 (1988).
[11] N. Beier, K. Jaquet, K. Schnackerz, and L. M. G. Heilmeyer, Jr., *Eur. J. Biochem.* **176,** 327 (1988).

usually between 50 and 100 μl containing approximately 50 pmol of peptide) to a 5 × 60 mm test tube (Pyrex glass). Dry down in the SpeedVac.

8. Take batches of 10 test tubes and add 20 μl of the modification mixture. Seal each tube with Parafilm and incubate at 50° for 1 hr. Add 5μl of acetic acid to each tube after cooling down to room temperature.

9. Dry down using a strong vacuum (less than 0.01 mbar) to achieve dry samples in less than 1 hr.

10. Hydrolyze the samples using the gas-hydrolysis procedure and perform PTC-amino acid analysis using reversed-phase HPLC as described below.

11. Purify S-ethylcysteine-containing peptides to homogeneity. Repeat procedure, beginning with step 6.

12. Take purified phosphopeptides for sequence analysis.

As an example, screening for phosphoserine is shown on peptides generated by endoproteinase Lys-C digestion of the β subunit of phosphorylase kinase. Figure 3 shows S-ethylcysteine determination in the separated peptides with (A) and without (B) performic acid oxidation. The dramatic reduction of S-ethylcysteine-forming peptides is due to the prevention of β elimination at cysteine following peroxidation to cysteic acid.

PTC-Amino Acid Analysis

Protein and peptide hydrolysis in the gas-phase is carried out with 6 N HCl employing a newly designed hydrolysis vessel (see Fig. 1). A tight seal prevents penetration of oxygen after evacuation following flushing with argon, which is crucial for the success of the method.

Equipment

Autosampler (SP 876; Spectra Physics) equipped with a thermostat for automatic injection: The samples can be held at 6° for more than 24 hr without any decrease in the amount of PTC-amino acids

Microbore HPLC system (AB1 30; Applied Biosystems, Foster City, CA) equipped with the standard PTC column (220 × 1.2 mm) from ABI: Detector is set at 254 nm and 0.01 absorption units full scale. The column is thermostated at 34°

Recording integrator (D-2000; Hitachi, Tokyo, Japan)

Reagents

HCl (6 N) for amino acid analysis
Recrystallized phenol
Argon (99.999%)

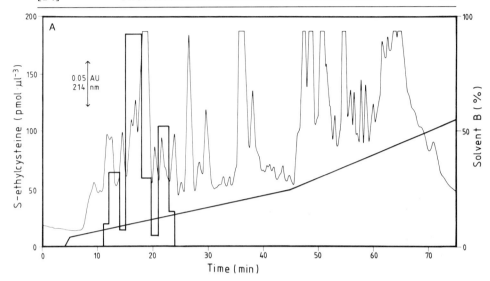

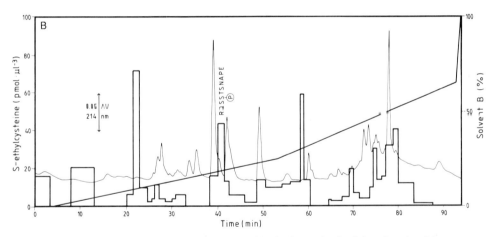

FIG. 3. (A and B) Screening for phosphopeptides in the β subunit of phosphorylase kinase. Isolated β subunit (10 nmol) is digested with endoproteinase Lys-C with (A) or without (B) prior performic oxidation. The digest is applied to a pH-stable reversed-phase HPLC column (4.6 × 250 mm; Vydac 228TP104). Separation is performed using a hexafluoroacetone/ ammonia/acetonitrile linear gradient at pH 8.6 as indicated; peptides are detected by their absorption at 214 nm (solid line). The open bars represent the amount of *S*-ethylcysteine found per 100 μl from each fraction. (C) Rechromatography of the main phosphopeptide-containing fraction shown in (A). After reduction of the volume under vacuum, all of the fraction eluting at 16 min in (A) is injected onto a Vydac C$_{18}$ column (4.6 × 150 mm, 218TP5415). Peptides are eluted with a trifluoroacetic acid/acetonitrile linear gradient as indicated by the broken line; they are detected by their absorption at 214 nm (solid line). Phosphopeptides identified are designated by K$_\beta$44/45 and K$_\beta$68/69, referring to the specific cleavage at lysine residues. The numbers identify the lysine residues in the whole sequence.[10]

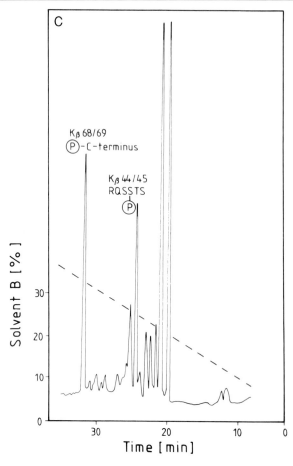

FIG. 3. (*continued*)

Conditioning solution 1: Ethanol, triethylamine, and water are mixed in a ratio of 2 : 2 : 1 (v/v)

Conditioning solution 2: Ethanol, triethylamine, and water are mixed in a ratio of 8 : 1 : 1 (v/v)

Coupling solution: Ethanol, triethylamine, water, and PITC are mixed in a ratio of 7 : 1 : 1 : 1 (v/v)

Sample buffer: 60 mM sodium acetate, pH 5, containing 1 mg/ml ethylenediaminetetraacetic acid

HPLC solvent A: 60 mM sodium acetate, pH 5.0

HPLC solvent B: 70% acetonitrile containing 30 m*M* sodium acetate, pH 4.6

The conditioning and coupling solutions must be freshly prepared.

Procedure

1. Put into the hydrolysis vessel up to 10 test tubes containing the modified or nonmodified peptides.
2. Add 400 µl of 6 *N* HCl and a crystal of phenyl beside the test tubes into the hydrolysis vessel.
3. Close the vessel, evacuate until the HCl starts boiling, then flush with argon. Repeat both steps two times. Seal the vessel under vacuum.
4. Carry out hydrolysis for 1 hr at 150°.
5. Transfer the test tubes into a new vessel and dry under high vacuum (less than 0.04 mbar).
6. Add 10 µl of conditioning solution 1 to each test tube and dry again under high vacuum (less than 0.04 mbar). Repeat this step.
7. Add 10 µl of conditioning solution 2 to each test tube. Flush with argon and incubate the mixture for 5 min at room temperature.
8. Add 10 µl of coupling solution to each test tube, flush with argon, and incubate the mixture for 5 min at 50°. Dry under high vacuum (less than 0.04 mbar).
9. Dissolve each sample in 30 µl of sample buffer. Transfer the solution into the sample vials of the autosampler and inject 20 µl of each sample into the ABI 130 for HPLC separation.

HPLC separation of the PTC-amino acids will be performed applying stepwise linear gradients starting with 6% solution B. In 10 min, solvent B rises to 30%, which is followed by a rise to 50% solvent B in the next 10 min. The flow rate is constant at 0.3 ml/min.

During performic acid oxidation tryptophan residues are destroyed. One of the PTC derivatives of the tryptophan oxidation products elutes at approximately the same time as PTC-*S*-ethylcysteine. Satisfactory separation is achieved by carefully adjusting the pH of solution A and optimizing the gradient. The pH of solution A is the most critical factor. Lowering pH shortens the elution time for *S*-ethylcysteine and is employed to optimize its separation from other components. However, the optimized pH must be evaluated with every new column; analysis under standard conditions must be repeated until optimal resolution is obtained. Changes in the steepness of the gradient and/or in the salt concentration of solution A have no marked effects. No more than 10 samples should be processed in

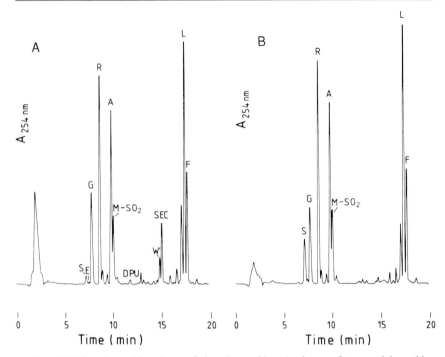

FIG. 4. PTC-amino acid analyses of phosphopeptides. A mixture of two model peptides [L-R-R-A-S(P)-L-G and L-W-M-R-F-A] containing 200 pmol of each is gas-phase hydrolyzed after performic acid oxidation (A and B). After S-ethylcysteine modification, the separation between the PtC derivatives of the oxidized tryptophan (W') and S-ethylcysteine is demonstrated in (A). As a control, the S-ethylcysteine modification is omitted and no disturbing signal is present at the position where PTC-S-ethylcysteine elutes (B). Amino acids are designated with the one-letter code. SEC, S-Ethylcysteine; M-SO$_2$, methionine sulfone. (C) The chromatogram of the PTC-amino acid analysis of 250 pmol oxidized and modified phosphopeptide K$_\beta$68/69 from the β subunit of phosphorylase kinase [see (C)].

1 batch to obtain high yield of S-ethylcysteine. Phosphokemptide (L–R–R–A–S(P)–L–G) is employed as external standard; the yield of S-ethylcysteine is taken to correct the S-ethylcysteine determination of unknown phosphopeptides or phosphoproteins. A tryptophan-containing peptide (L–W–M–R–F–A) is included in the phosphokemptide standard. It is employed to achieve an optimal separation between S-ethylcysteine and the tryptophan by-product which is demonstrated in Fig. 4A and B. Figure 4C shows as an example the chromatogram of PTC-amino acids obtained from the phosphopeptide K$_\beta$68/69 isolated from the β subunit of phosphorylase kinase identified by S-ethylcysteine modification (see Fig. 3C).

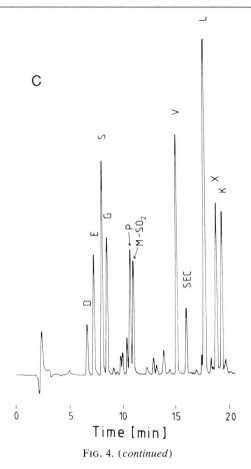

Fig. 4. (*continued*)

Sequence Analysis of Phosphopeptides

Generally, an unambiguous localization of phosphoserine in a sequence can be obtained by the *S*-ethylcysteine method. The most successful strategy is based on the following steps:

1. The phosphoprotein is digested following performic acid oxidation.
2. The resulting peptide mixture is separated and each fraction containing peptide(s) is analyzed for phosphoserine by the *S*-ethylcysteine method and amino acid analysis.
3. Fractions containing phosphoserine are tested for homogeneity, for example by capillary electrophoresis, or further separated in a second chromatographic step.

4. The phosphoserine residue(s) is then localized by sequence analysis with and without S-ethylcysteine modification.

Equipment

Gas-phase sequencer (model 470) online connected with a PTH analyzer (model 120), both from Applied Biosystems

Reagents

Modification mixture for sequence analysis: 80 μl ethanol, 65 μl of 5 N NaOH, 60 μl ethanethiol, 200 μl dimethyl sulfoxide, and 200 μl H₂O, which should be added as the last component

Sequence Analysis with S-Ethylcysteine Modification

Procedure

1. Add 50 μl of the modification mixture for sequence analysis to a test tube (5 × 50 mm) that contains an aliquot of the dried phosphopeptide.
2. Seal the test tube with Parafilm and incubate for 1 hr at 50°.
3. Cool to room temperature and add 10 μl acetic acid.
4. Apply the S-ethylcysteine-modified peptide as soon as possible onto the Polybrene-coated and precycled glass fiber disk. Dry thoroughly after each application in the cartridge chamber at elevated temperature.
5. Assemble the cartridge, dry the filter for a further 30 min with the argon dry function, and then start sequencing with the normal 03RPTH program.

Figure 5 gives an example for such a sequence analysis.

Sequence Analysis without S-Ethylcysteine Modification

During Edman degradation phosphoserine undergoes β elimination as soon as degradation reaches this phosphoamino acid, since the coupled PTC group increases the reactivity of the α hydrogen. As a result, even under the mild alkaline conditions in the gas-phase sequenator, β elimination of the phosphate group occurs after coupling of the PTC group. Thus, dehydroalanine is formed from phosphoserine, quantitatively. Following cleavage, this amino acid undergoes an addition reaction with dithiothreitol, which is present in the conversion reagent (25% trifluoroacetic acid), yielding a PTH-dithiothreitol adduct of dehydroalanine as the only end product. Equally important, no interruption of the Edman degradation process occurs.

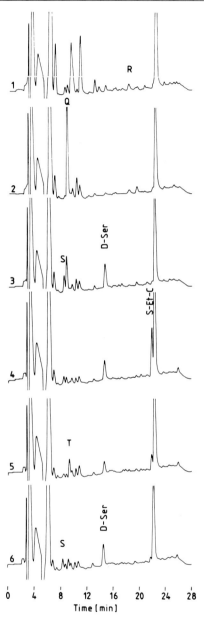

FIG. 5. Sequence analysis of the phosphopeptide $K_\beta 44/45$ after *S*-ethylcysteine modification. The oxidized and modified phosphopeptide $K_\beta 44/45$ (400 pmol) is analyzed using the gas-phase sequencer as described in Sequence Analysis with *S*-Ethylcysteine Modification. PTH-amino acid analyses of degradation steps 1 to 6 show that phosphoserine in cycle 4 is modified to *S*-ethylcysteine, whereas serines in cycles 3 and 6 are not modified. D-Ser, dithiothreitol adduct of dehydroalanine.

Procedure

Apply an aliquot of the phosphopeptide onto the pretreated glass fiber disk. Dry the sample, assemble the cartridge, and start sequencing as usual. Use the standard 03RPTH program.

Applying this procedure, phosphoserine can be detected during sequence analysis by the quantitative formation of the dithiothreitol adduct of dehydroalanine in the degradation step where phosphoserine was present formerly. The details about the chemical reactions which take place during Edman degradation are described in Meyer *et al.*[12]

An example for sequence analysis of a phosphoserine-containing peptide without modification is shown in Fig. 6A. Sequence analysis of the same peptide in the unphosphorylated form is shown in Fig. 6B. The chromatogram in cycle 2 of Fig. 6B shows the distribution of PTH-serine (ca. 25%) and PTH-dithiothreitol adduct of dehydroalanine (ca. 75%) that we get with the gas-phase sequenator if a serine residue is determined. However, in the case of a phosphoserine the PTH-dithiothreitol adduct of dehydroalanine is formed exclusively, as demonstrated in Fig. 6A.

Discussion

Phosphoserine Determination in Proteins and Peptides

The amount of phosphate determined by ashing of a protein known to contain exclusively phosphoserine agrees very well with the amount of phosphoserine determined by conversion to *S*-ethylcysteine. At present, there are only a limited number of proteins analyzed, but, according to our experience the *S*-ethylcysteine determination results systematically in a slightly higher phosphoserine content than the ashing procedure. Systematic deviations may arise from differences in the conversion rate found with the phosphokemptide standard and the conversion rate obtained with an unknown protein. Furthermore, the calculation of PTC-*S*-ethylcysteine is based on the calibration factor of PTC-methionine, which might not be the exact value.

Cysteine is the major source of error when small peptides are analyzed. This is not observed when the holoprotein is analyzed, at least with the proteins employed so far. However, as shown before this error can easily be circumvented by prior performic acid oxidation.

[12] H. E. Meyer, E. Hoffmann-Posorske, C. C. Kuhn, and L. M. G. Heilmeyer, Jr., *in* "Modern Methods in Protein Chemistry" (H. Tschesche, ed.), Vol. 3, p. 185. de Gruyter, Berlin and New York, 1988.

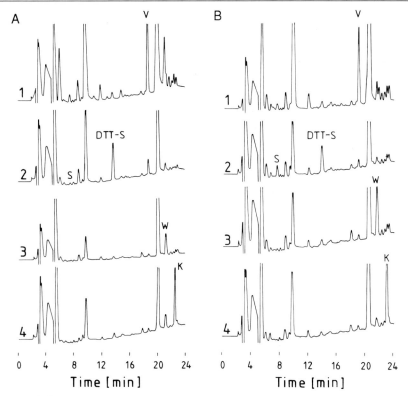

FIG. 6. Sequence analyses of the phospho- and dephosphopeptide $E_\beta 2$ without S ethyl cysteine modification. In (A), about 100 pmol of the phosphopeptide $E_\beta 2$ (E refers to glutamic acid residue number 2 of the whole β-subunit sequence) is applied onto the gas-phase sequenator and sequenced without S-ethylcysteine modification. The PTH chromatograms of the degradation steps 1 to 4 are shown. The chromatogram of cycle 2 demonstrates that the phosphoserine, present in this position, is quantitatively transformed to the dithiothreitol adduct of dehydroalanine (DTT-S). In (B), 60 pmol of the same peptide in the nonphosphorylated form is analyzed. A clear difference between the phosphoserine (A) and the serine in (B) can be seen. This demonstrates that phosphoserine can be located unambiguously in such a case without S-ethylcysteine modification in the gas-phase sequenator.

Sequence Analysis of Phosphoserine-Containing Peptides

In the spinning cup sequenator, Quadrol, a strongly basic buffer is employed for coupling. Thus, β elimination might occur immediately on base delivery before PITC coupling. Therefore, blockage of the degradation might occur due to rearrangement of dehydroalanine to pyruvate.[13]

[13] R. J. A. Grand, J. M. Wilkinson, and L. Mole, *Biochem. J.* **159,** 633 (1976).

When a ^{32}P-labeled phosphopeptide is degraded by the Edman procedure, often the released radioactivity is employed to localize phosphoserine. However, the released inorganic phosphate is strongly bound to the Polybrene support and, therefore, determination of inorganic [32P]phosphate in the elutate of the gas-phase sequencer is inadequate. Polybrene releases usually less than 1% of the radioactivity applied, which easily can lead to erroneous assignment of phosphoserine in the sequence. As an alternative, the glass fiber disk can be cut in several pieces of the same size and after each degradation step one piece is removed. Ultrasonic treatment in the presence of formic acid allows the extraction of [32P]phosphate quantitatively. Indeed, if the peptide is labeled with [32P]phosphate identification of inorganic phosphate in that cycle has been employed successfully to locate ^{32}P-labeled phosphoserine in the sequence.[14] This procedure is the method of choice when radioactively labeled peptides are available.

As a basis for a kind of diagonal technique, several authors[2,15] have suggested performing β elimination on the whole protein with subsequent addition of different nucleophiles. Digestion of the modified and unmodified protein followed by HPLC separation of the generated peptides should allow identification of phosphoserine-containing peptides not radioactively labeled. After β elimination of the phosphate group(s) and modification with a nucleophile the peptides formerly containing phosphoserine should elute quite differently from the nonmodified ones. In this way phosphopeptides should be identifiable. However, due to the intermediate formation of an optically inactive dehydroalanine during treatment with 0.5 M sodium hydroxide, racemization occurs at those positions where phosphoserine was located. The newly formed derivatives thereof are enantiomers. Therefore, peptides containing one phosphoserine residue will produce two diastereomers which will be separated by reversed-phase HPLC (two phosphoserine residues will produce four diastereomers, etc.). Furthermore, an appreciable amount of other amino acids will also be racemized during the sodium hydroxide treatment, which will result in a mixture of diastereomeric forms of each peptide. This makes this kind of diagonal technique impossible. The time course of phosphokemptide modification by alkali treatment demonstrates the formation of these products.[10] A similar effect was observed by Mega et al.[16] However, these authors were not aware of the racemization reaction. Additionally, modified proteins

[14] Y. Wang, C. J. Fiol, A. A. DePaoli-Roach, A. W. Bell, M. A. Hermodson, and P. J. Roach, Anal. Biochem. **174,** 537 (1988).
[15] T. G. Hastings and E. M. Reimann, FEBS Lett. **231,** 431 (1988).
[16] T. Mega, Y. Hamazume, Y.-M. Nong, and T. Ikenaka, J. Biochem. (Tokyo) **100,** 1109 (1986).

are incompletely digested by proteolytic enzymes since they need the optical integrity of the amino acids at which they split the polypeptide backbone. Separation of such a peptide mixture is an unresolvable task.

S-Ethylcysteine modification is very gentle if performed in the suggested manner. However, it does not work if the phosphoserine residue is the N-terminal or C-terminal amino acid. In both cases, *S*-ethylcysteine will not be formed; instead, blockage occurs if phosphoserine is the N-terminal amino acid; due to the rearrangement of the double bond in the dehydroalanine intermediate, pyruvate is formed. Likewise, ethylamine is formed from phosphoserine when it is located at the C-terminal position.

In some instances hydrolytic cleavage of the N-terminal amino acid occurs, depending on the individual peptide. Very seldom, hydrolytic cleavage of other sensitive peptides bonds is observed; it is recognized on sequence analysis without modification.

A low degree of tyrosine destruction results in a PTH by-product which elutes behind PTH-valine. Similarly, a low degree of lysine destruction forms a PTH by-product which elutes before PTH-alanine.

Peptides containing glycosylated serine residues also yield *S*-ethylcysteine when they are modified. In the gas-phase sequencer, using the non-modified peptide, deglycosylation occurs only partially. Therefore, the dithiothreitol adduct of dehydroalanine is obtained in a yield of 5–10% of that found normally with phosphoserine.

Identification of Phosphothreonine

Phosphothreonine cannot be modified like phosphoserine. Apparently, the double bond of the formed α-aminodehydrobutyric acid does not add the nucleophile ethanethiol. Likewise, catalytic dehydrogenation of the double bond is not useful, since the N terminus will be modified in such a manner that Edman degradation is no longer possible. Until now, there has been no way to identify phosphothreonine residues unambiguously. However, phosphothreonine forms by-products in the gas-phase sequencer that differ from those formed from threonine and this difference can be used to identify phosphothreonine. Phosphothreonine can be identified successfully in the N-terminal region of an 8-kDa protein isolated from *Chlamydomonas reinhardtii*.[17]

Acknowledgments

We thank Mr. H. Korte for the expert technical assistance and Mrs. Humuza for preparing the manuscript. This work was supported by the Deutsche Forschungsgemeinschaft and the Fonds der Chemischen Industrie.

[17] N. Dedner, H. E. Meyer, C. Ashton, and G. F. Wildner, *FEBS Lett.* **236**, 77 (1988).

[15] Solid-Phase Sequencing of ^{32}P-Labeled Phosphopeptides at Picomole and Subpicomole Levels

By RICHARD E. H. WETTENHALL, RUEDI H. AEBERSOLD, and LEROY E. HOOD

Introduction

Protein phosphorylation is usually restricted to the hydroxyamino acids, serine, threonine, and tyrosine, although phosphorylation of histidine can occur in some systems.[1,2] Specificity determinants for site-specific phosphorylation are usually provided by structural signals located within amino acid sequences adjacent to the phosphorylation site.[1] However, phosphorylation sites cannot be predicted with certainty from primary structure alone, hence the need for methodology to determine the sites experimentally. Edman degradation of phosphorylated serine and threonine residues generates unstable products which undergo β elimination of the phosphoryl group.[3] Indirect detection methods for the assignment of such phosphorylation sites are based on the detection of novel Edman degradation products generated from peptides in which phosphorylserine or phosphorylthreonine groups have been specifically modified; an example of this approach is the conversion of phosphoserine residues to the phenylthiocarbamyl-S-ethylcysteine derivative.[4] Such methods are particularly useful for the determination of *in vivo* phosphorylation sites,[5] but usually require quantities of phosphopeptides in excess of 20 pmol.

Assignment of phosphorylation sites can be based on the release of [32P]P$_i$ or other phosphorylated degradation products at the position of the phosphorylated residue during Edman degradation[6,7] or, alternatively, on the disappearance of phosphopeptides identified in Edman degradation

[1] A. M. Edelman, D. K. Blumenthal, and E. G. Krebs, *Annu. Rev. Biochem.* **56,** 567 (1987).
[2] N. Weigel, M. A. Kukuruzinska, A. Nakazawa, E. B. Waygood, and S. Roseman, *J. Biol. Chem.* **257,** 14477 (1982).
[3] J.-C. Mercier, F. Grosclaude, and B. Ribadeau-Dumas, *Eur. J. Biochem.* **23,** 41 (1971); R. Greenberg, M. L. Groves, and R. F. Peterson, *J. Dairy Sci.* **59,** 1016 (1976).
[4] H. E. Meyer, E. Hoffman-Posorske, H. Korte, and L. M. G. Heilmeyer, *FEBS Lett.* **204,** 61 (1986).
[5] K. Swiderek, K. Jaquet, H. E. Meyer, and L. M. G. Heilmeyer, Jr., *Eur. J. Biochem.* **176,** 335 (1988).
[6] D. Rylatt and P. Cohen, *FEBS Lett.* **98,** 71 (1979).
[7] R. E. H. Wettenhall and P. Cohen, *FEBS Lett.* **140,** 263 (1982).

reaction mixtures by electrophoresis[8,9] or high-performance liquid chromatography (HPLC)[10] analyses. However, with standard automated sequencing protocols in which peptides are immobilized within a matrix of the polycationic compound Polybrene, the ionic interaction between the phosphorylated Edman degradation products, including P_i, and Polybrene prevents efficient extraction of these products from the reaction chamber.[7,11] To overcome this problem, we have developed a method for the identification of ^{32}P-labeled phosphorylation sites during automated sequencing employing covalent attachment of phosphopeptides to glass fiber disks covalently modified with aminophenyl groups. The covalent attachment procedure is based on the method originally developed for the covalent attachment of peptides to aminopolystyrene beads[12,13] that we have modified for use in high-sensitivity gas/liquid solid-phase sequenators with aminophenyl-coated glass disks.[14]

Preparation of Aminophenyl-Glass Fiber Paper

Reagents

Circular (11-cm diameter) Whatman GF/F glass filter papers (Whatman, Ltd., Maidstone, England)
Trifluoroacetic acid (TFA), Sequenal grade (Pierce Chemical Company, Rockford, IL)
Aminophenyltriethoxysilane (mixed isomers; Petrach, Bristol, PA)/acetone/H_2O (2 : 50 : 48, v/v/v); freshly prepared immediately prior to use
Acetone, analytical grade
Double glass-distilled H_2O (required for preparation of all aqueous reagents used in this and all subsequent sections)

[8] J. Y. Henderson, A. J. G. Moir, L. A. Fothergill, and J. E. Fothergill, *Eur. J. Biochem.* **114**, 439 (1981).
[9] A. Aitken, T. Bilham, P. Cohen, D. Aswad, and P. Greengard, *J. Biol. Chem.* **256**, 3501 (1981).
[10] Y. Wang, C. J. Fiol, A. A. DePaoli-Roach, A. W. Bell, M. A. Hermodson, and P. J. Roach, *Anal. Biochem.* **174**, 537 (1988).
[11] W. D. Annan, W. Manson, and J. A. Nimmo, *Anal. Biochem.* **121**, 62 (1982).
[12] J. J. L'Italien and R. A. Laursen, *J. Biol. Chem.* **256**, 8992 (1981).
[13] J. J. L'Italien and J. E. Strickler, *Anal. Biochem.* **127**, 198 (1982).
[14] R. Aebersold, N. Nika, G. D. Pipes, R. E. H. Wettenhall, S. M. Clarke, L. E. Hood, and S. B. H. Kent, *in* "Methods in Protein Sequence Analysis" (B. Wittmann-Liebold, ed.), p. 79. Springer-Verlag, Berlin, 1989; R. Aebersold, G. D. Pipes, R. E. H. Wettenhall, H. Nika, and L. E. Hood, *Anal. Biochem.* **187**, 56 (1990).

Procedure

Acid Etching. Acid etching of the glass filters is required to expose more silanol groups and, hence, to enhance the reactivity of the glass toward the silylating reagent aminophenyltriethoxysilane.[14] Whatman GF/F filters are incubated at room temperature in trifluoroacetic acid (TFA) for 1 hr in a fume hood.[15] To ensure thorough etching, the filter container is gently agitated using an orbital shaker. After 1 hr, the filters are dried on Whatman #3 paper in a fume hood.[15] *Caution:* Fume hood containment and the use of Neoprene full-length gloves and apron and face shield must be used for all steps involving concentrated TFA because of its extremely corrosive nature.

Derivatization of Glass Filters Based on Method of Aebersold et al.[14] One or two acid-etched GF/F filters are sealed in a plastic bag containing 50 ml of the 2% aminophenyltriethoxysilane solution. Particular care must be taken to eliminate air bubbles from the bag prior to sealing. The sealed bags are incubated for 16 hr at 37° with gentle movement on a rocking platform. After the derivatization reaction, the filters are washed for 5 min in 200 ml of acetone on an orbital shaker. Washing is repeated 10 times. The washed sheets are placed on Whatman #1 paper and dried and the silane linkages cured in an oven for 45 min at 110°. The aminophenyl glass filters, having a pale purple–brown appearance and an amino group loading in the range of 12–24 nmol/mg of glass,[14] can be stored in covered glass containers at room temperature for periods in excess of 6 months, without any detectable loss of peptide-coupling ability.

Covalent Coupling of Phosphopeptides to Aminophenyl-Derivatized Glass Disks

Reagents

Aminophenyl glass (AϕG) disks (see above)

EDC-coupling buffer: *N*-Ethyl-*N'*-(3-dimethylaminopropyl)carbodiimide (EDC), 5 mg/ml in 160 m*M* 2-(4-morpholino)ethanesulfonic acid (MES) buffer, pH 4.0, containing 0.4 *M* guanidinium hydrochloride. This buffer must be freshly prepared just prior to use

Synthetic peptide 1, NH$_2$-Val-Gln-Ala-Ala-Ile-Asp-Tyr-Ile-Asn-Gly-COOH,[14] and peptide 2, NH$_2$-Ala-Lys-Arg-Arg-Arg-Leu-Ser-Ser-

[15] R. Aebersold, G. D. Pipes, H. Nika, L. E. Hood, and S. B. H. Kent, *Biochemistry* **27**, 6860 (1988).

Leu-Arg-Ala(NH$_2$) (ribosomal protein S6.229–239 analog),[16,17] are chemically synthesized using modified reaction protocols in an ABI 430A peptide synthesizer (Applied Biosystems, Foster City, CA) and the HF-cleaved peptide purified by reversed-phase HPLC (Vydac C$_4$ column; Separations Group, Hesperia, CA) using a linear gradient of acetonitrile in aqueous TFA (0.1%, v/v) to elute the peptides.[18] Peptide 3, NH$_2$-Ile-Glu-Gly-Arg-Ile-Leu-Leu-Ser-Glu-Leu-Ser-Arg-Arg(NH$_2$), a synthetic analog of the rabbit protein synthesis initiation factor eIF-2α41–52, is prepared as described previously[19]

Tryptic phosphopeptides of known structures used for the development of coupling and subsequent sequencing protocols are generated by trypsin digestion of the S6.229–239 analog[17] or the parent rat liver 40S ribosomal protein S6 phosphorylated with the purified catalytic subunit of beef heart cAMP-dependent protein kinase[16,20]; the tryptic derivative of the eIF-2α41–52 synthetic analog[19] phosphorylated with eIF-2α kinase activity prepared from rabbit reticulocytes.[21] The specific radioactivity of the [γ-^{32}P]ATP used in phosphorylation reactions is in the range 1–200 × 10^3 dpm/pmol. The individual tryptic peptides are purified by reversed-phase HPLC as described previously.[17,22]

Procedure

The coupling method is based on the previously described procedure for the coupling of peptides through their α-carboxyl groups to the NH$_2$ groups of aminopolystyrene beads[12,13] and, more recently, modified by Aebersold et al. for coupling to aminophenyl-derivatized glass fiber disks.[14]

The glass filters are cut into circular disks 1 cm in diameter. Phosphopeptide samples for application to disks should contain > 1 fmol of peptide and > 1000 cpm ^{32}P Cerenkov radiation per phosphorylation site. De-

[16] R. E. H. Wettenhall and F. J. Morgan, *J. Biol. Chem.* **259,** 2084 (1984); R. E. H. Wettenhall, H. P. Nick, and T. Lithgow, *Biochemistry* **27,** 170 (1988).

[17] C. House, R. E. H. Wettenhall, and B. E. Kemp, *J. Biol. Chem.* **262,** 772 (1987).

[18] S. B. H. Kent, *Annu. Rev. Biochem.* **57,** 957 (1988).

[19] W. Kudlicki, R. E. H. Wettenhall, B. E. Kemp, R. Szyszka, G. Kramer, and B. Hardesty, *FEBS Lett.* **215,** 16 (1987).

[20] H. P. Nick, R. E. H. Wettenhall, F. J. Morgan, and M. T. W. Hearn, *Anal. Biochem.* **148,** 93 (1985).

[21] G. Kramer, J. M. Cimadevilla, and B. Hardesty, *Proc. Natl. Acad. Sci. U.S.A.* **73,** 3078 (1976).

[22] R. E. H. Wettenhall, W. Kudilicki, G. Kramer, and B. Hardesty, *J. Biol. Chem.* **261,** 12444 (1986).

pending on the properties of the peptide, the lyophilized material can be dissolved in acetonitrile/H_2O (30 : 70, v/v), acetonitrile/TFA/H_2O (30 : 20 : 50, v/v/v), or acetonitrile/H_2O (30 : 70, v/v) containing 0.1% (w/v) trimethylamine; post-HPLC fractions in aqueous acetonitrile TFA (0.1%) buffer can be applied directly to the disk; in addition, samples can be applied to the disks in dilute sodium dodecyl sulfate (SDS) or guanidine hydrochloride-containing buffers.[14]

Peptide solutions are spotted onto the glass disks and dried with a warm stream of air from a hair dryer. In the case of subpicomole peptide samples, a carrier peptide is usually included in the peptide solution to serve as an internal sequencing standard and to aid in the quenching of side reactions that might compromise the NH_2 terminus of the phosphopeptide during the coupling and Edman degradation reactions. We usually use 200 pmol of the synthetic peptide 1[14] for this purpose. Where multiple applications of dilute peptide solutions (e.g., post-HPLC fractions) are required, care must be taken to ensure even distribution of the peptide on the disk by drying the filters between applications and to avoid washing of the previously applied peptide to the edges of the disks.

The AϕG-coupling reaction is initiated by applying 45 μl of the EDC-coupling buffer to the dried disk, transferring the disk to a glass scintillation vial, capping the vial, and incubating it at 37° for 1 hr. After incubation, the total radioactivity applied to the disk can be determined by counting of Cerenkov radiation in a liquid scintillation counter. The disks are then rinsed in 200 ml of double-distilled H_2O for 1 min and the ^{32}P-labeled peptide remaining bound to the disks quantitated by Cerenkov counting. Typically, the coupling efficiencies of tryptic phosphopeptides by this measurement are 40–80% of the phosphopeptide applied to the disk, with peptides ending (COOH terminal) in arginine usually coupling more efficiently than those ending in lysine.[14] Studies by L'Italien and Strickler[13] and Aebersold et al.,[14] using comparable coupling reaction conditions, have shown that peptides ending with neutral (e.g., glycine) or acidic amino acids (glutamic or aspartate) are also generally more efficiently coupled (yields > 80%). Thus, the method is suitable for the analyses of phosphopeptides generated with other proteases such as staphylococcal V8 protease, chymotrypsin, elastase, thermolysin, or subtilisin.

A portion of the phosphopeptide bound to the disk (up to 30% depending on the batch of AϕG paper) may be coupled to residual free aminophenyltriethoxysilane. The majority of such material is extracted with the methanol (S2) and methanol/H_2O (90 : 10, v/v) (S3) washes during the "begin" cycle (Table I). Any remaining noncovalently bound material is removed during the early sequencing cycles and is usually particularly evident at the first sequencing cycle unless extended S2 and/or S3 washes

TABLE I
PHOSPHOPEPTIDE SEQUENCING CYCLES[a]

Step	Normal cycle: reaction cartridge steps	Time (min)		Step	"Begin" cycle: reaction cartridge steps	Time (min)	
N1	R3 deliver	14.0	(2.0)	B1	Argon dry	5.0	
N2	S3 deliver (extraction)	1.5		B2	R3 deliver	7.0	(1.3)
N3	Argon transfer/collect	2.0		B3	S3 deliver (extraction)	1.5	
N4	S2 deliver (wash)	1.0	(1.0)	B4	Argon transfer/collect	2.0	
N5	R3 deliver	5.0	(2.0)	B5	S2 deliver	2.0	(2.0)
N6	S3 deliver (extraction)	1.5		B6	R2 deliver	2.0	(0.5)
N7	Argon transfer/collect	0.5		B7	R1 deliver	0.15	(1.0)
N8	S3 deliver (wash)	2.0	(0.2)	B8	R2 deliver	10.0	(2.0)
N9	S2 deliver (wash)	2.0	(2.0)	B9	S1 deliver	4.0	(1.0)
N10	R2 deliver	2.0	(0.5)	B10	S2 deliver	2.0	(2)
N11	R1 deliver	0.15	(0.9)	B11–B20[b]			
N12	R2 deliver	8.0	(0.5)				
N13	R1 deliver	0.1	(0.9)				
N14	R2 deliver	8.0	(0.5)				
N15	S1 deliver (wash)	2.0	(0.5)				
N16	R1 deliver	0.15	(0.9)				
N17	R2 deliver	6.0	(1.0)				
N18	S1 deliver (wash)	5.0	(0.5)				
N19	S2 deliver (wash)	2.0	(2.0)				

[a] The sequencing protocol is a modification of that originally developed for the reaction cartridge functions of gas/liquid-phase sequenators (Edman degradation)[23] using a no-vacuum program. Sequencing commences with the "begin" cycle (steps B1–B20) followed by the required number of normal cycles (steps N1–N19). Only the principal steps are numbered with the times for the following argon dry steps indicated in parentheses. Individual solvents (S1–S3) and reagents (R1–R3) are as defined in the text. Solvents are delivered to waste where wash is indicated, and directly to the fraction collector where extraction and argon transfer/collect functions are indicated. In the cases of steps N2 and N6, S3 is delivered in pulses of 0.5, 0.25, 0.25, and 0.5 min with 0.3-min pauses between S3 pulses to ensure efficient extraction of any phosphorylated Edman degradation products in the minimum volume (in the Caltech instruments the fraction collector allows a collection volume of ca. 0.3 ml; normally, separate 0.3-ml fractions are collected at step N2 + N3 and N5 + N6). Where required, conversions are carried out in methanolic HCl (1 M) for 15 min at 50°, following lyophilization of sequencer fractions in a Savant vacuum concentrator. Following conversion, the samples are redried and analyzed for PTH-amino acids as described in the text.

[b] Steps B11–20 are equivalent to steps N10–N19 in the normal cycle.

TABLE II
SEQUENCE ANALYSIS OF TRYPTIC PHOSPHOPEPTIDE DERIVATIVES OF
RIBOSOMAL S6 SEQUENCE S6.233–238[a]

| | | Recovery of ^{32}P radioactivity | | |
| | | | Diphospho S6.233–238 (cpm $\times$ 10^{-2}) | |
Cycle No.	Amino acid identified	Monophospho S6.233–238 (cpm $\times$ 10^{-3})	Normal	Extended dry
1	R	70.8 (68.2)	39.7 (40.2)	23.3 (5.5)
2	L	27.6 (40.4)	12.1 (16.0)	10.3 (5.8)
3	S	559.2 (114.5)	42.0 (18.1)	30.9 (8.4)
4	S	61.5 (50.5)	39.8 (18.4)	34.6 (9.2)
5	L	35.2 (34.5)	21.3 (16.8)	24.4 (6.5)
6	R	12.9 (17.3)	12.4 (14.1)	16.6 (4.8)
7	—	8.9 (11.0)	8.3 (9.0)	11.6 (4.0)

[a] The previously defined S6.233–238 mono- and diphosphopeptides were prepared as tryptic derivatives of the ribosomal protein S6 analog, S6.229–239 (referred to as peptide 2 in text), phosphorylated with the catalytic subunit of beef heart cAMP-dependent protein kinase.[17,20] The sequence analyses were as described in Table I except that, in the monophosphopeptide analysis (sample size 1.6 $\times$ 10^6 cpm), the initial methanol wash (step B5) was extended for an additional 5 min to reduce the level of noncovalently bound peptide and, in the extended dry analysis of the diphosphopeptide (38,000 cpm), the times for step N4 and the following argon dry step were both increased to 2 min. The ^{32}P radioactivity recovered after the primary cleavage step (N1) is given, with the values in parentheses referring to the radioactivity recovered after the second cleavage step (N5). The bursts of counts showing phosphorylation sites are underlined. After 10 cycles, the residual ^{32}P radioactivities remaining on the disks were 114,000, 9000, and 9810 cpm for the monophospho- diphospho- (normal dry), and diphospho- (extended dry) peptides, respectively. Individual amino acides are denoted by the single-letter code (R, Arg; L, Leu; S, Ser); the individual amino acid residues were determined by HPLC analysis for the monophosphorylated derivative only.

are employed during the begin cycle (Table II; compare monophospho- and diphosphopeptide analyses). The washed phosphopeptide-coupled disks can be stored in capped vials at −20° for at least 1 month without loss of sequence performance.

Automated Gas/Liquid-Phase Edman Degradation and Product Analysis

Reagents

Reagents are obtained from Applied Biosystems (Foster City, CA) or prepared in house at the California Institute of Technology (Caltech).

Phenyl isothiocyanate (5%, w/v, in heptane) (R1)
Aqueous trimethylamine (12.5%, w/v) (R2)
Trifluoroacetic acid (R3)
Methanolic HCl (R4) (1 M HCl in anhydrous methanol)
Ethyl acetate/heptane (50 : 50, v/v) (S1)
Methanol (S2)
Methanol/H_2O (90 : 10, v/v) containing 2 mM sodium phosphate buffer, pH 6.0 (S3)
Acetonitrile containing 10 mg dithiothreitol (Calbiochem, San Diego, CA) per 200 ml (S4)

Sequenator Configuration

The procedure for automated Edman degradation of phosphopeptides was developed and the data presented in this communication were obtained primarily by using a modified Caltech gas/liquid-phase sequenator.[23] The general principle of operation of these instruments is similar to commercially available gas/liquid-phase instruments (e.g., Applied Biosystems 470A or 477A sequenators). The major modifications specifically required for the phosphopeptide sequencing protocol are adjustments to the plumbing and controller program of the instrument to allow the required solvent delivery protocol and the collection of the anilinothiazolinone amino acid derivatives (ATZ derivatives) and any [^{32}P]P$_i$ directly from the valve block outlet of the reaction cartridge. The ATZ program and the availability of solvent delivery options in the ABI 477A sequenators allow these instruments to be easily modified for this purpose. In the ABI 470A and 477A sequenators, it is desirable to use an external fraction collector to allow sufficiently large volumes of S3 (>0.5 ml) for efficient extraction and transfer of phosphorylated Edman degradation products to the fraction collector tubes. Argon gas, flow rates, pressures (similar to those used in standard gas-phase sequenators[23]), and solvent delivery times must be adjusted for individual instruments to achieve efficient transfers.

Sequencing Procedure

The phosphopeptide sequencing protocol, which incorporated the optimal conditions for the recovery of [^{32}P]P$_i$ and other phosphorylated Edman degradation products from the sequenator, is summarized in Table I. In the case of phosphorylserine and phosphorylthreonine residues, the ^{32}P

[23] R. M. Hewick, M. W. Hunkapiller, L. E. Hood, and W. J. Dreyer, *J. Biol. Chem.* **256**, 7990 (1981); S. B. H. Kent, L. E. Hood, R. Aebersold, D. Teplow, L. Smith, V. Farnsworth, P. Cartier, W. Hines, P. Hughes, and C. Dodd, *BioTechniques* **5**, 314 (1987).

radioactivity is recovered mainly as inorganic phosphate as a result of the β elimination reaction.[3] Measurement of the ^{32}P release at each step in the sequencer cycle at positions of phosphorylated amino acid residues indicates that the radioactivity remains predominantly covalently bound to the AϕG disk until the cleavage step.[24]

Cleavage. Separate S3 transfers in methanol/H$_2$O (9 : 1, v/v) containing 2 mM sodium phosphate, pH 7.0, into fraction collector tubes are carried out after each of the two acid (R3) cleavage steps to ensure effective extraction of phospho derivatives from the cartridge (Table I). The conditions were optimized with [^{32}P]P$_i$ directly spotted on to an AϕG disk and subjected to normal sequenator operation to give overall recoveries of about 80%.[24] The second cleavage step is important for completing the cleavage process, usually with an additional 10 to 15% of the phospho cleavage products being recovered (Table II). A methanol (S2) wash followed by an argon-dry step are introduced after each S3 transfer to ensure drying of the disk. The conditions for drying must be carefully optimized: if insufficient, appreciable hydrolysis of the peptide-bound phosphoryl group occurs during the second cleavage phage, whereas extended drying can result in lower recoveries of ^{32}P and a more pronounced lag (see comparison of analyses of identical AϕG disks coupled with diphosphorylated S6.233–238 in Table II).

Coupling. The most efficient coupling, particularly at positions of phosphorylated residues, is achieved with a triple-coupling protocol (Table I), although effective sequencing of short phosphopeptides can be achieved with the normal double-coupling protocol. In most instruments, due to limitations in the number of solvent reservoirs and program options, it is necessary to combine the ethyl acetate and heptane solvents (referred to as S2 and S1, respectively, in standard Applied Biosystems 470A and 477A protocols); the new solvent is referred to as S1 in the phosphopeptide sequencing protocol (Table I). Thus, the disk is washed with a mixture of ethyl acetate and heptane (S1, 50 : 50, v/v) after the coupling steps instead of the separate ethyl acetate and heptane washes normally employed. These are followed by a methanol wash (S2) to facilitate drying prior to the cleavage step.

Conversion. The ATZ-amino acid derivatives are collected directly from the cartridge valve block and, where necessary, manually converted to the corresponding phenylthiohydantoin (PTH)-amino acids by heating in methanolic HCl (R4) at 50° for 15 min.

PTH-Amino Acid and ^{32}P Radioactivity Analyses. Following conversion, the residues are dried [Savant (Hicksville, NY) vacuum concentra-

[24] R. E. H. Wettenhall, R. Aebersold, T. Hunter, and L. E. Hood, in preparation.

tor], the tubes placed in scintillation vials for direct quantitation of Cerenkov ^{32}P radiation in a liquid scintillation counter, and subsequently
processed for PTH-amino acid analysis by reversed-phase HPLC. The
PTH-amino acid data presented in this communication (Tables III and IV)
were obtained using an IBM Cyano column (IBM Instruments, Walling-
ford, CT) eluted with aqueous acetonitrile/methanol/sodium acetate
buffer, as described previously.[25]

Applications and Critical Comment

The sequence analyses of the tryptic phosphopeptides relating to ribo-
somal protein S6 generated bursts of ^{32}P radioactivity at the positions
of previously defined phosphorylation sites for the peptides Arg-Leu-
Ser(PO$_4$)-Ser-Leu-Arg (Table II), Arg-Leu-Ser(PO$_4$)-Ser(PO$_4$)-Leu-Arg
(Table II), and Arg-Leu-Ser(PO$_4$)-Ser(PO$_4$)-Leu-Arg-Ala-Ser(PO$_4$)-Thr-
Ser-Lys (Table III). The analyses of samples of 50, 2, and 150 pmol of the
monophospho- (Table II), diphospho- (Table II), and triphospho- (Table
III) peptide, respectively, are described. Samples containing as little as 1
fmol of peptide and 1000 dpm of ^{32}P radioactivity yield bursts of ^{32}P
radioactivity at the site of phosphorylation, but the recoveries of radioactivity are relatively low compared with samples of >2 pmol; improved
recoveries are obtained in subpicomole analyses with the inclusion of
carrier peptide (we use 200 pmol of peptide 1) in the coupling reaction.

The effectiveness of the method is critically dependent on the use of
aqueous methanol as S3 for improving the recovery of P$_i$ and any other
phosphorylated Edman degradation products. The inclusion of sodium
phosphate buffer (2 mM, pH 7) in S3 and the bypassing of the conversion
flask are also important modifications for reducing the losses due to the
absorption of ^{32}P-labeled products to internal surfaces within the sequencer.[24] For peptides having uncomplicated structures and single phosphorylation sites close to their NH$_2$ termini, the ^{32}P released at the phosphorylation site can be up to 30% of the radioactivity in the original sample,
with <15% of the phosphorylation site signal carried over at the next cycle
(for example, the monophosphopeptide in Table II). Appreciably greater
sequencing lags are apparent when sequencing through multiple sites,
reflecting inefficient Edman degradation. An example of this is the pronounced lag observed at cycle 5 following the adjacent phosphorylserines
in the analyses of the diphospho- (Table II) and triphosphopeptides (Table
III); despite this lag, a burst of counts corresponding to the position of the

[25] M. W. Hunkapiller and L. E. Hood, this series, Vol. 91, p. 486.

TABLE III
SEQUENCE ANALYSIS OF TRYPTIC TRIPHOSPHOPEPTIDE ISOLATED FROM
RIBOSOMAL PROTEIN S6[a]

Cycle No.	Edman degradation products[b]		
	Amino acid identified	Yield of PTH-amino acid (pmol)	^{32}P recovered (cpm)
1	R	68 (4)	280
2	L	75 (5)	70
3	S′	20 (22)	570
4	S′	22 (11)	727
5	L	36 (14)	410
6	R	31 (11)	171
7	A	36 (15)	88
8	–	—	420
9	T	9 (5)	199
10	S	6 (4)	147
11	K	8 (6)	84

[a] Tryptic fragment of ribosomal protein S6 was isolated from HPLC-purified S6 originating from rat liver 40S ribosomal subunits phosphorylated *in vitro* with the catalytic subunit of beef heart cAMP-dependent protein kinase, as described.[20] The tryptic peptide corresponded to the previously characterized triphosphopeptide Arg-Leu-Ser(PO$_4$)-Ser(PO$_4$)-Leu-Arg-Ala-Ser(PO$_4$)-Thr-Ser-Lys (see text).[16]

[b] Yields of PTH-amino acids were determined by quantitative HPLC analysis using standard PTH-amino acids (Applied Biosystems). Values are given for the recoveries of amino acids at each cycle (collected at steps N3 + N4); the value for the corresponding amino acid at the subsequent cycle is provided in parentheses as an indication of the sequencing lag. In the case of serine residues, PTH-serine adducts (denoted as S′), but not authentic PTH-serine, were identified at positions 3 and 4 as would be expected for phosphorylated serine residues. Serine-related PTH signals were not quantitated at cycle 8, the position of the third phosphoryl serine, because of insufficient material. Individual amino acids are denoted by the single-letter code (A, Ala; K, Lys; T, Thr). ^{32}P radioactivity released at each cycle was quantitated by counting of Cerenkov radiation of sequencer fractions prior to HPLC analysis. Based on the ^{32}P radioactivity (12,880 cpm) of known specific activity, ca. 150 pmol of the triphosphopeptide was coupled to the AφG disk; of the radioactivity, 3035 cpm was extracted at the begin cycle S3 step (B5, Table I) and 2514 cpm remained bound to the disk after normal sequencer cycles. The bursts of counts showing phosphorylation sites are underlined.

third phosphorylation site was clearly apparent in the latter case in cycle 8 (Table III).

The recovery of only 30% of the ^{32}P originating from phosphorylation sites reflects a combination of losses due to partial NH_2 blocking of the peptide, hydrolysis of the phosphoryl and/or silanyl bonds during sequencing, and immobilization of the phosphoryl groups on the AϕG disks. Additional factors are peptide insolubility or nonideal interaction of the peptide with the glass disk that interferes with Edman degradation. The extent of interference appears to be influenced by the solvent used to apply the peptide to the disk. This is apparent in analyses of peptides requiring high concentrations of TFA for dissolving the lyophilized material, where the residual ^{32}P radioactivity remaining on the disk several cycles after sequencing through phosphorylation sites can be 50% or more of the starting material. The possibility of partial coupling of phosphoryl groups directly to the AϕG disks is suggested by the observation that about 35% of the phosphorylated S6.229–239 (C-terminal amide) peptide remained bound to the disk following the coupling reaction and extensive washing in water and 100% TFA, although some of this may have been due to strong noncovalent interactions with the glass disk.

The presence of guanidinium hydrochloride in the coupling buffer improves initial sequencing yields and overall recoveries of phosphate at phosphorylation sites with some peptides. The beneficial effects of this modification are illustrated with the peptide Ile-Leu-Leu-Ser-Glu-Leu-Ser(PO$_4$)-Arg[19] (Table IV), which is difficult to keep in solution in mildly acidic conditions,[22] such as those used in the coupling reaction.

Quantitative PTH-amino acid analysis has shown initial yields of about 30% of the starting material in analyses of various phosphopeptides in the low picomole range. For example, in the analysis of the S6 triphosphopeptide a yield of 68 pmol was obtained from the 150 pmol of starting material (Table III). The results of quantitative PTH-amino acid analyses also have indicated that the general performance of the sequencing protocol (Table I) with a variety of phosphopeptides (e.g., Table III), and other peptides[14,24] is comparable with conventional gas/liquid-phase sequencing protocols.[23]

Phosphopeptide sequencing applications are effectively limited to the analyses of phosphorylation sites within 25 residues of the NH_2 termini of phosphopeptides. This is mainly because of the gradual decline in repetitive yields in long sequencing runs due to the hydrolysis of the silanyl and phosphoryl linkages.[14,24] There is also the partial failure of Edman degradation after sequencing through phosphorylation sites,[3,11] as illustrated by the twofold reduction in yield of PTH-amino acids following cycles 3, 4, and 8 in the analyses of the S6 triphosphopeptide derivative (Table III).

TABLE IV

SEQUENCE ANALYSIS OF TRYPTIC FRAGMENT OF PHOSPHORYLATED SYNTHETIC ANALOG
OF eIF-2α SUBUNIT: EFFECT OF GUANIDINIUM HYDROCHLORIDE[a]

Cycle No.	Amino acid identified	Yield of PTH-amino acid (pmol)			^{32}P recovered (cpm)
		Guanidinium hydrochloride concentration (M)			
		0	0.4	2.0	0.4
1	I	17	32	34	5110
2	L	12	30	33	1607
3	L	16	27	31	601
4	(S)[b]	—	—	—	549
5	E	—	3	2	4174
6	L	3	13	16	2141
7	(S)[b]	—	—	—	4663
8	R	—	8	7	2200
9	—	—	—	—	966

[a] Tryptic fragment of the synthetic peptide Ile-Glu-Gly-Arg-Ile-Leu-Leu-Ser-Glu-Leu-Ser-Arg[19] was isolated by reversed-phase HPLC[22] and coupled to AϕG disks in the presence of various concentrations of guanidinium hydrochloride. Details of the sequence analysis are as described in Tables I and II. E, glutamic acid.

[b] PTH-serine adducts were identified at cycles 4 and 7 but the levels were too low for accurate quantitation.

The hydrolysis and subsequent release of aminophenyl groups are sometimes indicated by a trace of brownish material in dried sequencer fractions. The background hydrolysis of phosphoryl and, possibly, aminophenyl groups is contributed to by incomplete drying after the first cleavage step (N1, Table I). Hydrolysis of phosphoryl groups during the second cleavage step, as indicated by the radioactivity released, can be reduced by extending the drying steps (N3 and N4, Table I) preceding the second cleavage step (Table II). However, even under these stringent drying conditions background hydrolysis (up to 5% of the peptide-bound phosphate per cycle) can occur and a more pronounced lag is apparent (illustrated with the diphosphopeptide analyses in Table II).

Apart from the ability to recover phosphorylated derivatives, particular advantages of this solid-phase sequencing method are the low backgrounds in the HPLC analyses of PTH-amino acid and the fact that the COOH-terminal residue can be identified.[14] Another advantage is the improved recoveries of basic amino acid derivatives as well as authentic PTH-serine and PTH-threonine residues. Disadvantages include the partial coupling of internal acidic residues (see below) and the relatively low recoveries of amide (50 to 60% hydrolysis to the parent acidic residues) and tyrosine residues compared with those we obtain with Polybrene-based supports.

We usually observe appreciable coupling of internal carboxyl groups (50–90%), despite the use of coupling reactions conditions similar to those shown previously to favor the selective coupling of C-terminal carboxyl groups.[13] This is reflected in low yields of PTH-glutamyl and PTH-aspartyl residues and can result in false radioactive signals due to the premature release of phosphopeptide where coupling occurred only at internal carboxyl positions. This is illustrated with the analysis of the Ile-Leu-Leu-Ser-Glu-Leu-Ser(PO$_4$)-Arg peptide (Table IV, see burst of counts at cycle 5 due to premature release of phosphopeptide). ^{32}P radioactivity released in this way can be distinguished from ^{32}P-labeled Edman degradation products by high-voltage electrophoresis on thin-layer chromatography plates, isoelectrofocusing in polyacrylamide gels, or reversed-phase HPLC analyses.

The method is also effective in the identification of phosphorylthreonine and phosphoryltyrosine, residues. In these cases, the general sequencing characteristics and recoveries of phosphorylated Edman degradation products[24] are similar to those observed with phosphoserines; phosphorylthreonine derivatives also undergo β elimination of the phosphoryl group, whereas PTH-phosphoryltyrosine is stable and can be detected as such by HPLC.[24] Phosphorylhistidine cannot be identified by this method due to instability of the phosphoryl bond (complete hydrolysis occurs during the ''begin'' cycle).[24]

Acknowledgments

This work was supported by funds from the National Science Foundation Biological Instrumentation Programme and by research grants from the Monsanto Corporation and the National Health and Medical Research Council of Australia. We wish to thank Chin Sook Dim for the preparation of sequencing reagents; Steve Clark, Petros Arakelian, and Charles Spence for the construction of the gas/liquid-phase sequenator; and Richard Szyszka for supplying one of the preparations of phosphorylated eIF-2α41–52. We are indebted to Stephen B. H. Kent for his generous support and valuable comments.

[16] Identification of Phosphorylation Sites in Peptides Using a Microsequencer

By PETER J. ROACH and YUHUAN WANG

Principle of Method

As examples of protein phosphorylation proliferate, identifying the specific phosphorylated residues in peptides becomes a more frequent task. Analysis of phosphorylation sites has not kept pace with advances in amino acid sequencing in large part because the standard Edman degradation procedures do not yield stable phenylthiohydantoin (PTH) derivatives of phosphoserine or phosphothreonine residues, the derivatives of Edman degradation breaking down during the acid cleavage step to release inorganic phosphate. In automated sequencers, the inorganic phosphate is poorly extracted into the organic phase, thus complicating the identification of cycles at which it is released. Many early identifications of phosphorylation sites using spinning cup machines were based on extremely low or undisclosed recoveries of phosphate in eluates from cycles of the sequencer. With gas-phase instruments, the situation is even worse and virtually no phosphate is released from the sample filter. Out of this background, we developed the present method,[1-3] which is based on analyzing the free inorganic phosphate that remains associated with the filter, not the eluate, to assess cycles of the automatic Edman degradation at which phosphate is released from the peptide. Quantitation of the amount of free inorganic phosphate measures the cumulative release of phosphate from phosphoserine or phosphothreonine residues up to the interrupt cycle and can be used to judge which residues were phosphorylated. The following equipment is required: (1) pair of scissors (or scalpel), (2) Gasphase or pulsed liquid-phase microsequencer, and (3) means to separate and to quantitate phosphate and phosphopeptide.

Preparation and Loading of Sample

Peptide samples are prepared no differently than for routine amino acid sequence determination. In our work and in what we see as the most

[1] Y. Wang, C. J. Fiol, A. A. DePaoli-Roach, A. W. Bell, M. A. Hermodson, and P. J. Roach, *Anal. Biochem.* **174,** 537 (1988).
[2] Y. Wang, A. W. Bell, M. A. Hermodson, and P. J. Roach, *J. Biol. Chem.* **261,** 16909 (1986).

likely application of the method, the peptide to be analyzed contains [^{32}P]phosphate, introduced by the action *in vitro* of a protein kinase, and is available in quantities and purity adequate for sequence analysis. The method is described below with this situation in mind. However, there is no reason why protein labeled metabolically with ^{32}P could not be similarly analyzed. Likewise, there is no theoretical reason why quantitation of phosphate and phosphopeptide must depend on the use of ^{32}P, although radioactivity determinations are both simple and sensitive when available. With a radioactive sample, monitoring recoveries at different stages is easy. One can also measure the cycles of phosphate release from an impure sample or a sample of peptide too small to permit sequence identification. Provided the peptide sequence is known and no other phosphopeptides are present to interfere with the analysis, information on the phosphorylated residues can still be obtained.

The phosphopeptide is applied to the sample filter with special attention given to obtaining as even a loading as possible over the area of the filter. Thereafter, the filter is carefully cut into the required number of sectors either with a pair of scissors or a scalpel. The number depends on the chosen strategy, as discussed below, and we have cut the filter into as many as eight pieces. The radioactivity loaded on each piece of filter can be determined by Cerenkov counting. If obtaining an even loading is problematic, an alternative that we have used successfully is to separate the filter segments initially and then to load equal amounts of peptide onto each segment. The segments are then assembled in the sample holder, on top of a clean, intact filter. Most of our experiments have used trifluoro-acetic acid (TFA)-treated precycled glass fiber filters coated with Poly-brene in Applied Biosystems (Foster City, CA) machines. In preliminary trials, the technique appears to work also with the peptide support disks used in the Porton Instruments (Tarzana, CA) machine. There would also appear to be no inherent reason why samples on polyvinylidene difluoride (PVDF) filters[4] could not be processed similarly, but we have not tried this.

Analysis of Phosphate Release

The sequencing run is then initiated and interrupted at the end of predetermined cycles. At the interrupt, a filter segment is removed and the sequencing run continued. In some instances, we have replaced the

[3] C. J. Fiol, J. H. Haseman, Y. Wang, P. J. Roach, R. W. Roeske, M. Kowalczuk, and A. A. DePaoli-Roach, *Arch. Biochem. Biophys.* **267,** 797 (1988).
[4] P. Matsudeira, *J. Biol. Chem.* **262,** 10035 (1987).

segment with blank filter and in other cases we have not, with no apparent effect on the results. Thus, any changes in reagent flows occasioned by removal of a filter piece do not seem to matter. Analyses of PTH-amino acids monitor the performance of the sequence run and permit confirmation of the sequence. The yields must be corrected for the physical loss of a portion of the starting sample as each segment of the filter is removed. We have performed sequence runs of this type with an Applied Biosystems model 470 gas-phase sequencer and an Applied Biosystems model 477A sequencer run in pulsed liquid-phase mode. As noted above, we have also used a Porton Instruments model PI2090.

The filter segments can be analyzed by Cerenkov counting as a check for loss of radioactivity from the filter. Inorganic phosphate will almost certainly remain on the filter, but one cannot exclude loss of phosphopeptide, as this can happen with any peptide under unfavorable circumstances. Inorganic phosphate and peptides are then extracted from the filter. Essentially any method which does not cause release of phosphate from the peptide is acceptable, provided it is compatible with the subsequent analysis. Since the technology of microsequencing has sought to stabilize the association of peptide with the sample filter, strong conditions are usually required. We usually extract three times into 50% formic acid with sonication and obtain ^{32}P recoveries of >90%. This value can depend on the filter used as well as the particular peptide. In trials, we have also successfully extracted peptides with 2% sodium dodecyl sulfate (SDS) and with acetic acid.

Inorganic phosphate is then separated from the shortened peptides. We use several techniques, including reversed-phase high-performance liquid chromatography, and thin-layer electrophoresis. Strictly speaking, it would be sufficient to extract and analyze only the free inorganic phosphate from the filter. In our opinion, however, the analysis is better if both free phosphate and phosphopeptide can be quantitated. Phosphate release can then be expressed as the percentage of total phosphate, allowing simple correction for any differences in the amount of the sample on different filter pieces.

Strategy

To exemplify the strategy involved, data are presented for the first phosphopeptide that we analyzed, a peptide containing a site of cyclic AMP-dependent protein kinase phosphorylation. We now know that the peptide is 35 residues long and located close to the NH$_2$ terminus of rabbit liver glycogen synthase. At the time, we knew the sequence of the first 28 residues and, from phosphoamino acid analysis, we knew that P-Ser, but

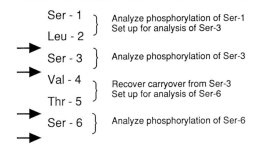

Ser - 1 } Analyze phosphorylation of Ser-1
          Set up for analysis of Ser-3

Leu - 2

→ Ser - 3 } Analyze phosphorylation of Ser-3

→ Val - 4 } Recover carryover from Ser-3
           Set up for analysis of Ser-6

Thr - 5

→ Ser - 6 } Analyze phosphorylation of Ser-6

→

FIG. 1. Example of a strategy to select cycles for analysis of phosphate release. Arrows indicate interrupts at which analyses were performed.

not P-Thr, was present in the peptide after the action of cyclic AMP-dependent protein kinase. Serines were present at residues 1, 3, and 6, with a Thr residue at position 5 (Fig. 1). In selecting cycles for analysis (i.e., interruption of the sequence run and withdrawal of a filter segment[5]), the fundamental strategy is to analyze before and after candidate phosphorylated residues. If the succeeding residues cannot be phosphorylated, then one can avoid analyzing at every cycle. Also, if the nature of the phosphorylated residue is known, then other cycles may be omitted. For example, in this case, we elected not to analyze before cycle 5, a Thr, since we knew from other experiments that it was not modified. Filter segments were removed after cycles 2, 3, 5, and 6. ^{32}P-Labeled peptide was extracted from the filters using 50% formic acid with recoveries of ^{32}P around 90%. The analysis of free phosphate was achieved by reversed-phased HPLC, taking advantage of the protocol that had been worked out during the initial purification of the peptide. The ^{32}P-labeled inorganic phosphate was found in the pass-through of the column with the remaining phosphopeptide eluting much later (around 60 min in an 80-min gradient). We demonstrated, by thin-layer electrophoresis, that the flow-through fractions contained a single radioactive species that comigrated with inorganic phosphate. Thus, the radioactivity in the pass-through was a measure of the cumulative phosphate release from the peptide up to the point of filter removal.

The analysis gave very clear results (Table I). In the starting peptide sample and after cycle 2, virtually no free phosphate was present.[6] A

[5] This practice, *sequendum interruptum*, originated in the laboratory of Mark A. Hermodson, Department of Biochemistry, Purdue University, West Lafayette, Indiana.

[6] It is good practice to check that, at the time of analysis, all of the ^{32}P is still associated with the peptide. We have observed partial release of phosphate even with storage of some peptides at −70°.

TABLE I
ANALYSIS OF LIVER GLYCOGEN SYNTHASE PHOSPHOPEPTIDE[a]

Cycle	Residue	Yield of PTH derivative (pmol)	^{32}P on filter segment (pmol)	^{32}P after HPLC (pmol)	Free phosphate (pmol)	Phosphate release (% of total)	
						Cumulative	Per cycle(s)
1	Ser	480					
2[b]	Leu	470	115	88	1	1	1
3[b]	Ser		131	104	48	46	45
4	Val	230					47
5[b]	Thr	165	109	86	80	93	
6[b]	Ser	150	94	76	73	97	4

[a] Approximately 75,000 cpm was loaded totally on the filter.
[b] Cycles at which filters were removed.

significant amount, 45% of the total radioactivity, was present as free phosphate after cycle 3, indicating that Ser-3, but not Ser-1, was significantly phosphorylated. After cycle 5, 93% of the radioactivity was present as free phosphate. Since the residues at cycles 4 and 5 could not have been phosphorylated, the additional phosphate release must have been due to carryover from cycle 3 and must have been associated with Ser-3. By cycle 6, also a Ser, 97% of the phosphate was released. Thus, we could unequivocally assign Ser-3 as the major phosphorylation site. Furthermore, since almost all the phosphate was released by cycle 6, we could exclude the presence of other phosphorylation sites in the COOH terminus of the peptide. This was important information since we did not know the complete sequence of the peptide at the time of the analysis and there could have been other possible phosphorylation sites. A simple graphic analysis, expressing the percentage of total radioactivity present as free phosphate at different cycles, is shown in Fig. 2.

Advantages and Disadvantages of Method

The reader will find in this volume reports[7-10] on other, perhaps more sophisticated, approaches to identifying phosphorylation sites and most likely the future lies in developing such new chemistries and physical

[7] H. E. Meyer, E. Hoffmann-Posorske, and L. M. G. Heilmeyer, Jr., this volume [14].
[8] P. Cohen, B. Gibson, and C. F. B. Holmes, this volume [13].
[9] E. H. R. Wettenhall, R. H. Aebersold, and L. E. Hood, this volume [15].
[10] H. E. Meyer, E. Hoffmann-Posorske, A. Donella-Deana, and H. Korte, this volume [17].

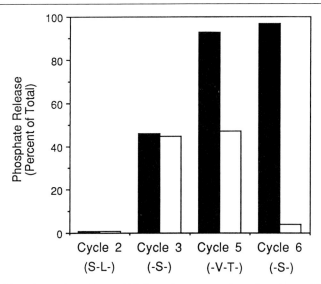

FIG. 2. Phosphate release from rabbit liver glycogen synthase phosphopeptide using data from Table I. The solid bars represent the cumulative release of ^{32}P and the open bars the release after the previous cycle interrupt. S, Serine, L, leucine; V, valine; T, threonine.

analytical methods. However, it is probably still true that no method is quite at the point of being routinely preferred in all situations. We view our method as a simple approach that is so easy to perform that it may often be worthwhile trying. Its chief merit is its simplicity and its dependence on readily available equipment. From a technical standpoint, the main advantage of the method over some of the alternatives is that high recoveries of phosphate can be achieved. In addition, when applied to peptides phosphorylated *in vitro* with ^{32}P, sensitive detection of phosphate is possible. The minimum number of counts per minute (cpm) necessary will depend in large part on the method chosen to separate and to analyze phosphate and phosphopeptides, but we feel that 1000 cpm/filter segment should be sufficient in most cases.

The principal drawbacks to the method are those shared by any technique relying on repeated Edman degradations. The repetitive yield of a given sequencing run will impose a limit on how far from the NH_2 terminus a residue can be in order to be susceptible to the analysis. Furthermore, the presence of P-Ser itself appears to have a severe effect on the efficiency of the Edman degradation.[1] Although this can often be tolerated in peptides with a single phosphorylation site, it can be problematic when multiple phosphorylation sites are involved. In this regard, techniques that convert the phosphoserine to a residue for which the recovery of the PTH derivative is higher, such as S-ethylcysteine,[7] have an advantage. The degree to which carryover complicates the assignment of the cycles of phosphate release is also very much dependent on the sequence. The worst case, for any analysis based on Edman degradation, is where candidate residues are closely spaced together. Nonetheless, our application of the method has included successful analysis of the sequences -S(P)-E-S-S and -S(P)-S(P)-; [S, serine; E, glutamic acid; S(P), phosphoserine].

Acknowledgments

This work is supported in part by National Institutes of Health Grant DK-27221 and the Indiana University Diabetes Research and Training Center, DK-20542. We are indebted to Mark Hermodson and Alex Bell for their encouragement, advice, and friendship in the development of this method, which was implemented first in the Department of Biochemistry, Purdue University, West Lafayette, Indiana.

[17] Sequence Analysis of Phosphotyrosine-Containing Peptides

By Helmut E. Meyer, Edeltraut Hoffmann-Posorske, Arianna Donella-Deana, and Horst Korte

Phosphorylation of distinct tyrosine residues is an essential step in the response on stimulation of cells with a variety of hormones and growth factors. Thus, phosphorylation occurs at special target proteins to transduce the outer cellular signal(s) into intracellular actions.[1] Whereas extraordinary developments have been made over the last few years to identify tyrosine kinases[2,3] and tyrosine-specific phosphatases,[4] little improvement has been achieved regarding the sequence analysis of phosphotyrosine-containing peptides. In contrast to the other O-phosphorylated amino acids, i.e., serine phosphate and threonine phosphate,[5,6] no viable methods are so far available allowing phosphotyrosine to be identified

[1] Y. Yarden and A. Ulrich, *Annu. Rev. Biochem.* **57,** 443 (1988).
[2] T. Hunter and M. Sefton, *Proc. Natl. Acad. Sci. U.S.A.* **77,** 1311 (1980).
[3] S. E. Shoelson and R. Kahn, *in* "Insulin Action" (B. Drazinin, S. Melmed, and D. LeRoith, eds.), p. 23. Alan R. Liss, New York, 1989.
[4] N. K. Tonks and H. Charbonneau, *Trends Biochem. Sci.* **14,** 497 (1989).
[5] H. E. Meyer, E. Hoffmann-Posorske, H. Korte, and L. M. G. Heilmeyer, Jr., *FEBS Lett.* **204,** 61 (1986).
[6] H. E. Meyer, E. Hoffmann-Posorske, and L. M. G. Heilmeyer, Jr., this volume [14].

after partial hydrolysis[2] or during Edman degradation without radioactive labeling. However, ^{32}P labeling does permit the assignment of phosphorylated tyrosine residues and methods and examples are given in the literature.[2,7-10]

A new, solid-phase sequencing technique for microsequence analysis has been described.[11,12] However, the detection of ^{32}P radioactivity is the only criterion for localizing the phosphorylated amino acids.

We have established two different methods which permit the identification of phosphotyrosine after partial hydrolysis or during sequence analysis in the lower picomolar range. After partial hydrolysis of proteins and peptides phosphotyrosine can be determined by capillary electrophoresis as its phenylthiohydantoin (PTH) derivative. This can be used as a nonradioactive screening method for phosphotyrosine-containing proteins and peptides.

Using solid-phase sequencing and subsequent ion-exchange high-performance liquid chromatography or capillary electrophoresis,[13] phosphotyrosine is determined as its phenylthiohydantoin derivative by measuring the light adsorption at 269 or 261 nm. As little as 1 pmol of phenylthiohydantoin phosphotyrosine can be quantitatively identified without the need of prior radioactive labeling.

Determination of Phosphotyrosine in Phosphotyrosine-Containing Peptides and Proteins by Capillary Electrophoresis

The procedure consists of the following steps:

1. Partial hydrolysis of phosphotyrosine-containing proteins and peptides[2] and derivatization of liberated amino acids to their PTH derivatives[14]
2. Analysis by capillary electrophoresis

[7] M. F. White, S. E. Shoelson, H. Keutmann, and C. R. Kahn, *J. Biol. Chem.* **263,** 2969 (1988).
[8] H.-U. Häring, M. Kasuga, M. F. White, M. Crettaz, and C. R. Kahn, *Biochemistry* **23,** 3298 (1984).
[9] L. Stadtmauer and O. M. Rosen, *J. Biol. Chem.* **261,** 10000 (1986).
[10] J. A. Cooper, B. M. Sefton, and T. Hunter, this series, Vol. 99, p. 387.
[11] R. H. Aebersold, H. Nika, G. D. Pipes, R. E. H. Wettenhall, S. M. Clark, L. E. Hood, and S. B. H. Kent, *in* "Methods in Protein Sequence Analysis" (B. Wittmann-Liebold, ed.), p. 79. Springer-Verlag, Berlin, 1988.
[12] R. E. H. Wettenhall, R. H. Aebersold, L. E. Hood and S. B. H. Kent, this volume [15].
[13] H. E. Meyer, E. Hoffmann-Posorske, H. Korte, A. Donella-Deana, A.-M. Brunati, L. A. Pinna, J. Coull, J. Perich, R. M. Valerio, and R. Basil Johns, *Chromatographia* **30,** 691 (1990).
[14] HPLC, High-performance liquid chromatography; PTH, phenylthiohydantoin; ATZ, anilinothiazolinone; PITC, phenyl isothiocyanate; MES, 4-morpholinoethanesulfonic acid; EDC, 1-ethyl-3-(dimethylaminopropyl)carbodiimide.

Partial Hydrolysis of Phosphotyrosine-Containing Proteins and Peptides and Derivatization of Liberated Amino Acids to Their PTH Derivatives

Reagents

HCl (6 N)
Recrystallized phenol
Argon (99.999%)
Conditioning solution 1: Mix ethanol, triethylamine, and water in a ratio of 2 : 2 : 1 (v/v)
Conditioning solution 2: Mix ethanol, triethylamine, and water in a ratio of 8 : 1 : 1 (v/v)
Coupling solution: Mix ethanol, triethylamine, water, and PITC in a ratio of 7 : 1 : 1 : 1 (v/v)
Aqueous trifluoroacetic acid (25%)

Procedure

1. Place the peptide or protein samples (0.02 to 1 nmol) in 5 × 60 mm test tubes (Pyrex glass) and dry under vacuum.

2. Put up to 10 test tubes into the hydrolysis vessel,[6] add 10 μl of 6 N HCl to each test tube, and add 400 μl of 6 N HCl and a crystal of phenol beside the test tubes.

3. Close the vessel, evacuate until the HCl starts boiling, then flush with argon. Repeat both steps twice. Seal the vessel under vacuum.

4. Carry out partial hydrolysis for 2 hr at 110°.

5. Transfer the test tubes into a new vessel and dry under high vacuum.

6. Add 10 μl of conditioning solution 1 to each test tube and dry again under high vacuum (less than 0.04 mbar). Repeat this step.

7. Add 10 μl of conditioning solution 2 to each test tube. Flush with argon and incubate the mixture for 5 min at room temperature.

8. Add 10 μl of coupling solution to each test tube, flush with argon, and incubate the mixture for 5 min at 50°. Dry immediately thereafter under high vacuum (less than 0.04 mbar).

9. Add 50 μl of ethyl acetate to each test tube. Gently rotate the tubes (do not vortex) and decant each wash. Dry briefly under vacuum. Repeat this step.

10. Add 50 μl of 25% trifluoroacetic acid, flush with argon, and seal with Parafilm. Incubate for 30 min at 50°. Dry down under vacuum.

Capillary Electrophoresis

Equipment

Capillary electrophoresis system (model 270A; Applied Biosystems, Foster City, CA)

Reagents

Sodium citrate buffer (20 mM, pH 2.5) for running the capillary electrophoresis

Aqueous acetonitrile (20%, v/v) containing 0.35% trifluoroacetic acid

Sample solution: 50 mM sodium chloride

Procedure

1. Dissolve the dried samples in 3–50 μl of sample solution.
2. Perform the capillary electrophoresis at −25 kV, 30°, and 20 min running time. Perform UV detection at 261 nm.

Figure 1 shows the results of PTH-phosphoamino acid determination by capillary electrophoresis. On the left-hand side, 1 nmol of two model peptides (Fig. 1A and B), containing phosphotyrosine, was taken for partial hydrolysis and PTH derivatization. In both cases, a clear signal at 15 min demonstrates the presence of PTH-phosphotyrosine. The different yields in PTH-phosphotyrosine are due to the different amounts of phosphotyrosine liberated from the individual phosphopeptides, depending on the amino acids surrounding the phosphotyrosine residue.

On the right-hand side, a standard mixture (Fig. 1C) containing the PTH derivatives of cysteic acid, phosphoserine, phosphothreonine, and phosphotyrosine is analyzed by capillary electrophoresis. All four amino acid derivatives are well separated. As expected, PTH-phosphoserine is not stable and therefore, not applicable for phosphoserine determination of phosphoserine-containing peptides.

The other PTH derivatives are quite stable and we have found that this method is especially well suited for determination of phosphothreonine in phosphothreonine-containing peptides.[15]

The detection limit of this method is about 1 pmol of phosphotyrosine in 5 μl of sample solution. Therefore this procedure is very useful to detect phosphotyrosine-containing proteins and peptides. It can be used as a nonradioactive screening method, allowing the purification of phosphotyrosine-containing peptides from a proteolytic digest, a prerequisite to localize *in vivo* phosphorylation sites by sequence analysis.

[15] H. E. Meyer, C. Liedtke, M. Heber, A. Donella-Deana, H. Korte, and E. Hoffmann-Posorske, in preparation.

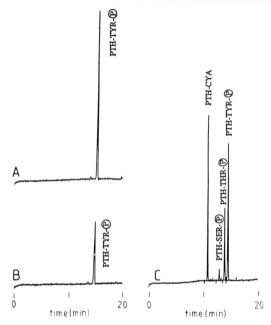

FIG. 1. Determination of PTH-phosphoamino acids by capillary electrophoresis. *Left:* Peptides L-R-R-A-Y(P)-L-G (A) or D-R-V-Y(P)-I-H-P-F (B) (1 nmol) is partially hydrolyzed with 6 *N* HCl at 110° for 2 hr. The liberated amino acids were coupled with PITC and transformed to the more stable PTH derivatives. After drying, the samples are dissolved in 50 μl 20% aqueous acetonitrile containing 0.35% trifluoroacetic acid and 50 m*M* sodium chloride. Thirty nanoliters is analyzed by capillary electrophoresis. Detection is at 261 nm; run time is 20 min at −25 kV and 30°; buffer is 20 m*M* sodium citrate, pH 2.5. *Right:* Cysteic acid, phosphoserine, phosphothreonine, and phosphotyrosine (1 nmol each) are transferred to their PTH derivatives. Sample is dissolved in 50 μl 20% aqueous acetonitrile. Thirty nanoliters is analyzed by capillary electrophoresis. Detection is at 261 nm; run time is 20 min at −20 kV and 30°; buffer is 20 m*M* sodium citrate, pH 2.5.

Sequence Analysis of Phosphotyrosine-Containing Peptides

The procedure outlined below consists of the following steps:

1. Attachment of phosphotyrosine-containing peptide(s) to Sequelon-AA membrane
2. Sequence analysis using either Applied Biosystems model 470 gas-phase sequencer modified for solid-phase sequencing or MilliGen/Biosearch (Burlington, MA) ProSequencer 6600
3. Determination of PTH-phosphotyrosine[14] by
   a. Reversed-phase high-performance liquid chromatography (HPLC)
   b. Ion-exchange HPLC

c. Capillary electrophoresis

d. Measurement of ^{32}P radioactivity

Attachment of Phosphotyrosine-Containing Peptide(s) to Sequelon-AA Membrane

This procedure follows the method given by MilliGen/Biosearch with slight modifications.

Reagents

Sequelon-AA membrane (MilliGen/Biosearch)

Two different peptides[15a] are used for the following experiments: D-R-V-Y-[^{32}P]I-H-P-F and L-R-R-A-Y(P)-L-G, which is not radioactively labeled. The first peptide is from Serva (Westbury, NY) and is enzymatically phosphorylated as described.[16] The other peptide is synthesized as described[17]

Acetonitrile, HPLC quality (Aldrich, Milwaukee, WI)

4-Morpholinoethanesulfonic acid (MES) (Serva)

1-Ethyl-3-(dimethylaminopropyl)carbodiimide (EDC) (Sigma, St. Louis, MO)

Solution of 0.1 M sodium hydroxide pro analysis (Roth, Karlsruhe, Germany) freshly prepared in HPLC water (Aldrich)

Solution A: 30% aqueous acetonitrile (the concentration of acetonitrile can be varied dependent on solubility of the peptide; the given concentration is useful for most peptides)

Solution B: 0.1 M MES, pH 5.0, containing 15% acetonitrile (pH is adjusted with the 0.1 M sodium hydroxide solution before the addition of acetonitrile). This solution can be stored frozen at $-20°$

Procedure

1. Add 20 μl of solution A to the dry phosphotyrosine-containing peptide (1–2 nmol) and mix (sonicate) to dissolve the peptide.

2. Place a Sequelon-AA disk on a piece of aluminum foil or plastic film.

3. Apply the peptide solution to the disk.

4. Allow the disk to thoroughly dry at room temperature.

[15a] Single-letter abbreviations for amino acids: A, Alanine; D, aspartic acid; F, phenylalanine; G, glycine; H, histidine; I, isoleucine; L, leucine; P, proline; R, Arginine; Y, tyrosine; V, valine.

[16] A. M. Brunati and L. A. Pinna, *Eur. J. Biochem.* **172,** 451 (1988).

[17] R. M. Valerio, P. F. Alewood, R. B. Johns, and B. E. Kemp, *Int. J. Pep. Protein Res.* **33,** 428 (1989).

5. To 0.1 ml of solution B add 1.0 mg of EDC.

6. Carefully apply 10 μl of the solution B containing EDC to the Sequelon-AA disk onto which the peptide has been dried.

7. Allow the reaction to proceed for 20 min at room temperature.

8. Using the Applied Biosystems model 470 gas-phase sequencer, the disk must be washed with 0.5 ml water and 0.5 ml methanol to remove the reagents.

The sample is now ready for sequence analysis.

Sequence Analysis

Application of Applied Biosystems Model 470 Gas-Phase Sequencer Modified for Solid-Phase Sequencing. ATZ-phosphotyrosine, which is insoluble in butyl chloride, was found to be soluble in trifluoroacetic acid, allowing the transfer from the reaction cartridge to the conversion flask almost quantitatively. Therefore, we have exchanged the extraction solvent S3 (butyl chloride) for trifluoroacetic acid. This allows one to perform the splitting procedure with liquid trifluoroacetic acid, which is also used for the transfer of the ATZ derivatives of the cleaved amino acids. A further recommendation is to exchange the vent valves in the formerly butyl chloride position S3 for trifluoroacetic acid-resistant ones. In addition, needed changes in the sequence control program are described below. Performing these changes, solid-phase sequencing of phosphotyrosine-containing peptides with the model 470 gas-phase sequenator is possible.

Equipment

Gas-phase sequencer (model 470) online connected with a PTH analyzer (model 120), both from Applied Biosystems

Modifications for Solid-Phase Sequencing

Solution S3 (butyl chloride) is exchanged for 100% trifluoroacetic acid
In the first cycle, the standard 03CBGN program (from Applied Biosystems) is used, but S2 (ethyl acetate) is employed instead of solution S3 (now 100% trifluoroacetic acid!)
The following degradation cycles (03CYP) are newly designed to get optimum results for solid-phase sequencing. The cycle listing is shown in Table I

Procedure

1. Place the Sequelon-AA disk with the coupled phosphotyrosine-containing peptide into the cartridge as usual. Cover the membrane with a trifluoroacetic acid-pretreated glass fiber disk and close the cartridge.

2. Start sequence analysis using the program 03RYPP (see Table I).

TABLE I

CYCLE LISTING OF OPTIMIZED DEGRADATION PROGRAM 03CYP FOR SOLID-PHASE
SEQUENCING WITH MODEL 470 GAS-PHASE SEQUENATOR[a]

Step	Cartridge function[b]	Flask function	Time (sec)
1	Deliver R2	Pause	100
2	Prep R1	Load R4	12
3	Deliver R1	Argon dry	8
4	Argon dry	Pause	50
5	Deliver R2	Pause	300
6	Deliver R1	Pause	4
7	Argon dry	Pause	50
8	Deliver R2	Pause	300
9	Deliver R1	Pause	4
10	Argon dry	Pause	50
11	Deliver R2	Pause	300
12	Argon dry	Line purge	60
13	Deliver S1	Pause	100
14	Deliver S2	Argon dry	240
15	Argon dry	Argon dry	120
16	Deliver R3	Argon dry	20
17	Prep S3	Argon dry	6
18	Deliver S3 (midpoint)	Argon dry	21
19	Pause	LC start	10
20	Pause	Collect	60
21	Pause	Load S4	6
22	Pause	Argon dry	4
23	Pause	Load S4	6
24	Pause	Argon dry	4
25	Pause	Pause	540
26	Pause	Collect	14
27	Pause	Load S4	6
28	Pause	Argon dry	4
29	Pause	Load S4	6
30	Pause	Argon dry	4
31	Pause	Collect	40
32	Pause	Deliver S4	30
33	Pause	Argon dry	5
34	Pause	FC step	2
35	Pause	Empty	40
36	Pause	Line purge	10
37	Transfer with S3	Pause	20
38	Pause	Argon dry	2
39	Transfer with Ar	Pause	20
40	Prep S3	Argon dry	6
41	Deliver S3	Argon dry	21
42	Pause	Argon dry	30
43	Transfer with S3	Pause	70
44	Pause	Pause	10
45	Transfer with Ar	Pause	30
46	Argon dry	Argon dry	30
47	Deliver S2	Argon dry	60
48	Argon dry	Pause	120

[a] Total time for cycle: 49 min 16 sec.

[b] S1, Heptane; S2, ethyl acetate; S3, trifluoroacetic acid delivered as liquid; S4, 20%
acetonitrile in water; R1, 5% phenyl isothiocyanate in heptane; R2, 10% trimethylamine
in water; R3, trifluoroacetic acid delivered in gaseous form; R4, 25% trifluoroacetic acid
in water. Run program 03RYPP consists of one cycle 03CBGN (S2 is used instead of
S3) and an optional number of cycles 03CYP.

Application of MilliGen/Biosearch ProSequencer 6600 Solid-Phase Sequencer. The ProSequencer is especially designed for solid-phase sequencing and uses liquid trifluoroacetic acid as cleavage reagent and transfer solvent. This instrument is used without any modification.

Equipment

Solid-phase sequencer model ProSequencer 6600 connected to an external fraction collector (model 6400), both from MilliGen/Biosearch

Procedure

1. Place the Sequelon-AA disk containing the phosphotyrosine-containing peptide into the cartridge as usual.
2. Start sequence analysis employing the standard degradation program.

Determination of PTH-Phosphotyrosine

Following Edman degradation, different analysis methods are employed to achieve positive identification of the resulting PTH-phosphotyrosine. PTH-phosphotyrosine standard is prepared according to Tarr's method for large peptides,[18] with slight modifications as outlined below.

Preparation of PTH-Phosphotyrosine Standard

Reagents

Phosphotyrosine (Sigma)
Argon (99.999%) (Linde, Witten, Germany)
Ethanol p.a. (Merck, Darmstadt, Germany), HPLC water (Aldrich), triethylamine, sequence grade (Pierce, Rockford, IL), phenyl isothiocyanate, sequence grade (Beckman, Munich, Germany), ethyl acetate, 20% (v/v) acetonitrile in water, and 25% (v/v) trifluoroacetic acid, all sequence grade (Applied Biosystems)
Conditioning solution 1: Ethanol, triethylamine, and water mixed in a ratio of 2 : 2 : 1 (v/v)
Conditioning solution 2: Ethanol, triethylamine, and water mixed in a ratio of 8 : 1 : 1 (v/v)
Coupling solution: Ethanol, triethylamine, water, and phenyl isothiocyanate mixed in a ratio of 7 : 1 : 1 : 1 (v/v)

[18] G. E. Tarr, *in* "Methods of Protein Microcharacterization" (J. E. Shively, ed.), p. 155. Humana Press, Clifton, New Jersey, 1986.

The conditioning and coupling solutions must be freshly prepared every time.

Procedure

1. Dry down (10 times) 10 nmol of phosphotyrosine in 5 × 60 mm glass test tubes.
2. Add 10 μl of conditioning solution 1 to each test tube and dry again under high vacuum (less than 0.04 mbar). Repeat this step.
3. Add 10 μl of conditioning solution 2 to each test tube. Flush with argon and incubate the mixture for 5 min at room temperature.
4. Add 10 μl of coupling solution to each test tube, flush with argon, and incubate the mixture for 5 min at 50°. Dry under high vacuum (less than 0.04 mbar).
5. Add 50 μl of ethyl acetate to each test tube. Gently rotate tube (do not vortex) and decant each wash. Dry briefly under vacuum. Repeat this step.
6. Add 50 μl of 25% trifluoroacetic acid, flush with argon, and seal with Parafilm. Incubate for 30 min at 50°. Dry down under vacuum.

These PTH-phosphotyrosine standards can be stored frozen under argon at −20° for at least 1 month. For use, the content of one tube is dissolved in 20% acetonitrile containing 0.35% trifluoroacetic acid and 50 mM sodium chloride. This solution is stable for at least 1 day at room temperature.

Reversed-Phase HPLC

Equipment

For reversed-phase PTH-amino acid analyses the model 120 PTH analyzer from Applied Biosystems is used with both sequenators. Only the Applied Biosystems sequencer is connected on line to this analyzer

Chromatograms are recorded using a calculating integrator C-R3A from Shimadzu (Columbia, MD)

Reagents

Gradient solutions A and B are prepared according to the recommendations of Applied Biosystems.

Gradient solution A: 5% (v/v) tetrahydrofuran solution in water containing 50–100 mM sodium acetate (pH ~4)

Gradient solution B: Acetonitrile supplied with 500 nmol of dimethylphenylthiourea per liter

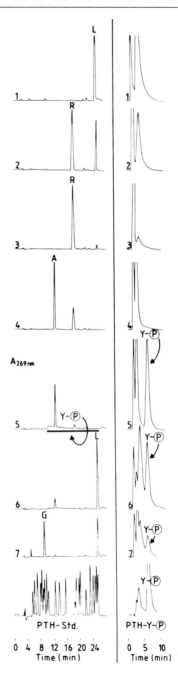

Procedure

1. The standard gradient program is used for separation of the PTH-amino acids.

2. For sequence analysis with the Applied Biosystems sequencer, 40% of the PTH samples is used for the reversed-phase HPLC.

3. For sequence analysis with the MilliGen/Biosearch sequencer, 10% of the PTH samples is manually injected onto the PTH analyzer.

4. Ten percent of the samples is taken for ^{32}P determination of radioactivity directly (in the case of labeled peptides); the residual parts are dried down for other determination methods (see below).

The Applied Biosystems sequencer is used for the sequence analysis of the phosphotyrosine-containing peptide L-R-R-A-Y(P)-L-G, which is not radioactively labeled. The results of reversed-phase HPLC method are seen on the left-hand side in Fig. 2. The chromatogram of degradation step 5 demonstrates the different chromatographic behavior of PTH-phosphotyrosine compared to the other standard PTH-amino acids in the reversed-phase HPLC system. Thus, only a slight increase in the UV signal starting at 8 min elution time and extending until the end of the chromatogram demonstrates the presence of PTH-phosphotyrosine (marked by a filled bar under the chromatogram and an arrow). This specific signal

---

FIG. 2. Sequence analysis of the phosphotyrosine-containing peptide L-R-R-A-Y(P)-L-G. The phosphopeptide (2.2 nmol) is used for covalent attachment to the Sequelon-AA membrane; 1.5 nmol (about 60% yield) is covalently bound according to the repetitive yield calculation after sequence analysis. Solid-phase sequence analysis is performed with the modified Applied Biosystems gas-phase sequencer. *Left:* The chromatograms of the usual reversed-phase HPLC separation are shown, performed with 50% of each sample. In cycle 5, where PTH-phosphotyrosine is expected, only an increase in the baseline (marked by a horizontal line and an arrow) is visible. The other peaks, at 12 and 16.5 min, are due to the carryover of alanine and arginine, respectively, from the two previous cycles. *Right:* The chromatograms of the ion-exchange HPLC are presented. These analyses are performed with a 25% aliquot of each sample. In degradation step 5 a specific signal at 6 min demonstrates the presence of PTH-phosphotyrosine (marked by an arrow), which elutes from the column after all the other nonphosphorylated PTH-amino acids. The chromatogram at bottom, left shows the separation of a PTH-amino acid standard mixture containing 62.5 pmol of 19 PTH-amino acids normally found in proteins. At bottom, right the ion-exchange HPLC separation of a PTH-phosphotyrosine standard is shown, which was prepared as described before, containing 100 pmol of PTH-phosphotyrosine. The peak at 2.5 min is due to the presence of some unidentified UV-positive material emerging in front of the chromatogram. PTH-phosphotyrosine elutes as a symmetrical peak at 6-min elution time. A flow rate of 200 μl/min is chosen for both chromatographic separations. UV absorbance is measured at 269 nm. The one-letter code is used for the amino acids. Y-P, PTH-phosphotyrosine. All other experimental details are as given in the text.

allows the identification of phosphotyrosine in this position only with an appreciable amount of experience.

When the MilliGen/Biosearch sequencer is used for sequence analysis of the ^{32}P-labeled peptide D-R-V-Y-[^{32}P]I-H-P-F the chromatogram of degradation step 4 shows no identifiable PTH-derivative (Fig. 3). Even the slight rise of the UV-signal which could be observed with the other peptide (see above and Fig. 2) is not detectable. However, an unambiguous identification is possible with the ion-exchange HPLC or capillary electrophoresis, as will be shown in the following section.

Ion-Exchange HPLC

Equipment

LiChroGraph L-6210 inert HPLC pump (Merck-Hitachi, Darmstadt, Germany); for UV detection, a Merck-Hitachi variable-wavelength detector (model L-4000) is used. As an alternative, the inert HPLC pump model 2249 (Pharmacia) is employed. For UV detection, a Kratos (Karlsruhe, Germany) Spectroflow 773 variable-wavelength detector is used

Inert manual injection valve (model 9125) is used for sample injection in conjunction with the Merck-Hitachi equipment; the inert injection valve model 2154 (Valco) is used with the alternative system

Glass-lined microbore column (2 × 100 mm) (No. 201005); SGE, Weiterstadt, Germany) filled with 10-μm Mono Q (Pharmacia) is used for separation of PTH-phosphotyrosine

Calculating integrator (D 2000 from Merck-Hitachi) to record chromatograms

Reagents

Solvent for separation consists of 10 mM potassium phosphate, pH 3.0, containing 100 mM potassium chloride and 25% (v/v) acetonitrile, and is freshly prepared every day and filtered through a GVWP 0.22-μm membrane (Millipore, Bedford, MA) under vacuum. This solution is held under a helium atmosphere. Potassium phosphate and potassium chloride are from Baker (Phillipsburg, NJ) and acetonitrile is from Aldrich

Procedure

1. Dissolve the dried samples from sequence analysis in 50–150 μl of separation solvent containing 20% acetonitrile.
2. Take a separate aliquot and apply onto the Mono Q column.
3. Perform isocratic separation at a flow rate of 200 μl/min.

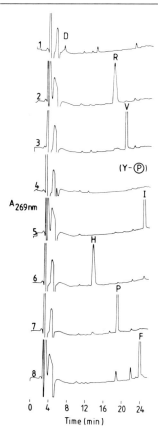

FIG. 3. Sequence analysis of the phosphotyrosine-containing peptide D-R-V-[^{32}P]Y-I-H-P-F. ^{32}P-labeled phosphopeptide (1.8 nmol) is taken for the attachment procedure to Sequelon-AA; 1.2 nmol is covalently bound to the support and taken for sequence analysis with the MilliGen/Biosearch ProSequencer. Ten percent of each fraction, collected after each degradation cycle, is manually injected onto the model 120 PTH-amino acid analyzer. In cycle 4 PTH-[^{32}P]phosphotyrosine is expected. No signal for this amino acid is visible. All other details are as given in the legend to Fig. 2.

4. Set the detector to 269 nm.

5. Optionally, collect fractions of 200 μl for radioactive measurement in the case of ^{32}P-labeled samples.

Inert pumps were selected since, in accordance with our experience, phosphorylated compounds will be absorbed to all Fe^{3+}-containing surfaces, making the determination of such compounds in the lower picomolar range impossible. Therefore, capillaries, columns, and other parts of the HPLC device are made from noniron materials, too.

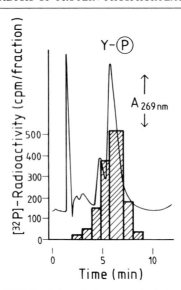

FIG. 4. Ion-exchange HPLC of degradation step 4 of the [^{32}P]phosphopeptide. This analysis is performed with a 72% aliquot of the sample. A specific UV signal at 6 min demonstrates the presence of PTH-phosphotyrosine. At the indicated time intervals, the eluates are collected to determine the ^{32}P radioactivity. The hatched bars represent the amount of radioactivity found for each fraction. The maximum of ^{32}P radioactivity coelutes with the maximum of the UV signal at about 6 min. It confirms the identity between PTH-phosphotyrosine and ^{32}P radioactivity. All other details are as described in the legend to Fig. 2.

The results of ion-exchange HPLC are shown on the right-hand side of Fig. 2 and in Fig. 4. As can be seen in Fig. 2 (right), a clear signal emerges in the chromatogram of degradation step 5 at 6 min elution time, which is identical to that of the PTH-phosphotyrosine standard (lowest chromatogram in Fig. 2, right). This peak positively identifies phosphotyrosine as the fifth amino acid in this peptide. The detection limit of this method is as low as 1 pmol.

Capillary Electrophoresis

The equipment and reagents for PTH-phosphotyrosine analysis by capillary electrophoresis are the same as described above.

Procedure

1. Dissolve the dried 50% aliquot from the sequence analyzer in 3–50 µl of 20% aqueous acetonitrile containing 0.35% trifluoroacetic acid and 50 mM sodium chloride.

2. Perform the capillary electrophoresis at -25 kV, 30°, with a 20-min running time. Perform detection at 261 nm.

A typical result of capillary electrophoresis is shown in Fig. 5, using the D-R-V-Y(P)-I-H-P-F peptide. A clear, sharp signal of PTH-phosphotyrosine in cycle 4, where phosphotyrosine is expected, demonstrates the presence of this amino acid. The advantage of this method is the high selectivity for amino acids that are negatively charged at pH 2.5. Therefore, only PTH-phosphotyrosine can be detected in this example. All the other nonmodified amino acids are running in the opposite direction under the applied conditions and will not be detected (see PTH standard in Fig. 5).

Measurements of ^{32}P Radioactivity

Equipment

LKB WALLAC model 1209 RACKBETA liquid scintillation counter (Bromma, Sweden)

Reagent

Scintillation solvent: Prepare from 24 g Permablend (Hewlett-Packard, Zurich, Switzerland) and 375 g naphthalene dissolved in 3000 ml dioxane; both are kind gifts from Hoechst (Frankfurt, Germany)

Procedure

1. Put 10% of each PTH sample after sequence analysis or 200-μl aliquots of the ion-exchange chromatography into a 5-ml liquid scintillation vial (Baker)
2. Supply with 2 ml of scintillation solvent.
3. Count the ^{32}P radioactivity.

The results of the measurement of ^{32}P radioactivity are presented in Figs. 4 and 6. Figure 6 shows the sequence analysis of the ^{32}P-labeled phosphopeptide D-R-V-[^{32}P]Y-I-H-P-F performed with the MilliGen/Biosearch solid-phase sequencer. Of the 1.8 nmol of the peptide (containing 11500 cpm of ^{32}P radioactivity) taken for the coupling procedure, 67% (7650 cpm) is covalently attached at the Sequelon-AA membrane. A sharp increase in the radioactivity starting at cycle 4 demonstrates the presence of [^{32}P]phosphotyrosine in this position. A 50% yield of ^{32}P radioactivity is calculated from the results of degradation steps 4 to 7. This is in agreement with the recovered amount of PTH-phosphotyrosine determined by the UV absorption during ion-exchange HPLC. The extent of carryover of the radioactivity visible in the following cycles is comparable to that of the PTH-amino acids shown in Fig. 3. Figure 4 shows the ion-

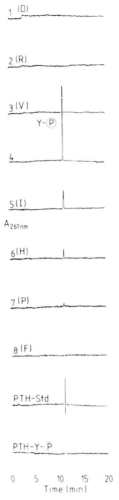

FIG. 5. Capillary electrophoresis of PTH-amino acids of the peptide D-R-V-[^{32}P]Y-I-H-P-F. Fifty percent of each degradation step collected by the built-in fraction collector of the sequenator was transferred into 500-μl Eppendorf vials, dried under vacuum in a SpeedVac, and dissolved in 20 μl of 20% (v/v) aqueous acetonitrile containing 0.35% (v/v) trifluoroacetic acid and 50 mM sodium chloride. Conditions: injection time (vacuum), 10 sec; 30°; capillary length, 72 cm (50 cm to detector); 50-μm i.d.; voltage = -25 kV; buffer, 20 mM sodium citrate, pH 2.5; detector rise time, 1 sec; detection at 261 nm, 0.008 AUFS.

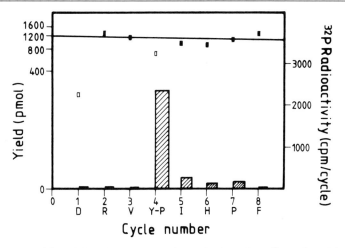

FIG. 6. Repetitive yield calculation and determination of the ^{32}P radioactivity of the peptide D-R-V-[^{32}P]Y-I-H-P-F. The repetitive yield is calculated by plotting the cycle number against the yield of each PTH-amino acid. Only the amounts of PTH-amino acids denoted by a filled square are taken for the calculation of the regression line. The amounts of the PTH-amino acids are calculated from the results of the reversed-phase HPLC chromatograms. PTH-phosphotyrosine was quantified from the results of the ion-exchange HPLC chromatograms. The hatched bars represent the amount of ^{32}P radioactivity found in each degradation step. Of the 6000 cpm covalently attached to the solid-phase support about 2500 cpm are recovered in cycle 4. The yield of ^{32}P radioactivity found in cycle 4 to 7 is about 50%.

exchange HPLC chromatogram of Edman degradation step 4 of the same peptide. It demonstrates that the ^{32}P radioactivity coelutes with PTH-phosphotyrosine.

Discussion

For the first time, it is possible to detect phosphotyrosine-containing peptides and proteins nonradioactively after partial hydrolysis by capillary electrophoresis of PTH-phosphotyrosine. Using these methods, phosphotyrosine-containing peptides can be purified from a proteolytic digest of a phosphotyrosine-containing protein. This allows the localization of the phosphotyrosine residue by sequence analysis as described in this chapter.

Employing the unmodified gas-phase sequenator, the yield of PTH-phosphotyrosine is usually between 0 and 2%, depending on the water content of the butyl chloride used as transfer solution from the cartridge to the conversion flask. Thus, the phosphorylation sites in the insulin receptor protein were identified only by the moderate rise in the radioactivity at those positions where phosphotyrosine residues are located.[7,9] In addition, the poor solubility of ATZ-phosphotyrosine in butyl chloride leads to a high overlap during sequence analysis, which renders an unambiguous identification difficult. These difficulties are overcome by the application of liquid trifluoroacetic acid (instead of butyl chloride), in which the phosphotyrosine derivatives are relatively soluble. Thus, taking trifluoroacetic acid as the transfer solvent, a >50% yield of PTH-phosphotyrosine can be achieved using either the modified Applied Biosystems gas-phase sequencer or the MilliGen/Biosearch solid-phase sequencer. This is a lower yield than found for the standard PTH-amino acids, which are soluble in trifluoroacetic acid almost quantitatively (see Fig. 6). However, trifluoroacetic acid is more efficient than the buffered methanol/water solution that was suggested before[11] and a higher recovery should be obtained by optimization of the transfer procedure from the reaction chamber to the conversion flask.

Another problem is the application of a suitable solid-phase support that allows a high efficient covalent coupling of the phosphotyrosine-containing peptide. This aim can be achieved with the Sequelon-AA membrane, developed by MilliGen/Biosearch. An optimized attachment protocol allows the coupling procedure to be performed with an overall efficiency of over 60%. Using these improved methods, the sequence analysis of a [^{32}P]phosphotyrosine-containing peptide becomes possible with low amounts of radioactivity (< 1000 cpm), allowing the identification of the phosphotyrosine residues unambiguously (Fig. 6).

The more significant advantage is the possibility of localizing phosphotyrosine during sequence analysis of phosphotyrosine-containing peptides that are not radioactively labeled. This is achieved by the development of ion-exchange HPLC and capillary electrophoresis methods that both allow the quantitative determination of PTH-phosphotyrosine.

These new tools for phosphotyrosine determination and sequence analysis open the door to further experiments allowing the study of *in vivo* regulation of tyrosine phosphorylation/dephosphorylation.

Acknowledgments

We wish to thank Dr. Ralf Nendza (Pharmacia) for the kind gift of Mono Q material; Dr. Dogan (Merck) for making the inert Merck-Hitachi pump available; Dr. Coull and Prof. Köster (MilliGen/Biosearch) for the Sequelon-AA membranes; Dr. Bovens (MilliGen/Biosearch) for making the ProSequencer available; Prof. Johns (University of Melbourne) for a sample of the peptide L-R-R-A-Y(P)-L-G. Many thanks are due to Prof. Pinna (University of Padua) for helpful discussions and to Prof. L. M. G. Heilmeyer, Jr. for support during the time of this study. This study was supported by the Deutsche Forschungsgemeinschaft, the Fonds der Chemischen Industrie, and the Italian CNR, Target Project on Biotechnology and Bioinstrumentation.

[18] Synthesis of *O*-Phosphoserine- and *O*-Phosphothreonine-Containing Peptides

By JOHN W. PERICH

The increased recognition of protein phosphorylation in many physiological processes has resulted in the need for efficient chemical methods for the synthesis of phosphorylated serine- and threonine-containing peptides for use as model substrates. Prior to 1980, phosphoserine [Ser(P)]-containing peptides were generally prepared by a "global phosphorylation" approach[1,2] that involved (1) the phosphorylation of a protected Ser-containing peptide with dibenzyl or diphenyl phosphorochloridate in pyridine followed by (2) hydrogenolytic deprotection. Although this approach has been successfully used for the preparation of many simple Ser(P)-containing peptides,[1] this approach is limited since the phosphorylation or deprotection steps are often incomplete and lead to major synthetic problems. Since 1980, however, these problems were overcome by the development of an alternative approach that involved the use of protected Boc-Ser(PO$_3$R$_2$)-OH derivatives in the *tert*-butyloxycarbonyl (Boc) mode of peptide synthesis.[3] This chapter describes the synthetic methods used for the preparation of Ser(P)- and phosphothreonine [Thr(P)] containing peptides by the use of (1) Boc-Ser(PO$_3$Ph$_2$)-OH and Boc-Thr(PO$_3$Ph$_2$)-OH in the Boc mode of peptide synthesis for the synthesis of protected Ser(PO$_3$Ph$_2$)- or Thr(PO$_3$Ph$_2$)-containing peptides followed by (2) their deprotection using modified hydrogenation conditions.

Equipment and Reagents

The synthesis of Ser(P)- and Thr(P)-peptides requires a laboratory equipped with a rotary evaporator, a high vacuum line, a fume hood, a hydrogenation apparatus, and a high-performance liquid chromatography (HPLC) instrument. In addition, the availability of nuclear magnetic resonance (NMR) spectroscopy (^{1}H, ^{13}C, and ^{31}P) and fast atom bombardment (FAB) mass spectrometry facilities is considered necessary for the proper analysis of protected amino acids and peptides.

[1] G. Folsch, *Sven. Kem. Tidskr.* **79**, 38 (1967).
[2] A. W. Frank, *CRC Crit. Rev. Biochem.* **16**, 51 (1984).
[3] J. W. Perich, Ph.D. Thesis, University of Melbourne, Melbourne, Australia (1986).

Synthesis of Ser(P)- and Thr(P)-Containing Peptides

The most suitable procedure for the synthesis of Ser(P)- and Thr(P)-containing peptides is accomplished by (1) the use of Boc-Ser(PO$_3$Ph$_2$)-OH or Boc-Thr(PO$_3$Ph$_2$)-OH in the Boc mode of peptide synthesis for the synthesis of protected Ser(PO$_3$Ph$_2$)- or Thr(PO$_3$Ph$_2$)-peptides followed by (2) the hydrogenolytic cleavage (platinum) of the phenyl phosphate groups.[4,5] This approach permits a phosphorylated serine residue to be incorporated at any specific site in a peptide during the peptide synthesis procedure, and is suitable for the synthesis of multiple Ser(P)-containing peptides and mixed Ser/Ser(P)-containing peptides.

The procedure for the synthesis of Ser(P)- and Thr(P)-containing peptides is outlined as follows: (1) preparation of Boc-Ser(PO$_3$Ph$_2$)-OH and Boc-Thr(PO$_3$Ph$_2$)-OH, (2) synthesis of protected Ser(PO$_3$Ph$_2$)- or Thr(PO$_3$Ph$_2$)-containing peptides, (3) peptide deprotection (including cleavage of phosphate-protecting groups), and (4) characterization of Ser(P)- and Thr(P)-containing peptides.

Preparation of Boc-Ser(PO$_3$Ph$_2$)-OH and Boc-Thr(PO$_3$Ph$_2$)-OH

The synthesis of Boc-Ser(PO$_3$Ph$_2$)-OH and Boc-Thr(PO$_3$Ph$_2$)-OH is accomplished by a three-step procedure which involves (1) initial protection of the carboxyl group of Boc-Ser-OH as its 4-nitrobenzyl ester, (2) phosphorylation of the hydroxyl group of Boc-Ser-ONBzl or Boc-Thr-ONBzl by the use of diphenyl phosphorochloridate in pyridine, and (3) hydrogenolytic cleavage of the 4-nitrobenzyl group from the carboxyl terminus (see Fig. 1).[6] Protection of the carboxy terminus prior to the phosphorylation step is necessary, since a carboxylic acid group readily reacts with diphenyl phosphorochloridate and leads to by-products. In this case, the 4-nitrobenzyl group is a suitable protecting group since it can be readily introduced to Boc-Ser-OH or Boc-Thr-OH without side-chain protection and can be cleaved in high yield by palladium-catalyzed hydrogenolysis or sodium dithionite reduction. The phosphorylation of Boc-Ser-ONBzl and Boc-Thr-ONBzl is straightforward and is accomplished by the use of diphenyl phosphorochloridate in pyridine. The cleavage of the 4-nitrobenzyl group from Boc-Ser(PO$_3$Ph$_2$)-ONBzl or Boc-Thr(PO$_3$Ph$_2$)-ONBzl is best performed by palladium-catalyzed hydrogenolysis using a hydrogen column apparatus operated at 1 atm of pressure.

[4] J. W. Perich, P. F. Alewood, and R. B. Johns, *Tetrahedron Lett.* **27**, 1373 (1986).
[5] J. W. Perich and R. B. Johns, *J. Org. Chem.* **53**, 4103 (1989).
[6] J. W. Perich, P. F. Alewood, and R. B. Johns, *Aust. J. Chem.* **44**, 233 (1991).

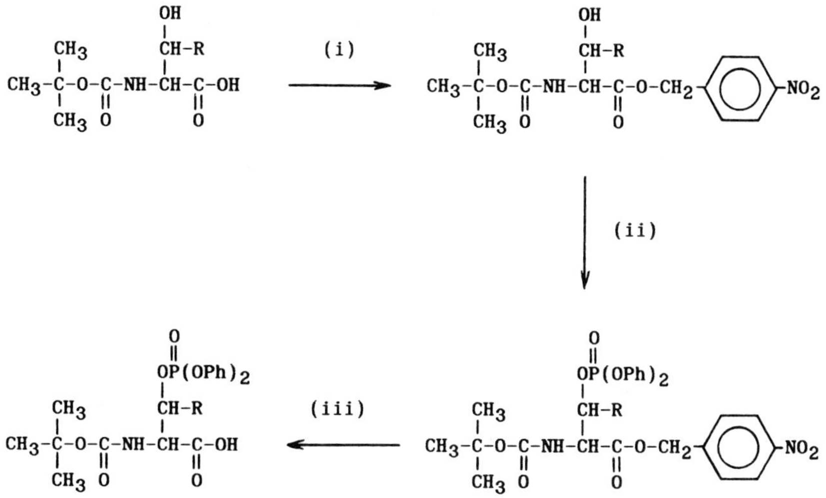

$$\underset{\underset{CH_3}{|}}{\overset{\overset{CH_3}{|}}{CH_3-C}}-O-\underset{\overset{\parallel}{O}}{C}-NH-\underset{\overset{|}{CH-R}}{\overset{|}{\overset{OH}{|}}}-\underset{\overset{\parallel}{O}}{C}-OH \quad \xrightarrow{(i)} \quad \underset{\underset{CH_3}{|}}{\overset{\overset{CH_3}{|}}{CH_3-C}}-O-\underset{\overset{\parallel}{O}}{C}-NH-\underset{\overset{|}{CH-R}}{\overset{\overset{OH}{|}}{}}-\underset{\overset{\parallel}{O}}{C}-O-CH_2-\langle\text{aryl}\rangle-NO_2$$

(ii)

$$\underset{\underset{CH_3}{|}}{\overset{\overset{CH_3}{|}}{CH_3-C}}-O-\underset{\overset{\parallel}{O}}{C}-NH-\underset{\overset{|}{CH-R}}{\overset{\overset{OP(OPh)_2}{\overset{|}{O}}}{}}-\underset{\overset{\parallel}{O}}{C}-OH \quad \xleftarrow{(iii)} \quad \underset{\underset{CH_3}{|}}{\overset{\overset{CH_3}{|}}{CH_3-C}}-O-\underset{\overset{\parallel}{O}}{C}-NH-\underset{\overset{|}{CH-R}}{\overset{\overset{OP(OPh)_2}{\overset{|}{O}}}{}}-\underset{\overset{\parallel}{O}}{C}-O-CH_2-\langle\text{aryl}\rangle-NO_2$$

FIG. 1. Synthesis of Boc-Ser(PO$_3$Ph$_2$)-OH and Boc-Thr(PO$_3$Ph$_2$)-OH. (i) 4-Nitrobenzyl bromide, triethylamine, ethyl acetate (80°, 6 hr); (ii) (PhO)$_2$P(O)Cl/pyridine [20°, 4 hr (R = H) or 16 hr (R = CH$_3$)]; (iii) H$_2$, 10% (w/v) palladium on charcoal, 5% (v/v) acetic acid/ethyl acetate.

Carboxyl Esterification Procedure[7]

1. 4-Nitrobenzyl bromide (125 mmol) is added to a solution of Boc-Ser-OH or Boc-Thr-OH (100 mmol) and triethylamine (125 mmol) in ethyl acetate (150 ml).

2. The solution is refluxed for 6 hr.

3. After cooling, water (60 ml) is added and the solution transferred to a separating funnel.

4. The aqueous phase is discarded and the organic phase successively washed with 1 M HCl (two times, 50 ml each) and 5% (w/v) NaHCO$_3$ (two times, 50 ml each) dried, (Na$_2$SO$_4$), and then filtered.

5. The solvent is then evaporated on a rotary evaporator under reduced pressure.

While Boc-Thr-ONBzl is obtained as a thick oil, Boc-Ser-ONBzl is obtained as a white solid and can be further purified by recrystallization from ethyl acetate/ligroine 60–80°. In the esterification procedure, particular care should be used in the handling of 4-nitrobenzyl bromide since this reagent is extremely corrosive and causes severe skin and eye irritation.

[7] P. F. Alewood, J. W. Perich, and R. B. Johns, *Synth. Commun.* **12**, 821 (1982).

Phosphorylation Procedure[6]

1. A solution of diphenyl phosphorochloridate (12.0 mmol) in tetrahydrofuran (5 ml) is added to a solution of Boc-Ser-ONBzl or Boc-Thr-ONBzl (10.0 mmol) in pyridine (10 ml) at 20°.

2. After stirring for 4 hr at 20° (for Ser) or overnight at 4° (for Thr), water (2 ml) is added and the solution stirred for 15 min.

3. Ethyl acetate (60 ml) (for Ser) or diethyl ether (60 ml) (for Thr) is added and the organic phase washed successively with 1 M HCl (three times, 30 ml each), 5% $NaHCO_3$ (two times, 30 ml each), 1 M HCl (once with 30 ml), dried (Na_2SO_4), and filtered.

4. The solvent is then evaporated on a rotary evaporator under reduced pressure.

5. The crude Boc-Ser(PO_3Ph_2)-ONBzl is recrystallized from ethyl acetate/lingroine 60–80°.

While complete phosphorylation of the primary hydroxyl group of Boc-Ser-ONBzl is effected after 2 hr, the slower phosphorylation of the secondary hydroxy group of Boc-Thr-ONBzl requires the phosphorylation solution to be stirred at 4° overnight for complete hydroxyl phosphorylation. In the case of Boc-Thr(PO_3Ph_2)-ONBzl, diethyl ether is the solvent of choice for product isolation since this low-polarity solvent permits the complete removal of diphenyl hydrogen phosphate (formed by hydrolysis of excess diphenyl phosphorochloridate) from the organic phase by sodium bicarbonate extraction. The use of ethyl acetate, dichloromethane, or chloroform is not recommended since sodium diphenyl phosphate cannot be completely extracted from these organic phases by base extraction. As Boc-Ser(PO_3Ph_2)-ONBzl is insoluble in diethyl ether, the isolation of this product is best performed by the use of ethyl acetate for product isolation followed by recrystallization from ethyl acetate/diethyl ether to removal diphenyl hydrogen phosphate from the crude isolated solid.

Hydrogenation Procedure[6]

1. Boc-Ser(PO_3Ph_2)-ONBzl or Boc-Thr(PO_3Ph_2)-ONBzl (10.0 mmol) is dissolved in 5% acetic acid/ethyl acetate (50 ml) and 10% palladium on charcoal (0.5 g) added.

2. The hydrogenation column is charged with hydrogen and the solution vigorously stirred until hydrogen uptake ceases.

3. The catalyst is removed by gravity filtration through filter paper [Whatman (Clifton, NJ) No. 1] and the solvent evaporated under reduced pressure.

4. The residue is dissolved in diethyl ether (60 ml) and the solution

transferred to a separating funnel and washed with 1 M HCl (once with 30 ml).

5. The organic phase is extracted with 5% $NaHCO_3$ (three times, 15 ml each) and the combined base extractions then washed with diethyl ether (once with 15 ml).

6. The aqueous phase is then acidified to pH 1 with 2 M HCl and the aqueous solution then extracted with dichloromethane (three times, 30 ml each).

7. The solvent is then evaporated on a rotary evaporator under reduced pressure.

While Boc-Thr(PO_3Ph_2)-OH is obtained as a thick oil, Boc-Ser (PO_3Ph_2)-OH is obtained as a clear oil which solidifies on storage and can be recrystallized from diethyl ether. Also, both products can be converted to their dicyclohexylamine (DCHA) derivatives and purified, if necessary, by recrystallization.

Synthesis of Protected Ser(PO_3Ph_2)- or Thr(PO_3Ph_2)-Containing Peptides

The protected Ser(PO_3Ph_2)-containing peptide is assembled by standard chemical procedures used in the Boc mode of solution-phase peptide synthesis. In solution-phase synthesis, the mixed anhydride condensation procedure[8] is generally the method of choice for the coupling of protected amino acids to peptides. An advantage in the use of phenyl phosphate groups is that, depending on the peptide sequence and its length, protected Ser(PO_3Ph_2)-containing peptides are generally soluble in common organic solvents (such as ethyl acetate and dichloromethane) and thereby peptide isolation and synthesis are facilitated. Also, the N-methylamine group is used for the peptide carboxyl terminus because peptide N-methylamides are generally obtained as solids and have good solubility properties.

Procedure

1. Dissolve Boc-Ser(PO_3Ph_2)-OH (1.4 Eq) in tetrahydrofuran and cool the solution to $-20°$ using dry ice/acetone.

2. Add a solution of N-methylmorpholine (1.4 Eq) in tetrahydrofuran.

3. Add a solution of isobutyl chloroformate (1.3 Eq) in tetrahydrofuran so that the temperature of the coupling solution is maintained at $-20°$.

4. After an activation period of 3 min, a solution of the peptide trifluoroacetate (1.0 Eq) and N-methylmorpholine (1.0 Eq) in tetrahydrofuran

[8] J. Meienhofer, in "The Peptides: Analysis, Synthesis, Biology" (E. Gross and J. Meienhofer, eds.), Vol. 1, Chapter 6. Academic Press, New York, 1983.

(or dichloromethane or *N*-methylpyrrolidone, depending on the solubility of the peptide) is added at $-20°$.

5. After a coupling period of 2 hr, 5% $NaHCO_3$ (5 ml) is added and the solution stirred for a further 30 min.

6. Transfer the solution to a separating funnel with the use of a suitable solvent (60 to 100 ml) (the selection of diethyl ether, ethyl acetate, dichloromethane, or chloroform being determined by the solubility of the peptide in the solvent). (*Note:* In some cases, a low-solubility peptide is isolated by aqueous precipitation.)

7. The organic phase is washed with 5% $NaHCO_3$ (two times, 30 ml each), 1 M HCl (two times, 30 ml each), the organic phase evaporated under reduced pressure, and then dried under high vacuum.

Peptide Deprotection (Including Cleavage of Phosphate-Protecting Groups)

The cleavage of the phenyl phosphate groups from $Ser(PO_3Ph_2)$- and $Thr(PO_3Ph_2)$-containing peptides is readily achieved by hydrogenolysis over platinum. While early phenyl reductions using catalytic quantities were often incomplete,[1] complete phenyl reduction is effected by the use of (1) 50% (v/v) CF_3CO_2H/CH_3CO_2H as hydrogenation solvent and (2) molar equivalents of platinum oxide per phenyl group.[4-6] Also, preliminary palladium-catalyzed cleavage of peptides containing benzylic groups (benzyl ether and ester, benzyloxycarbonyl, etc.) is unnecessary because the platinum-mediated hydrogenation cleaves the benzylic protecting groups simultaneously.

Hydrogenation Procedure[5,6]

1. The protected $Ser(PO_3Ph_2)$-containing peptide (generally 0.1 mmol) is dissolved in 50% CF_3CO_2H/CH_3CO_2H (4 ml) and 83% platinum oxide (1.1 mEq of PtO_2/phenyl group) added.

2. The hydrogenation column is charged with hydrogen and the solution vigorously stirred until hydrogen uptake ceases.

3. The platinum is removed by gravity filtration through filter paper (Whatman No. 1) and the solvent evaporated under reduced pressure. (*Note:* The platinum is also washed with water and the water washings separately evaporated under reduced pressure.)

4. The residue is triturated with diethyl ether (three times, 30 ml each) and the solid residue then dried under high vacuum.

Summary

A feature of this synthetic approach is that since protected peptides are obtained in high yield and the hydrogenolytic deprotection procedure proceeds cleanly, the isolated Ser(P)- and Thr(P)-containing peptides are often obtained in high purity (>99.5%) and, in most cases, do not require any further HPLC or anion-exchange chromatographic purification.

The advantages of the above phenyl phosphate-based protection system is that this approach permits the synthesis of (1) large, complex Ser(P)- and Thr(P)-containing peptides [such as Ac-Glu-Ser(P)-Leu-Ser(P)-Ser(P)-Ser(P)-Glu-Glu-NHMe], (2) mixed Ser/Ser(P)-containing peptides [such as H-Ser-Ser(P)-Ser(P)-NHMe · trifluoroacetic acid (TFA) and H-Ser(P)-Ser-Ser(P)-NHMe · TFA], and (3) multiple Ser(P)- and Thr(P)-containing peptides [such as H-Ser(P)-Ser(P)-Ser(P)-Glu-Glu-NHMe · TFA and H-Thr(P)-Thr(P)-Thr(P)-Glu-Glu-NHMe · TFA].

Characterization of Ser(P)- and Thr(P)-Containing Peptides

The characterization of Ser(P)- and Thr(P)-containing peptides is accomplished by the use of NMR spectroscopy (^{13}C and ^{31}P) and FAB mass spectrometry. In the ^{13}C NMR spectrum of these peptides, the C_α and the C_β carbons of the Ser(P) and the Thr(P) residues are observed as phosphorus-coupled doublet signals with coupling constants varying from 3 to 9 Hz. In the case of FAB mass spectrometry, greater information is obtained in the positive mode with the use of aqueous acetic acid/glycerol as the matrix phase (thioglycerol can also be used) and either argon or xenon as the ionization gas.[9] Apart from high-intensity pseudomolecular ions being obtained for Ser(P)- and Thr(P)-containing peptides, this spectrometry technique is also useful in establishing complete phenyl cleavage after hydrogenolytic deprotection. The interpretation of the FAB mass spectra for multiple Ser(P)- and Thr(P)-containing peptides can often be complicated due to the high-molecular-weight mass spectral region containing ion peaks due to extensive sodium and potassium complexation. Also, trace amounts of platinum in the peptide can give rise to ion peaks of +96 mass units due to platinum complexation of the peptide in the matrix. However, the observation of platinum–complex ions can be overcome by HPLC purification of the peptide.

Solid-Phase Synthesis of Ser(P)- and Thr(P) Peptides

The solid-phase synthesis of Ser(P)- and Thr(P)-containing peptides is also possible by the use of Boc-Ser(PO_3Ph_2)-OH in the Merrifield mode[10] of solid-phase peptide synthesis.

[9] R. B. Johns, P. F. Alewood, J. W. Perich, A. L. Chaffee, and J. K. MacLeod, *Tetrahedron Lett.* **27**, 4791 (1986).
[10] G. Baramy and R. B. Merrifield, *in* "The Peptides: Analysis, Synthesis, Biology" (E. Gross and J. Meienhofer, eds), Vol. 2, Chapter 1. Academic Press, New York, 1983.

Procedure

1. Prepare Boc-Ser(PO₃Ph₂)-OH as described above.

2. Assemble peptide according to the method described by Baramy and Merrifield,[10] using 3 Eq of the Boc-amino acid for dicyclohexycarbodiimide (DCC)/1-hydroxybenzotriazole (HOBt) couplings and 50% CF_3CO_2H/CH_2Cl_2 for Boc cleavage.

3. The Ser(PO₃Ph₂)-containing peptide is cleaved from the resin using method a, b, or c.

    a. A suspension of the peptide–resin in dry dimethylformamide (DMF) (30 ml) containing palladium acetate (0.3 g) is hydrogenated at 60 psi for 24 hr (60°). The catalyst is removed by gravity filtration (Whatman No. 1) and the solvent evaporated under reduced pressure.

    b. Dry hydrogen bromide is bubbled into a suspension of the peptide–resin in trifluoroacetic acid (10 ml) for 90 min (20°). The solvent is evaporated under reduced pressure and the residue triturated with diethyl ether (two times, 30 ml each) and dried under high vacuum.

    c. Hydrogen fluoride is distilled into a Kel-F vessel (Peninsula, CA) containing the peptide–resin and anisole (10%) and the solution allowed to stand at 0° for 45 min. The HF is evaporated under reduced pressure and the residue is triturated with diethyl ether (two times, 30 ml each) and dried under high vacuum.

4. The isolated peptide residue is hydrogenated with platinum oxide as described above.

5. The solvent is evaporated under reduced pressure and the residue is triturated with diethyl ether and then dried under high vacuum.

6. The peptide is purified by semipreparative reversed-phase HPLC or anion-exchange chromatography.

This procedure was used by Perich *et al.*[11] for the synthesis of Glu-Ser(P)-Leu using palladium acetate-mediated hydrogenation (step 3,a above) or HBr/CF₃CO₂H (step 3,b above) for the cleavage of the Ser(PO₃Ph₂)-containing tripeptide from the polystyrene resin. Also, Arendt *et al.*[12] prepared Leu-Arg-Arg-Ser(P)-Leu-Gly using liquid hydrogen fluoride (step 3,c above) for the cleavage of a Ser(PO₃Ph₂)-containing peptide from a polystyrene resin (or phenylacetamidomethyl resin) support.

---

[11] J. W. Perich, R. M. Valerio, and R. B. Johns, *Tetrahedron Lett.* **27,** 1377 (1986).

[12] A. Arendt, K. Palczewski, W. T. Moore, R. M. Caprioli, J. H. McDowell, and P. A. Hargrave, *Int. J. Pept. Protein Res.* **33,** 468 (1989).

Synthesis of Ser(P)-Containing Peptides through Benzyl
   Phosphate Protection

   The synthesis of Ser(P)- and Thr(P)-containing peptides is also possible by the use of benzyl phosphate protection. A feature of this approach is that the benzyl phosphate groups are quantitatively cleaved by palladium-catalyzed hydrogenolysis. Three variations of this approach are briefly described.

   Using Boc-Ser(PO$_3$Bzl$_2$)-OH.[6,13] Boc-Glu(OBzl)-Ser(PO$_3$Bzl$_2$)-Leu-OBzl was prepared in 81% yield[6,14] by the use of Boc-Ser(PO$_3$Bzl$_2$)-OH in peptide synthesis with formic acid used for cleavage of the Boc group from the Boc-dipeptide.[6] However, the use of Boc-Ser(PO$_3$Bzl$_2$)-OH in peptide synthesis is limited due to the acid sensitivity of benzyl phosphate groups; approximately 50, 10, and 1% benzyl loss occurs after a 60-min treatment with 4 M HCl/dioxane, 1 M HCl/acetic acid, or formic acid, respectively.

   Using Ppoc-Ser(PO$_3$Bzl$_2$)-OH. By changing to the more acid-labile 2-phenylisopropyloxycarbonyl group (Ppoc), Boc-Glu(OBzl)-Ser(PO$_3$Bzl$_2$)-Leu-OBzl was prepared in 94% yield by the use of Ppoc-Ser(PO$_3$Bzl$_2$)-OH in peptide synthesis with 0.5 M HCl/dioxane used for cleavage of the Ppoc group from the Ppoc-dipeptide. However, in general applications, the use of Ppoc-Ser(PO$_3$Bzl$_2$)-OH in peptide synthesis requires subsequent peptide extension of the Ser(PO$_3$Bzl$_2$)-containing peptide to adopt the Ppoc or Bpoc mode of peptide synthesis.

   Using Boc-Ser(PO$_3$BrBzl$_2$)-OH. The 4-bromobenzyl group can also be used for phosphate protection and takes advantage of the fourfold increase in acid stability of the 4-bromobenzyl group over the benzyl group in 1 M HCl/acetic acid or formic acid. For example, Boc-Glu(OBzl)-Ser(PO$_3$BrBzl$_2$)-Leu-OBzl was prepared in 94% yield by the use of Boc-Ser(PO$_3$BrBzl$_2$)-OH in peptide synthesis and formic acid for the cleavage of the Boc group from the Boc-dipeptide.

[13] P. F. Alewood, J. W. Perich, and R. B. Johns, *Aust. J. Chem.* **37,** 429 (1984).
[14] P. F. Alewood, J. W. Perich, and R. B. Johns, *Tetrahedron Lett.* **25,** 987 (1984).

[19] Synthesis of O-Phosphotyrosine-Containing Peptides

By JOHN W. PERICH

It is only recently that tyrosine kinases have been identified and that protein phosphorylation of tyrosine residues has been recognized in numerous cellular processes.[1,2] As past syntheses of synthetic phosphotyrosine [Tyr(P)]-containing peptides were inadequate,[3] the investigation of more efficient chemical methods was undertaken in 1980 so as to provide suitable model Tyr(P)-containing peptide substrates for biochemical study. As a result of such studies, it is now possible to prepare complex Tyr(P)-containing peptides by the use of protected tyrosine derivatives in butyloxycarbonyl (Boc)- or 9-fluorenylmethyloxycarbonyl (Fmoc)-peptide synthesis.

Both Boc-Tyr(PO_3Bzl_2)-OH and Boc-Tyr(PO_3Me_2)-OH have been used in peptide synthesis for the preparation of Tyr(P)-containing peptides. In the case of benzyl phosphate protection, Perich and Johns[4] prepared Tyr(P)-Leu-Gly-OH by the use of Boc-Tyr(PO_3Bzl_2)-OH for the solution-phase synthesis of Boc-Tyr(PO_3Bzl_2)-Leu-Gly-OBzl followed by quantitative deprotection by palladium-catalyzed hydrogenolysis in 40% (v/v) CF_3CO_2H/CH_3CO_2H. However, due to the sensitivity of benzyl phosphate groups to the acid conditions used for the removal of the Boc group (i.e., 4 M HCl/dioxane and 40% $CF_3CO_2H/CH_2Cl_2^{4,5}$), this derivative is therefore limited to the synthesis of N-terminal Tyr(PO_3Bzl_2)-containing peptides. In a solid-phase approach (Merrifield, see Ref. 14), Gibson et al.[6] prepared Arg-Tyr(P)-Val-Phe by the use of Boc-Tyr(PO_3Bzl_2)-OH with peptide–resin cleavage and peptide deprotection effected by a final HF/10% (v/v) anisole deprotection step. However, the yields of Tyr(P)-peptides are low with this approach since extensive dephosphorylation of the Tyr(P) residue occurs in liquid hydrogen fluoride.

The methyl phosphate-protected derivative, Boc-Tyr(PO_3Me_2)-OH, has proved to be better suited to peptide synthesis and has been used for

---

[1] P. J. Blackshear, A. C. Nairn, and J. F. Kuo, *FASEB J.* **2**, 2957 (1988).

[2] G. Carpenter, *Annu. Rev. Biochem.* **56**, 881 (1987).

[3] A. W. Frank, *CRC Crit. Rev. Biochem.* **16**, 51 (1984).

[4] J. W. Perich and R. B. Johns, *J. Org. Chem.* **54**, 1750 (1989).

[5] R. M. Valerio, P. F. Alewood, R. B. Johns, and B. E. Kemp, *Int. J. Pept. Protein Res.* **33**, 428 (1989).

[6] B. W. Gibson, A. M. Falick, A. L. Burlingame, L. Nadasdi, A. C. Nguyen, and G. L. Kenyon, *J. Am. Chem. Soc.* **109**, 5343 (1987).

the solution-phase synthesis of Asn-Glu-Tyr(P)-Thr-Ala and Pro-Tyr(P)-Val.[7,8] Depending on the sequence of the peptide, the cleavage of methyl phosphate groups is possible by the use of either (1) 33% HBr/acetic acid, (2) bromotrimethylsilane/CH_3CN, (3) 1 M bromotrimethylsilane–thioanisole/CF_3CO_2H/m-cresol, (4) trifluoromethanesulfonic acid/CF_3CO_2H/thioanisole/m-cresol (1 : 5 : 3 : 1), or (5) 1 M trimethylsilyl triflate–thioanisole/CF_3CO_2H/m-cresol.

The preparation of Tyr(P)-containing peptides can be accomplished by (1) the use of Boc-Tyr(PO_3Me_2)-OH in Boc solution or solid-phase peptide synthesis or (2) the use of Fmoc-Tyr(PO_3Me_2)-OH in Fmoc/polyamide solid-phase peptide synthesis.

Preparation of Tyr(P)-Containing Peptides by Boc-Peptide Synthesis

The procedure for the synthesis of Tyr(P)-containing peptides by Boc-peptide synthesis is outlined as follows: (1) preparation of Boc-Tyr (PO_3Me_2)-OH, (2) synthesis of protected Tyr(PO_3Me_2)-containing peptides (solution and solid phase), and (3) peptide deprotection (including cleavage of methyl phosphate groups).

The synthesis of the protected derivative, Boc-Tyr(PO_3Me_2)-OH, is accomplished from Boc-Tyr-OH by a simple three-step procedure which involves (1) initial esterification of Boc-Tyr-OH with 4-nitrobenzyl bromide or 2-(bromomethyl)anthraquinone, (2) phosphoro triester or "phosphite triester" phosphorylation of the tyrosyl hydroxyl group, followed by (3) hydrogenolytic or reductive cleavage of the 4-nitrobenzyl or 2-methyleneanthraquinonyl (Maq) ester group.

Preparation of Boc-Tyr(PO_3Me_2)-OH (Fig. 1)

1. Carboxyl Esterification Procedure[8,9]

1. 4-Nitrobenzyl bromide or 2-(bromomethyl)anthraquinone (12.5 mmol) is added to a solution of Boc-Tyr-OH (10 mmol) and triethylamine (12.5 mmol) in ethyl acetate (30 ml).

2. The solution is refluxed for 6 hr.

3. After cooling, water is added and the organic phase successively washed with 1 M HCl and 5% $NaHCO_3$, dried (Na_2SO_4), filtered, and the solvent then evaporated under reduced pressure.

[7] R. M. Valerio, J. W. Perich, E. A. Kitas, P. F. Alewood, and R. B. Johns, *Aust. J. Chem.* **42**, 1519 (1989).

[8] E. A. Kitas, J. W. Perich, R. B. Johns, and G. W. Tregear, *J. Org. Chem.* **55**, 4181 (1990).

[9] J. Jentsch and E. Wunsch, *Justus Liebigs Ann. Chem.* **97**, 2490 (1964).

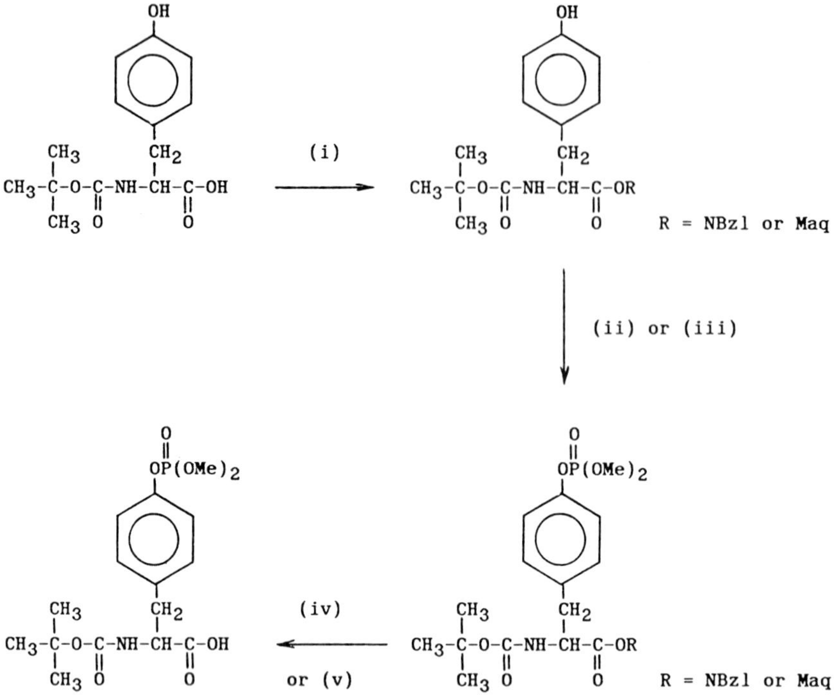

FIG. 1. Preparation of Boc-Tyr(PO$_3$Me$_2$)-OH. (i) 4-Nitrobenzyl bromide or 2-(bromomethyl)anthraquinone, triethylamine, ethyl acetate (80°, 6 hr); (ii) NaH (30 min), then (MeO)$_2$P(O)Cl (30 min, 20°); (iii) (MeO)$_2$PNEt$_2$/1H-tetrazole (20 min, 20°), then MCPBA (0°, 10 min); (iv) H$_2$, 10% palladium on charcoal, 5% acetic acid/ethyl acetate; and (v) Na$_2$S$_2$O$_4$/Na$_2$CO$_3$ (1 hr, 50°).

The phosphorylation of Boc-Tyr-ONBzl or Boc-Tyr-OMaq is achieved by its initial treatment with 1 Eq of sodium hydride at −60° followed by *in situ* treatment of the resultant sodium phenoxide with dimethyl phosphorochloridate. In this phosphorylation, the generation of the phenoxide intermediate gives an orange solution which, after addition of dimethyl phosphorochloridate, becomes colorless due to rapid phosphorylation of the sodium phenoxide intermediate. As this phosphorylation step is water sensitive, it is necessary that dry tetrahydrofuran or dioxane is used. An advantage in the use of the 2-methylanthraquinone group is that this protecting group generally leads to solid products; for example, Boc-Tyr(PO$_3$Me$_2$)-ONBzl is isolated as a yellow oil while Boc-Tyr(PO$_3$Me$_2$)-OMaq is obtained as a yellow solid and can be purified by crystallization.

2a. Phosphoro Triester Phosphorylation Procedure[5,8]

1. Sodium hydride is washed with pentane (three times, 5 ml each) and dried under a stream of dry nitrogen.

2. Sodium hydride (5.0 mmol) is suspended in dry tetrahydrofuran (5 ml) (or dioxane) and cooled to 10°.

3. A solution of Boc-Tyr-ONBzl or Boc-Tyr-OMaq (4.0 mmol) in dry tetrahydrofuran (5 ml) (or dioxane) is added to the hydride suspension with vigorous stirring.

4. After stirring for 30 min at 20°, the solution is cooled to 10° and dimethyl phosphorochloridate (6.0 mmol) is added in one portion to the bright orange solution (the solution becomes colorless).

5. After stirring for 30 min at 20°, water (1 ml) is added and the solvents are evaporated under reduced pressure.

6. The residue is dissolved in ethyl acetate (30 ml) and the organic phase washed successively with 1 M HCl (10 ml) and 5% $NaHCO_3$ (10 ml). The organic phase is dried (Na_2SO_4), filtered, and the solvent then evaporated under reduced pressure.

7. In the case of Boc-Tyr(PO_3Me_2)-OMaq, the solid is triturated with diethyl ether and then recrystallized from ethyl acetate/ligroine 60–80° (mp 149–150°).

This derivative can also be prepared by the "phosphite triester" phosphorylation of Boc-Tyr-ONBzl or Boc-Tyr-OMaq with (1) $(Meo)_2PNEt_2$/ 1H-tetrazole followed by (2) m-chloroperoxybenzoic acid (MCPBA) oxidation. It should be noted that dimethyl N,N-diethylphosphoramidite is particularly pungent and that its synthesis uses dangerous reagents and therefore should only be performed by experienced personnel. Also, as phosphite triester phosphorylation reagents are very sensitive to water, all reagents and tetrahydrofuran should be thoroughly dried. If possible, tetrahydrofuran should be freshly distilled from a potassium/benzophenone still.

2b. Phosphite Triester Phosphorylation Procedure[8]

1. Boc-Tyr-ONBzl or Boc-Tyr-OMaq (5.0 mmol) and dimethyl N,N-diethylphosphoramidite (5.5 mmol) are dissolved in dry tetrahydrofuran (5 ml) at 20°.

2. 1H-Tetrazole (16.5 mmol) is added in one portion to the stirred solution at 20°.

3. After stirring for 20 min, the solution is cooled to −40° and a solution of 85% m-chloroperoxybenzoic acid (6.0 mmol) in CH_2Cl_2 (12 ml) or 14% $tert$-butyl hydroperoxide (3.8 ml, 6.0 mmol) is added so that the temperature of the reaction solution is kept below 0°.

4. After stirring for 10 min, a solution of 10% $Na_2S_2O_5$ (5 ml) is added to the solution at 0°.

5. The solution is transferred to a separating funnel using ethyl acetate (30 ml) and the organic phase washed successively with 5% $NaHCO_3$ (15 ml) and 1 M HCl (15 ml). The organic phase is then dried (Na_2SO_4), filtered, and the solvent then evaporated under reduced pressure.

6. In the case of Boc-Tyr(PO_3Me_2)-OMaq, the solid is triturated with diethyl ether and then recrystallized from ethyl acetate/ligroine 60–80° (mp 149–152°).

3a. Hydrogenation Procedure[5]

1. Boc-Tyr(PO_3Me_2)-ONBzl or Boc-Tyr(PO_3Me_2)-OMaq (10.0 mmol) is dissolved in 5% acetic acid/methanol (50 ml) and 10% palladium on charcoal (0.5 g) added.

2. The hydrogenation column is charged with hydrogen and the solution vigorously stirred until hydrogen uptake ceases.

3. The catalyst is removed by gravity filtration through filter paper [Whatman (Clifton, NJ) No. 1] and the solvent evaporated under reduced pressure.

4. The residue is dissolved in diethyl ether (60 ml) and the organic phase washed with 1 M HCl (30 ml).

5. The organic phase is extracted with 5% $NaHCO_3$ (three times, 15 ml each) and the aqueous phase then washed with diethyl ether (15 ml).

6. The aqueous phase is then acidified to pH 1 with 2 M HCl and the aqueous solution then extracted with dichloromethane (three times, 30 ml each).

7. The solvent is then evaporated under reduced pressure.

In the case where laboratories lack a hydrogenation apparatus, the synthesis of Boc-Tyr(PO_3Me_2)-OH can be accomplished by the sodium dithionite reduction of the 2-methylanthraquinone-protecting group. The methylanthraquinone group is preferred over the use of the 4-nitrobenzyl group in this route since, unlike dithionite reduction of the 4-nitrobenzyl group, dithionite reduction of the methylanthraquinone group does not proceed with the formation of polymeric by-products.

3b. Sodium Dithionite Reduction Procedure[8]

1. Boc-Tyr(PO_3Me_2)-OMaq (8.5 mmol) is dissolved in acetonitrile (50 ml) and a solution of sodium dithionite (34.0 mmol) and sodium carbonate (34.0 mmol) in hot water (25 ml) added.

2. The solution is vigorously stirred at 50° for 1 hr, cooled at 20°, and then acidified to pH 1 with 1 M HCl.

3. The acetonitrile is then evaporated under reduced pressure.

4. Diethyl ether (60 ml) is then added and the aqueous phase discarded.

5. The organic phase is washed with 1 M HCl; (30 ml) and the organic phase extracted with 5% NaHCO$_3$ (three times, 15 ml each). The combined base extracts are combined and the aqueous phase then washed with diethyl ether (15 ml).

6. The aqueous phase is then acidified to pH 1 with 2 M HCl and the aqueous solution then extracted with dichloromethane (three times, 30 ml each).

7. The solvent is then evaporated under reduced pressure and the thick oil dried under high vacuum.

In addition, Boc-Tyr(PO$_3$Me$_2$)-OH can also be prepared by a one-pot procedure which uses *in situ tert*-butyldimethylsilyl protection of the carboxyl terminus of Boc-Tyr-OH followed by phosphite triester phosphorylation of the tyrosyl hydroxyl group using dimethyl N,N-diethylphosphoramidite.[10]

One-Pot Procedure

1. Boc-Tyr-OH (3.0 mmol) is dissolved in tetrahydrofuran (9 ml) and solutions of N-methylmorpholine (3.0 mmol) in tetrahydrofuran (1 ml) and *tert*-butyldimethylchlorosilane (3.0 mmol) in tetrahydrofuran (1 ml) are added, respectively.

2. After 2 min at 20°, a solution of dimethyl N,N-diethylphosphoramidite (1.0 mmol) in tetrahydrofuran (THF) (1 ml) is added followed by the addition of 1H-tetrazole (9.0 mmol) in one portion.

3. After 20 min at 20°, the reaction solution is cooled to $-40°$ and a solution of iodine (4.0 mmol) in tetrahydrofuran/water (3 : 1, 4 ml) added.

4. After 10 min at 20°, a solution of 10% Na$_2$S$_2$O$_5$ (10 ml) is added and the solution stirred for a further 10 min.

5. The solution is transferred to a separating funnel using diethyl ether (40 ml) and the aqueous phase discarded.

6. The organic phase is washed with 10% Na$_2$S$_2$O$_5$ (30 ml) and then extracted with 5% NaHCO$_3$ (three times, 15 ml each).

7. The combined base extracts are washed with diethyl ether (30 ml) and then acidified to pH 1 with 30% HCl.

8. The organic extracts are combined, dried (Na$_2$SO$_4$), filtered, and evaporated under reduced pressure to give Boc-Tyr(PO$_3$Me$_2$)-OH as a thick oil.

---

[10] J. W. Perich and R. B. Johns, *Synthesis* p. 701 (1989).

Solution-Phase Synthesis of Tyr(P)-Containing Peptides

The isobutoxycarbonyl mixed anhydride coupling procedure[11] is the method of choice for the addition of protected amino acids and Boc-Tyr(PO₃Me₂)-OH to peptides. This coupling method is favored since it is simple, rapid, gives high product yields, and uses readily available reagents. In addition to the mixed anhydride coupling procedure, the dicyclohexylcarbodiimide (DCC)/1-hydroxybenzotriazole (HOBt) and active ester coupling procedures can also be used.

Procedure

1. Dissolve Boc-Tyr(PO₃Me₂)-OH (1.4 Eq) in tetrahydrofuran and cool the solution to −20°, using dry ice/acetone cooling.
2. Add solution of *N*-methylmorpholine (1.4 Eq) in THF.
3. Add solution of isobutyl chloroformate (1.3 Eq) in THF so that the temperature of the coupling solution is maintained at −20°.
4. After an activation period of 3 min, a solution of the peptide trifluoroacetate (1.0 Eq) and *N*-methylmorpholine (1.0 Eq) in either tetrahydrofuran (or dichloromethane or *N*-methylpyrrolidone, depending on the solubility of the peptide) is added at − 20°.
5. After a coupling period of 1.5 to 2 hr, 5% NaHCO₃ is added and the solution stirred for a further 30 min.
6. The solution is transferred to a separating funnel, using a suitable solvent (the selection of diethyl ether, ethyl acetate, dichloromethane, or chloroform being determined by the solubility of the peptide in the solvent). (*Note:* In some cases, the peptide is isolated by aqueous precipitation.)
7. The organic phase is washed with 5% NaHCO₃ (two times, 30 ml each), 1 *M* HCl (two times, 30 ml each), the organic phase evaporated under reduced pressure, and then dried under high vacuum.

Peptide Deprotection (Including Cleavage of Methyl Phosphate Groups)

The deprotection of protected Tyr(PO₃Me₂)-containing peptides is performed in accordance with standard peptide deprotection procedures.[12] The cleavage of the methyl phosphate groups is accomplished by acidolytic

[11] J. Meienhofer, *in* "The Peptides: Analysis, Synthesis, Biology" (E. Gross and J. Meienhofer, eds), Vol. 1, Chapter 6. Academic Press, New York, 1983.
[12] H. Yajima and N. Fujii, *in* "The Peptides: Analysis, Synthesis, Biology" (E. Gross and J. Meienhofer, eds), Vol. 5, Chapter 2. Academic Press, New York, 1983.

or silylitic treatment and, if possible, can be monitored by ^{31}P nuclear magnetic resonance (NMR) spectroscopy.

Procedure

Method 1[5]: The peptide (0.1 mmol) is dissolved in 33% HBr/acetic acid (5 ml) and left at 20° for 24 hr (or as determined by ^{31}P NMR spectroscopy to be complete). The solvent is evaporated under reduced pressure, the residue triturated with diethyl ether, and then purified by reversed-phase high-performance liquid chromatography (RP-HPLC) or anion-exchange chromatography.

Method 2[7]: The peptide (0.1 mmol) is dissolved in 10% bromotrimethyl-silane/CH$_3$CN (20 ml) and left at 20° for 5 hr (or as determined by ^{31}P NMR spectroscopy to be complete). The solvent is evaporated under reduced pressure, the residue triturated with diethyl ether, and then purified by RP-HPLC or anion-exchange chromatography.

Method 3[7,8,13]: The peptide (0.1 mmol) is dissolved in 1 M bromotri-methylsilane and thioanisole in CF$_3$CO$_2$H (3 ml) containing m-cresol (10 mEq/peptide) at 0° and left at 20° for 12 hr (or as methyl cleavage was determined by ^{31}P NMR spectroscopy to be complete). The solvent is evaporated under reduced pressure and cold diethyl ether is added. The precipitated residue is washed with diethyl ether and then purified by RP-HPLC or anion-exchange chromatography.

Method 4[7,8,13]: The peptide (0.1 mmol) is dissolved in trifluorometh-anesulfonic acid/CF$_3$CO$_2$H/thioanisole (or dimethyl sulfide)/m-cresol (1 : 5 : 3 : 1) (3 ml) and left at 20° for 1 hr (or as methyl cleavage was determined by ^{31}P NMR spectroscopy to be complete). The solvent is evaporated under reduced pressure and cold diethyl ether is added. The precipitated residue is washed with diethyl ether and then purified by RP-HPLC or anion-exchange chromatography.

Method 5[8]: The peptide (0.1 mmol) is dissolved in 1 M trimethylsilyl triflate and thioanisole in CF$_3$CO$_2$H (3 ml) containing m-cresol (10 mEq/peptide) and left at 20° for 24 hr (or as methyl cleavage was determined by ^{31}P NMR spectroscopy to be complete). Methanol is added to the solution and solvent is evaporated under reduced pressure. Cold diethyl ether is added and the precipitated residue is washed with diethyl ether and then purified by RP-HPLC or anion-exchange chromatography.

[13] E. A. Kitas, J. W. Perich, R. B. Johns, and G. W. Tregear, *Tetrahedron Lett.* **29,** 3591 (1988).

Boc/Solid-Phase Synthesis of Tyr(P)-Containing Peptides

The synthesis of Tyr(P)-containing peptides is possible by the use of Boc-Tyr(PO$_3$Me$_2$)-OH in the Merrifield solid-phase methodology.[14]

Boc Method

1. Prepare Boc-Tyr(PO$_3$Me$_2$)-OH as outlined above.
2. Assemble the peptide according to the method described by Merrifield, using a polystyrene support, 40% CF$_3$CO$_2$H/CH$_2$Cl$_2$ for Boc cleavage, and DCC/HOBt amino acid couplings performed with 3 Eq of the Boc-amino acid.
3. Peptide–resin cleavage: Dry hydrogen bromide is bubbled into a suspension of the peptide–resin in trifluoroacetic acid (10 ml) for 90 min (at 20°). The solvent is evaporated under reduced pressure, the residue dissolved in 10% acetic acid/water, and the solution filtered (Whatman No. 1) to remove resinous material. The solvent is then evaporated under reduced pressure, the residue triturated with diethyl ether (two times, 30 ml), and then dried under high vacuum.
4. Methyl phosphate cleavage: The above isolated solid is dissolved in 33% HBr/acetic acid (5 ml) and the solution kept at 20° for 15 hr.
5. The solvent is evaporated under reduced pressure, the residue triturated with diethyl ether, and the residue then dried under high vacuum.
6. The peptide is purified by semipreparative reversed phase HPLC or anion-exchange chromatography.

In the first demonstration of this method, Valerio *et al.*[5,15] prepared Leu-Arg-Arg-Ala-Tyr(P)-Leu-Gly with the use of nitro protection for the arginine side-chain groups; the nitro groups were cleaved by palladium acetate-mediated hydrogenolysis in acetic acid (60 psi, 24 hr).

Fmoc/Solid-Phase Synthesis of Tyr(P)-Containing Peptides

The procedure for the synthesis of Tyr(P)-containing peptides by the use of Fmoc-Tyr(PO$_3$Me$_2$)-OH in Fmoc-polyamide solid-phase synthesis[16] is outlined as follows: (1) preparation of Fmoc-Tyr(PO$_3$Me$_2$)-OH, (2) solid-phase synthesis of protected Tyr(PO$_3$Me$_2$)-containing peptides, and (3) peptide deprotection (including cleavage of methyl phosphate groups).

[14] G. Baramy and R. B. Merrifield, *in* "The Peptides: Analysis, Synthesis, Biology" (E. Gross and J. Meienhofer, eds), Vol. 2, Chapter 1. Academic Press, New York, 1983.
[15] R. M. Valerio, P. F. Alewood, R. B. Johns, and B. E. Kemp, *Tetrahedron Lett.* **25,** 2609 (1984).
[16] E. Atherton and R. C. Sheppard, "Solid Phase Peptide Synthesis—A Practical Approach." IRL Press at Oxford Univ. Press, Oxford, 1989.

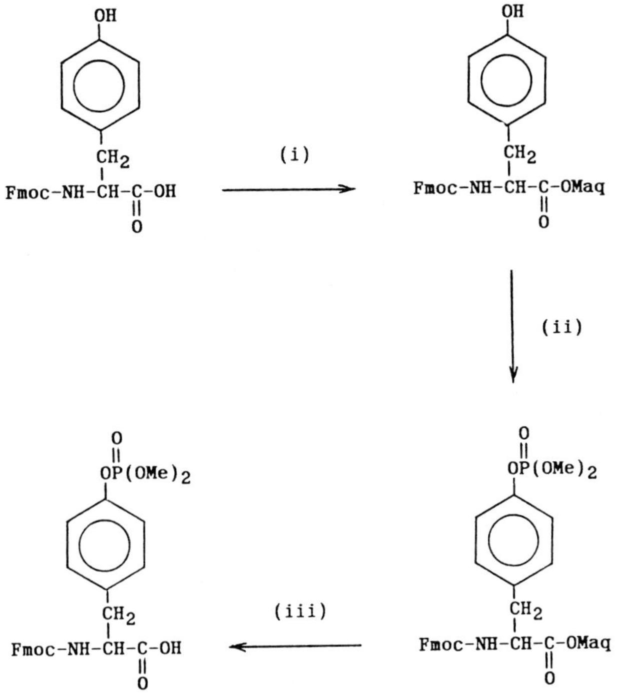

FIG. 2. Preparation of Fmoc-Tyr(PO₃Me₂)-OH. (i) 2-(Bromomethyl)anthraquinone, triethylamine, ethyl acetate (80°, 6 hr); (ii) (MeO)₂PNEt₂/1*H*-tetrazole (20 min, 20°), then MCPBA (0°, 10 min); and (iii) Na₂S₂O₄/Na₂CO₃ (1 hr, 50°).

Preparation of Fmoc-Tyr(PO₃Me₂)-OH.[17] Phosphite triester phosphorylation procedure (Fig. 2)

1. Fmoc-Tyr-OMaq (5.0 mmol) and dimethyl *N,N*-diethylphosphoramidite (5.5 mmol) are dissolved in dry tetrahydrofuran (5 ml) at 20°.

2. 1*H*-Tetrazole (16.5 mmol) is added in one portion to the stirred solution at 20°.

3. After stirring for 30 min at 20°, the solution is cooled to −40° and a solution of *m*-chloroperoxybenzoic acid (6.0 mmol) in CH₂Cl₂ (12 ml) or 14% *tert*-butyl hydroperoxide (3.8 ml) is added so that the temperature of the reaction solution is kept below 0°.

4. After stirring for 10 min at 0°, a solution of 10% Na₂S₂O₅ (5 ml) is added to the reaction solution at 0°.

5. The solution is transferred to a separating funnel using ethyl acetate

[17] E. A. Kitas, J. W. Perich, J. D. Wade, R. B. Johns, and G. W. Tregear, *Tetrahedron Lett.* **30**, 6229 (1989).

(30 ml) and the organic phase washed successively with 5% NaHCO₃ (15 ml) and 1 M HCl (15 ml). The organic phase is then dried (Na₂SO₄), filtered, and the solvent then evaporated under reduced pressure.

6. The solid is triturated with diethyl ether and then dried under high vacuum.

Reduction procedure

1. Fmoc-Tyr(PO₃Me₂)-OMaq (4.25 mmol) is dissolved in acetonitrile (25 ml) and a solution of sodium dithionite (17.0 mmol) and sodium carbonate (17.0 mmol) in hot water (12.5 ml) added.

2. The solution is vigorously stirred at 50° for 1 hr, cooled to 20°, and then acidified to pH 1 with 1 M HCl.

3. The acetonitrile is then evaporated under reduced pressure.

4. Diethyl ether (60 ml) is then added and the aqueous phase discarded.

5. The organic phase is washed with 1 M HCl (30 ml) and the organic phase extracted with 5% NaHCO₃ (three times, 15 ml each). The combined base extracts are combined and the aqueous phase then washed with diethyl ether (15 ml).

6. The aqueous phase is acidified to pH 1 with 2 M HCl and the aqueous solution then extracted with dichloromethane (three times, 30 ml each).

7. The solvent is then evaporated on a rotary evaporator under reduced pressure.

8. The isolated oil is dried under high vacuum to give a light white honeycomb solid.

Solid-Phase Synthesis of Protected Tyr(PO₃Me₂)-Containing Peptides.[17] The solid-phase synthesis is performed using a kieselguhr-polydimethylacrylamide resin that is functionalized with the acid-labile 4-hydroxymethylphenoxyacetic acid linkage and can include a β-alanine internal reference. Due to variations in commercially available solid-phase peptide synthesizers, the synthesis of peptides should be performed in accordance with the manufacturer's recommendations.

Procedure

1. Prepare Fmoc-Tyr(PO₃Me₂)-OH as outlined above.

2. Assemble peptide according to the method described by Atherton and Sheppard,[16] using a polyamide support, the use of 3 Eq of the Fmoc-amino acid for BOP/HOBt or DCC/HOBt couplings, and 20% piperidine/dimethylformamide (DMF) for Fmoc cleavage.

3. Peptide-resin cleavage and methyl cleavage: The peptide resin (0.25 mmol) is suspended in 1 M bromotrimethylsilane and thioanisole in CF₃CO₂H (5 ml) containing m-cresol (10 mEq) at 0° and stirred for 16 hr

at 4°. The solvent is evaporated under reduced pressure, the residue dissolved in 10% (v/v) acetic acid/water and the solution filtered (Whatman No. 1) to remove resinous material. The solvent is then evaporated under reduced pressure, the residue triturated with diethyl ether (two times, 30 ml each), and then dried under high vacuum.

4. The peptide is purified by semipreparative reversed phase HPLC or anion-exchange chromatography.

This approach has been used for the synthesis of Arg-Leu-Ile-Glu-Asp-Asn-Glu-Tyr(P)-Thr-Ala-Arg-Gln-Gly in 70% yield, using BOP/HOBt (3 Eq) as the coupling procedure. While the repeated piperidine cleavage of the Fmoc group also causes partial methyl phosphate cleavage ($t_{1/2} = 7$ min), this side reaction does not seem to effect peptide extension.

Summary

As a consequence of developments in peptide synthesis, it is now well established that Tyr(P)-containing peptides can be prepared in high yield by the use of Boc-Tyr(PO_3Me_2)-OH in Boc solution or solid-phase peptide synthesis or Fmoc-Tyr(PO_3Me_2)-OH in Fmoc/solid-phase peptide synthesis. It is considered that with further developments in solid-phase synthesis techniques and the use of alternative phosphate-protecting groups, the synthesis of large and complex Tyr(P)-containing peptides will be routine.

[20] Measurement of Stoichiometry of Protein Phosphorylation by Biosynthetic Labeling

By BARTHOLOMEW M. SEFTON

Determination of the extent of phosphorylation of proteins that can be obtained only in trace amounts or in impure form is not possible by chemical means. The stoichiometry of phosphorylation of such proteins can, however, be inferred if the protein of interest can be labeled biosynthetically. In the most general sense, this is accomplished by labeling cells to equilibrium with a radioactive amino acid and with [^{32}P]P_i, measurement of the specific activity of total cellular protein and intracellular ATP, isolation of the protein of interest by immunoprecipitation, quantification of the protein from the recovery of amino acid label, and calculation of the number of moles of phosphate present in the protein from the recovery of ^{32}P. This approach has been used to quantify the phosphorylation of p60[v-src],[1] the cytoskeletal proteins vinculin[2] and talin,[3] and phospholipase C-γ.[4] Alternative approaches are described in [21] and [22] in this volume.

General Considerations

Measurements of this sort are never, however, as simple as was just described. Labeling cellular protein to true equilibrium with a radioactive amino acid is often difficult because the average half-life of cellular protein, 40 to 50 hr,[5] is long relative to the length of time that many kinds of cells can tolerate radiolytic damage. This is especially a problem in the case of hematopoietic cells. As a result, in cells labeled for relatively short periods, the specific activity of a protein with a short half-life will be greater than that of bulk cellular protein, and the abundance of the protein calculated from recovered radioactivity will overestimate the true amount of the protein isolated.

The specific activity of the cellular ATP pool has been estimated to reach equilibrium with that of the medium in as little as 6 hr after the addition of medium containing [^{32}P]P$_i$ to cells.[6] Given the often rapid turnover of phosphates on proteins, this renders the labeling of the phosphoryl moieties of phosphoproteins to equilibrium much simpler than the labeling of the polypeptide backbone. However, not all phosphates in proteins undergo rapid turnover. Phosphate at Thr-701 of the large T antigen of SV40 virus has a much longer half-life than that of the phosphates bound to serine in the molecule.[7] Estimation of the extent of phosphorylation by labeling cells with [^{32}P]P$_i$ for 6 hr would probably underestimate the occupancy of this site.

An important point ot keep in mind is the variable tolerance of cells to radiolytic damage. We found chicken fibroblasts to be relatively tolerant of labeling with [^{35}S]methionine or [^{32}P]P$_i$ for 48 hr,[1] but murine lymphocytes to be killed by labeling with [^{35}S]methionine for a little as 18 hr.[8] The viability or integrity of labeled cells can be assessed microscopically.

[^{35}S]Methionine is the obvious choice as a radioactive label for protein, but the use of a mixture of radioactive amino acids will reduce possible

[1] B. M. Sefton, T. Patschinsky, C. Berdot, T. Hunter, and T. Elliott, J. Virol. 41, 813 (1982).
[2] B. M. Sefton, T. Hunter, E. H. Ball, and S. J. Singer, Cell (Cambridge, Mass.) 24, 165 (1981).
[3] J. E. DeClue and G. S. Martin, Mol. Cell. Biol. 7, 371 (1987).
[4] J. Meisenhelder, P.-G. Suh, S. G. Rhee, and T. Hunter, Cell (Cambridge, Mass.) 57, 1109 (1989).
[5] M. J. Weber, Nature (London), New Biol. 235, 58 (1972).
[6] C. Colby and G. Edlin, Biochemistry 9, 917 (1970).
[7] K.-H. Scheidtmann, A. Kaiser, and G. Walter, in "Protein Phosphorylation" (O. M. Rosen and E. G. Krebs, eds.), p. 1273. Cold Spring Harbor Lab., Cold Spring Harbor, New York, 1981.
[8] T. R. Hurley and B. M. Sefton, Oncogene 4, 265 (1989).

errors due to an unusual amino acid composition of a protein of unknown sequence. Use of cells at a less than maximal cell density should increase labeling significantly in that their rate of protein synthesis will be maximal. If the cells require frequent changes of medium, the labeling medium can be replaced with fresh medium containing new label without affecting the measurement.

Accuracy of the approach described here depends on the abundance of the protein of interest being the same in the culture labeled with a radioactive amino acid as in the culture labeled with $[^{32}P]P_i$. There is real value, therefore, in using exactly the same labeling medium for both the radioactive amino acid and $[^{32}P]P_i$. This avoids the possibility of different levels of the protein of interest in cultures growing at different rates in labeling medium lacking amino acids on one hand and phosphate on the other. In many cases, it should in fact be possible to achieve sufficient incorporation of label into the protein of interest by labeling in complete medium. The stoichiometry of phosphorylation of p60$^{v\text{-}src}$, a protein composing only 0.06 to 0.08% of the total protein in Rous sarcoma virus-transformed chick cells and which could be immunoprecipitated with an efficiency of only 25%,[9] could be measured by biosynthetic labeling in Dulbecco–Vogt medium containing as much as 20% of the normal concentration of methionine and 100% of the normal level of phosphate.[1]

In the case of abundant cellular proteins, such as vinculin[2] or the SV40 virus large T antigen,[10] for which high-titer antibodies are available, biosynthetic labeling with a radioactive amino acid is not necessary. The stoichiometry of phosphorylation of such proteins can be esimated simply by staining the gel used to fractionate the immunoprecipitate from the ^{32}P-labeled cells with either Coomassie Blue or silver and estimating the amount of the specific protein on the gel by comparison with the staining of known amounts of several purified proteins subjected to electrophoresis on the same gel. Alternatively, the gel-fractionated protein can be transferred electrophoretically to nitrocellulose and the filter stained with India ink.[11]

Note that this general approach does not depend on knowledge of the absolute efficiency of immunoprecipitation. All that it requires is that the efficiency of precipitation of the $[^{35}S]$methionine-labeled protein be the same as that of the ^{32}P-labeled protein.

It is critical to wash cells labeled with radioactive amino acids once or twice with Tris-buffered saline or phosphate-buffered saline prior to the measurement of the specific activity of cellular protein, as this removes

[9] J. E. Buss and B. M. Sefton, *J. Virol.* **53**, 7 (1985).
[10] G. Walter and P. J. Flory, *Cold Spring Harbor Symp. Quant. Biol.* **44**, 165 (1980).
[11] K. Hancock and V. C. W. Tsang, *Anal. Biochem.* **133**, 157 (1983).

bound serum proteins. Otherwise, the amount of protein in the cells will be overestimated seriously in the protein assay. The technique described here employs the Lowry protein assay. Other similar techniques should work equally well.

Traditionally, we have measured the [35S]methionine present in an immunoprecipitated protein by dissolving gel pieces in perchloric acid and hydrogen peroxide.[1,4] This is to eliminate quenching of the radioactivity in the protein embedded in the dried gel. While this procedure works, it is tedious in that the pieces of gel are often very slow to dissolve and the mixture of perchloric acid and hydrogen peroxide excites scintillation fluid chemically. As a result, the chemiluminescence must often be allowed to decay before the incorporation can be determined accurately.

The specific activity of cellular protein is determined by precipitation of cellular protein with trichloroacetic acid (TCA) and collection of the precipitate on a filter. Accurate calculation of the recovery of the protein of interest requires that the counting efficiency of the precipitated cell protein on the surface of a filter be the same as that of the protein from the dissolved gel piece. The filter should therefore also be incubated in hydrogen peroxide and perchloric acid.

An alternative procedure, which we have not actually employed to determine the stoichiometry of phosphorylation, would be to transfer the gel-fractionated proteins electrophoretically to nitrocellulose, identify the protein of interest by autoradiography, and count the relevant piece of membrane in scintillation fluid. In this procedure, the TCA-precipitated total cellular protein collected on a nitrocellulose filter and the immunoprecipitated protein would be counted under identical conditions.

The latter procedure is obviously less time consuming than is the dissolving of the dried gel pieces and is not affected by chemiluminescence. The only obvious drawback to this approach is the fact that the efficiency of transfer of proteins from polyacrylamide gels to nitrocellulose is never 100% and varies from protein to protein. Since the general approach described here requires equal recoveries of the protein labeled with radioactive amino acids and phosphate, the protein labeled with [32P] should also be transferred to nitrocellulose for measurement of the radioactivity in the immunoprecipitated protein. The use of immobilized proteins in both cases would correct for unpredictable inefficiencies of transfer to the nitrocellulose membrane.

Determination of Stoichiometry of Phosphorylation by
    Biosynthetic Labeling

1. If labeling adherent cells, set up four identical dishes of cells. Count one at the time of cell lysis to determine the appropriate amount of antibody

to be used for immunoprecipitation. Label two with [^{35}S]methionine—one for immunoprecipitation, one for determination of the specific activity of total cellular protein. Label the fourth with [^{32}P]P$_i$. Incubate all of the cultures in identical medium.

2. If labeling suspension cells, set up only three identical cultures—one to count, one to label with [^{35}S]methionine, and another to label with [^{32}P]P$_i$. A portion of the [^{35}S]methionine-labeled cells can be used to determine the specific activity of cellular protein and another portion used for immunoprecipitation. Incubate all of the cultures in identical medium.

Determination of Recovery of Protein of Interest

1. Label one (suspension cells) or two (adherent cells) cultures of cells with [^{35}S]methionine or other radioactive amino acid for at least 48 hr. Fifty to 100 μCi/ml is often sufficient. Change the medium to fresh medium containing radioactive label if necessary.

2. Count the cells present in an unlabeled culture. Lyse an appropriate number of labeled cells and isolate the protein of interest by immunoprecipitation.

3. Wash either a duplicate dish of labeled adherent cells, or an equivalent aliquot of suspension cells, with Tris-buffered or phosphate-buffered saline and dissolve the cells in 2 ml of Lowry C solution.[12,13] Dissolution can take 30 to 60 min at room temperature and can be monitored microscopically.

4. Remove a small portion of the Lowry C solution and precipitate cellular proteins with 25% trichloroacetic acid. Collect the precipitate on a Millipore (Bedford, MA) or glass fiber filter and measure incorporation of [^{35}S]methionine into acid-precipitable material by scintillation counting.

5. Determine the concentration of protein in the remaining solution using a standard Lowry assay.[12]

6. Calculate the specific activity of cellular protein.

7a. Fractionate the immunoprecipitate by SDS–polyacrylamide gel electrophoresis, dry the gel without fixation, and localize the protein of interest by autoradiography. Excise the piece of the gel containing the protein being examined. For determination of background, excise an equivalent portion of gel from either a lane containing an immunoprecipitate prepared with nonspecific serum or from a region of the experimental lane that contains a representative background level of radioactivity.

[12] O. H. Lowry, N. J. Rosebrough, A. L. Farr, and R. J. Randall, *J. Biol. Chem.* **193**, 265 (1951).

[13] G. S. Martin, S. Venuta, M. Weber, and H. Rubin, *Proc. Natl. Acad. Sci. U.S.A.* **68**, 2739 (1971).

7b. Alternatively, transfer the gel-fractionated proteins electrophoretically to nitrocellulose, identify the band of interest by autoradiography, excise the piece containing the protein of interest, and measure the radioactivity by scintillation counting using background samples equivalent to those described above.

8. Immerse the gel pieces and, separately, the filter on which the TCA precipitate was collected, in 0.4 ml of 60% (v/v) perchloric acid and 0.8 ml 30% (v/v) H_2O_2 and incubate at 60° until the gel pieces dissolve—approximately 5 hr. This is best done in a glass scintillation vial. Add Aquasol (New England Nuclear, Boston, MA) or Ecolume (ICN, Costa Mesa, CA) and determine the amount of radioactivity in each sample by scintillation counting. The apparent background level of radioactivity in the samples may well be unacceptably high at first. This is due to chemical excitation of the fluor. It will, however, disappear if the samples are held for 24 to 48 hr.

9. From the radioactivity present in the specific protein band and the specific activity of cellular protein, calculate the amount of the protein of interest in the immunoprecipitate.

Determining Moles of Phosphate in Protein of Interest

1. Label one set of cells with $[^{32}P]P_i$. Labeling for 48 hr is optimal. Labeling for 6 hr is probably adequate. Five hundred to one thousand microcuries per milliliter should be more than sufficient. Change the medium to fresh medium containing radioactive label if necessary.

2. Remove and save the labeling medium. Determine the concentration of ^{32}P in the medium by counting a suitably diluted aliquot by scintillation counting. Calculate the specific activity of the labeling medium from this number and the calculated concentration of phosphate in the labeling medium. Assume that the cells have not depleted the medium of phosphate significantly. It can also be assumed that the phosphate concentration of undialyzed calf serum is 1.5 mM.

3. Lyse a known number of cells and isolate the protein of interest by immunoprecipitation, using the same ratio of antibody to cell lysate used for the precipitation of the $[^{35}S]$methionine-labeled protein.

4a. Fractionate the immunoprecipitate by sodium dodecyl sulfate (SDS)–polyacrylamide gel electrophoresis, dry the gel, and locate the band of interest by autoradiography. If analyzing an abundant protein, stain the gel with either Coomassie Blue or silver and estimate, by comparison with standards, the amount of the protein of interest in the gel. Excise the piece of the gel containing the protein being examined. For determination of background, excise an equivalent portion of gel from either a lane containing an immunoprecipitate prepared with nonspecific serum or from a region of the experimental lane that contains a representative background level of radioactivity.

4b. If the [^{35}S]methionine-labeled protein was quantified by transfer of the gel-fractionated protein to nitrocellulose (see step 7b above, determination of recovery of protein), also transfer the ^{32}P-labeled protein to nitrocellulose prior to counting.

5. Determine the amount of label in the protein of interest by scintillatiion counting. Dissolving of the gel pieces in perchloric acid and hydrogen peroxide is probably not essential for ^{32}P-labeled proteins, because the extent of quenching is much less than is the case with ^{35}S.

6. Assuming that the phosphate residues in the protein of interest have the same specific activity as that of the phosphate in the labeling medium, calculate the number of moles of phosphate recovered in the protein by immunoprecipitation.

7. Calculate the stoichiometry of phosphorylation from the number of moles of protein and moles of phosphate isolated by immunoprecipitation.

[21] Estimation of Phosphorylation Stoichiometry by Separation of Phosphorylated Isoforms

By Jonathan A. Cooper

The stoichiometry of phosphorylation of a protein is of central importance in assessing the possible significance of the phosphorylation. Although it is theoretically possible that enzymatic function could be regulated by the rate of phosphorylation or rate of dephosphorylation of an enzyme, generally it is the phosphorylation state (i.e., the fraction of molecules that are phosphorylated relative to those that are not) that determines activity. For example, in the absence of allosteric effectors, phosphorylation switches glycogen phosphorylase from an inactive state to an active state, with concomitant change in oligomerization state from dimer to tetramer.[1] The actual rates of phosphate esterification (by protein kinases) and hydrolysis (by phosphatases) are important only inasmuch as their balance determines the extent of phosphorylation of the substrate.

There are two primary ways of assessing phosphorylation stoichiome-

[1] N. B. Madsen, in "The Enzymes" (P. D. Boyer and E. G. Krebs, eds.), Vol. 17, p. 366. Academic Press, Orlando, Florida, 1986.

try. One is radioactive, labeling the phosphoprotein with [32P]P$_i$ and an amino acid label, and factoring in the specific activities of the two labels to calculate the molar ratio.[2] The alternative method is to physically separate the phosphoprotein and dephosphoprotein (or peptide derivatives of them), then estimate their relative quantities. Here we describe how protein phosphorylation stoichiometry can be assessed by two-dimensional (2D) gel electrophoresis, and also how the stoichiometry of phosphorylation of a particular residue can be assessed by 2D peptide mapping. The procedures used are in widespread use, and we give only a few examples of the many variations that are possible.

Two-Dimensional Gel Electrophoresis Involving Isoelectric Focusing[3,4]

Two-dimensional gel electrophoresis with isoelectric focusing as the first dimension separates a phosphoprotein from its dephospho form because the isoelectric point (pI) of the phosphoprotein is decreased by the acidic phosphate group. Indeed, forms of the protein having increasing numbers of phosphate groups are commonly separated from each other, forming a series of spots moving across the gel toward the acidic end. Quantifying the protein in each of the spots allows a direct estimate of the proportion of molecules containing zero, one, two, or more phosphate residues. The procedure requires knowledge of the coordinates of the dephosphoprotein and of each of its phosphorylated derivatives, and a means of quantifying the amount of protein present at each of these positions in a 2D gel.

Identification and Quantitation of Phosphorylated and Nonphosphorylated Isoforms

Total Cell Proteins

Some abundant phosphoproteins can be detected when total [32P]P$_i$-labeled proteins are analyzed on a 2D gel (Fig. 1). Their abundance can be estimated as follows.[5] Total cell proteins are labeled with [35S]methionine or cysteine, or with 3H- or 14C-labeled amino acids, and separated by 2D gel electrophoresis. A sample of [32P]P$_i$-labeled proteins from the same source is analyzed in parallel, and a third gel is loaded with a mixture of

[2] B. M. Sefton, this volume [20].
[3] P. H. O'Farrell, *J. Biol. Chem.* **250,** 4007 (1975).
[4] J. I. Garrells, *J. Biol. Chem.* **254,** 7961 (1979).
[5] J. A. Cooper and T. Hunter, *J. Biol. Chem.* **258,** 1108 (1983).

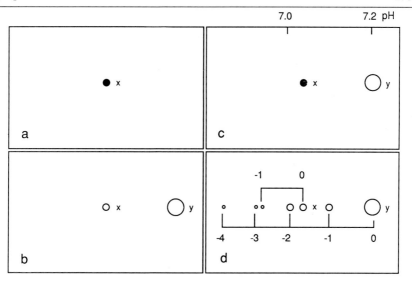

FIG. 1. Steps in assessing phosphorylation stoichiometry from 2D gels. Illustration of model autoradiographs of 2D gels (acidic proteins on left) of the following samples: (a) ^{32}P-labeled proteins, (b) [^{35}S]methionine-labeled proteins, (c) mixture of samples (a) and (b), (d) partially carbamylated [^{35}S]methionine-labeled sample. Spots x and y are the phospho and dephospho forms of the same protein; filled symbols represent ^{32}P radioactivity and open symbols ^{35}S radioactivity. The mixing experiment (gel c) allows the identification of spot x in gel (b) as the phospho form. Peptide map comparisons confirm that spot y in gel (b) is closely related to spot x, and is presumably the nonphosphorylated precursor. The relative abundances of spots x and y, and hence the phosphorylation stoichiometry, are determined from gel (b). In gel (d), 0, -1, -2, -3, and -4 represent charged isoforms of the phospho- and dephosphoprotein created by partial carbamylation. Using the carbamylation series as a ruler, phosphoprotein x is judged to contain 1.7 negative charges relative to apoprotein y, which at this pH (7.0, measured under the 9.5 M urea conditions used for isoelectric focusing) is close to that predicted for a single phosphate group.

the two samples. The 2D gel containing both ^{32}P and amino acid labels can be autoradiographed (for ^{35}S or ^{14}C and ^{32}P) or fluorographed (for ^{3}H and ^{32}P), then exposed through several layers of aluminum foil, with an intensifying screen, for ^{32}P only (there may be a small contribution from ^{35}S or ^{14}C which will be evident if the parallel gel containing amino acid label only is exposed under the same conditions). Once the ^{3}H-, ^{14}C-, or ^{35}S-labeled spot corresponding to the phosphoprotein has been identified, excising it and determining its radioactivity allows its abundance to be estimated, relative to total cell protein radioactivity [from trichloroacetic acid (TCA) precipitation].

The abundance of the corresponding nonphosphorylated protein also

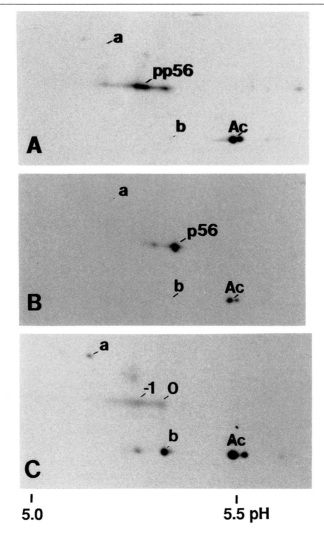

FIG. 2. Analysis of phosphorylation stoichiometry by 2D gel electrophoresis. p56^{lck} was immunoprecipitated from LSTRA cells labeled with [^{35}S]methionine.[7] Before 2D gel electrophoresis, the immunoprecipitate was treated in three different ways. (A) Immunoprecipitate analyzed directly. Mixing with a [^{32}P]P$_i$-labeled sample showed that the spots designated pp56^{lck} were phospho-p56^{lck}. Spots a, b, and actin (Ac) are reference proteins. (B) The immunoprecipitate was treated with acid phosphatase (0.1 μg/μl, in 40 mM PIPES, pH 6, 1 mM DTT, 20 mM leupeptin, 20 μg/ml aprotinin) for 20 min at 30° before analysis. This collapses the pp56^{lck} spots into a single, nonphosphorylated form (p56^{lck}). Similar treatment of a [^{32}P]P$_i$-labeled sample removed essentially all of the radioactivity (data not shown). (C) Immunoprecipitate was treated with acid phosphatase, then carbamylated prior to electrophoresis as described in Table II (except that one-third volumes of the sample were incubated

requires knowledge of its coordinates on the 2D gel. In the absence of data gained by use of specific antibodies, the usual recourse is guesswork. The challenge is to identify a 2D gel spot that has the properties expected for the dephospho form of the phosphoprotein under study. The dephospho-protein is usually of similar (or slightly smaller) apparent molecular mass and more basic pI, and should have a very similar peptide map. A search through several candidate 2D gel spots may be needed. One clue is to look for a protein whose abundance decreases under conditions of cell treatment where the abundance of the phospho form increases (i.e., when phosphorylation would decrease the pI of the protein). A structural rela-tionship between the phosphoprotein and apoprotein must be confirmed by peptide mapping the two species (both labeled with the same radioactive amino acid), cut individually from 2D gels.[5] The peptide maps should be identical, unless one of the labeled peptides happens to contain a phosphorylation site. Once both the phospho- and dephosphoprotein have been identified as spots on a 2D gel of proteins labeled with radioactive amino acids, then their relative abundances may be determined by cutting and counting or by densitometry.

Immunoprecipitates

Less abundant proteins can also be studied following biosynthetic labeling, if they can be separated from other cell proteins, by immunopre-cipitation or other purification procedures. As described above for total cell proteins, ^{32}P- and [^{35}S]methionine-labeled immunoprecipitated pro-teins can be electrophoresed separately and as a mixture, and the ^{35}S-labeled phosphorylated and nonphosphorylated forms identified and quan-tified. This procedure has been used to monitor the phosphorylation state of p60$^{c\text{-}src}$ [6] and p56^{lck} (Fig. 2).[7] Estimates of phosphorylation stoichiometry involving any purification procedure such as immunoprecipitation must

[6] K. L. Gould and T. Hunter, *Mol. Cell. Biol.* **8**, 3345 (1988).
[7] J. D. Marth, J. A. Cooper, C. S. King, S. F. Zeigler, D. A. Tinker, R. W. Overell, E. G. Krebs, and R. M. Perlmutter, *Mol. Cell. Biol.* **8**, 540 (1988).

in urea at 98° for 1, 2, and 5 min before recombining). The major carbamylated form of p56^{lck}, presumably the −1 form, has the same pI as the major phosphoform present in the original sample (A). At this pH, a phosphate group has a single negative charge. Therefore most of the p56^{lck} molecules in immunoprecipitates from LSTRA cells have a single phosphate group. Because 2D tryptic peptide mapping reveals two major phosphorylation sites, it appears that most p56^{lck} molecules are phosphorylated at one site or the other, a few lack phosphate, and a few are phosphorylated at both sites.

be interpreted with caution, unless it is known that the purification procedure does not discriminate on the basis of phosphorylation state.

Nonradioactive Methods

Very abundant proteins can be detected on 2D gels by staining with Coomassie Blue or silver. The phosphorylated forms can be identified by autoradiography of a stained gel on which ^{32}P-labeled proteins have been analyzed. Less abundant proteins can also be studied if an antiserum appropriate for immunoblotting is available. ^{32}P-Labeled and unlabeled proteins are subjected to 2D gel electrophoresis on separate gels, blotted, and probed with the antiserum. The relative quantities of the phospho and dephospho forms of the antigen can be estimated from the relative intensities of immunostaining.[5] As with immunoprecipitation, immunoblotting results need to be interpreted with caution, since the antibody may react differently with phospho and dephospho antigen. Also, controls for the linearity of immunoblot intensity with quantity of antigen are needed.

pI Shift Caused by Phosphorylation

The magnitude of the pI shift caused by a single phosphate group varies according to the polypeptide of interest. There are two reasons for this. First, the charge contribution of a phosphate group depends on the ambient pH. The phosphate esters of serine, threonine, and tyrosine have two acidic protons, with pK_a values in H_2O of <2 and about 5.8, respectively.[8] In 9.5 M urea solution, the second pK_a is approximately 6.5. Therefore, under typical conditions for isoelectric focusing in high concentrations of urea, adding a phosphate group to a basic protein (pI in urea >7.5) adds two negative charges ($-OPO_3^{2-}$), but adding a phosphate group to an acidic protein (pI in urea <5.5) adds only a single negative charge ($-OPO_3H^-$). For pI values close to the pK_a for phosphate, the Henderson–Hasselbach equation can be used to calculate the charge shift:

$$pH = pK_a + \log([A^-]/[HA])$$

$$pH - pK_a = \log([A^-]/[HA])$$

$$[A^-]/[HA] = 10^{(pH-pK_a)}$$

the average charge on a phosphate group is $-(1 + X)$, where

[8] J. A. Cooper, B. M. Sefton, and T. Hunter, this series, Vol. 99, p. 387.

TABLE I

ESTIMATED pK_a VALUES OF IONIZING GROUPS IN PROTEINS[a]

Ionizing group	pK_a	Charge	
		Below pK_a	Above pK_a
α-COOH	3.7	0	−1
Asp and Glu, COOH	4.7	0	−1
His, imidazole	6.5	+1	0
α-NH$_2$	7.8	+1	0
Cys, sulfhydryl	8.3	0	−1
Tyr, phenol	10.0	0	−1
Lys, ε-NH$_2$	10.2	+1	0
Arg, guanidyl	<12	+1	0

[a] From H. R. Mahler and E. H. Cordes, "Biological Chemistry," 1st ed., pp. 53 and 267. Harper & Row, New York, 1966.

$$X = [A^-]/([HA] + [A^-]) = 1/[10^{(pK_a - pH)} + 1]$$

Second, the effect on pI of adding an acidic group to a polypeptide depends on the buffering effect of other ionizing groups on the polypeptide. It is possible to have two polypeptides with the same pI, but different buffering capacity. This is commonly seen with polypeptides of different sizes. Generally, the larger the polypeptide the greater the number of ionizing side chains that act to buffer the pI against the effect of an added phosphate group. Therefore, larger proteins are shifted less in pI than small proteins.

The effect of phosphorylation on the pI of a protein of known amino acid composition can be predicted using the Henderson–Hesselbach equation for each of the ionizing side chains, making the simplifying assumption that under denaturing conditions each side chain ionizes with its usual pK_a (Table I). The theoretical pI can then be calculated by an iterative procedure, recalculating the net charge on the polypeptide at different pH values until the pH at which the net charge is zero is found. If this calculation is performed for both the dephosphoprotein and its phospho form, then the charge shift caused by phosphorylation can be predicted. This prediction is unreliable for several reasons. First, urea appears to alter the pK_a values of ionizing groups, presumably because it affects the hydration states of the acid and conjugate base (notice that the water content of a 9.5 M urea solution is only about 50%!). Different pK_a values may be affected to different degrees. Second, any remaining secondary structure in a protein could mask some ionizing groups and, even if fully denatured, adjacent ionizing residues could cause shifted pK_a values.

Third, posttranslational modifications of many varieties can alter the ionization properties of side chains. All these considerations mean that predicting pI by calculation is only approximate.

In actuality, a reliable assignment of the phosphate content of a given isoform of a protein may be made only if a calibration curve is generated for the protein of interest.[9] The effects on pI of adding sequential single negative charges to the protein need to be determined. A simple way of doing this is to heat the nonphosphorylated protein (either total cell proteins, purified proteins, or an immunoprecipitate) with urea under conditions where one or more amino groups on the protein (primarily ε-amino groups of lysine residues, which are normally fully positively charged at pH <9) become carbamylated (no charge at any pH).[10] By combining samples treated with urea for different periods of time, a single sample containing the protein with different degrees of carbamylation can be made (Table II). A 2D gel of this sample provides a calibration series. Using this as a ruler, the mobility shift induced by phosphorylation can then be translated into an effective charge change (Fig. 1). Allowing for the pK_a of the phosphate group in the pH range of interest, the observed mobility shift can then be attributed to the addition of one, two, or more phosphate residues.

Other Two-Dimensional Procedures

Any separation procedure that resolves the various phosphorylated forms of the protein may be adaptable for estimation of phosphorylation stoichiometry. A classical example is the quantitation of ribosomal protein S6 phosphorylation following 2D separation in which gels containing urea at pH 8.6 form the first dimension.[11] Under these conditions, each phosphate group adds about two negative charges to the S6 protein, decreasing its mobility toward the cathode.

One-Dimensional Analyses

For some proteins, phosphorylation leads to a considerable mobility shift on SDS–PAGE. In such cases, 2D gel electrophoresis may be unnecessary. For example, SDS–PAGE of immunoprecipitates of [^{35}S]methionine-labeled p60$^{c\text{-}src}$ shows that phorbol diester treatment of the cells

[9] R. A. Steinberg, P. H. O'Farrell, U. Friedrich, and P. Coffino, Cell (Cambridge, Mass.) 10, 381 (1977).
[10] D. Bobb and B. H. J. Hofstee, Anal. Biochem. 40, 209 (1971).
[11] J. Martin-Perez and G. Thomas, Proc. Natl. Acad. Sci. U.S.A. 80, 926 (1983).

TABLE II
PROCEDURES FOR GENERATING CARBAMYLATION SERIES[a]

Total cell proteins
1. Wash cells
2. Lyse 3–4 × 10⁶ cells in 100 μl warm lysis buffer [0.3% (w/v) SDS, 1% (w/v) dithio-threitol (DTT), 20 mM Tris-HCl, pH 8.0 (at 4°), 1 mM EDTA]. Scrape and transfer to 1.5-ml snap-cap tube
3. Add 10 μl of 0.5 mg/ml RNase A, 1 mg/ml DNase I, 50 mM MgCl$_2$, 0.5 M Tris-HCl, pH 8.0 (at 4°)
4. Leave 2 to 5 min at 0°, then transfer to dry ice
5. Lyophilize
6. At this stage, sample preparation for 2D gels routinely involves dissolving the lyophilized proteins in 100 μl of 2DSB (9.95 M urea, 4% Nonidet P-40, 2% ampholytes, pH 6–8, 0.1 M DTT, 0.0015% Phenol Red). For carbamylation, the lyophilized proteins are dissolved instead in 100 μl of 9.95 M urea. Remove 40 μl to a second tube, on ice. Place the first tube at 98° and at 0.5, 2, and 5 min[b] remove 20-μl samples to the second tube. Freeze and lyophilize. Dissolve in 60 μl of double-strength 2DSB lacking urea. The final volume will be about 100 μl, due to the presence of urea in the dried sample
Immunoprecipitates
Prepare and wash immunoprecipitate as usual. Perform an additional wash to remove any detergents. Remove as much residual liquid as possible after the final spin. Lyophilize. Continue at step 6, above

[a] Based on 2D gel electrophoresis sample preparation procedures of Garrels,[4] and B. R. Franza (personal communication).
[b] Different heating times may be required depending on the proteins and concentrations of reducing agents. Urea will react with many nucleophiles at high temperature. The complete 2DSB solution contains 0.1 M DTT, which reacts preferentially with urea. Therefore longer heating times are needed if the sample is dissolved in 2DSB prior to heating, in order to achieve a similar extent of carbamylation of proteins in the sample.

causes most of the p60$^{c\text{-}src}$ molecules to run with reduced mobility, suggesting that phosphorylation by protein kinase C is almost quantitative.[12] Probing immunoblots of cell lysates with antibody to the retinoblastoma gene product (RB) shows that RB undergoes a mobility shift in S and G$_2$ phase cells. This mobility shift is obliterated by phosphatase, suggesting cell cycle-specific phosphorylation of most of the RB molecules in the cell.[13] Although SDS–PAGE techniques are more convenient than 2D gel electrophoresis, the former procedure does not readily permit an estimate of the mole ratio of phosphate to protein to be determined.

[12] K. L. Gould, J. R. Woodgett, J. A. Cooper, J. E. Buss, D. Shalloway, and T. Hunter, *Cell* (*Cambridge, Mass.*) **42**, 849 (1985).
[13] K. Buchkovich, L. Duffy, and E. Harlow, *Cell* (*Cambridge, Mass.*) **58**, 1097 (1989).

Peptide Mapping

Assessing phosphorylation stoichiometry has little meaning unless the number of sites phosphorylated is also determined. For example, a protein may be phosphorylated to 100% stoichiometry, with every molecule containing a single phosphate, but if five sites contain equal amounts of phosphate, then each is phosphorylated only in 20% of molecules. This would make it difficult to evaluate the significance of any one of the phosphorylations. In the case of p36 (calpactin I), transformation by Rous sarcoma virus (RSV) causes about 12% of molecules to be singly phosphorylated, but this phosphate is found on both serine and tyrosine residues, at approximately equal levels. Therefore the stoichiometry of tyrosine phosphorylation is about 6%.[5,14]

Phosphopeptide mapping of individual forms of a phosphoprotein, separated on the basis of charge or pI, can also be useful to test whether phosphorylation of one site precedes phosphorylation of others, i.e., whether phosphorylations are sequential. For example, Martin-Perez and Thomas showed that less phosphorylated forms of S6 contained phosphate at a subset of the sites phosphorylated in highly phosphorylated forms of S6, suggesting an ordered filling of the sites.[11]

Peptide mapping can also be used to determine directly the stoichiomtery of phosphorylation of individual sites. This is possible because, just as 2D gel electrophoresis separates proteins of different charges, some peptide mapping procedures resolve peptides of different charges, so the phospho and dephospho forms of a particular peptide can be separated. If the peptides are labeled with an amino acid label, such as [^{35}S]methionine or a single ^{3}H-labeled amino acid, and provided the phosphopeptide of interest contains the labeled residues, then the relative quantities of the phospho and dephospho forms of the peptide can be determined. Of course, it is necessary to identify which of the many peptides derived from the protein corresponds to the phosphopeptide, and which to the dephosphopeptide. Appropriate markers need to be analyzed. For proteins of known sequence, the dephosphopeptide can be synthesized, and position on a peptide map detected chemically (e.g., with fluorescein). The peptide may also be phosphorylated *in vitro* with a suitable kinase and radioactive ATP, and the position of the phosphopeptide on the map determined by radioactivity detection. The amino acid-labeled protein may then be digested to generate peptides, resolved, and the amounts of radioactivity migrating with the two markers determined. One control is to dephosphorylate the protein enzymatically (with phosphatase) prior to

[14] E. Erikson and R. L. Erikson, *Cell (Cambridge, Mass.)* **21**, 829 (1980).

proteolysis and peptide mapping. In this map, only the dephosphopeptide should be detected. These procedures have been used to estimate the extents of p60[v-src] phosphorylation at Tyr-416[15] and of SV40 large T antigen phosphorylation at Thr-124.[16]

Acknowledgment

Supported by Public Health Service Grant CA41072.

[15] B. M. Sefton, T. Patschinsky, C. Berdot, T. Hunter, and T. Elliott, *J. Virol.* **41,** 813 (1982).
[16] K.-H. Scheidtmann, A. Kaiser, A. Carbone, and G. Walter, *J. Virol.* **38,** 59 (1981).

[22] Microchemical Determination of Phosphate in Proteins Isolated from Polyacrylamide Gels

By VICTOR D. VACQUIER and GARY W. MOY

We designed the following methods to determine the phosphate content of proteins from cells which cannot be labeled to equilibrium with radioactive phosphate. We describe a simple method to electroelute and recover protein bands from polyacrylamide gels. The method of analysis for phosphate is as previously described.[1] We have used these procedures to determine the loss of 15 phosphate groups from sea urchin sperm guanylate cyclase when the sperm contacts the jelly coat of the egg,[2] and to study the phosphorylation of sea urchin sperm histone H3 induced by egg jelly.[3]

Protein

All glassware is soaked overnight in 6 N HCl, rinsed in glass-distilled water, and air dried. Protein is precipitated by making samples 10% in trichloroacetic acid (TCA) from a 100% (w/v) stock (4°). The 10,000 g (10 min) pellet of TCA-insoluble protein is collected and washed twice with an excess of 90% (v/v) acetone. The final pellet is collected by centrifugation at 3000 g for 10 min and dissolved in 1 ml 10% (w/v) sodium dodecyl sulfate (SDS). It is important that chemically pure SDS be used in all these procedures, because impure SDS contains a substantial quantity of

[1] J. E. Buss and J. T. Stull, this series, Vol. 99, p. 7.
[2] V. D. Vacquier and G. W. Moy, *Biochem. Biophys. Res. Commun.* **137,** 1148 (1986).
[3] V. D. Vacquier, D. C. Porter, S. H. Keller, and M. Ackerman, *Dev. Biol.* **133,** 111 (1989).

phosphate. Protein concentration is determined by the Lowry method[4] using bovine serum albumin as a standard. The protein mixture to be separated by SDS–polyacrylamide gel electrophoresis is reprecipitated by adding 9 ml 100% acetone and the pellet collected by centrifugation at 10,000 g for 10 min. The pellet is dissolved at 2.0 mg protein/ml Laemmli sample buffer[5] containing 4% SDS and 5% (v/v) mercaptoethanol. The dissolved sample is then placed in a boiling water bath for 5 min.

Electrophoresis

The bands to be cut from the gel must separate cleanly in the gel system used. In some cases we continue the run for twice the time needed to run the tracking dye off the gel. Laemmli gels[5] containing 0.1% SDS are used. The gel dimensions are 15 × 15 cm and 1.5 mm thick. The 5% stacking gel is poured without a comb. If molecular weight standards are needed one narrow lane is made by putting a wedge at one end of the stacker before it polymerizes. We load 500–2000 μg of total protein on such gels and run the gels according to Laemmli.[5] Gloves rinsed in distilled water are worn at all times. The protein runs as a single curtain and excellent resolution is achieved for the sea urchin sperm proteins we have studied. Following electrophoresis, one glass plate is removed and the gel is carefully lowered into 200 ml of 2 M KCl in a Pyrex baking dish. The protein bands appear as clear bands against a background of whitish gel representing the precipitation of SDS by K[+].[6] The bands are best visualized by indirect illumination, which is done by suspending the dish with the gel about 15 cm above a light box with a square of black paper directly on the light box. Optimal conditions for visualizing the clear protein bands can be easily determined by altering the size of the black paper square and the distance of the gel from the light box. The duration of time the gel is in KCl is critical; if it is left too long in KCl the clear bands of protein will also turn white. When the clear bands of protein appear against the white background (usually by 2 min) the gel is transferred to a glass plate and the protein bands of choice cut out with a razor blade as strips of polyacrylamide gel. The slices are washed twice in water and stored frozen.

[4] O. H. Lowry, N. J. Rosebrough, A. L. Farr, and R. J. Randall, *J. Biol. Chem.* **193,** 265 (1951).
[5] U. K. Laemmli, *Nature (London)* **227,** 680 (1970).
[6] L. P. Nelles and J. R. Bamburg, *Anal. Biochem.* **73,** 522 (1976).

Electroelution

We usually start electroelution with the combined cutout strips from 5 to 10 gels. A 5-ml volume of gel slices is suspended in 20 ml of elution buffer[2] [25 mM Tris base, 192 mM glycine, 0.15% pure SDS purchased from Bio-Rad Laboratories (Richmond, CA) Cat. #161-0301] and the tube contents placed in a Spectrapor (Spectrum Medical, Los Angeles, CA) dialysis bag of 25-mm diameter. The bag is dialyzed overnight versus at least 2 liters of elution buffer. The bag is placed in a cylindrical chamber (14 cm high and 11 cm in diameter) with one electrode at the bottom and the other electrode on the flat Plexiglas lid. Plastic tubes with small holes in their sides are glued over each electrode so that it is impossible for the bag to contact the electrode. The chamber is filled with elution buffer to 1 cm from the top (1.7 liters elution buffer) and a current of 160 mA is applied to the chamber for 12 hr (room temperature). Following the first electroelution, the liquid in the bag is removed, 20 ml fresh elution buffer is added, and the electroelution is repeated. Following the second electroelution the 40 ml of elution buffer is centrifuged at 10,000 g and the supernatant dialyzed against 5-liter portions of distilled water with four changes at 12-hr intervals. The contents of the bag are placed in four 30-ml Corex tubes and lyophilized to complete dryness. The residue of SDS-protein is extracted twice with 10 ml of 90% acetone by vortexing and collecting the insoluble protein by centrifugation. The protein pellet is partially dried by vacuum evaporation for 1 min and then dissolved in a small volume of 0.1% pure SDS in water (the volume used depends on the size of the acetone pellet). Protein content is then determined.[4] A sample of the final eluted protein is electrophoresed on a gel that is stained with silver to be certain of the purity and the molecular weight of the protein.

Phosphate Determination

The exact method previously described by Buss and Stull[1] is used to assay different volumes of the protein in 0.1% SDS. Blanks of those volumes of 0.1% SDS without protein must be run in each assay. Centrifuge the phosphate reagent solution at 30,000 g for 20 min instead of filtering it through paper.[1] The phosphate reagent is discarded after 48 hr. The KH_2PO_4 is dried at 110° for 24 hr before weighing it to prepare a standard phosphate solution.[1] The samples are placed in acid-washed 10 × 75 mm glass tubes, evaporated, and carefully ashed as previously described.[1] After ashing the samples, the volumes, reagents, and conditions for the assay are as described.[1] The standard curve with KH_2PO_4 is

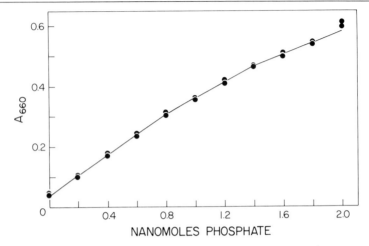

FIG. 1. Standard curve for phosphate using KH_2PO_4.[1]

close to the curve of Buss and Stull,[1] but extends to 2.0 nmol phosphate (Fig. 1), whereas their curve is only to 1.0 nmol. The points are very reproducible and the assay method is simple to perform.

From nine gels of 800 μg protein/gel of cavitated sperm membrane vesicles (CMV)[7] we recovered 686 μg of the M_r 150,000 form of guanylate cyclase. We used 12–24 μg/assay and determined that there were 2.57 ± 0.42 phosphates per molecule of cyclase ($n = 19$). From four gels of 800 μg protein/gel we isolated 209 μg of the M_r 160,000 form of the enzyme. Assays of 3–15 μg/tube yielded a value of 17.95 ± 1.24 phosphates per molecule of cyclase ($n = 16$).[2]

[7] G. E. Ward, G. W. Moy, and V. D. Vacquier, *J. Cell Biol.* **103**, 95 (1986).

[23] Production of Phosphorylation State-Specific Antibodies

By ANDREW J. CZERNIK, JEAN-ANTOINE GIRAULT, ANGUS C. NAIRN, JONATHAN CHEN, GRETCHEN SNYDER, JOHN KEBABIAN, and PAUL GREENGARD

The importance of protein phosphorylation in the regulation of a wide variety of cellular processes has been clearly established. The ability to detect and to quantitate changes in the state of phosphorylation of specific substrate proteins is of great utility in the study of their functional significance. Standard methods for measuring the state of protein phosphorylation in intact cell preparations utilize prelabeling with [^{32}P]P$_i$ or "back phosphorylation." Although these methods have contributed valuable information for many test systems, they do have several practical and theoretical limitations.[1] The degree of sensitivity and selectivity afforded by immunochemical methodology makes it an attractive alternative for detecting changes in the state of phosphorylation of specific proteins. For example, antibodies against phosphotyrosine-containing proteins have been widely utilized.[2] Phosphorylation state-dependent monoclonal antibodies against a variety of cytoskeletal proteins have also been produced and characterized.[3–7] Each of these monoclonal antibodies (MAb) had been isolated from a large pool derived from immunization protocols in which the specific targeting of phosphorylated epitopes was not the primary objective.

A more specific approach was used to prepare serum antibodies that distinguished between the phospho and dephospho forms of G substrate,[8] a protein which is localized to cerebellar Purkinje cells and is a substrate for cGMP-dependent protein kinase. A synthetic heptapeptide, Arg-Lys-Asp-Thr-Pro-Ala-Leu, which corresponds to the sequence surrounding the two phosphorylated threonyl residues in the holoprotein, served as antigen, after cross-linking to keyhole limpet hemocyanin (KLH). The resulting antibodies against the peptide–KLH conjugate were specific for the dephospho form of G substrate. Phospho-specific antibodies were prepared by using holo-G substrate, purified in the fully phosphorylated form, as the antigen. Despite the success with G substrate, attempts to produce phospho-specific polyclonal antibodies by immunization with the phospho form of other holoproteins have, in general, failed to work. Two major factors can account for the lack of success of this protocol. First, most phosphorylated proteins are thought to undergo rapid dephosphorylation, regardless of the route of immunization, which would result in the

[1] E. J. Nestler and P. Greengard, "Protein Phosphorylation in the Nervous System." Wiley, New York, 1984.

[2] J. Y. J. Wang, *Anal. Biochem.* **172**, 1 (1988).

[3] L. A. Sternberger and N. H. Sternberger, *Proc. Natl. Acad. Sci. U.S.A.* **80**, 6126 (1983).

[4] F. C. Luca, G. S. Bloom, and R. B. Vallee, *Proc. Natl. Acad. Sci. U.S.A.* **83**, 1006 (1986).

[5] I. Grundke-Iqbal, K. Iqbal, Y.-C. Tung, M. Quinlan, H. M. Wisniewski, and L. I. Binder, *Proc. Natl. Acad. Sci. U.S.A.* **83**, 4913 (1986).

[6] K. S. Kosik, L. K. Duffy, M. M. Dowling, C. Abraham, A. McCluskey, and D. J. Selkoe, *Proc. Natl. Acad. Sci. U.S.A.* **81**, 7941 (1984).

[7] F. M. Davis, T. Y. Tsao, S. K. Fowler, and P. N. Rao, *Proc. Natl. Acad. Sci. U.S.A.* **80**, 2926 (1983).

[8] A. C. Nairn, J. A. Detre, J. E. Casnellie, and P. Greengard, *Nature (London)* **299**, 734 (1982).

loss of the desired epitope. Second, holoproteins generally contain several highly immunogenic epitopes at other locations in the protein, decreasing the probability that clonal dominance for an epitope containing the phosphorylation site will be obtained.

We have used the phosphorylated and nonphosphorylated forms of synthetic phosphorylation site peptides to produce phosphorylation state-specific antibodies for selected sites in two neuronal phosphoproteins. Phospho- and dephosphospecific polyclonal antibodies were produced[9] against site 1 of synapsin I, a synaptic vesicle-associated protein that is implicated in the regulation of neurotransmitter release. Monoclonal and polyclonal phospho-specific antibodies were generated[10,11] against the cAMP-dependent phosphorylation site of DARPP-32, a dopamine- and cAMP-regulated phosphoprotein that, in its phosphorylated state, is an inhibitor of protein phosphatase 1. The methods used for these proteins should have general application in the production of phosphorylation state-specific antibodies to substrates whose sites of phosphorylation have been established.

Antibody Production: Use of Synthetic Peptides as Phosphorylation State-Specific Epitopes

Step 1: Selection of Synthetic Peptides

The specific targeting of antibody production to a phosphorylated epitope is accomplished by the use of a short synthetic peptide, corresponding to the amino acid sequence surrounding the phosphorylation site(s) in the protein substrate. While prior knowledge of the particular sequence is required, advances in various aspects of protein microsequencing techniques[12,13] have vastly improved the ability to identify phosphorylation sites directly. In addition, the characterization of consensus sequences for a number of protein kinases[14] has enabled rational predictions to be made concerning the location of putative phosphorylation sites by searching for

[9] A. J. Czernik, J.-A. Girault, and P. Greengard, *J. Cell Biol.* **109,** 215a (1989).

[10] J. Chen, Z.-J. Huang, J. Kebabian, J.-A. Girault, A. J. Czernik, and P. Greengard, *FASEB J.* **2,** A550 (1988).

[11] G. Snyder, J.-A. Girault, J. Chen, A. J. Czernik, J. A. Nathanson, J. Kebabian, and P. Greengard, *J. Neurosci.* (submitted for publication).

[12] M. W. Hunkapiller, J. E. Strickler, and K. J. Wilson, *Science* **226,** 304 (1984).

[13] D. W. Speicher, *in* "Techniques in Protein Chemistry" (T. E. Hugli, ed.), p. 24. Academic Press, San Diego, California, 1989.

[14] P. J. Blackshear, A. C. Nairn, and J. F. Kuo, *FASEB J.* **2,** 2957 (1988).

homology within the deduced amino acid sequence of cloned protein substrates. A short peptide, encompassing the sequence surrounding the phosphorylatable residue, is chosen for synthesis. The synthetic peptide should include any apparent consensus site features, such as the basic residues found on the amino-terminal side of sites phosphorylated by cAMP-dependent protein kinase, or the acidic residues found on the carboxyl-terminal side of sites for casein kinase II. Peptides of 10 to 12 residues in length appear to be optimal. This size is short enough to maximize the probability of including the phosphorylatable residue within the epitope, while minimizing the chances of providing other, phosphorylation-independent epitopes. Peptides of this length are usually efficient substrates for *in vitro* phosphorylation with purified kinase, which is a requirement in step 2. To better approximate their structure in the native protein, peptides are normally synthesized with a free amino terminus and an amidated carboxyl terminus. However, this can vary depending on the type of coupling chemistry one plans to use (step 3). Other amino acid residues, not present in the native sequence, can also be included at either end of the synthetic peptide. For example, addition of a tyrosyl residue would facilitate subsequent labeling with ^{125}I, or addition of a cysteinyl residue would permit coupling to the carrier protein via the sulfhydryl group.

The 11-mer, Tyr-Leu-Arg-Arg-Arg-Leu-Ser-Asp-Ser-Asn-Phe-amide, was chosen for the synapsin I experiments. This peptide encompasses Ser-9,[15,16] the residue in bovine and rat synapsin I that is phosphorylated by both cAMP dependent protein kinase and calcium/calmodulin-dependent protein kinase I. The DARPP-32 peptide, Ile-Arg-Arg-Arg-Arg-Pro-Thr-Pro-Ala-Met-amide, chosen for immunization is a 10-mer encompassing the cAMP-dependent phosphorylation site (Thr-34)[17,18] of bovine DARPP-32. The peptides were synthesized by standard solid-phase methodology and were purified by preparative reversed-phase high-performance liquid chromatography (HPLC). Purity was estimated to be >95% by analaytical HPLC and amino acid analysis, and was verified by mass spectrometry.

[15] A. J. Czernik, D. T. Pang, and P. Greengard, *Proc. Natl. Acad. Sci. U.S.A.* **84,** 7518 (1987).

[16] T. C. Sudhof, A. J. Czernik, H.-T. Kao, K. Takei, P. A. Johnston, A. Horiuchi, S. D. Kanazir, M. A. Wagner, M. S. Perin, P. De Camilli, and P. Greengard, *Science* **245,** 1474 (1989).

[17] H. C. Hemmings, Jr., K. R. Williams, W. H. Konigsberg, and P. Greengard, *J. Biol. Chem.* **259,** 14486 (1984).

[18] K. R. Williams, H. C. Hemmings, Jr., M. B. LoPresti, W. H. Konigsberg, and P. Greengard, *J. Biol. Chem.* **261,** 1890 (1986).

Step 2: Phosphorylation of Synthetic Peptides

The phosphorylated form of the synthetic peptide is prepared by *in vitro* phosphorylation using an appropriate purified protein kinase. The catalytic subunit of cAMP-dependent protein kinase is used to phosphorylate the synapsin I peptide and the DARPP-32 peptide. A typical reaction mixture contains 0.5 mM synapsin I peptide, 2 mM ATP, 75 mM HEPES, pH 7.4, 15 mM $MgCl_2$, 0.1 mM EGTA, 0.25 mM dithiothreitol (DTE), and 2 μg/ml kinase in a volume of 5 ml. The reaction mixture is incubated at 30° for 10 hr, with a second aliquot of enzyme added after 4 hr. The reaction is stopped with 2.5 ml glacial acetic acid. The excess ATP is removed by batch adsorption to Dowex AG 1-XB anion-exchange resin (Bio-Rad, Richmond, CA). Resin (2 ml) is added to the acidified reaction mixture, stirred for 1 hr at 4°, and then poured into a small column. The flow-through and a 4-ml wash with 30% (v/v) acetic acid are pooled, lyophilized, and redissolved in 0.1% (v/v) trifluoroacetic acid (TFA). The phosphorylated form of the peptide is separated from the remaining unphosphorylated peptide (which typically represented <5% of the total in reactions using catalytic subunit) by reversed-phase HPLC. Aliquots of the redissolved peptide are applied to a Vydac C_{18} column (0.46 × 25 cm) (Separations Group, Hesperia, CA) and eluted with a shallow gradient of increasing concentration of acetonitrile (CH_3CN) in 0.085% TFA (0–5 min, 0% CH_3CN; 5–10 min, 0–13.3% CH_3CN; 10–30 min, 13.3–18.9% CH_3CN) at a flow rate of 1 ml/min. Peptide elution is monitored by UV absorbance at 214 nm. Phosphorylated peptides normally elute about 1 min earlier in the gradient than the unphosphorylated form. The peak fractions are pooled and lyophilized. In some instances, a tracer amount of [γ-^{32}P]ATP is added to the phosphorylation reaction mixture to monitor the stoichiometry of phosphorylation, and the recovery of the ^{32}P-labeled phosphopeptide is quantitated by Cerenkov counting. A similar protocol is used to prepare the phosphorylated form of the DARPP-32 peptide, with the HPLC gradient elution conditions adjusted accordingly.

Step 3: Peptide Conjugation and Immunization

In order to enhance antigenicity, the phosphorylation site peptides are coupled to a carrier protein prior to immunization, using a molar ratio of 25 : 1 [(phospho)peptide : carrier]. The coupling procedure is similar to the general methods described for the preparation of anti-peptide antibodies.[19,20] The dephospho and phospho forms of the synapsin I peptide (1.25

[19] J. P. Briand, S. Muller, and M. H. V. Van Regenmortel, *J. Immunol. Methods* **78**, 59 (1985).

[20] E. Harlow and D. Land, "Antibodies: A Laboratory Manual." Cold Spring Harbor Lab., Cold Spring Harbor, New York, 1988.

μmol) and the phospho form of the DARRP-32 peptide are each coupled to *Limulus* hemocyanin (50 mg; Sigma, St. Louis, MO). As a precaution, the hemocyanin is assayed for the presence of contaminating phosphatase activity by incubation (60 min) with samples of ^{32}P-labeled holoprotein followed by precipitation with 10% TCA and centrifugation. The loss of acid-precipitable counts into the supernatant would provide evidence for phosphatase activity. The samples of hemocyanin [as well as bovine serum albumin (BSA) or bovine thryoglobulin] tested were devoid of phosphatase activity. The hemocyanin and peptide are dissolved in 2.5 ml of 125 mM sodium phosphate buffer, pH 7.4. Coupling is carried out at 4° by the dropwise addition of 1.5 ml of 0.2% glutaraldehyde over the course of 20 min with constant stirring, followed by an additional incubation for 100 min. Excess glutaraldehyde is reduced by the addition of 150 μl sodium borohydride (12.5 mg/ml). The use of sodium borohydride should be avoided if essential disulfide bonds are to be maintained in the peptide. The reaction mixture is dialyzed overnight against 5 liters of 10 mM sodium phosphate buffer, pH 7.4/154 mM NaCl, divided into four aliquots, and stored frozen at $-20°$. Other cross-linking reagents, such as carbodiimide, bisdiazotized benzidine, or *m*-maleimidodibenzoyl-*N*-hydroxysuccinimide ester, can be used[19] as required by the particular structure of the peptide to be conjugated. Although the efficiency of coupling can vary, we have not found this to be a significant problem. The amount of peptide used here provides sufficient material to immunize and boost two rabbits or eight mice.

Step 4: Antibody Production

Once the peptide conjugates have been prepared, the investigator must decide between monoclonal and polyclonal antibody production. The overall advantages and disadvantages of these procedures have been described elsewhere.[20] Thus far, both techniques have worked equally well. Given the relative simplicity of polyclonal antibody production, we use rabbits first.

Polyclonal Antibodies. Rabbits are immunized by multiple intradermal injections of the peptide–carrier conjugate (150–200 μg peptide) in Freund's complete adjuvant. Subsequent booster injections of the conjugate (100 μg peptide) are given in incomplete adjuvant at 21 and 42 days. After the first boost, serum samples are obtained from weekly bleedings and stored at $-20°$. Preimmune serum is obtained from each animal prior to the initial injection.

Sera from rabbits injected with the phosphorylated and nonphosphorylated forms of the conjugated synapsin I peptide are screened by a dot-immunoblotting procedure (see the next section for screening protocols). Positive sera that appear to be specific for either the phospho or dephospho forms of synapsin I at phosphorylation site 1 are identified. Positive serum samples are further purified.

The dephospho-specific synapsin I antibodies (G-143) are purified using a dephosphosynapsin I peptide affinity column. The dephosphorylated form of the synapsin I peptide (25 mg) is coupled to 1.5 g activated CH-Sepharose 4B (Pharmacia/LKB, Piscataway, NJ) following the instructions provided by the manufacturer. Ethanolamine (1 M, pH 8) is used to block any remaining reactive groups on the resin. Serum samples (25–40 ml) are pooled, filtered through a 0.22-μm Millex-GV syringe filter (Millipore, Bedford, MA) and applied to the peptide column at room temperature. To reduce potential proteolysis of the coupled peptide and the antibodies, a cocktail of protease inhibitors is added to the serum prior to loading (final concentrations: benzamidine, 25 mM; EDTA and EGTA, 1 mM each; leupeptin and antipain, 20 μg/ml each; pepstatin and chymostatin, 2 μg/ml each; phenylmethylsulfonyl fluoride, 100 μM; aprotinin, 100 units/ml). After the serum had been applied, the column is washed with six bed volumes of each of the following buffers: (1) 50 mM Tris, pH 7.5/1 M NaCl/0.05% Tween 20, (2) 50 mM sodium borate, pH 8.5/500 mM NaCl/0.05% Tween 20, (3) 50 mM sodium acetate, pH 5.5/500 mM NaCl/0.05% Tween 20, (4) 50 mM HEPES, pH 7.6/500 mM NaCl. The bound antibodies are then eluted with 4.6 M MgCl$_2$. The MgCl$_2$ elutate is diluted with an equal volume of H$_2$O and dialyzed four times against 5 liters of 10 mM HEPES, pH 7.6/154 mM NaCl. After dialysis, the antibodies are concentrated to approximately 1 mg/ml with the aid of a Centriprep-30 concentrator (Amicon, Danvers, MA) or by use of polyethylene glycol flakes (Aquacide III, Calbiochem, San Diego, CA). Affinity-purified antibodies are stored at 4° in the presence of 0.02% sodium azide or stored frozen in aliquots at $-20°$.

An IgG pool containing phosphospecific synapsin I antibodies (G-257) is obtained using a protein A-Sepharose fast protein liquid chromatography (FPLC) column (Pharmacia/LKB). A filtered serum sample (20 ml), which contains the protease inhibitors described above, is adjusted to 100 mM glycine, pH 8.5, 3 M NaCl and applied to the protein A-Sepharose column. The IgG fraction is eluted with a decreasing salt/pH gradient to 100 mM glycine, pH 3. Protein elution is monitored by UV absorption at 280 nm. The peak fractions are rapidly neutralized with 1 M Tris base and then passed through a dephosphosynapsin I peptide-CH-Sepharose column to adsorb any dephospho-specific antibodies that might have been present in the IgG pool. The flow-through from the peptide column, containing the phospho-specific antibodies, is then dialyzed four times against 5 liters of

10 mM HEPES, pH 7.6, 154 mM NaCl. Aliquots are stored at 4° in the presence of 0.02% sodium azide or stored frozen at −20°.

Only phosphospecific antibodies to synapsin I are present in the IgG pool from G-257 serum. In the event that serum samples contain a mixture of dephospho- and phosphospecific antibodies, the IgG pool can be applied to a column having the phosphorylated form of the peptide as the ligand. The phospho-specific antibodies should bind to the column and be recovered by the MgCl$_2$ elution as described above. This procedure would also remove other unrelated antibodies that are present in the IgG pool. Antibodies that react with a 50-kDa protein band on immunoblots of rat brain homogenate are removed by using phosphopeptide affinity purification of the G-257 IgG pool. If a phosphopeptide affinity column is used, precautions should be taken to minimize the potential dephosphorylation of the phosphopeptide by the addition of phosphatase inhibitors, such as 50 mM NaF and 20 mM orthophosphate, and performing all steps except the MgCl$_2$ elution at 4°. This approach is also used to purify anti-phospho-DARPP-32 antibodies.

Monoclonal Antibody Production. Monoclonal antibodies specific for the phosphorylated form of DARPP-32 (at Thr-34) are prepared by standard techniques[20] using BALB/c mice and Sp 2/0 mouse myeloma cells. Additional antibody is produced in ascites fluid by injection of cells from selected hybridoma cultures into pristane-primed BALB/c mice. Screening procedures are described in the following section.

Screening Procedures for Identification of Phosphorylation State-Specific Antibodies

We have employed either dot-immunobinding, enzyme-linked immunosorbent assays (ELISAs), or immunoprecipitation assays to screen for phosphorylation state-specific antibody production. For primary screening, the choice of assay depends on the number of samples to be tested and the availability of sufficient quantities of the dephospho and phospho forms of the peptide or holoprotein. When peptides are used for the primary screening, secondary tests using the holoprotein should be performed to confirm the results. Dot-immunobinding and ELISAs are simple and sensitive; they can handle large numbers of samples, such as those produced during monoclonal screening. Immunoprecipitation assays are generally performed as a secondary screen.

Dot-Immunobinding Assay

Dot-blot assays are used to identify phospho- and dephospho-specific polyclonal antibodies for phosphorylation site 1 in synapsin I. Serially diluted aliquots of purified phosphosynapsin I (0.1–50 ng/2 μl), which have been phosphorylated *in vitro* at site 1 to a stoichiometry of 0.9 mol phosphate/mol protein with the catalytic subunit of cAMP-dependent protein kinase,[15] are spotted onto strips of pure nitrocellulose, together with identical amounts of dephosphosynapsin I. Aliquots of phosphosynapsin I that have been phosphorylated *in vitro* at sites 2 and 3 to a stoichiometry of 1.9 mol/mol by calcium/calmodulin-dependent protein kinase II[15] are also spotted as an additional control. Assays are performed in 50 mM Tris, pH 7.5, 200 mM NaCl, 0.05% Tween 20 containing 2% nonfat dried milk. Nitrocellulose strips are preincubated in buffer, incubated with aliquots of antisera (1 : 100 dilution) for 90 min, washed, and incubated for 90 min with ^{125}I-labeled protein A. Strips are then washed thoroughly, dried, and subjected to autoradiography. Initial screening with this method identifies positive sera that are affinity purified as described above. Assays employing antibodies affinity purified from each antiserum are shown in Fig. 1. During the initial screening of sera with this procedure, the possibility that the serum samples might contain phosphatase activity is tested by incubation of serum (1 : 50 dilution) with dot blots of ^{32}P-labeled synapsin I, followed by autoradiography and Cerenkov counting. No significant reduction in the total radioactivity present in any spot is observed, indicating that no phosphatase activity is detected toward ^{32}P-labeled synapsin I at either site 1 or at sites 2 and 3. Note that the phosphospecific antibody, G-257, is selective for the presence of a phosphoryl group at site 1, and does not recognize the phosphorylated form of synapsin I at sites 2 and 3. Conversely, the dephospho-specific antibody, G-143, does not react with synapsin I that has been phosphorylated at site 1. This type of assay is most useful when the number of samples to be screened is low, as is typically the case for the production of polyclonal rabbit antisera. It is important to prepare holoprotein samples that have been stoichiometrically phosphorylated, since the presence of any dephospho form of the holoprotein in the sample would complicate the detection of phospho-specific antibodies. If the *in vitro* phosphorylation stoichiometry is poor, or if the available quantities of holoprotein are limiting, then the phosphorylated and unphosphorylated forms of the synthetic peptide can be used for screening. To obviate detection of carrier-specific antibodies, each form of the peptide is conjugated to a carrier protein (e.g., BSA) other then that used for the immunization. It is suggested that a different cross-linking reagent be used as well, since antibodies to the linker can produce false positives in this type of assay (glutaraldehyde is particularly immunogenic). The peptide conjugates are then used for dot blots as described above.

NANOGRAMS SPOTTED

50   25   5   1   0.1

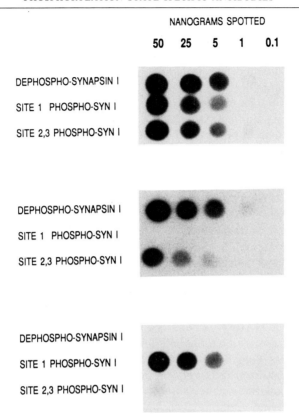

DEPHOSPHO-SYNAPSIN I

SITE 1 PHOSPHO-SYN I

SITE 2,3 PHOSPHO-SYN I

DEPHOSPHO-SYNAPSIN I

SITE 1 PHOSPHO-SYN I

SITE 2,3 PHOSPHO-SYN I

DEPHOSPHO-SYNAPSIN I

SITE 1 PHOSPHO-SYN I

SITE 2,3 PHOSPHO-SYN I

FIG. 1. Identification of phosphorylation state-specific antibodies for site 1 in synapsin I. Antibodies were tested for phosphorylation state specificity after spotting equal amounts of the three forms of synapsin I onto nitrocellulose. Site 1 of synapsin I was phosphorylated with the catalytic subunit of cAMP-dependent protein kinase; sites 2 and 3 were phosphorylated with Ca^{2+}/calmodulin-dependent protein kinase II. Antibody binding was detected with ^{125}I-labeled protein A. *Top:* G-116 (1 : 1000 dilution). This antibody, which recognizes an epitope in the COOH-terminal portion of the molecule,[16] demonstrates the equal reactivity of a phosphorylation state-independent antibody with the three forms of synapsin I. *Middle:* G-143 (1 : 300 dilution). *Bottom:* G-257 (1 : 300 dilution).

ELISA

ELISA[20] are most useful when the primary screening requirements are large, as is the case for monoclonal antibody production. Monoclonal antibodies specific for the phosphorylated form of DARPP-32 at Thr-34 are identified with a modified ELISA protocol. Microtiter plates (Nunc II, Roskilde, Denmark) are coated overnight with aliquots (1 μl/well) of the phosphorylated form of the DARPP-32 peptide conjugated to BSA. After blocking with 1% (w/v) BSA in phosphate-buffered saline (PBS), 100-μl aliquots of fusion hybrid supernatants (or mouse sera at 1 : 50 dilution) are added and incubated for 30 min at 37°. The plates are washed 3 times with PBS containing 1% BSA and 0.1% Tween 80. Goat anti-mouse IgG conjugated to horseradish peroxidase (100 μl of a 1/100 dilution, Kirkegaard and Perry, Gaithersburg, MD) is added to each well and incubated for an additional 30 min at 37°. The plates are then washed three times with 1% BSA and 0.1% Tween 80. An aliquot (50 μl) of substrate solution containing a 1 : 1 mixture of hydrogen peroxide and 2,2-azinodi(3-ethyl-benzthiazoline sulfonate) (ATBS; Kirkegard and Perry) is added to each well and the resulting color reaction quantified after 30 min using an SLT ER400 EIA reader (SLT Laboratory Instruments, Hillsborough, NC).

Approximately 2000 clones were initially screened with the phosphory-lated DARPP-32 peptide and about 500 were positive. The positive clones were then tested for specificity against the phospho and the dephospho forms of native DARPP-32. The protocol for these experiments is essen-tially the same as that used for the initial screening except that the plates are coated with aliquots (approximately 100 ng) of dephosphorylated or phosphorylated DARPP-32. Nearly 250 of these clones reacted preferably with with phosphorylated DARPP-32, and 27 of these were selected for further analysis. One of these clones (Mab-23) was shown to react strongly with the phosphopeptide but not the dephosphopeptide of DARPP-32 with the ELISA and was further assessed by dot-immunobinding assay and immunoprecipitation of ^{32}P-labeled phospho-DARPP-32.

Immunoprecipitation Assay

Immunoprecipitation is useful as a secondary screening procedure to assess and to quantify further the selectivity of the phosphorylation state-specific antibodies identified in the primary screen. However, we have observed that some antibodies that work well for immunoblotting do not have the capability to immunoprecipitate the holoprotein. Results demon-strating the relative specificity of the G substrate antibodies are shown in Fig. 2. Dephospho-G substrate and phospho-G substrate were purified from rabbit cerebellum as described.[8,21] Dephospho-G substrate and phos-pho-G substrate were iodinated using ^{125}I-labeled Bolton–Hunter reagent (New England Nuclear, Boston, MA). Precipitation of immune complexes of ^{125}I-labeled dephospho-G substrate or ^{125}I-labeled phospho-G substrate was performed using protein A-bearing *Staphylococcus aureus* cells

[21] D. W. Aswad and P. Greengard, *J. Biol. Chem.* **256**, 3487 (1981).

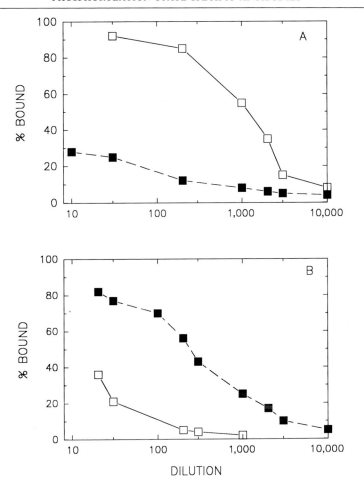

FIG. 2. Immunoprecipitation of dephospho-G substrate and phospho-G substrate by dephospho- and phosphoselective antibodies. Serial dilutions of each antibody were used to immunoprecipitate the radiolabeled phospho- and dephospho forms of G substrate. □, ^{125}I-Labeled dephospho-G substrate; ■, ^{125}I-labeled phospho-G substrate. (A) Dephosphospecific antibody; (B) phosphospecific antibody. The phosphorylated form of G substrate was prepared as described,[8] using cGMP-dependent protein kinase. The percentage bound represents the counts per minute (cpm) precipitated divided by the total G substrate (cpm) added to each assay. Each data point represents the mean of duplicate determinations. (Adapted from Nairn *et al.*[8])

(SAC). Before immunoprecipitation, SAC was prepared as described,[22] except that the buffer (NET buffer) used contained 150 mM NaCl, 15 mM EDTA, 50 mM Tris-HCl, pH 7.4, 50 mM NaF, 0.02% sodium azide, and 1% Nonidet P-40. ^{125}I-Labeled substrate in 100 μl NET buffer was "precleared" with 10 μl of SAC. After incubating for 20 min at 0°, the sample was centrifuged and the supernatant containing the precleared ^{125}I-labeled G substrate was removed and diluted to required concentrations (1000–2000 cpm μl⁻¹) in NET buffer. Immunoprecipitation of ^{125}I-labeled G substrate was performed using 96-well microtiter plates. Serum (10 μl, diluted as indicated in NET buffer), and ^{125}I-labeled G substrate (10 μl) were added to 80 μl NET buffer and the samples were incubated for 2 hr at 4°. SAC (20 μl) was then added and the samples were incubated for an additional 15 min before the assay was stopped by centrifugation for 15 min at 1000 g using the Cooke Microtiter system (Dynatech, Baton Rouge, LA). Supernatants were removed by inversion. The pellets were resuspended and washed with 100 μl of NET buffer. Bound radioactivity was measured, either directly in the washed pellets, or following SDS–PAGE and autoradiography, using a γ counter (type 2/200; Micromedic Systems, Horsham, PA). The dephospho-specific antibody precipitated ^{125}I-labeled dephospho-G substrate in preference to ^{125}I-labeled phospho-G substrate. With a serum dilution of 1 : 1000, approximately 55% of the dephospho-G substrate was precipitated, compared with only 6% for the phospho-G substrate (Fig. 2A). Conversely, the phosphospecific antibody displayed a significantly greater affinity for ^{125}I-labeled phospho-G substrate than for ^{125}I-labeled dephospho-G substrate (Fig. 2B). At a 1 : 200 dilution of serum, 50% of the phospho-G substrate was precipitated under conditions in which less than 5% of the dephospho form was precipitated. Similar procedures were used to characterize the selectivity of the phosphospecific monoclonal antibodies against phospho-DARPP-32.

Characterization of Phosphorylation State-Specific Antibodies

The utility of phospho- and dephosphospecific antibodies has been demonstrated in a variety of biochemical and physiological experiments. The state of cAMP-dependent phosphorylation of synapsin I during an *in vitro* reaction was measured by quantitative immunoblotting with antibodies G-257 and G-143, and compared to results obtained by direct measure-

[22] S. E. Goelz, E. J. Nestler, B. Chehrazi, and P. Greengard, *Proc. Natl. Acad. Sci. U.S.A.* **78**, 2130 (1981).

SYNAPSIN I PHOSPHORYLATION
TIME COURSE (SEC)

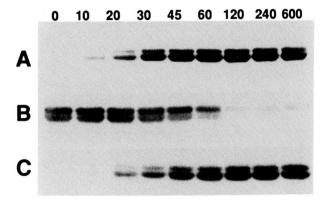

FIG. 3. Measurement of the state of phosphorylation of synapsin I using quantitative immunoblotting procedures. The state of site 1 phosphorylation of synapsin I was measured by quantitative immunoblotting with antibodies G-257 and G-143 and compared to the values obtained by direct measurement of ^{32}P incorporation from [γ-^{32}P]ATP. Purified synapsin I (7.5 μM) was phosphorylated with the catalytic subunit of cAMP-dependent protein kinase (5 nM) in the presence of 50 mM HEPES, pH 7, 10 mM MgCl$_2$, 0.1 mM DTT, and 150 μM ATP at 30°. At various times, aliquots were removed and added to an SDS–containing solution. Two reactions were run in parallel, one containing only unlabeled ATP and the other with tracer amounts of [γ-^{32}P]ATP. All samples were subjected to SDS–PAGE using 7.5% gels. Gels containing the ^{32}P-labeled samples were dried and subjected to autoradiography. The nonradioactive samples were transferred[23] to nitrocellulose, and then the membranes were incubated sequentially with affinity-purified G-257 or G-143, and ^{125}I-labeled protein A, followed by autoradiography of the blots. (A) Phosphosynapsin I as detected by immunoblotting with the phospho-specific antibody G-257 and ^{125}I-labeled protein A; (B) dephosphosynapsin I as detected by immunoblotting with the dephospho-specific antibody G-143 and ^{125}I-labeled protein A; (C) phosphosynapsin I as detected by autoradiography of ^{32}P incorporation from [γ-^{32}P]ATP.

ment of ^{32}P incorporation (Fig. 3). The values obtained with the phospho-specific antibody G-257 correlated well with those obtained by measurement of ^{32}P incorporation, while an inverse correlation was observed with the dephosphospecific antibody, G-143 (Fig. 4). Using a nonequilibrium radioimmunoassay procedure, similar results were obtained in experiments using G substrate.[8]

The ability of the phospho-specific synapsin I antibodies to detect changes in the state of phosphorylation of the protein in intact cell systems in response to pharmacological manipulations was also characterized (Fig.

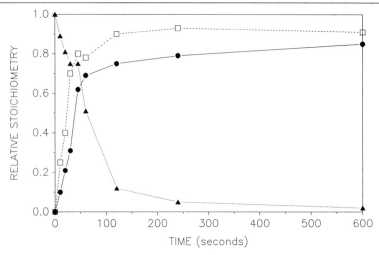

FIG. 4. Comparison of stoichiometry of synapsin I phosphorylation as determined by phosphorylation state-specific antibodies and ^{32}P incorporation. The results obtained in Fig. 3 were quantitated and depicted graphically. ^{32}P incorporation into synapsin I was determined by Cerenkov counting of the gel piece after Coomassie Blue staining and excision from the dried gel. Synapsin I bands were detected with ^{125}I-labeled protein A and quantitated by γ counting of the nitrocellulose pieces after visualization by autoradiography. The stoichiometry of phosphorylation at site 1 reached 0.9 mol phosphate/mol protein as measured by ^{32}P incorporation, and this value was used to normalize the relative values obtained from the immunoblots. □, Phosphosynapsin I as determined by ^{32}P incorporation; ●, phosphosynapsin I as determined with antibody G-257; ▲, dephosphosynapsin I as determined by antibody G-143.

5).[23–25] Treatment with forskolin increased synapsin I phosphorylation 3.6-fold over control levels in mouse cortical reaggregate cultures, and 1.3-fold in rat hippocampal slice preparations. Additional bands which cross-reacted with G-257 antibodies were observed on the autoradiogram. Two of the bands corresponded to synapsins IIa and IIb, two additional members of the synapsin family of phosphoproteins. The relative increase in the forskolin-stimulated phosphorylation of synapsin IIb was similar to that observed for synapsin I. Synapsins I and II share a large degree of sequence homology,[16] including the sequence surrounding the NH$_2$-terminal phosphorylation site, where 9 of 11 residues are identical. There-

[23] J.-A. Girault, I. Shalaby, N. Rosen, and P. Greengard, *Proc. Natl. Acad. Sci. U.S.A.* **85**, 7790 (1988).
[24] H. Towbin, T. Staehelin, and J. Gordon, *Proc. Natl. Acad. Sci. U.S.A.* **76**, 4350 (1979).
[25] H. C. Hemmings, Jr., J.-A. Girault, K. R. Williams, M. B. LoPresti, and P. Greengard, *J. Biol. Chem.* **264**, 7726 (1989).

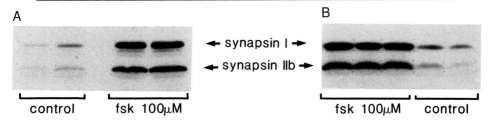

FIG. 5. Measurement of forskolin-stimulated phosphorylation of synapsins I and II in (A) reaggregate cell cultures and (B) tissue slice preparations. Rotation-mediated reaggregate cultures of mouse striatum were prepared as described.[23] Experiments were performed with aggregates that were maintained in culture for 3 weeks. Aggregates from several flasks were pooled, washed in buffer (124 mM NaCl, 5 mM KCl, 25.9 mM NaHCO₃, 1.4 mM MgSO₄, 1.2 mM KH₂PO₄, 1.5 mM CaCl₂, 10 mM D-glucose, 25 mM HEPES, pH 7.4), and then aliquoted into individual tubes. After a 1-hr preincubation, forskolin (fsk; 100 µM final concentration from a 10 mM stock in 95% ethanol) was added. After a 10-min incubation the buffer was removed and the aggregates were frozen in liquid N₂. The aggregates were later sonicated in hot 1% SDS, and the samples were assayed for protein content by the BCA assay (Pierce, Rockford, IL). Equal amounts of protein (150–200 µg) were subjected to SDS–PAGE using 12.5% gels and transferred to nitrocellulose.[24] Rat striatal slice preparations were prepared as described,[25] and the experiments were carried out in a fashion similar to that with the reaggregates. The nitrocellulose blots were incubated with affinity-purified G-257 (1 : 300 dilution) followed by ¹²⁵I-labeled protein A. Immunoreactive synapsin bands were detected by autoradiography, excised, and quantitated by γ counting.

fore, it was not unexpected that the use of the synthetic synapsin I peptide would produce antibodies that cross-reacted with synapsin II. A similar cross-reactivity was observed with G-143, the dephosphospecific antibody. Purification of the G-257 IgG pool by phosphopeptide affinity chromatography eliminated the appearance of cross-reactive bands other than synapsin II on the immunoblots.

Results obtained using a phosphospecific monoclonal antibody for DARPP-32 with an intact cell preparation are shown in Fig. 6. Treatment of the striatal cultures with forskolin produced a large increase in the amount of phospho-DARPP-32 detected. The antibodies also reacted with two bands of higher molecular weight which were not recognized by other antibodies to DARPP-32. The identity of these additional bands has not been established. They could represent proteins having epitopes resembling the phosphorylated form of DARPP-32, although they did not appear to be sensitive to treatment with forskolin. In other experiments, phospho-DARPP-32 antibodies also exhibited cross-reactivity with the phospho form of phosphatase inhibitor-1, a protein which contains a nine-residue sequence identical to DARPP-32 in the region surrounding the cAMP-dependent site of phosphorylation in each protein.[25]

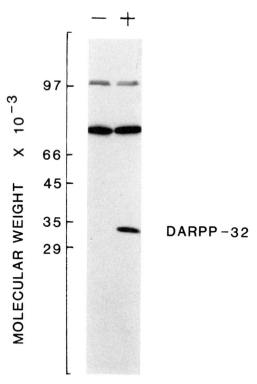

FIG. 6. Measurement of forskolin-stimulated phosphorylation of DARPP-32 in striatal reaggregate cell cultures. Striatal reaggregate cultures were prepared, treated, and processed as described in the legend to Fig. 5, with a 6-min period of incubation with forskolin (50 μM). Equal amounts of protein (270 μg) were subjected to SDS–PAGE and transferred to nitrocelluose.[24] Phospho-DARPP-32 was detected with monoclonal antibody Mab-23 and ^{125}I-labeled protein A overlay, followed by autoradiography.

General Utility of Method: Advantages, Limitations, and Applications

A flow diagram of the general procedure is shown in Fig. 7. In principle, it should be possible to prepare dephospho- and phosphospecific antibodies for any phosphoprotein. The use of small synthetic peptides improves the chance of targeting antibody production to epitopes of the native protein containing phosphorylation sites. The consensus sequences for several protein kinases include either basic or acidic amino acids in proximity to the phosphoryl acceptor residue, and these hydrophilic residues generally improve the immunogenic potential of small peptides. Large

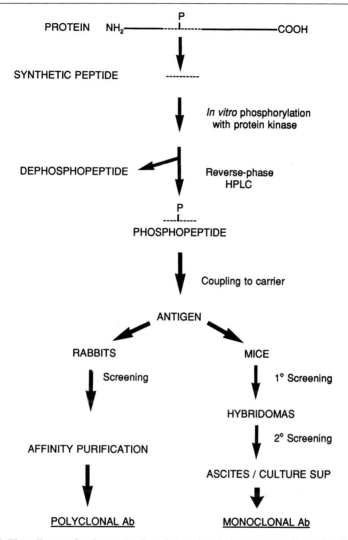

FIG. 7. Flow diagram for the production of phosphorylation state-specific antibodies. The phosphorylation site in the protein of interest can be determined by direct microsequencing techniques or predicted from sequence information based on homology with known substrate consensus sequences. To prepare dephospho-specific antibodies, the synthetic peptide can be directly conjugated to the carrier protein. For other details, see text.

amounts of the phosphorylated peptide antigens can be used for immunization, and it appears that small phosphopeptides are more resistant to dephosphorylation than the corresponding holoprotein.[26] Although phosphothreonyl residues are thought to be more resistant to phosphatase activity, we have produced phospho- and dephosphospecific antibodies to both phosphothreonyl-containing sites (G substrate and DARPP-32) and phosphoseryl-containing sites (synapsin I). Since antibodies that are specific for phosphotyrosine itself have been prepared by a number of laboratories, it should be possible to produce phosphorylation state-specific antibodies that recognize selected substrates for protein-tyrosine kinases, although we have not yet attempted this in our own laboratory.

One limitation of this procedure is the potential for cross-reactivity with other unrelated substrates, due to the homology between a particular phosphorylation site sequence and the consensus sequence for the specific protein kinase. This is not usually a problem for immunoblotting assays, as the cross-reacting species can usually be resolved by SDS–PAGE prior to assay. However, cross-reactivity of the antibodies would restrict their use in other types of quantitative assays and in immunocytochemistry.

Phosphorylation state-specific antibodies can provide a direct means to quantitate the regulation of the state of phosphorylation of a protein in intact cells in response to various physiological and pharmacological manipulation. The immunochemical techniques are generally simple enough to perform and would avoid several of the methodological pitfalls, such as proteolysis, denaturation, and recovery, which can restrict the use of back phosphorylation assays. When phosphospecific antibodies are used in combination with other phosphorylation-independent antibodies (to measure the total amount of protein), and known standards of the phospho (or dephospho) forms of the protein, it is possible to quantitate the *in vivo* stoichiometry of phosphorylation. This value can help to assess the physiological relevance of the reaction. Direct measurement of phosphorylation state avoids the problems associated with the estimation of the *in vivo* specific activity of intracellular ATP pools as well as the health risks and disposal problems associated with the high amounts of $[^{32}P]P_i$ required for prelabeling experiments. The technique also allows assays to be performed on tissue samples derived from *in vivo* studies with intact animals. If the antibodies recognize the native form of the protein, they can also be used to inhibit the phosphorylation and dephosphorylation reactions in cell-free systems and could be introduced into intact cell systems. Cellular fractionation and immunocytochemical studies using

[26] H. C. Hemmings, Jr., A. C. Nairn, J. Elliot, and P. Greengard, *J. Biol. Chem.* **265,** 20369 (1990).

these antibodies could provide evidence for phosphorylation-dependent changes in the subcellular distribution of the protein. All of these approaches can be useful in the elucidation of the functional significance of a particular phosphoprotein.

Acknowledgments

We thank Dr. Shelley Halpain for providing rat hippocampal slices tissue, Dr. John Haycock for comments on the manuscript, and Catherine Belleville for excellent technical assistance. This work was supported by U.S. Public Health Service Grants MH 39327 and MH 40899 and by cooperative agreement CR 813826 with the U.S. Environmental Protection Agency.

Section II

Protein Kinase Inhibitors

A. Peptide Inhibitors
Articles 24 and 25

B. Chemical Inhibitors
Articles 26 through 32

[24] Pseudosubstrate-Based Peptide Inhibitors

By Bruce E. Kemp, Richard B. Pearson, and Colin M. House

Many protein kinases exist as catalytically inactive or latent forms unless activated by the binding of a ligand or by covalent modification such as phosphorylation. It is now clear that the inactive form of a number of protein kinases results from the presence of an autoinhibitory region that masks the catalytic activity.[1] We term this type of control intrasteric regulation since an internal segment of the protein structure acts as an autoinhibitor. Studies on several of these protein kinases have demonstrated that the autoinhibitory regions act by mimicking the features of the protein substrate phosphorylation site of the enzyme.[1,2] Any structure that folds back into the active site to turn a protein kinase off using substrate-like recognition groups can be termed a pseudosubstrate. Subsequent removal of the pseudosubstrate region from the active site is required for expression of protein kinase activity. While this is normally achieved by binding of an activator it can also be mimicked experimentally by proteolysis or site-directed mutagenesis to give a constitutively active form of the enzyme. Thus far all pseudosubstrate structures have been found outside the conserved protein kinase catalytic domain, which includes the ATP and putative protein substrate-binding regions.[1] The pseudosubstrate region may be present on a separate subunit, as in the case of the regulatory subunits R_I and R_{II} of the cyclic AMP-dependent protein kinase[3] and the cyclic AMP-dependent protein kinase inhibitor. Alternatively, the pseudosubstrate region may reside on either the amino-terminal side of the catalytic domain as is seen for protein kinase C[5] or on the carboxyl-terminal side as observed for the myosin light-chain kinase (MLCK).[4] Synthetic peptides corresponding to the pseudosubstrate regions of several protein kinases have been found to act as inhibitors.[4-6] Furthermore, substitution of a phosphorylatable residue in the appropriate position in

---

[1] B. E. Kemp, R. B. Pearson, C. House, P. J. Robinson, and A. R. Means, *Cell. Signalling* **1**, 303 (1989).

[2] T. R. Soderling, *J. Biol. Chem.* **265**, 1823 (1990).

[3] S. S. Taylor, J. A. Buechler, and W. Yonemoto, *Annu. Rev. Biochem.* **59**, 971 (1990).

[4] B. E. Kemp, R. B. Pearson, V. Guerriero, I. C. Bagchi, and A. R. Means, *J. Biol. Chem.* **262**, 2542 (1987).

[5] C. House and B. E. Kemp, *Science* **238**, 1726 (1987).

[6] M. E. Payne, Y. L. Fong, T. Ono, R. J. Colbran, B. E. Kemp, T. R. Soderling, and A. R. Means, *J. Biol. Chem.* **263**, 7190 (1988).

the pseudosubstrate peptide may permit it to act as a substrate.[5] Where synthetic peptides can mimic the function of the pseudosubstrate regions these regions are referred to as pseudosubstrate prototopes.[5] Pseudosubstrate-based peptide inhibitors can be used to study the physiological roles of their respective protein kinases in extracts or in cells by microinjection, use of permeabilized cell, or expression in transfected cells. Since there may be overlaps in the specificity of pseudosubstrate peptides it is necessary to include several types of controls in such studies. These include an inactive pseudosubstrate analog peptide as a negative control and, if available, the constitutively active form of the protein kinase as a positive control.

Kinetics of Inhibition

Pseudosubstrate inhibitors may be particularly potent, with K_i values in the nano- to micromolar range. Two approaches have been taken to measure inhibitory constants. For potent inhibitors it is necessary to use Dixon plots to estimate K_i values because a significant portion of the enzyme is present as the enzyme–inhibitor complex.[7,8] The less rigorous, but useful, alternative is to use IC_{50} values that represent the concentration of inhibitor required to cause 50% inhibition of phosphate transfer at a protein or peptide substrate concentration equivalent to the K_m value. In some respects an IC_{50} value is more appropriate when there is uncertainty about the precise mechanism of inhibition. Care should be taken to ensure that other components of the reaction mixture, including ATP and Mg^{2+} concentrations, are optimal. Pseudosubstrate peptide inhibitors may act either as peptide substrate and/or ATP antagonists (see Calmodulin-Dependent Protein Kinases). This complicates the interpretation of the mechanism of inhibition operating and emphasizes the need for three-dimensional structural information to fully understand how pseudosubstrates interact with the active site.

Synthetic peptides will readily bind to glass or plastic surfaces. This becomes a problem with potent synthetic peptides tested at low concentration, where a significant portion of the peptide may bind to plastic unless precautions are taken. The same behavior is seen with dilute enzyme solutions.[9] The inclusion of 0.1% Tween 80 (w/v) in the dilution buffer is usually sufficient to block peptide binding to plastic surfaces. For example, the calmodulin antagonist peptide ARRKWQKTGHAVRAIGRLSS has an IC_{50} of 33.3 nM when diluted into H_2O or 10.6 nM when diluted into

[7] B. E. Kemp, H. C. Cheng, and D. A. Walsh, this series, Vol. 159, p. 173.
[8] D. A. Walsh, K. L. Angelos, S. M. Van Patten, D. B. Glass, and L. P. Garetto, *in* "Peptides and Protein Phosphorylation" (B. E. Kemp, ed.), p. 43. Uniscience CRC Press, Boca Raton, Florida, 1990.
[9] R. B. Pearson, C. House, and B. E. Kemp, *FEBS Lett.* **145**, 327 (1982).

buffer containing Tween 80.[9a] Note, however, that some protein kinases, such as protein kinase C, may be adversely affected by including high concentrations of detergent in the reaction mixture.

Preparation of Pseudosubstrate Peptide Inhibitors

The Merrifield solid-phase synthesis chemistry, peptide purification, and characterization for pseudosubstrate peptide inhibitors are the same as those used for the design of peptide substrates (see this series, Vol. 200, [10]). For an earlier account of the synthesis of oligopeptides for the study of cyclic nucleotide-dependent protein kinases, see Glass.[10]

Identification of Pseudosubstrate Regions

Pseudosubstrate sequences identified for a number of protein kinases are listed in Table I.[11–37] In the case of cyclic AMP-dependent protein kinase the pseudosubstrate region is found on the separate regulatory

[9a] Single-letter abbreviations for amino acids: A, Alanine; R, arginine; N, asparagine; D, aspartic acid; C, cysteine; Q, glutamine; E, glutamic acid; G, glycine; H, histidine; I, isoleucine; L, leucine; K, lysine; M, methionine; F, phenylalanine; P, proline; S, serine; T, threonine; W, tryptophan; Y, tyrosine; V, valine.

[10] D. B. Glass, this series, Vol. 99, p. 119.

[11] C. H. Clegg, G. G. Cadd, and G. S. McKnight, *Proc. Natl. Acad. Sci. U.S.A.* **85**, 3703 (1988).

[12] R. Mutzel, M. L. Lacombe, M. N. Simon, J. De Gunzburg, and M. Vernon, *Proc. Natl. Acad. Sci. U.S.A.* **84**, 6 (1987).

[13] J. D. Scott, M. B. Glaccum, M. J. Zoller, M. D. Uhler, D. M. Helfman, G. S. McKnight, and E. G. Krebs, *Proc. Natl. Acad. Sci. U.S.A.* **84**, 5196 (1987).

[14] T. Toda, S. Cameron, P. Sass, M. Zoller, J. D. Scott, B. McMullen, M. Hurwitz, E. G. Krebs, and M. Wigler, *Mol. Cell. Biol.* **7**, 1371 (1987).

[15] J. D. Scott, E. H. Fischer, K. Takio, J. G. Demaille, and E. G. Krebs, *Proc. Natl. Acad. Sci. U.S.A.* **82**, 572 (1985).

[16] K. Takio, R. D. Wade, S. B. Smith, E. G. Krebs, K. A. Walsh, and K. Titani, *Biochemistry* **23**, 4207 (1984).

[17] W. Wernet, V. Flockerzi, and F. Hofmann, *FEBS Lett.* **251**, 191 (1989).

[18] D. Kalderon and G. M. Rubin, *J. Biol. Chem.* **264**, 10738 (1989).

[19] S. Ohno, H. Kawasaki, S. Imajoh, K. Suzuki, M. Inagaki, H. Yokokura, T. Sakoh, and H. Hidaka, *Nature (London)* **325**, 161 (1987).

[20] K. Kubo, S. Ohno, and K. Suzuki, *FEBS Lett.* **223**, 138 (1987).

[21] L. Coussens, P. J. Parker, L. Rhee, T. L. Yang-Feng, E. Chen, M. D. Waterfield, U. Francke, and A. Ullrich, *Science* **233**, 859 (1986).

[22] A. Rosenthal, L. Rhee, R. Yadegari, R. Paro, A. Ullrich, and D. V. Goeddel, *EMBO J.* **6**, 433 (1987).

[23] S. Ohno, Y. Akita, Y. Konno, S. Imajoh, and K. Suzuki, *Cell (Cambridge, Mass.)* **53**, 731 (1988).

[24] Y. Ono, T. Fujii, K. Ogita, U. Kikkawa, K. Igarashi, and Y. Nishizuka, *J. Biol. Chem.* **263**, 6927 (1988).

TABLE I
AUTOREGULATORY PSEUDOSUBSTRATE SEQUENCES

Protein kinase[a]	Sequence	Ref.[b]
cAMP-PK R_I subunit (88–107)		
Bovine, human, pig[c]	VVKGRRRRGAISAEVYTEED	3
Mouse	VVKARRRRGGVSAEVYTEED	11
Dictyostelium	NNITRKRRGAISSEPLGDKP	12
cAMP-PK R_{II} subunit (86–105)[d]		
Bovine	PGRFDRRVSVCAET	3
Pig	PSKFTRRVSVCAET	3
Rat	PAKFTRRVSVCAET	13
Yeast (136–155)	HFNAQRRTSVSGET	14
PKI (5–26)	TTYADFIASGRTGRRNAIHDIL	15
cGMP-PKα (54–67)	GPRTTRAQGISAEP	16[e]
cGMP-PK β (70–83)	GEPRTKRQAISAEP	17
Drosophila		
G1 gene (137–150)	MPAAIKKQGVSAES	18
G2–T1 gene (474–485)	QNFRQRALGISAEP	18
Protein kinase C		
α (15–31)	DVANRFARKGALRQKNV	19
$β_I$ and $β_{II}$ (15–31)	ESTVRFARKGALRQKNV	20
γ (14–30)	GPRPLFCRKGALRQKVV	21
Drosophila (24–40)	NKMKSRLRKGALKKNV	22
ε (149–165)	ERMRPRKRQGAVRRRVH	23[f]
δ (137–153)	AMFPTMNRRGAIKQAKI	24
ζ (109–125)	EKAESIYRRGARRWRKL	24
Lung	FTRKRQRAMRRVHQ	g
Yeast (391–407)	QLMGGLHRHGAIINRKE	25
Acidic protein 14-3-3 (45–61)	LSVAYKNVVGARRSSWR	26[h]
Cam-II PK[i]		
α (280–307)	CMHRQETVDCLKKFNARRKLKGAILTTM	27
β (281–308)	MMHRQETVECLKKFNARRKLKGAILTTM	28
γ (281–308)	MMHRQETVECLKKFNARRKLKGAILTTM	29
δ (281–308)	MMHRQETVDCLKKFNARRKLKGAILTTM	30
Cam PK-Gr (53–80)	NFVHMDTAQKKQEFNARRKLKAAVKAVV	31[j]
Calspermin (2–24)	MDTAQKKQEFNARRKLKAAVKAVV	32
Myosin light chain kinase		
Smooth muscle (787–810)	SKDRMKKYMARRKWQKTGHAVRAI	33
Skeletal muscle (570–593)	SQRLLKKYLMKRRWKKNFIAVSAA	34
Twitchin (5421–5443)	SRYTKIRDSIKTKYDAWPEPLPP	35[k]
Phosphorylase kinase γ subunit[l]		
Rabbit (332–353)	VIRDPYALRPLRRLIDAYAFRI	36
Mouse (333–354)	VIRDPYALPPLRRLIDAYAFRI	m
Human (336–357)	LLRDPYALRSVRHLIDNCAFRL	37

[a] cAMP-PK R_I subunit, cyclic AMP-dependent protein kinase regulatory subunit type I; cAMP-PK R_{II} subunit, cyclic AMP-dependent protein kinase regulatory subunit type II; PKI, cyclic AMP-

subunit, whereas in other protein kinases the autoregulatory region is contiguous with the amino acid sequence of the catalytic domain and may act intramolecularly. The recognition of pseudosubstrate regions has depended very much on a knowledge of the substrate specificity of protein kinases,[38] as well as the determination of their amino acid sequences[39] and proximity to allosteric regulator sites (see below). Once a putative pseudosubstrate sequence has been located outside the catalytic domain of the kinase in question, synthetic peptides corresponding to this region can be tested as inhibitors. While pseudosubstrate inhibitors are often potent (K_i < 10 μM) this is not always the case, as the corresponding regions in the cyclic GMP-dependent and cyclic AMP-dependent protein kinases are not.[40] Several experimental approaches can be used to confirm that the identified region is responsible for intrasteric regulation of the protein kinase. These include generation of proteolytic fragments of the enzyme that have been rendered constitutively active by removal of the pseudosubstrate region. Alternatively, the pseudosubstrate region can be removed by site-directed mutagenesis and the recombinant protein

---

dependent protein kinase inhibitor (Walsh inhibitor); Cam-II PK, calmodulin-dependent protein kinase II (α, β, γ, and δ forms). The pseudosubstrate residues corresponding to the substrate phosphate acceptor sites are shown in bold.

[b] Numbers refer to text footnotes.

[c] Sequence conserved for bovine, human, pig, and rat.

[d] The core pseudosubstrate sequence RRVS[95]VCAE is conserved for bovine, mouse, pig, and rat; Ser-95 is autophosphorylated.

[e] Thr-58 is autophosphorylated and is widely thought to occupy the active site of the enzyme. Inspection of the aligned sequences suggest Gly-62 occupies the active site as the pseudosubstrate. The corresponding *Drosophila* genes do not have a corresponding autophosphorylatable Thr residue.

[f] The ε isoform of protein kinase C is also called the *n* form.

[g] N. Bacher, Y. Zisman, E. Berent, and E. Livneh, *Mol. Cell. Biol.* **11,** 126 (1991).

[h] Protein 14-3-3 isolated from brain activities tyrosine hydroxylase.[26] The inhibition of protein kinase C by this protein has been proposed to be due to the pseudosubstrate sequence 45–61 but not proven.

[i] Cam-II PK, calmodulin-dependent protein kinase II, also known as the multifunctional calmodulin-dependent protein kinase. Thr-286 is autophosphorylated and widely thought to occupy the active site; however, inspection of other pseudosubstrate sequences suggests that the GAIL sequence may be a more likely candidate to occupy the active site prior to activation by calmodulin.

[j] Cam PK-Gr, brain granulocyte calmodulin-dependent protein kinase, which has homology with the testis protein calspermin.[32]

[k] Twitchin is the product of the *unc*-22 gene of *Caenorhabditis elegans*. It is not known if the sequence shown is autoinhibitory.

[l] The pseudosubstrate sequence shown for phosphorylase kinase was derived by inspection of the amino acid sequence; no experimental evidence is available for this proposal.

[m] J. C. Chamberlain, unpublished, EMBO Protein Data Base.

expressed. Antibody directed at the pseudosubstrate region may also activate the enzyme.[41] Note that technical difficulties will vary with each protein kinase and the approach used may need to be modified accordingly. Comparison of the sequences listed in Table I indicates some of the common features that may be used in identifying pseudosubstrate regions. Alanine is the most common residue occupying the phosphate acceptor site but Gly, Thr, and Ser can also occur in some cases. The frequent use of Ala is not surprising, given its structural similarity to Ser. The residue in the phosphate acceptor site is frequently followed by a hydrophobic residue (Val, Ile, or Leu): In this regard MLCK is interesting (Table I) because sequence comparisons suggest that the Ala-Val sequence should occupy the active site; however, proteolysis experiments have shown that these residues are not essential for maintaining the enzyme in the inactive form, only the nearby basic residues (see below). It should be emphasized that scanning the target protein kinase sequence for a pseudosubstrate region may not always yield a signle unequivocal site. For instance, the cyclic AMP-dependent protein kinase contains adjacent Arg residues in each of the regulatory subunits $R^{92}RGAISA^{98}$ and $R^{234}RIIV^{238}$ for R_I and

[25] D. A. Levin, F. O. Fields, R. Kunisawa, J. M. Bishop, and J. Thorner, Cell (Cambridge, Mass.) 62, 213 (1990).
[26] T. Ichumura, T. Isobe, T. Okuyama, N. Takahashi, K. Araki, R. Kuwano, and Y. Takahashi, Proc. Natl. Acad. Sci. U.S.A. 85, 7084 (1988).
[27] R. M. Hanley, A. R. Means, T. Ono, B. E. Kemp, K. E. Burgin, N. Waxham, and P. T. Kelly, Science 237, 293 (1987).
[28] M. K. Bennett and M. B. Kennedy, Proc. Natl. Acad. Sci. U.S.A. 84, 1794 (1987).
[29] T. Tobimatsu, I. Kameshita, and H. Fujisawa, J. Biol. Chem. 263, 16082 (1988).
[30] T. Tobimatsu and H. Fujisawa, J. Biol. Chem. 264, 17907 (1989).
[31] C. Ohmstede, K. F. Jensen, and N. E. Sayhuon, J. Biol. Chem. 264, 5866 (1989).
[32] T. Ono, G. R. Slaughter, R. G. Cook, and A. R. Means, J. Biol. Chem. 264, 2081 (1989).
[33] N. J. Olson, R. B. Pearson, D. S. Needleman, M. Y. Hurwitz, B. E. Kemp, and A. R. Means, Proc. Natl. Acad. Sci. U.S.A. 87, 2284 (1990).
[34] K. Takio, P. K. Blumenthal, A. M. Edelman, K. A. Walsh, E. G. Krebs, and K. Titani, Biochemistry 24, 6028 (1985).
[35] G. M. Benian, J. E. Kiff, N. Neckelmann, D. G. Moerman, and R. H. Waterston, Nature (London) 342, 45 (1989).
[36] E. M. Reimann, K. Titani, L. H. Ericsson, R. D. Wade, E. H. Fischer, and K. A. Walsh, Biochemistry 23, 4185 (1984).
[37] S. K. Hanks, Mol. Endocrinol. 3, 110 (1989).
[38] B. E. Kemp, D. B. Bylund, T. S. Huang, and E. G. Krebs, Proc. Natl. Acad. Sci. U.S.A. 72, 3448 (1975).
[39] S. Hanks, A. M. Quinn, and T. Hunter, Science 241, 42 (1988).
[40] D. B. Glass, S. B. Smith, J. Biol. Chem. 258, 14797 (1983).
[41] M. Makowske and O. M. Rosen, J. Biol. Chem. 264, 16155 (1989).

RRVSVR^{230}RILM^{234} for R$_{II}$. Both sequences contain Ile instead of Ala at the phosphate acceptor site, but this is followed by a second hydrophobic residue, I or L, analogous to other pseudosubstrate structures. These sites occur in the cyclic AMP-binding domains of the respective regulatory subunits and it is clear from proteolysis experiments that they do not interact with the catalytic domain.[3] A second feature of the pseudosubstrate structures listed in Table I is the presence of a phosphorylation site at or near the proposed pseudosubstrate regulatory site for at least one member of each protein kinase type listed. These include cyclic AMP-dependent protein kinase (Ser-95 for the R$_{II}$ subunit), cyclic GMP-dependent protein kinase α form (Thr-58), calmodulin-dependent protein kinase II (Thr-286), the β form of protein kinase C (Ser-16 and Thr-17), and smooth muscle MLCK (Ser-814 and Ser-828). The presence of autophosphorylation sites in or near the proposed pseudosubstrate regulatory regions poses questions as to their significance and whether a phosphorylatable residue lies in the active site in the latent form of the enzyme. For cyclic AMP-dependent protein kinase regulatory subunit R$_{II}$ Ser-95 probably does occupy the active site and the phosphorylated regulatory subunit does inhibit the catalytic subunit, albeit with reduced affinity.[42] For protein kinase C the autophosphorylation sites are clearly distinct from the pseudosubstrate site[43] and no single site is phosphorylated stoichiometrically; however, the situation with calmodulin-dependent protein kinase II and cyclic GMP-dependent protein kinase is more complex (see below). While autophosphorylation near the pseudosubstrate site may merely reflect its proximity to the active site (see below) it seems reasonable to expect that a number of protein kinases that are themselves regulated by protein phosphorylation, such as those in the receptor-linked ribosomal protein S6 phosphorylation cascade, will turn out to have regulatory phosphorylation sites near their pseudosubstrate regions. This expectation is based on the observation that the pseudosubstrate regions of the calmodulin-dependent protein kinases and protein kinase C are in close proximity to their binding sites for calmodulin and phorbol ester, respectively. At present it is not known what proportion of protein kinases is regulated by pseudosubstrate structures. So far these have been identified by knowing the protein kinase sequence outside the catalytic domain, the substrate specificity, and that the enzyme exists in an inactive latent form. Recognition of the substrate specificity determinants of a protein kinase is not necessarily straightforward, as there is considerable variability in phosphorylation site sequences even for protein kinases whose substrate

[42] R. Randel Aldao and O. M. Rosen, *J. Biol. Chem.* **252,** 7140 (1977).
[43] A. J. Flint, R. D. Paladini, and D. E. Koshland, *Science* **249,** 408 (1990).

specificity has been exhaustively investigated. This variability adds to the uncertainty in identifying pseudosubstrate structures. Nevertheless, as the mechanisms operating for different classes of protein kinases are determined, it will become increasingly possible to recognize the pseudosubstrate motifs of related family members by inspecting the amino acid sequence alone, as has been done for protein kinase C. The following examples detail how the pseudosubstrate region have been identified for individual enzymes and illustrate potential uses of pseudosubstrate peptides. We have not considered the regulation of tyrosine protein kinases by pseudosubstrate mechanisms because of insufficient information. However, Cooper[44] has considered this and suggested that phosphorylation of Tyr-527 in pp60^{src} may be acting as a pseudoproduct inhibitor rather than a pseudosubstrate.

Cyclic Nucleotide-Dependent Protein Kinases

The pseudosubstrate region in the cyclic AMP-dependent protein kinase regulatory subunits (Table I) was historically referred to as the hinge region.[3] A battery of evidence derived from proteolysis studies, chemical modification, and site-directed mutagenesis now support the concept that this is the regulatory region responsible for maintaining the catalytic subunit in an inactive form in the holoenzyme.[3] Identification of Arg as an important specificity determinant for the cyclic AMP-dependent protein kinase[38] and the synthesis of peptide substrates and inhibitors[45] paved the way for understanding the mechanism of inhibition of the enzyme by the regulatory subunit[46] and the protein kinase inhibitor.[8]

Protein kinase inhibitor (PKI) contains a pseudosubstrate region between residues 15 and 22, RTGRRNAI (Table I). The 18-residue synthetic peptide, TTYADFIASGRTGRRNAI, corresponding to the region PKI (5–22), has a K_i of approximately 3 nM (Table II).[47–50] This is the most potent and specific pseudosubstrate inhibitor identified for any protein

[44] J. A. Cooper, in "Peptides and Protein Phosphorylation" (B. E. Kemp, ed.), p. 85. Uniscience CRC Press, Boca Raton, Florida, 1990.

[45] B. E. Kemp, E. Benjamini, and E. G. Krebs, Proc. Natl. Acad. Sci. U.S.A. 73, 1038 (1976).

[46] J. D. Corbin, P. H. Sugden, L. West, D. A. Flockhart, T. M. Lincoln, and D. McCarthy, J. Biol. Chem. 253, 3997 (1978).

[47] C. House and B. E. Kemp, Cell. Signalling 2, 187 (1990).

[48] R. J. Colbran, M. K. Smith, Y. L. Fong, C. M. Schworer, and T. R. Soderling, J. Biol. Chem. 264, 4800 (1989).

[49] C. J. Foster, S. A. Johnston, B. Sunday, and F. C. A. Gaeta, Arch,. Biochem. Biophys. 280, 397 (1990).

[50] M. Ikebe, Biochem. Biophys. Res. Commun. 168, 714 (1990).

TABLE II
SYNTHETIC PSEUDOSUBSTRATE INHIBITOR PEPTIDES

Protein kinase	Sequence[a]	K_i/IC$_{50}$ (μM)	Ref.[b]
cAMP-PK	TTYADFIASGRTGRRNAIHD	0.003	8
cGMP-PK			
α form	PRTTRAQGISAEP	2100	c
β form	PRTKRQAISAEP	900	c
Protein kinase C	RFARKGALRQKNV	0.13	47
Cam-II PK	MHRQETVDCLKKFNARRKLKGAILTTMLA	2.7	48
	LKKFNARRKLKGAILTTMLA	24	6
MLCK smooth			
(787–807)	SKDRMKKYMARRKWQKTGHAV	1.0	d
(783–807)	AKKLSKDRMKKYMARRKWQKTGHAV	0.013	49[e]
(783–804)	AKKLSKDRMKKYMARRKWQKTG	0.9	d
(783–804)	AKKLSKDRMKKYMARRKWQKTG	0.025	50[e]

[a] Residues corresponding to the phosphate acceptor sites are shown in bold.
[b] Numbers refer to text footnotes.
[c] P. J. Robinson, B. Michell, K. I. Mitchelhill, and B. E. Kemp, unpublished (1990).
[d] Determined in the presence of calmodulin with intact MLCK.
[e] IC$_{50}$ values determined with a calmodulin-independent proteolysed form of the MLCK.

kinase. Detailed structure/function studies by Walsh and colleagues have delineated the parts of the peptide sequence that contribute to its potency.[7,8,51] While the pseudosubstrate sequence RRNAI is necessary for recognition by the cyclic AMP-dependent protein kinase active site the additional basic residue Arg-15 and the amino-terminal extension TTYAD-FIAS contribute substantially to potency.[8] The Ile at position 22 acts as an important hydrophobic anchor and its substitution with Gly decreases potency more than 100-fold. As noted above, this is of interest in that almost all pseudosubstrate sequences listed in Table I contain a hydrophobic residue Leu, Val, or Ile at the $n + 1$ position, where n is the equivalent of the phosphate acceptor site. By making judicious substitutions with nonstandard amino acids the potency of the PKI pseudosubstrate peptide inhibitor has been further increased.[51] The PKI analog 5–22 containing 1-naphthylalanine in place of Phe-10 has a sevenfold greater potency with a $K_i = 0.39$ nM. This observation further emphasizes the important contribution surrounding residues may play in optimizing the spatial arrangement of residues involved directly in the interaction with the active site. In the absence of three-dimensional information about the inhibitor

[51] D. B. Glass, H. C. Cheng, L. Mende-Mueller, J. Reed, and D. A. Walsh, J. Biol. Chem. **264,** 8802 (1989).

peptide–enzyme complex it is not possible to specify precisely the spatial arrangements of residues required. Once this is known it may explain why alanine analogs of peptide substrates corresponding to a number of phosphorylation site sequences have been poor inhibitors. Knowledge of the pseudosubstrate structure–function relationships also provides clues for making more potent substrates. The PKI (14–22) S_{21} analog is a potent substrate for the cyclic AMP-dependent protein kinase with a K_m of 0.11 μM compared to a K_i of 0.04 μM for the PKI (14–22) parent peptide or a K_m of 4.7 μM for the kemptide.

The PKI pseudosubstrate prototope has been exploited in the design of synthetic genes that permit the controlled expression of the inhibitor peptide in recipient transfected cells.[52] It should be possible to extend this approach to other protein kinases and even place additional address motif sequences onto the ends of the peptide to permit membrane targeting by myristylation or organelle localization for enzymes such as pyruvate dehydrogenase kinase. Despite the polarity of the PKI peptide small amounts appear to gain access to the cytoplasm of cells incubated in millimolar concentrations of the peptide.[53] The potency of the peptide permits almost a 10^6-fold concentration gradient between the 1 mM extracellular concentration and the 3 nM K_i. It should also be possible to enhance the lipophilicity of the PKI peptide but this might result in localizing the peptide to the cell membrane and actually hinder access to the cytoplasm. Cleavable ester groups can be used to provide temporary protection of carboxyl groups to allow compounds access to the cytoplasm; however, there is no comparable means of masking amine or guanidino groups. One approach may be to design bifunctional reagents with a cleavable ester at one end and a sulfate or phosphate group at the other end capable of forming an intramolecular ion pair with a basic residue. The only successful attempt reported so far of designing a selectively protected protein kinase pseudosubstrate inhibitor with the capacity to enter the cell has been an inhibitor of the insulin receptor tyrosine protein kinase, (hydroxy-2-naphthalenylmethyl)phosphonic acid, used as an (acyloxyl)methyl prodrug with an IC_{50} of approximately 10 μM.[54]

Cyclic GMP-dependent protein kinase is an example of an enzyme whose pseudosubstrate region has been reported but for which the corresponding peptide binds poorly to the enzyme.[40] Given the increasing importance of this enzyme in smooth muscle relaxation and atriopeptin

[52] J. R. Grove, D. J. Price, H. M. Goodman, and J. Avruch, *Science* **238**, 530 (1987).
[53] W. Buchler, U. Walter, B. Jastoroff, and S. M. Lohmann, *FEBS Lett.* **228**, 27 (1988).
[54] R. Saperstein, P. P. Vicario, H. V. Strout, E. Brady, E. E. Slater, W. J. Greenlee, D. L. Ondeyka, A. A. Patchett, and D. G. Hangauer, *Biochemistry* **28**, 5694 (1989).

action, the development of an inhibitor of comparable specificity and potency of the PKI peptide would be highly desirable. Cyclic GMP-dependent protein kinase regulation was studied in the laboratories of Corbin[55] and Gil[56] and from their work came the initial concept that this enzyme was regulated by a pseudosubstrate mechanism. Thr-58 is autophosphorylated in the cyclic GMP-dependent protein kinase hinge region.[16] However, the initial suggestion that this residue occupies the active site analogous to Ser-95 for R_{II} may not be correct. Inspection of the range of pseudosubstrate sequences (Table I) suggests that, in fact, Gly-62 is likely to occupy the phosphate acceptor site. In the β isoform of cyclic GMP-dependent protein kinase[17] Ala-78 occupies the corresponding position to Gly-62. We have found that substitution of the corresponding pseudosubstrate peptides with Ser (β isoform Ala-78 or the α isoform Gly-62) renders them substrates at these positions. If Gly-62 occupies the active site then autophosphorylation on Thr-58 would be a secondary event following release of the pseudosubstrate from the active site in the α form. Activation of the enzyme by cyclic GMP leads to autophosphorylation on Thr-58 whereas up to six sites can be phosphorylated when cyclic AMP is used to activate the enzyme.[57] This suggests that a large segment of the sequence surrounding the hinge region is in proximity to the active site and that conformational changes induced by cyclic nucleotide binding dictate when residues are accessible for autophosphorylation rather than any of these residues being in the active site prior to activation. Inspection of the *Drosophila* cyclic GMP-dependent protein kinase gene sequences in this region reveals an absence of an autophosphorylation site corresponding to Thr-58.

Calmodulin-Dependent Protein Kinases

The pseudosubstrate prototope for smooth muscle MLCK was identified on the basis of homology with the substrate phosphorylation site sequence around Ser-19 in the myosin light chains.[4] The role of this region in autoregulation of the enzyme has been proved by using proteolysed forms of the enzyme[33,58,59] and more recently by recombinant protein expression.[59a] The MLCK pseudosubstrate prototope overlaps the calmodulin-binding region.[4] Thus the corresponding synthetic peptide can

[55] T. M. Lincoln, D. A. Flockhart, and J. D. Corbin, *J. Biol. Chem.* **253**, 6002 (1978).
[56] G. N. Gill, *J. Cyclic Nucleotide Res.* **3**, 153 (1977).
[57] A. Aitken, B. A. Hemmings, and F. Hofmann, *Biochim. Biophys. Acta* **790**, 219 (1984).
[58] R. B. Pearson, R. E. H. Wettenhall, A. R. Means, D. J. Hartshorne, and B. E. Kemp, *Science* **241**, 970 (1988).
[59] M. Ikebe, S. Maruta, and S. Reardon, *J. Biol. Chem.* **264**, 6967 (1989).
[59a] M. Itoh, V. Guerriero, X. Chen, and D. J. Hartshome, *Biochemistry*, in press (1991).

TABLE III
SUBSTRATE ANTAGONIST VERSUS CALMODULIN ANTAGONISM FOR MYOSIN LIGHT-CHAIN
KINASE PSEUDOSUBSTRATE

A[783]KKLSKDRMKKYMARRKWQKTGHAVRAIGRLSS[815] [a]

	IC$_{50}$ (μM)	
Sequence	Substrate antagonist activity	Calmodulin antagonist activity
SKDRMKKYMARRKWQKTGHAV	1.0	0.54
AKKLSKDRMKKYMARRKWQKTG	0.9	0.046
ARRKWQKTGHAVRAIGRLSS	3	0.011
RRKWQKTGHAV	3	260
RRKWQK	3	181
RRKLQK	13.5	1700
KRRWKKNFI AV[b]	1.8	20.9
KRRWKK	1.6	6
KRRFKK	0.5	93

[a] Chicken gizzard MLCK sequence.[33] Residues corresponding to the substrate acceptor sites are shown in bold.
[b] Rabbit skeletal MLCK sequence.[34]

inhibit the enzyme by acting both as a calmodulin antagonist and a substrate antagonist. Structure–function studies using synthetic peptides have made it possible to design pseudosubstrate inhibitors in which the calmodulin antagonist activity is reduced (Table III) while maintaining the substrate antagonist potency. These peptides may be used to probe the role of the MLCK in systems such as smooth muscle cells, skinned muscle fibers, and nonmuscle cells without affecting other calmodulin-dependent processes.

Extending the smooth muscle pseudosubstrate peptide toward either the amino terminus or the carboxyl terminus did not change the substrate antagonist activity, but strongly enhanced calmodulin antagonist activity (Table III). This however, does not mean that calmodulin binding extends toward the amino terminus in the parent protein because it is known that tryptic cleavage of smooth muscle MLCK at Arg-808 destroys calmodulin binding.[58] The fact that peptide 783–804 is also a potent calmodulin antagonist illustrates the need for caution in using the potency of peptide action alone to define a functional region in a protein. In Table III it can be seen that the basic residue-rich segment, RRKWQK is a potent substrate antagonist and a poor calmodulin antagonist (IC$_{50}$ 181 μM). The related peptide KRRWKK has greater potency as a calmodulin antagonist (IC$_{50}$ 6 μM), but this can be reduced to 93 μM by substituting the Trp with Phe.

Gaeta *et al.*[60] have developed pseudosubstrate inhibitors of the MLCK by incorporating nonnatural amino acids into the sequence around the phosphorylation site Ser-19 in the myosin light chains, SNVF. The most potent inhibitor resulted from the substitution of Phe-22 with 1,8-diphenyl-5-(3-phenylpropyl)-3-octene. Approaches such as these may yield compounds that have a greater capacity for acting on intact cells than the basic residue-rich sequences. The inhibition of MLCK by peptides corresponding to the pseudosubstrate region is complex. For instance, the peptide RRKWQKTHGAV, which acts as a peptide substrate antagonist (IC_{50} 3 μM) and a poor calmodulin antagonist (IC_{50} 260 μM), also has ATP antagonist activity (IC_{50} 13 μM). The ATP concentration employed in assays of the substrate antagonist activity of the pseudosubstrate peptides is 250 μM and at this concentration ATP antagonist effects will be greatly reduced. The calmodulin-independent MLCK derived by proteolysis[56,57] has also been used to estimate the potency of peptide substrate antagonist activity, thereby avoiding the confounding effect of the pseudosubstrate peptide binding to calmodulin. Several MLCK pseudosubstrate peptides have IC_{50} values in the low nanomolar range (Table II) when tested with the calmodulin-independent form of the enzyme.

The MLCK pseudosubstrate peptide has been used in microinjection experiments to inhibit muscle contraction in individual smooth muscle cells.[61] An important control in these experiments was the demonstration that the injection of the constitutively activated proteolytic fragment of the MLCK caused calcium-independent contraction and that this could be blocked with microinjection of the pseudosubstrate peptide.

The calmodulin-dependent protein kinase II autoregulatory region has been studied in several laboratories. In the initial study it was shown that the peptide Cam kinase II (290–309), LKKFNARRKLKGAILTTMLA, was an inhibitor, competitive with respect to peptide substrate and non-competitive toward ATP.[6] Subsequently it was found that amino-terminal extension of this sequence from residue 290 to residue 281, MHRQETVD-CLKKFNARRKLKGAILTTMLA, increased potency 10-fold (K_i 0.2 μM) but altered the mechanism of inhibition to become competitive toward ATP[48] and non competitive with respect to peptide substrate. Thr-286 in the sequence RQETV is a site of autophosphorylation that renders the enzyme partially independent of calmodulin after autophosphorylation.[2,48] Soderling and colleagues have proposed that Thr-286 is an im-

[60] F. C. A. Gaeta, L. S. Lehman De Gaeta, T. P. Kogan, Y. S. Or, C. Foster, and M. Czarniecki, *J. Med. Chem.* **33**, 964 (1990).
[61] T. Itoh, M. Ikebe, G. J. Kargacin, D. J. Hartshorne, B. E. Kemp, and F. S. Fay, *Nature* (*London*) **338**, 164 (1989).

portant component of the autoregulatory region of calmodulin-dependent protein kinase II.[62] Using a proteolysed calmodulin-independent form of the enzyme it was found that the peptide 281–309, MHRQET[286]VDCLKK-FNARRKLKGAILTTMLA, inhibited activity and that this was reversed with calmodulin.[62] Since phosphorylation of this peptide at Thr-286 was strongly enhanced with the addition of calmodulin these authors suggested that Thr-286 fulfills an inhibitory role that is overcome by conformational changes induced by calmodulin. An alternative explanation derived by inspecting the sequence comparisons in Table I is that the sequence GAIL with its proximal Ala and Ile may be the actual pseudosubstrate region that can be removed by adding calmodulin, thus allowing phosphorylation of Thr-286. The sequence RRKLKGAIL is not precisely like a substrate in that it does not have the consensus motif RXXS or RXXT.[63] But it should be noted that only 13 of the 20 known phosphorylation sites in the substrates of this enzyme actually conform to this consensus motif (see Vol. 200 of this series, [3]). It seems reasonable that the increased inhibitory potency obtained by increasing the peptide length from 290–309 to 281–309 may reflect the influence of the amino-terminal extension on the potency of the region RRKLKGAIL, analogous to the effect of the amino-terminal region of the PKI peptide (see above). The suggestion that calmodulin causes a conformational change that switches Thr-286 from being an inhibitor to being a substrate seems less likely.[62] Note that the Cam PK-Gr isoform (brain granulocyte calmodulin-dependent protein kinase; Table I) does not contain an Arg equivalent to Arg-283 and therefore raises further doubts about interpreting Thr-286 as having a direct role in occupying the active site in the latent form of the enzyme. Thus autophosphorylation of Thr-286 is viewed as having a secondary regulatory role occurring after the primary activation step. Whatever the correct mechanism, these considerations clearly illustrate the difficulty in defining precisely where the autoinhibitory region is located and emphasize the critical need for three-dimensional structural information. Peptide and enzyme mutagenesis studies can provide considerable information but not necessarily resolve these questions unequivocally.

Phosphorylase kinase is a multisubunit enzyme, $(\alpha,\beta,\gamma,\delta)_4$, where the α and β subunits are regulatory, γ is the catalytic subunit, and δ is calmodulin. The regulation of this enzyme is complex and there is no direct evidence yet of a pseudosubstrate regulatory mechanism operating. The isolated $\gamma\delta$ complex is calcium dependent and the isolated γ subunit is

[62] R. J. Colbran, Y.-Liang Fong, C. M. Schworer, and T. R. Soderling, *J. Biol. Chem.* **263**, 18145 (1988).

[63] B. E. Kemp and R. B. Pearson, *Trends Biochem. Sci.* **15**, 342 (1990).

TABLE IV
STRUCTURE–FUNCTION RELATIONSHIPS FOR PROTEIN KINASE C
PSEUDOSUBSTRATE PROTOTOPE

S^{16}TVRFARKGALRQKNVHEVKN^{36} [a]

Sequence		IC_{50} (μM)	Ref.[b]
(19–27)	RFARKGALR	6.2	5
(19–31)	RFARKGALRQKNV	0.13	47
(19–36)	RFARKGALRQKNVHEVKN	0.18	5
(16–31) K^{17}	SKVRFARKGALRQKNV	0.075	47
(19–31) A^{22}	RFAAKGALRQKNV	81	47
(19–31) retroinversion[c]	RFARKGALRQKNV	3	67
(19–31) S^{25} retroinversion[c]	RFARKGSLRQKNV	5	67

[a] Amino acid sequence of the β isoform of protein kinase C. Residues in the pseudosubstrate corresponding to the substrate phosphate acceptor site are shown in bold.

[b] Numbers refer to text footnotes.

[c] Retroinversion peptide contains all D-amino acids synthesized in reverse order so that the side-chain spatial orientation is maintained. The D-Ser-25-containing retroinversion peptide is not a substrate.

activated by the addition of calmodulin, albeit modestly.[64,65] These findings raise the possibility that the γ subunit contains a pseudosubstrate region. One candidate is the sequence **RDPYALR**, which is present in the carboxyl-terminal end of the γ subunit overlapping a recently reported calmodulin-binding region.[66] This sequence resembles the phosphorylation site sequence in phosphorylase, **RKQIS(P)VR** with Ala in place of the phosphate acceptor site Ser-14 and Leu at the $n + 1$ position as seen in many pseudosubstrate regions.

Protein Kinase C

The pseudosubstrate autoregulatory region in protein kinase C (see Table IV)[67] was located by inspecting the amino-acid sequence of the phospholipid diacylglycerol-binding domain. Substrate specificity studies in several laboratories[68–70] had shown that basic residues on both sides

[64] S. M. Kee and D. J. Graves, *J. Biol. Chem.* **261,** 4732 (1986).

[65] J. R. Skuster, K. F. Chan, and D. J. Graves, *J. Biol. Chem.* **257,** 2203 (1980).

[66] M. Dasgupta, T. Honeycutt, and D. K. Blumenthal, *J. Biol. Chem.* **264,** 17156 (1989).

[67] A. Ricouart, A. Tartan, and C. Sergheraet, *Biochem. Biophys. Res. Commun.* **165,** 1382 (1989).

[68] C. House, R. E. H. Wettenhall, and B. E. Kemp, *J. Biol. Chem.* **262,** 772 (1987).

[69] J. R. Woodgett, K. L. Gould, and T. Hunter, *Eur. J. Biochem.* **161,** 177 (1986).

[70] I. Yasuda, A. Kishimoto, S. Tanaka, M. Tominaga, A. Sakurai, and Y. Nishizuka, *Biochem. Biophys. Res. Commun.* **166,** 1220 (1990).

of the serine or threonine acceptor site can favor phosphorylation. The sequence RFARKGALRQKNV, corresponding to residues 19–31 of the α isoenzyme of protein kinase C, was observed and the corresponding synthetic peptide found to act as a potent competitive inhibitor (IC_{50} 0.13 μM).[47] In this case the phosphate acceptor site is occupied by alanine at position 25. Substitution of the Ala-25 with Ser results in a correspondingly potent substrate. Pseudosubstrate sequences occur in all protein kinase C isoenzyme forms thus far sequenced with absolute conservation of Arg-22 and Gly-Ala from man to yeast. Structure/function studies of the protein kinase C pseudosubstrate peptide[47] supported the concept that the basic residues were important. Substitution of Arg-22 with Ala increased the IC_{50} over 600-fold to 81 μM. However, substitution of other nonbasic residues, such as Leu-26 and Gly-24, also reduced inhibitor potency by more than 10-fold.[47] These results indicate that the potency of the protein kinase C pseudosubstrate results from the contribution of a number of residues in a manner analogous to the cyclic AMP-dependent protein kinase inhibitor peptide (see above). There is now support from several other types of experiments that the region 19–31 is indeed the autoregulatory pseudosubstrate region for protein kinase C. These include activation of the enzyme by anti-pseudosubstrate peptide antibodies,[41] mutagenesis experiments that cause activation of the enzyme by deletion of the pseudosubstrate region,[71] and site studies which have identified Ser-16 and Thr-17 as autophosphorylation sites, thereby inferring the proximity of the pseudosubstrate region (19–31) to the active site.[43] The latter studies on autophosphorylation of protein kinase C have important implications for understanding the relationships between autophosphorylation and pseudosubstrate control. Of the six autophosphorylation sites on protein kinase C two are in proximity to the pseudosubstrate region, two are on the amino-terminal side of the catalytic domain in proximity to the ATP-binding region, and two occur in the carboxyl-terminal region outside the catalytic domain. Since only approximately 1 mol of phosphate is incorporated per mole of enzyme it suggests that any one of the six possible sites may be autophosphorylated on activation of protein kinase C. One interpretation of these observations is that activation of the enzyme following removal of the pseudosubstrate region from the active site permits autophosphorylation of any Ser or Thr residues proximal to the active site.

    Using a different approach O'Brian et al.[72] have found that the

[71] C. J. Pears, G. Kour, C. House, B. E. Kemp, and P. J. Parker, Eur. J. Biochem. **194**, 189 (1991).

[72] C. A. O'Brian, N. E. Ward, R. M. Liskamp, D. E. De Bont, and J. H. van Boom, Biochem. Pharmacol. **39**, 49 (1990).

N-myristylated analog of a peptide substrate KRTLR inhibits histone phosphorylation catalyzed by protein kinase C with an IC_{50} of 75 μM. Further studies examining a range of acylated peptide inhibitors may provide more potent compounds. Aitken *et al.*[73] reported the presence of a pseudosubstrate-like sequence, KNVVGARR, in a family of brain acidic proteins that act as potent inhibitors of protein kinase C. As yet it is not known whether the corresponding synthetic peptide is a potent inhibitor. Indeed, since these acidic proteins are thought to activate tyrosine hydroxylase by the acidic residue-rich carboxyl-terminal region binding to a basic residue-rich region of tyrosine hydroxylase,[26] it is conceivable that this acidic region and not the pseudosubstrate sequence is important for inhibition of protein kinase C. The activities of many protein kinases are substantially enhanced following one or two steps of purification from crude extracts, suggesting the presence of endogenous inhibitors. One interpretation of this observation is that there may be a number of pseudosubstrate inhibitors analogous to the brain acidic proteins that act as inhibitors for a variety of protein kinases. Walsh and colleagues[74] have characterized two such protein kinase C inhibitors, one a zinc-binding protein. Neither of these proteins appears to contain an identifiable pseudosubstrate sequence that could account for their inhibition of protein kinase C.

Concluding Remarks

In discussing protein kinase pseudosubstrate structures we have attempted to challenge some views held regarding their identification in order to illustrate the problems in this area of research. In terms of the practical use of pseudosubstrate peptides the central question is how specific are they and can they be used effectively in crude systems or in intact cells? Smith *et al.*[75] have made the important point that many protein kinases utilize basic residues as substrate specificity determinants and that one can get significant overlap in the inhibitory specificity of pseudosubstrate peptide inhibitors. For example, the MLCK peptide AKKLSK-DRMKKYMARRKWQKTG was just as good an inhibitor of protein kinase C as MLCK itself. Nevertheless, provided appropriate controls are used,[61] and the limitations recognized, the lack of absolute specificity may not be an overriding obstacle to their use. In addition, substitution with

[73] A. Aitken, C. A. Ellis, A. Harris, L. A. Sellers, and A. Toker, *Nature (London)* **344**, 594 (1990).
[74] J. D. Pearson, D. B. DeWald, W. R. Mathews, N. M. Mozier, H. A. Zürcher-Neely, R. L. Heinrikson, M. A. Morris, W. D. McCubbin, J. R. McDonald, E. D. Fraser, H. J. Vogel, C. M. Kay, and M. P. Walsh, *J. Biol. Chem.* **265**, 4583 (1990).
[75] M. K. Smith, R. J. Colban, and T. R. Soderling, *J. Biol. Chem.* **265**, 1837 (1990).

natural and unnatural amino acids may be used to increase the specificity and potency of the pseudosubstrate peptides. The situations where pseudosubstrate peptides are most needed are in crude systems and living cells, which are also the situations where questions of specificity will be greatest.

The protein kinase C pseudosubstrate peptide has been used in a number of studies[32,76-78a] to explore the role of this enzyme in a variety of physiological processes. Generally the level of peptide PKC (19–31) required to inhibit the enzyme in permealized cell preparations is in the micromolar range, substantially higher than that required to inhibit peptide substrate phosphorylation with purified enzyme. The polybasic nature of the pseudosubstrate peptide may mean that the free concentration is substantially lower due to binding to membrane phospholipids. Of all the pseudosubstrate peptides the cyclic AMP-dependent protein kinase inhibitor peptide PKI (5–22) has been used most extensively (see [25] in this volume). The potency and specificity of this peptide have greatly facilitated its use and provide greater confidence in the conclusions drawn. It seems likely that the expansion in the use of pseudosubstrate peptides will continue into the foreseeable future.

[76] D. R. Alexander, J. M. Hexham, S. C. Lucas, J. D. Graves, D. A. Cantrell, and M. J. Crumpton, *Biochem. J.* **260**, 893 (1989).
[77] S. G. Rane, M. P. Walsh, J. R. McDonald, and K. Dunlap, *Neuron* **3**, 239 (1989).
[78] T. Eichholtz, J. Alblas, M. van Overveld, W. Moolenaar, and H. Plough, *FEBS Lett.* **261**, 147 (1990).
[78a] M. M. Bosma and B. Hille, *Proc. Natl. Acad,. Sci. U.S.A.* **86**, 2943 (1989).

[25] Utilization of the Inhibitor Protein of Adenosine Cyclic Monophosphate-Dependent Protein Kinase, and Peptides Derived from It, as Tools to Study Adenosine Cyclic Monophosphate-Mediated Cellular Processes

By Donal A. Walsh and David B. Glass

The inhibitor protein of the cAMP-dependent protein kinase (PKI) is a useful tool to evaluate the physiological function of the cAMP-dependent protein kinase. This chapter presents considerations from our current state of knowledge that need to be taken into account when using PKI for these purposes.

The cAMP-dependent protein kinase consists of regulatory (R) and catalytic (C) subunits, and is activated by the following reaction process:

$$R_2C_2 + 4cAMP \rightarrow R_2cAMP_4 + 2C$$

There are two primary species of regulatory subunit, designated I and II, with each exhibiting multiple isoforms; there are also three isoforms of catalytic subunit.[1-4] Each potential species of holoenzyme is most likely activated by the indicated mechanism. All current knowledge indicates that the activation occurs via the formation of a ternary intermediate (R + C + cAMP), that this ternary intermediate is catalytically inactive, and that dissociation of the catalytic subunit from the regulatory subunit is essential for activation.[5] Inactivation of catalytic subunit by regulatory subunit is a consequence of it acting as a pseudosubstrate. The regulatory subunit binds to the protein-binding site of the catalytic subunit, and, in consequence, blocks protein–substrate binding. This mechanism is similar to what has been observed with many protein kinases.[6]

PKI, as isolated from rabbit skeletal muscle, has been extensively investigated.[7] It is a 75-amino acid protein, with the bulk of its active site being contained in the amino-terminal residues 6 to 22. The synthetic peptide PKI(6–22)amide, of sequence TYADFIASGRTGRRNAI, exhibits ~20% of the inhibitory potency of the native protein, with a K_i of 1.6 nM.[7-9] Like regulatory subunit, PKI (and peptides derived from it) inactivates the catalytic subunit of the protein kinase by binding to the protein–substrate site of the catalytic site. Established major recognition determinants on the inhibitor protein for PKI catalytic subunit interaction include the side chains of Phe-10; Arg-15, -18, and -19; and Ile-22.[7,8,10] Additional binding efficacy is provided by the presence of Gly-17 and Asn-20, and also by some component of structure believed to be a helical segment in the region

[1] M. D. Uhler, D. F. Carmichael, D. C. Lee, J. C. Chrivia, E. G. Krebs, and G. S. McKnight, Proc. Natl. Acad. Sci. U.S.A. 83, 1300 (1986).

[2] M. D. Uhler, J. C. Chrivia, and G. S. McKnight, J. Biol. Chem. 261, 15360 (1986).

[3] M. O. Showers and R. A. Maurer, J. Biol. Chem. 261, 16288 (1986).

[4] S. J. Beebe, O. Oyen, M. Sandberg, A. Froysa, V. Hansson, and T. Jahnsen, Mol. Endocrinol. 4, 465 (1990).

[5] C. E. Cobb, A. H. Beth, and J. D. Corbin, J. Biol. Chem. 262, 16566 (1987), and references quoted therein.

[6] T. R. Soderling, J. Biol. Chem. 265, 1823 (1990).

[7] D. A. Walsh, K. L. Angelos, S. M. Van Patten, D. B. Glass, and L. P. Garetto, in "Peptides and Protein Phosphorylation" (B. E. Kemp, ed.), p. 43. Uniscience CRC Press, Boca Raton, Florida, 1990.

[8] D. B. Glass, H.-C. Cheng, L. Mende-Mueller, J. Reed, and D. A. Walsh, J. Biol. Chem. 264, 8802 (1989).

[9] D. B. Glass, L. J. Lundquist, B. M. Katz, and D. A. Walsh, J. Biol. Chem. 264, 14579 (1989).

[10] J. D. Scott, M. B. Glaccum, E. H. Fischer, and E. G. Krebs, Proc. Natl. Acad. Sci. U.S.A. 83, 1613 (1986).

of approximately residues 6 to 12.[11] The recognition of PKI via the Arg-18-Arg-19 pair of basic amino acids and the hydrophobic residue, Ile-22, in the pseudosubstrate portion of the peptide mirrors the binding interactions of protein substrates to the protein kinase. However, the binding affinity of PKI for the protein kinase is at least 10,000-fold greater than that of any established protein substrate.

An extensive array of PKI-derived peptides has now been synthesized (Table I).[12] This set of analogs provides initial insights into possible alterations in the PKI peptide sequence and, hence, in overall structure that can be tolerated without losing inhibitory potency. In addition to straightforward structure–function studies, suitable changes in the structure of PKI(6–22)amide should be able to enhance the ability of the peptide to penetrate cells and increase its resistance to proteolytic degradation. In this regard, the NH_2 terminus of the peptide can be altered without loss of potency. However, the effects of N^{α}-acetylation or an NH_2-terminal sarcosine residue on permeability or metabolic stability have yet to be determined. Almost all of the analogs have been prepared as COOH-terminal carboxyamides, but other alterations at this end of the molecule have not been extensively examined. Either peptide end may be a suitable location to place a hydrophobic group that might enhance cell permeability. The amphiphilic α-helical segment from approximately residues 6 to 12 contains the important Phe-10 as a binding determinant. The most potent PKI(6–22) peptide synthesized so far contains naphthylalanine at this position,[9] a substitution that also substantially increases hydrophobicity. Substituting an azido-Phe residue in place of Phe-10 results in a useful photoaffinity analog of PKI[13] that is likely to lead to irreversible inhibition. The central portion of PKI(6–22)amide (i.e., residues 11–14) does not contain any essential recognition determinants but clearly provides essential spacing between Phe-10 and the pseudosubstrate domain. Substitutions of D-alanine, β-alanine, or sarcosine for Gly-14 are well tolerated,[14] and might be expected to lessen proteolysis of the involved peptide bonds. In the pseudosubstrate regions of PKI peptide, Arg-15, -18, and -19 have been replaced by Gly to produce much less active peptides and these have been used as suitable negative controls for the parent inhibitor peptide. The replacement of a single Arg residue with D-Arg results in a peptide that retains suitable inhibitory activity, but also

[11] J. Reed, J. S. de Ropp, J. Trewhella, D. B. Glass, W. K. Liddle, E. M. Bradbury, V. Kinzel, and D. A. Walsh, *Biochem. J.* **264,** 371 (1989).

[12] J. R. Grove, D. J. Price, H. M. Goodman, and J. Avruch, *Science* **238,** 530 (1987).

[13] B. M. Katz, L. J. Lundquist, D. A. Walsh, and D. B. Glass, *Int. J. Pept. Protein Res.* **33,** 439 (1989).

[14] D. A. Walsh, J. Trewhella, and D. B. Glass, *J. Cell Biol.* **107,** 490a (1988).

TABLE I
PKI INHIBITORY PEPTIDES[a]

Peptide sequence	K_i (nM)
Peptide Length	
T T Y A D F I A S G R T G R R N A I[a]	3.1
T Y A D F I A S G R T G R R N A I[a]	1.7
[c]T Y A D F I A S G R T G R R N A I[a]	2.5
Y A D F I A S G R T G R R N A I[a]	28
[c]Y A D F I A S G R T G R R N A I[a]	34
A D F I A S G R T G R R N A I[a]	90
D F I A S G R T G R R N A I[a]	97
G R T G R R N A I[a]	36
T T Y A D F I A S G R T G R R N A I H D[a]	2.3
F I A S G R T G R R N A I H D[a]	73
I A S G R T G R R N A I H D[a]	115
A S G R T G R R N A I H D[a]	119
G R T G R R N A I H D[a]	57
R T G R R N A I H D[a]	250
T G R R N A I H D[a]	1250
I A S G R T G R R N A I H D I L V S S A	800
G R T G R R N A I H D I L V S S A	75,000
R R N A I H D I L V S S A	150,000
N A I H D I L V S S A	nd
H D I L V S S A	nd
I L V S S A	nd
T D V E T T Y A D F I A S G R T G R R N A I H D I L V S S A S	4
T D V E T T Y A D F I A S G R T G R R N A I H D	4.8
Substitution of Threonine-5	
Sr T Y A D F I A S G R T G R R N A I[a]	6
Substitution of Threonine-6	
T **A** Y A D F I A S G R T G R R N A I[a]	7
Substitution of Tyrosine-7	
T T **A** A D F I A S G R T G R R N A I[a]	14
Substitution of Alanine-8	
T T Y **L** D F I A S G R T G R R N A I[a]	9
Substitution of Aspartate-9	
T T Y A **A** F I A S G R T G R R N A I[a]	8
T T Y A **G** F I A S G R T G R R N A I[a]	31
T T Y A **P** F I A S G R T G R R N A I[a]	68

(*continued*)

TABLE I (*continued*)

Peptide sequence	K_i (nM)
Substitutions of Phenylalanine-10	
T T Y A D **A** I A S G R T G R R N A I[a]	270
T Y A D **A** I A S G R T G R R N A I[a]	425
T Y A D **L** I A S G R T G R R N A I[a]	53
T Y A D **Y** I A S G R T G R R N A I[a]	17
T Y A D **W** I A S G R T G R R N A I[a]	1
T Y A D **dF** I A S G R T G R R N A I[a]	93
T Y A D **hF** I A S G R T G R R N A I[a]	115
T Y A D **gF** I A S G R T G R R N A I[a]	196
T Y A D **fW** I A S G R T G R R N A I[a]	2
T Y A D **tA** I A S G R T G R R N A I[a]	5
T Y A D **nA** I A S G R T G R R N A I[a]	0.4
T Y A D **zF** I A S G R T G R R N A I[a]	3
T Y A D **iF** I A S G R T G R R N A I[a]	7
T Y A D **cF** I A S G R T G R R N A I[a]	8
T Y A D **aF** I A S G R T G R R N A I[a]	15
T Y A D **nF** I A S G R T G R R N A I[a]	18
Substitution of Isoleucine-11	
T T Y A D F **A** A S G R T G R R N A I[a]	8
Substitutions of Alanine-12, Serine-13, and Glycine-14	
T T Y A D F I A L **I** R T G R R N A I[a]	130
T T Y A D F I A S **A** R T G R R N A I[a]	30
T T Y A D F I A S **dA** R T G R R N A I[a]	3
T T Y A D F I A S **Sr** R T G R R N A I[a]	20
T T Y A D F I ——— R T G R R N A I[a]	298
Substitution of Arginines-15, 18, and 19	
G **K** T G R R N A I H D[a]	370
G R T G **K** R N A I H D[a]	36,000
G R T G R **K** N A I H D[a]	4200
Y A D F I A S G **G** T G R R N A I H D I L V S S A	6600
Y A D F I A S G R T G **G** R N A I H D I L V S S A	20,000
Y A D F I A S G R T G R **G** N A I H D I L V S S A	120,000
Substitution of Threonine-16	
T T Y A D F I A S G R **Ui** G R R N A I[a]	2.3
G R **A** G R R N A I[a]	140

(*continued*)

TABLE I (*continued*)

Peptide sequence	K_i (nM)
Substitution of Glycine-17 and Asparagine-20	
T T Y A D F I A S G R T G R R **A** A I[a]	21
T T Y A D F I A S G R T **L** R R N A I[a]	26
G R T G R R **A** A I[a]	550
G R T G R R **G** A I[a]	110
G R T **L** R R N A I[a]	390
G R T **L** R R **A** A I[a]	4000
Substitution of Alanine-21	
I A S G R T G R R N **Sp**I H D I L V S S A	96,000
I A S G R T G R R N **U** I H D I L V S S A	19,000
Substitution of Isoleucine-22	
T T Y A D F I A S G R T G R R N A **L**[a]	11
T T Y A D F I A S G R T G R R N A **G**[a]	470
G R T G R R N A **L**[a]	180
G R T G R R N A **G**[a]	1400
Multiple Substitutions Residues 14–22	
G R T G R R N A I[a]	36
G R T G R R **A** A I[a]	550
G R T **L** R R N A I[a]	390
G R T **L** R R **A** A I[a]	4000
G R T G R R N A **L**[a]	180
A A A G R R N A **L**[a]	3800
G R T **L** R R **A** A **L**[a]	11,000
Substitution of Histidine-23	
Y A D F I A S G R T G R R N A I **G** D I L V S S A	1200
Substitution of Aspartate-24	
Y A D F I A S G R T G R R N A I H **E**	2
Multiple Substitutions	
I A **A** G R T G R R N A I H **E** I L V S S A	780
I A **A** G R T G R R N A I H **E**	800
Y A D F I A **A** G R **U** G R R N A I H D I L V **A A** A	12,000

[a] These data are taken from refs. 7–10 and 12, and unpublished work from our laboratory. Abbreviations: a, COOH-terminal amide; c, NH$_2$-terminal acetyl; dF, D-phenylalanine; hF, homophenylalanine; gF, phenylglycine; fW, tryptophan(formyl); tA, 2'-thienylalanine; nA, 1'-naphthylalanine; zF, 4'-azidophenylalanine; iF, 3',5'-diiodo-4'-aminophenylalanine; cF, 4'-Chlorophenylalanine; aF, 4'-aminophenylalanine; nF, 4'-nitrophenylalanine; dA, D-alanine; Sr, sarcosine; Sp, phosphoserine; U, aminobutyrate, Ui, isoaminobutyrate. A, alanine; R, arginine; N, asparagine; D, aspartic acid; C, cysteine; Q, glutamine; E, glutamic acid; G, glycine; H, histidine; I, isoleucine; L, leucine; K, lysine; M, methionine; F, phenylalanine; P, proline; S, serine; T, threonine; W, tryptophan; Y, tyrosine; V, valine.

exhibits an enhanced half-life after microinjection into cultured cells.[15] Finally, if Ala-21 in PKI(14–22)amide is replaced with Ser, the pseudosubstrate inhibitor is converted to a highly efficient substrate of the protein kinase.[8] The use of nonstandard amino acids in the design of PKI peptide analogs requires care in choosing activation and coupling schemes for peptide synthesis that are consistent with the solubilities, stabilities, and reactivities of the various protected amino acids. Much additional work will be required to delineate the optimum PKI pharmacophore, permeable to cells and metabolically stable, that would still recognize the active site of the protein kinase and mimic the interaction of PKI.[16]

A precise analysis has not yet been undertaken to evaluate whether PKI peptides have equal efficacies with each of the three catalytic subunit species that have now been identified from molecular cloning approaches.[1-4] The two forms of catalytic subunit that have been isolated by ion-exchange chromatography[17,18] from pure enzyme are inhibited equally by PKI.[19] One isoform of yeast catalytic subunit is inhibited by PKI(5–24) with a 32-fold lower potency as compared to the mammalian enzyme.[20]

In kinetic assays under steady state conditions the reaction mechanism of the cAMP-dependent protein kinase is ordered, with ATP binding first and protein substrate second.[21] A nucleotide-dependent interaction of PKI with the catalytic subunit is also favored which, in the presence of Mg^{2+}, is highly specific for ATP but has a broader nucleotide specificity when Mn^{2+} is substituted for Mg^{2+}.[22,23] It has been proposed that PKI binds optimally to a catalytic subunit–MgATP complex, in particular involving a specific conformation of the terminal ATP phosphodiester bond. Because of this, ATP has frequently been added when examining PKI–catalytic subunit interactions, and it may be helpful in this regard. However, the conclusions cited above have been derived primarily from steady state

[15] N. J. C. Lamb, J. Mery, J.-C. Cavadore, and A. Fernandez, *J. Cell Biol.* **107**, 493a (1988).
[16] D. B. Glass and D. A. Walsh, *in* "Therapeutic Peptides and Proteins: Formulation, Delivery and Targeting" (D. Marshak and D. Liu, eds.), p. 135. Cold Spring Harbor Lab., Cold Spring Harbor, New York, 1989.
[17] P. J. Bechtel, J. A. Beavo, and E. G. Krebs, *J. Biol. Chem.* **247**, 637 (1972).
[18] V. Kinzel, A. Hotz, N. Konig, M. Gagelmannm, W. Perin, J. Reed, D. Kobler, F. Hofmann, C. Obst, H. P. Gensheimer, D. Goldblat, and S. Shaltiel, *Arch. Biochem. Biophys.* **253**, 341 (1987).
[19] S. M. Van Patten, A. Hotz, V. Kinzel, and D. A. Walsh, *Biochem. J.* **256**, 785 (1988).
[20] M. J. Zoeller, J. Kuret, S. Cameron, L. Levin, and K. E. Johnson, *J. Biol. Chem.* **263**, 9142 (1988).
[21] S. Whitehouse, J. R. Feramisco, J. E. Casnellie, E. G. Krebs, and D. A. Walsh, *J. Biol. Chem.* **258**, 3693 (1983).
[22] S. Whitehouse and D. A. Walsh, *J. Biol. Chem.* **258**, 3682 (1983).
[23] S. M. Van Patten, W. H. Fletcher, and D. A. Walsh, *J. Biol. Chem.* **261**, 5514 (1986).

kinetic approaches that use low concentrations of protein. Such studies provide information only on the optimal kinetic route of interaction; it appears quite probable that ATP-independent PKI interaction with catalytic subunit can occur quite readily at higher protein concentrations.[7,24] Of note, catalytic subunit–regulatory subunit type I (C–R_I) interaction is also favored by the presence of ATP.[7,25] Thus the preferred interaction of R_I is like that of PKI and mirrors the sequential reaction sequence with protein substrate.

PKI can be used effectively to identify protein phosphorylation that is catalyzed by the cAMP-dependent protein kinase, and to distinguish such phosphorylation from that catalyzed by other enzymes. In consequence it can be used to dissect actions of cAMP, and distinguish them from those of other messengers. Its use in *in vitro* systems appears to allow the unequivocal identification of proteins that are phosphorylated by the cAMP-dependent protein kinase. With intact cells, the use of PKI can provide a high degree of definitive evidence concerning a role for cAMP and the cAMP-dependent protein kinase, but there are several cautions that must be exercised. The following considerations are of special concern.

1. Specificity of PKI. The inhibitor protein exhibits an extremely high degree of specificity for the cAMP-dependent protein kinase, and numerous other protein kinases are not affected. Even at a concentration of 0.1 mM (which is more than four orders of magnitude greater than that needed to inhibit the cAMP-dependent protein kinase) PKI does not inhibit the closely homologous cGMP-dependent protein kinase.[26] However, more caution is required in interpreting results when using PKI(5–22)amide and shorter peptides. Even though these peptides show a very high degree of selectivity for the cAMP-dependent protein kinase they can, nevertheless, inhibit the cGMP-dependent protein kinase when used at very high concentrations.[26] This can become problematic when, as expanded on below, the amount of PKI peptide used is dictated by the amount of protein kinase in the cell and not by its K_i value for the protein kinase. The K_i values of PKI(5–22)amide and PKI(14–22)amide for the cGMP-dependent protein kinase are 68 and 30 μM; these are 55,000 and 526 times higher, respectively, than those for inhibition of the cAMP-dependent protein kinase.

2. Potential for Dual Protein Kinase Messenger System. The actions of cAMP that are mediated by the cAMP-dependent protein kinase appear to be solely a consequence of the stimulation of protein phosphorylation.

[24] C. D. Ashby and D. A. Walsh, *J. Biol. Chem.* **247**, 6637 (1972).
[25] J. A. Beavo, P. J. Bechtel, and E. G. Krebs, *Adv. Cyclic Nucleotide Res.* **5**, 241 (1974).
[26] D. B. Glass, H.-C. Cheng, B. E. Kemp, and D. A. Walsh, *J. Biol. Chem.* **261**, 12166 (1986).

Despite several searches, there is currently no evidence that the regulatory subunit–cAMP complex, which is also produced by protein kinase dissociation, or free regulatory subunit which might be derived from it, have any physiological role (other than regulation of catalytic subunit kinase activity). Several specific binding proteins for regulatory subunit do exist,[27–30] but it is not known whether such interactions simply direct the holoenzyme to specific cellular locations, or are indicative of the regulatory subunit serving other purposes. The function of the dissociation of the protein kinase remains an enigma, especially since its close homolog, the cGMP-dependent protein kinase, demonstrates that there is no need for the protein kinase to actually dissociate in order to achieve activation of phosphotransferase activity. Until this issue is resolved there should be caution in interpreting effects of added PKI. There remains a possibility that the cAMP-dependent protein kinase is a two-messenger system and that upon dissociation both the catalytic and regulatory subunits have specific and different cellular functions. If so, addition of PKI would block cAMP-mediated protein phosphorylation but would not inhibit any cAMP-mediated activity that free regulatory subunit might manifest. In the presence of excess PKI it is in fact possible to promote dissociation of the protein kinase,[31] causing the formation of free, but inhibited, catalytic subunit and free regulatory subunit. Thus PKI could in fact promote a cAMP effect if it was mediated by free regulatory subunit.

3. *Potential for Multiple Mechanisms of cAMP Action.* The current state of knowledge suggests that essentially all actions of cAMP in mammalian cells are elicited by the cAMP-dependent protein kinase. One notable exception to this appears to be the observation of a direct cAMP-gated ion channel conductance in the cilia from olfactory cells,[32] i.e., an effect of cAMP that does not involve protein phosphorylation. While further data are necessary to fully dissect this latter system, it suggests that not all effects of cAMP in mammalian cells are mediated via the protein kinase. Thus inhibition by PKI would be diagnostic of cAMP action, but a lack of inhibition would not necessarily exclude cAMP as the mediator of an event.

4. *Relationship of Effective Concentration of PKI to K_i for cAMP-*

[27] W. E. Theurkauf and R. B. Vallee, *J. Biol. Chem.* **257,** 3284 (1982).
[28] P. Miller, U. Walter, W. E. Theurkauf, R. B. Vallee, and P. DeCamilli, *Proc. Natl. Acad. Sci. U.S.A.* **79,** 5562 (1982).
[29] S. M. Lohmann, P. DeCamilli, I. Einig, and U. Walter, *Proc. Natl. Acad. Sci. U.S.A.* **81,** 6723 (1984).
[30] R. B. Vallee, M. J. DiBartolomeis, and W. E. Theurkauf, *J. Cell Biol.* **90,** 568 (1981).
[31] C. D. Ashby and D. A. Walsh, *J. Biol. Chem.* **248,** 1255 (1973).
[32] T. Nakamura and G. H. Gold, *Nature (London)* **325,** 442 (1987).

Dependent Protein Kinase. The high affinity of PKI for the protein kinase can provide a misleading understanding of the amount needed to block protein kinase activity. The concentration of protein kinase catalytic subunit in cells is in the range of 0.3 to 1.0 μM.[33] Thus, although the K_i value of PKI for the protein kinase is 0.1 nM, it requires an intracellular concentration of 10,000-fold higher than the K_i value to achieve stoichiometric inhibition of the total cellular content of cAMP-dependent protein kinase. At such a concentration, PKI is still highly specific for inhibition of the protein kinase and, because of its very high affinity, only needs to be very slightly in excess of stoichiometric levels to achieve full inhibition. However, more caution is needed when using peptides such as PKI(5–22)-amide or PKI(14–22)amide. These also only need to be at concentrations essentially stoichiometric with the protein kinase for full inhibition, but the margin between this concentration and the amount that would inhibit the cGMP-dependent protein kinase is small. For example, a 30-fold excess of PKI(14–22)amide (i.e., ~30 μM) would inhibit the cGMP-dependent protein kinase by 50%.

Sources

The purification of PKI from rabbit skeletal muscle has been described in detail.[7] An alternate source is the *Escherichia coli* expression system described by Thomas *et al.*[34] employing an expression vector with a synthetic DNA coding insert constructed from the amino acid sequence of PKI. The protein product of this expression system is identical to native PKI in both activity and physical characteristics and differs only by the addition of a leader glycine and the absence of a blocked NH_2 terminus.

PKI(5–22)amide is available from Peninsula Laboratories (Belmont, CA) under the commercial name Wiptide. This is also a source for the shorter, less potent, PKI(14–24)amide. A broad range of substitution analogs of these peptides has been described encompassing five orders of magnitude in potency (Table I). Use of these allows a comparison of their potency in blocking a presumed cAMP-mediated event to their potency in inhibiting the protein kinase. This added correlation ensures that the effect observed is mediated by cAMP. In addition to the peptides themselves, Grove *et al.*[12] have described a synthetic plasmid constructed to produce PKI(1–31) within mammalian cells.

[33] F. Hofmann, P. J. Bechtel, and E. G. Krebs, *J. Biol. Chem.* **252,** 1441 (1977).
[34] J. Thomas, S. M. Van Patten, K. H. Day, J. Richardson, D. A. Walsh, and R. A. Maurer, *J. Biol Chem.* **265** (in press).

Applications

Direct Exposure of Cells to PKI Peptides

Buchler et al.[35] have reported that the direct exogenous addition to rat hepatocytes of PKI(5–24)amide (K_i for the protein kinase, 2.5 nM) or PKI(11–30)amide (K_i for the protein kinase, 800 nM) blocked the cAMP-dependent synthesis of mRNAs for tyrosine aminotransferase and phosphoenolpyruvate carboxykinase, but did not block non-cAMP-dependent mRNA synthesis. These results correlated well with other modulators of the cAMP response. Nearly complete inhibition was observed at a 1 mM external concentration of either peptide, with a half-maximal effect at near 0.5 mM; PKI(5–24)amide was very slightly more potent. The uptake of these two peptides by cells is unexpected, as is also the lack of any significant distinction between their efficacy. The uptake may represent a small degree of endocytosis and it is possible that quite variable results would be seen between different cell types. Attempts to correlate efficiency with intact cells with *in vitro* inhibitory constraints could be compromised by different rates of cellular uptake and of intracellular proteolysis.

Microinjection of PKI

The microinjection of PKI into cells has been an approach used successfully with several cell types where this methodology can be applied.[36–44] Generally to date an all-or-none approach has been used and an undetermined excess amount of PKI injected. This type of methodology, however, lends itself quite suitably to determining the correct amount of PKI, or PKI peptides, to be injected based upon an accurate determination of the amount of cAMP-dependent protein kinase present in the cell type

[35] W. Buchler, U. Walter, B. Jastorff, and S. M. Lohmann, *FEBS Lett.* **288,** 27 (1988).

[36] J. L. Maller and E. G. Krebs, *J. Biol. Chem.* **252,** 1712 (1977).

[37] J. C. Saez, D. C. Spray, A. C. Nairn, E. Hertzberg, P. Greengard, and M. V. L. Bennett, *Proc. Natl. Acad. Sci. U.S.A.* **83,** 2473 (1986).

[38] V. F. Castellucci, A. Nairn, P. Greengard, J. H. Schwartz, and E. R. Kandel, *J. Neurosci.* **2,** 1673 (1982).

[39] D. Huchon, R. Ozon, E. H. Fischer, and J. G. Demaille, *Mol. Cell. Endocrinol.* **22,** 211 (1981).

[40] W. B. Adams and I. B. Levitan, *Proc. Natl. Acad. Sci. U.S.A.* **79,** 3877 (1982).

[41] E. E. Bittar, G. Chambers, and E. H. Fischer, *J. Physiol. (London)* **333,** 39 (1982).

[42] E. E. Bittar and G. Chambers, *Comp. Biochem. Physiol. C* **80C,** 421 (1985).

[43] J. Nwoga and E. E. Bittar, *Comp. Biochem. Physiol. C* **74C,** 177 (1983).

[44] G. Bkaily and N. Sperelakis, *Am. J. Physiol.* **246,** H630 (1984).

being used. Intracellular destruction of peptides can be a significant problem.

PKI Delivery by Liposome Fusion

In order to overcome the limitation imposed by microinjection methodology (that of using only a few cells), an alternate approach has been devised, involving the fusion of PKI-containing liposomes with cells. Delivery of PKI to isolated cultured heart cells has been accomplished using unilamellar phosphatidylcholine vesicles,[45] and to an anti-N-CAM-treated AtT-20 tumor cell line using a more complex lipid vesicle targeted via bound protein A.[45,46] These methods would appear to be applicable to many other cell culture systems.

Delivery of PKI via Plasmid Expression

The expression systems devised by Grove et al.[12,47] and Day et al.[48] have considerable potential for delineating cAMP action. The expression vector of Grove et al.[12,47] contains an avian sarcoma virus long terminal repeat (ASV LTR) constitutive promoter, the nucleotide sequence reverse translated from the amino acid sequence of PKI(1–31) (plus an additional amino-terminal methionine), and the simian virus 40 (SV40) splice and polyadenylation signals. PKI(1–31) assayed with the protein kinase in vitro has a K_i of 4 nM. This sector has been successfully employed using transient transfection of JEG-3 cells, with the signal examined being cAMP-stimulated mRNA synthesis. An excellent negative control employed an alternate vector in which Arg-18 and Arg-19 of the PKI sequence had been replaced by glycine.

The expression vector of Day et al.[48] is similar, except that it synthesizes the full-length protein which has a higher degree of specificity for the protein kinase. It contains an RSV LTR constitutive promoter, a full-length synthetic reverse-translated amino acid sequence of PKI (plus an additional amino-terminal Met-Gly), and the SV40 splice and polyadenylation signals. This vector has been successfully employed using transient infection of GH_3 pituitary tumor cells, with the signal examined being the cAMP regulation of the prolactin gene, and also in CHO cells. These workers also utilized the negative control vector in which Arg-18 and Arg-

[45] T. Reisine, G. Rougon, and J. Barbet, J. Cell Biol. 102, 1630 (1986).
[46] T. Reisine, G. Rougon, J. Barbet, and H.-U. Affolter, Proc. Natl. Acad. Sci. U.S.A. 82, 8261 (1985).
[47] J. R. Grove, P. J. Deutsch, D. J. Price, J. F. Habener, and J. Avruch, J. Biol. Chem. 264, 19506 (1989).
[48] R. N. Day, J. A. Walder, and R. A. Maurer, J. Biol. Chem. 264, 431 (1989).

19 of the PKI sequence had both been replaced by glycine. Obtaining stably transformed cells was difficult, possibly because of the need for cAMP-dependent phosphorylation during the cell cycle. To do so may require that a PKI expression vector be devised which has a tightly regulated promoter.

Fluoresceinated PKI as Marker for Dissociated Catalytic Subunit of cAMP-Dependent Protein Kinase

As a current approach with fixed cells, the Fletcher and Byus laboratories have utilized fluroesceinated PKI to localize free catalytic subunit; details of this procedure have been presented in a previous volume of this series.[49] The use of PKI for these procedures allows the distinction between free (and active) catalytic subunit and that present as holoenzyme. Further development of these procedures, coupled to microinjection, would allow for the process of dissociation to be followed in intact cells.

Characterization of Protein Kinases in Vitro

This subject has been described in detail in previous volumes of this series.[50,51]

[49] C. V. Byus and W. H. Fletcher, this series, Vol. 159, p. 236.
[50] J. A. Traugh, C. D. Ashby, and D. A. Walsh, this series, Vol. 38, p. 290.
[51] S. Whitehouse and D. A. Walsh, this series, Vol. 99, p. 80.

[26] Use of Sphingosine as Inhibitor of Protein Kinase C

By Yusuf A. Hannun, Alfred H. Merrill, Jr., and Robert M. Bell

Protein kinase C functions as an important regulatory element of a nearly ubiquitous signal transduction system in mammalian cells; it plays critical roles in cell differentiation, tumor promotion, oncogenesis, and cell regulatory processes such as granule secretion and hormone release.[1] Protein kinase C exists as a family of closely related isoenzymes with differences in tissue distribution and intracellular localization.[2] Protein kinase C isoenzymes are physiologically regulated by diacylglycerol (DAG), which is generated from membrane phospholipids in response to

[1] Y. Nishizuka, *Science* **233**, 305 (1986).
[2] Y. Nishizuka, *Nature (London)* **334**, 661 (1988).

cell stimulation.[3] Because protein kinase C is activated pharmacologically by phorbol esters and related tumor promoters[4] and since protein kinase C is the predominant intracellular receptor for phorbol esters,[5] it has been implicated in mediating the numerous biological activities attributed to phorbol esters.[1]

A number of protein kinase C inhibitors have been identified, including phenothiazines,[6,7] adriamycin,[6,8] amiloride,[9] tamoxifen,[10] palmitoylcarnitine,[11] alkyllysophospholipid,[12] CP-46,665-1,[13] K-252,[14] staurosporine,[15] sangivamycin,[16] H-7,[17] aminoacridines,[18] and sphingosine.[19] While many of these inhibitors display weak activity against protein kinase C and/or have poor specificity, some have emerged as useful inhibitors, including sphingosine and H-7.

The identification of protein kinase C inhibitors allows their use as biochemical tools to probe the mechanism of *in vitro* regulation of the enzyme. Inhibitors can also be used to explore the cellular activities of protein kinase C and to determine the role of protein kinase C in mediating biological responses. Also, with inhibitors that are naturally occurring molecules (such as sphingosine), the question arises as to the possible

[3] R. M. Bell, *Cell* (*Cambridge, Mass.*) **45**, 631 (1986).

[4] M. Castagna, Y. Takai, K. Kaibuchi, K. Sano, U. Kikkawa, and Y. Nishizuka, *J. Biol. Chem.* **257**, 7847 (1982).

[5] J. E. Niedel, L. J. Kuhn, and G. R. Vandenbark, *Proc. Natl. Acad. Sci. U.S.A.* **80**, 36 (1983).

[6] B. C. Wise, D. B. Glass, C. H. Chou, R. L. Raynor, N. Katoh, R. C. Schatzman, R. S. Turner, R. F. Kibler, and J. F. Kuo, *J. Biol. Chem.* **257**, 8489 (1982).

[7] T. Mori, Y. Takai, R. Minakuchi, B. Yu, and Y. Nishizuka, *J. Biol. Chem.* **255**, 8378 (1980).

[8] Y. A. Hannun, R. J. Foglesong, and R. M. Bell, *J. Biol. Chem.* **264**, 9960 (1989).

[9] J. M. Besterman, W. S. May, H. LeVine, III, E. J. Cragoe, and P. Cuatrecasas, *J. Biol. Chem.* **260**, 1155 (1985).

[10] C. A. O'Brian, R. M. Liskamp, D. H. Solomon, and I. B. Weinstein, *Cancer Res.* **45**, 2462 (1985).

[11] N. Katoh, R. W. Wrenn, B. C. Wise, M. Shoji, and J. F. Kuo, *Proc. Natl. Acad. Sci. U.S.A.* **78**, 4813 (1981).

[12] D. M. Helfman, K. Barnes, J. M. Kinkade, Jr., W. R. Vogler, M. Shoji, and J. F. Kuo, *Cancer Res.* **43**, 2955 (1983).

[13] M. Shoji, W. R. Vogler, and J. F. Kuo, *Biochem. Biophys. Res. Commun.* **127**, 590 (1985).

[14] H. Kase, K. Iwahashi, S. Nakanishi, Y. Matsuda, K. Yamada, M. Takahashi, and C. Murakata, *Biochem. Biophys. Res. Commun.* **142**, 436 (1987).

[15] T. Tamaoki, H. Nomoto, I. Takahashi, Y. Kato, M. Morimoto, and F. Tomita, *Biochem. Biophys. Res. Commun.* **135**, 397 (1986).

[16] C. R. Loomis and R. M. Bell, *J. Biol. Chem.* **263**, 1682 (1988).

[17] H. Hidaka, M. Inagaki, S. Kawamoto, and Y. Sasaki, *Biochemistry* **23**, 5036 (1984).

[18] Y. A. Hannun and R. M. Bell, *J. Biol. Chem.* **263**, 5124 (1988).

[19] Y. A. Hannun, C. R. Loomis, A. H. Merrill, Jr., and R. M. Bell, *J. Biol. Chem.* **261**, 12604 (1986).

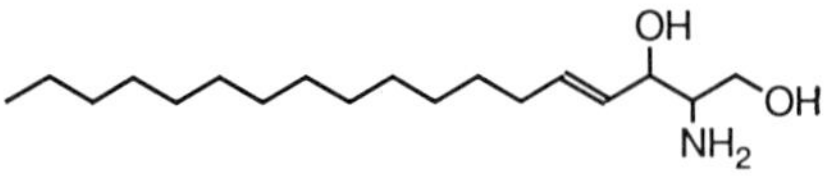

FIG. 1. Structure of sphingosine. Sphingosine is a naturally occurring amino base with a variable-length hydrocarbon chain (16–24 carbons) and a free amine with a pK_a of around 7.0.

physiological function of these molecules in negative regulation of protein kinase C activity.[20]

Sphingosine was found to be a potent inhibitor of protein kinase C *in vitro*.[19,21] Sphingosine (Fig. 1), a basic structural component of the sphingolipids, inhibited protein kinase C activity in different cells.[20] These results incriminated sphingosine and related lysosphingolipids as pathobiological agents in the sphingolipidoses[22] and led us to hypothesize a physiological function for sphingosine in the negative regulation of protein kinase C[22]; these aspects of sphingosine regulation of protein kinase C are reviewed elsewhere.[20] In this chapter, the rationale, guidelines, and limitations to the use of sphingosine as a cellular inhibitor of protein kinase C are presented.

Use of Sphingosine

A central issue in molecular cell biology is the dissection of biochemical pathways involved in signal transduction and cellular regulation. The determination of the role of protein kinase C in various biological processes has assumed critical significance and has been the subject of numerous investigations. The discovery of inhibitors of protein kinase C possessing high specificity and potency has emerged as a key goal; such inhibitors would permit the dissection of cellular activities involving protein kinase C.

The extensive use of sphingosine in different cellular systems has proved that sphingosine inhibits most, if not all, protein kinase C-mediated biological responses.[20,23] In this role, sphingosine acts as an antagonist of diacylglycerols and phorbol esters, a result expected from *in vitro* studies on the mechanism of inhibition of protein kinase C by sphingosine. These studies demonstrated that sphingosine is a competitive inhibitor with dia-

[20] Y. A. Hannun and R. M. Bell, *Science* **243**, 500 (1989).
[21] A. H. Merrill, Jr., A. Serenit, V. L. Stevens, Y. A. Hannun, R. M. Bell, and J. M. Kinkade, Jr., *J. Biol. Chem.* **261**, 12610 (1986).
[22] Y. A. Hannun and R. M. Bell, *Science* **235**, 670 (1987).
[23] A. H. Merrill, Jr. and V. L. Stevens, *Biochim. Biophys. Acta* **1010**, 131 (1989).

cylglycerol (DAG)/phorbol esters and prevents the interaction of these activators with the ternary complex of protein kinase C–phospholipid–calcium.[19] Sphingosine is equally potent against the three major isoforms of protein kinase C, namely the α, β, and γ isoforms.[24] Sphingosine has not been evaluated against other isoforms. Because sphingosine may affect targets other than protein kinase C (especially at higher concentration), interpretation of results on the effects of sphingosine on different cellular responses must be carefully evaluated. In this regard, inhibition of a cellular response by sphingosine is necessary, but insufficient, to demonstrate that this particular response is mediated by one of the three major isoforms of protein kinase C. That is, responses not inhibited by sphingosine are unlikely to be mediated by protein kinase C, while responses inhibited by sphingosine may involve protein kinase C and/or other targets. Therefore, additional tools must be used to dissect the role of protein kinase C in various processes (see Evaluation of Protein Kinase C-Mediated Biology).

Use of Sphingosine in Cell Systems

The chemical properties of sphingosine are well suited for studies with cells because it dissolves readily in water-miscible solvents [such as dimethyl sulfoxide (DMSO) and ethanol], can be further stabilized by formation of a complex with fatty acid-free serum albumin, and is rapidly taken up by cells.[21] However, because sphingosine is an amphipathic molecule with a positively charged amine at physiologic pH (pK_a appears to be near 7.0), special care should be taken in its delivery and in estimating its effective cellular concentration.

Method of Delivery

The use of sphingosine in DMSO solution can result in significant cytotoxicity.[25] Ethanol solutions appear to be appropriate for short-term cellular studies with sphingosine, such as in the examination of platelet responses[26] and neutrophil activation[27] that occur over a few minutes. To minimize cytotoxicity and for longer term cellular studies, sphingosine is most appropriately used in a 1 : 1 complex with serum albumin.[21,28]

[24] Y. Hannun and R. Bell, unpublished observations (1990).
[25] D. Pittet, K.-H. Krause, C. B. Wolheim, R. Bruzzone, and D. P. Lew, *J. Biol. Chem.* **262,** 10072 (1987).
[26] Y. A. Hannun, C. S. Greenberg, and R. M. Bell, *J. Biol. Chem.* **262,** 13620 (1987).
[27] E. Wilson, M. C. Olcott, R. M. Bell, A. H. Merrill, Jr., and J. D. Lambeth, *J. Biol. Chem.* **261,** 12616 (1986).
[28] J. D. Lambeth, D. N. Burnham, and S. R. Tyagi, *J. Biol. Chem.* **263,** 3818 (1988).

Solutions

Sources of Sphingosine. Commercially available sphingosine is prepared by the hydrolysis of different complex sphingolipids and is actually a mixture of several different long-chain bases, with *trans*-4-sphingenine as the major species. Since the other long-chain bases (mainly sphinganine and chain-length homologs of sphingenine) are also inhibitors of protein kinase C, this does not present a problem in most applications. Synthetic *erythro*-sphinganine (otherwise called dihydrosphingosine) is available in a very pure form with respect to chain length; however, it is a 50 : 50 mixture of D and L isomers. Sphingosine is also available as the sulfate salt, but this form is more difficult to solubilize. Chemically synthesized stereoisomers (sphingosine has four distinct stereoisomers) of sphingosine have also been evaluated for their effects against protein kinase C.[29]

Ethanol Solution. Sphingosine is dissolved in 95% (v/v) ethanol to a final concentration of 100 mM. Heating the mixture in hot water will aid the long-chain base to dissolve. Sphingosine is unstable in dilute solution; however, there is little loss over several months when stored in concentrated stocks (e.g., 100 mM). The ethanol solution can be used directly to deliver ethanol solutions of sphingosine or to form a 1 : 1 complex with serum albumin.

Bovine Serum Albumin Solution. A 1 mM solution of fatty acid-free bovine serum albumin is prepared in a suitable medium, such as tissue culture medium, Tyrode's solution, or other balanced salt solutions. Warming the mixture to 37° facilitates solubilization.

Sphingosine/Albumin Solution. To prepare a 1 mM solution of sphingosine in albumin, a 1 : 100 dilution is prepared. An appropriate amount of the sphingosine solution in ethanol is pipetted into the albumin solution to give the final concentration of 1 mM sphingosine. Sphingosine will initially come out of solution, forming a white precipitate. Sphingosine is redissolved by incubating the solution in a shaking water bath at 37° until the solution becomes clear. This may take several hours, depending on the quality of bovine serum albumin used. The albumin solution may be also dialyzed for special studies, since sphingosine is not lost following dialysis of the albumin complex (in Spectr/Por 4 membranes, Baxter, Atlanta, GA) against phosphate-buffered saline (three changes of approximately 10 times the sample volume each). Stock solutions of the sphingosine : albumin complex can be stored at −20° for at least several weeks.

[29] A. H. Merrill, Jr., S. Nimkar, D. Menaldino, Y. A. Hannun, C. Loomis, R. M. Bell, and S. R. Tyagi, *Biochemistry* **28**, 3138 (1989).

Estimating Effective Concentrations of Sphingosine:
Surface Dilution Considerations

The potency of sphingosine in inhibiting protein kinase C is primarily determined by its surface concentration. As with all amphipathic molecules, this is determined by two factors: (1) the partitioning of sphingosine into lipid bilayers; and (2) the ratio of the total mass of sphingosine present in solution to the total mass of lipids (present in membranes and in other lipid structures such as lipoproteins or lipid droplets[30]). In the absence of sphingosine-binding proteins, sphingosine will predominantly partition into cell membranes. Therefore, the effective cellular concentration of sphingosine ([sphingosine]$_{effective}$) will be primarily determined by the relative amount of cells and sphingosine:

$$[\text{Sphingosine}]_{effective} \propto \text{mass of sphingosine/total cells}$$
$$\propto [\text{sphingosine}]/[\text{cells}]$$

where [sphingosine] is the molar concentration of sphingosine and [cells] is the cell density (cells/volume).

In comparison, the effective concentrations of water-soluble agents, hormones, and growth factors are primarily determined by their molar concentrations irrespective of cell number or cell density. Therefore, the bulk molar concentration of sphingosine, independent of its ratio to cell density or cell number, may appear misleading in predicting the effectiveness of sphingosine in inhibiting protein kinase C. By decreasing the number of cells or by increasing the volume of the assay (i.e., by modulating cell density at a fixed molar concentration of sphingosine), the effective concentration of sphingosine will increase, and it will appear as a more potent molecule. By increasing cell density (with a fixed concentration of sphingosine), the effective concentration of sphingosine in membranes will decrease and, therefore, the dose range over which it inhibits protein kinase C will increase. Taking these parameters into consideration, a useful starting dose range for sphingosine to inhibit protein kinase C can be obtained for tissue culture experiments: sphingosine is effective at inhibiting protein kinase C activity in the 1–5 μM range using 10^6 cells/ml (this is equivalent to having 1–5 nmol sphingosine/10^6 cells).

Another complication in estimating the surface concentration of sphingosine arises from the presence of serum proteins in culture media or in the delivery of sphingosine in a 1 : 1 complex with albumin. Serum proteins will act as a buffer to bind sphingosine and, therefore, only a fraction of the added sphingosine is available to partition into cell membranes. For

---

[30] D. G. Robertson, M. DiGirolamo, A. H. Merrill, Jr., and J. D. Lambeth, *J. Biol. Chem.* **264**, 6773 (1989).

example, when sphingosine is delivered from an ethanol solution to washed human platelets, it is effective in inhibiting protein kinase C in the 1–20 μM range[19,31] (using 0.4 ml of 2.5 × 10⁸ platelets/ml). In the presence of serum proteins, as occurs in platelet-rich plasma, the effective concentration of sphingosine is in the 100–200 μM range[26] (with the same amount and concentration of platelets). When sphingosine is delivered in a 1 : 1 complex with serum albumin, the uptake of sphingosine from the culture medium is modulated. In the absence of sphingosine-binding serum proteins, nearly all of the added sphingosine is taken up by cells within minutes.[32] In the presence of equimolar bovine serum albumin, only 50% of the added sphinganine (dihydrosphingosine) was taken up from the culture medium over 6 hr.[21]

Monitoring Responses to Sphingosine

To optimize the use of sphingosine as a cellular inhibitor of protein kinase C, the effects of sphingosine on parameters of protein kinase C activation should be examined. Dose responses of sphingosine inhibition of cellular protein kinase C are then compared with dose responses for modulating the biologic response of interest. Monitoring the parameters listed below has proved useful.

Phorbol Dibutyrate Binding to Intact Cells. Sphingosine inhibits phorbol binding to cellular protein kinase C.[19,21] This inhibition is a prerequisite for inhibition of protein kinase C activation. Therefore, the effects of sphingosine on phorbol dibutyrate (PDBu) binding to intact cells is examined under conditions identical to those in which the cellular responses are measured. In particular, phorbol dibutyrate binding should be performed using the same number and concentration of cells, same buffer or media, and a concentration of [³H]PDBu close to the dissociation constant (approximately 20–50 nM); higher concentrations of PDBu are more difficult to displace with sphingosine.

Protein–Substrate Phosphorylation. A direct biochemical marker of protein kinase C activation or inhibition is obtained by monitoring the phosphorylation status of known protein kinase C substrates. A necessary and sufficient requirement for sphingosine to inhibit cellular protein kinase C activity is for sphingosine to inhibit protein kinase C-mediated substrate phosphorylation. Therefore, the ability of sphingosine to inhibit protein phosphorylation is monitored under conditions identical to those used in monitoring the physiological response of interest. For example, in human platelets, sphingosine inhibits platelet secretion in washed platelets at

---

[31] W. Khan and Y. Hannun, unpublished observations (1990).
[32] Y. Hannun, unpublished observations (1990).

concentrations of 1–20 μM. When platelets are labeled with ortho[^{32}P]-phosphate and then stimulated with agonist, sphingosine inhibits phosphorylation of a specific protein kinase C substrate (47 kDa) with an identical dose response to that required for inhibition of secretion.[31]

Structure–Function Relationship: Negative Controls

Structural features required for inhibition of protein kinase C by sphingosine have been elucidated.[29] These include a requirement for a hydrophobic character and the presence of a free amine. Inhibition is modulated by the length of the hydrocarbon tail of sphingosine, the stereospecificity of the molecule, the presence or absence of the *cis*-4,5 double bond, and substitutions at the 1-hydroxyl position. Acylation of the 2-amino group results in loss of inhibition of protein kinase C. *N*-Acetylsphingosine shares similar physical properties with sphingosine but does not inhibit protein kinase C. Therefore, *N*-acetylsphingosine is a useful control to evaluate the specificity of sphingosine in inhibiting protein kinase C.[19,26] Short-chain sphingosine analogs (e.g., C$_{11}$), which retain the free amine, are ineffective in inhibiting protein kinase C *in vitro* and in cells.[29] Therefore, they can also be used as additional controls.

Override by DAG/Phorbol Ester

Since sphingosine is a competitive inhibitor with respect to DAG or phorbol esters, these two molecules overcome sphingosine inhibition *in vitro* and in cell systems. Therefore, for a physiological response to be mediated by protein kinase C, its inhibition by sphingosine should be overcome by either diacylglycerol or phorbol esters. Two useful (cell-permeable) diacylglycerols are *sn*-1,2-dioctanoylglycerol (di-C$_8$) and 1-oleoyl-2-acetylglycerol (similar considerations should be applied when using these DAG analogs as when using sphingosine). In human platelets, di-C$_8$ overcomes sphingosine inhibition of secretion, aggregation, and 47-kDa phosphorylation.[26,32]

Limitations to Use of Sphingosine

Cytotoxicity

Cytotoxicity occurs at different concentrations in different cell types; sphingosine can be cytotoxic at rather low concentrations. This is particularly true when sphingosine is added to cells in ethanol or DMSO (instead of the albumin complex); nonetheless, these solvents have been used successfully when care was taken to determine the minimum effective does.[19,27] On the one hand, cytotoxicity may be a consequence of inhibition of protein kinase C, as suggested by concentration dependencies, structural specificity, the effects of long-chain bases on protein phosphorylation patterns (especially when wild-type cells have been compared to a cell line selected for partial resistance to the toxicity of long chain bases), and the apparent lack of perturbation of cellular acidic compartments, which could be another effect of these compounds.[33] On the other hand, cytotoxicity could occur through a variety of mechanisms unrelated to protein kinase C.

Metabolism

The use of sphingosine as a pharmacological tool is hampered by its potential to be metabolized into inactive compounds. Rapid metabolism of sphingosine to sphingosine phosphate occurs by the action of sphingosine kinase. For example, in platelets, sphingosine is rapidly taken up (within seconds) and metabolized (within 2–5 min) to sphingosine phosphate.[24] Preincubation of platelets with sphingosine is, therefore, restricted to 15–30 sec prior to addition of agonists or other effectors. In HL-60 cells, sphingosine is primarily metabolized by the addition of fatty acyl groups with the resulting formation of ceramide and subsequent incorporation into more complex sphingolipids.[21] This occurs with a half-life of 6 hr. Therefore, a single administration of sphingosine may not be optimal for inhibiting protein kinase C at later time points (e.g., after 12–24 hr). To circumvent this problem, sphingosine may be added at regular intervals (e.g., daily) to tissue culture cells.[34]

Specificity

Sphingosine is not exclusively an inhibitor of protein kinase C even though it is selective (i.e., it does not appear to inhibit the cAMP-dependent kinase or myosin light-chain kinase). Additional effects of sphingosine were first observed in studies of epidermal growth factor (EGF) receptor phosphorylation. The effects of sphingosine were not completely reversed by phorbol esters, and there were inconsistencies in the time course and stoichiometry of phosphorylation.[35] These studies suggested that sphingosine may stimulate a protein kinase C-independent pathway of protein

[33] V. L. Stevens, S. Nimkar, W. C. Jamison, D. C. Liotta, and A. H. Merrill, Jr., *Biochim. Biophys. Acta* **1051**, 37 (1990).

[34] V. L. Stevens, E. F. Winton, E. E. Smith, N. E. Owens, J. M. Kinkade, Jr., and A. H. Merrill, Jr., *Cancer Res.* **49**, 3229 (1989).

[35] M. Faucher, N. Girones, Y. A. Hannun, R. M. Bell, and R. Davis, *J. Biol. Chem.* **263**, 5319 (1988).

phosphorylation and may increase the affinity and number of cell-surface EGF receptors. It was also shown that 5 μM sphingosine increases the activity of the cytoplasmic tyrosine kinase domain of the EGF receptor to a level equal to or greater than that of the ligand-activated holoreceptor.[36] It was, therefore, suggested that sphingosine mimicks the effect of EGF to induce a conformation that is optimal for tyrosine kinase activity.

Long-chain bases are also mitogenic at concentrations lower than those necessary to inhibit phorbol ester binding.[37] Sphingosine inhibits the Na^+,K^+-ATPase with approximately the same dose response as for protein kinace C inhibition.[38] Long-chain bases have also been reported to affect several ion transport systems that are thought to be regulated by protein kinase C.[39] Sphingosine inhibits coagulation initiated by lipopolysaccharide stimulation of human monocytes. It also inhibits tissue factor activity *in vitro* by inhibiting factor VII binding.[40]

Some responses to sphingosine occur at concentrations significantly greater than required to inhibit protein kinase C. For example, sphingosine inhibits thyrotropin-releasing hormone binding to GH3 cells with an IC_{50} of 63 μM[41] (compared with an IC_{50} of 1–3 μM for inhibition of protein kinase C using a comparable number of cells). Sphingosine also inhibits a calmodulin-dependent kinase[42]; however, the cellular effects were achieved with only very high concentrations[42] presumably because sphingosine partitions in membranes whereas calmodulin kinases are soluble targets. Finally, sphingosine has been reported to inhibit c-src and v-src kinases, but at concentrations of 330 and 660 μM.[43] At high concentrations or in the absence of membranes, long-chain bases probably have nonspecific effects due simply to their charge and/or amphipathic nature. These effects could be avoided by working with a lower concentration range of sphingosine.

Evaluation of Protein Kinase C-Mediated Biology

Investigations on the possible involvement of protein kinase C in cellular processes have relied on a number of biochemical and pharmacologic tools and indicators. None of these tools, however, has emerged as infalli-

[36] P. Wedegaertner and G. Gill, *J. Biol. Chem.* **264,** 11346 (1989).

[37] H. Zhang, N. E. Buckly, K. Gibson, and S. Spiegel, *J. Biol. Chem.* **265,** 76 (1990).

[38] K. Oishi, B. Zheng, and J. F. Kuo, *J. Biol. Chem.* **265,** 70 (1990).

[39] R. J. Gillies, R. Martinez, J. M. Sneider, and P. B. Hoyer, *J. Cell. Physiol.* **139,** 125 (1989).

[40] P. R. Conkling, K. L. Patton, Y. A. Hannun, C. S. Greenberg, and J. B. Weinberg, *J. Biol. Chem.* **264,** 18440 (1989).

[41] I. Winicov and M. C. Gershengorn, *J. Biol. Chem.* **263,** 12179 (1988).

[42] A. B. Jefferson and H. Schulman, *J. Biol. Chem.* **263,** 15241 (1988).

[43] Y. Igarashi, S. Hakomori, T. Toyokuni, B. Dean, S. Fujita, M. Sugimoto, T. Ogawa, K. El-Ghendy, and E. Racker, *Biochemistry* **28,** 6796 (1989).

ble in proving a necessary and sufficient role for protein kinase C in modulating a particular cellular response. Therefore, a preferred approach to dissecting protein kinase C-mediated biology should rely on multiple and simultaneous evaluation of different parameters.

Measuring DAG Levels

At present DAG appears to be the primary endogenous activator of protein kinase C (although a role for unsaturated fatty acids and phosphatidylinositol 4,5-bisphosphate is suggested but not proved). Therefore, demonstration of a DAG response during cellular activation is a necessary requirement for the involvement of protein kinase C in that process. Most studies, however, have relied on measuring phosphatidylinositol (PI) turnover by examining phosphatidylinositol metabolism and inositol trisphosphate generation. But since diacylglycerol could be generated by the breakdown of other phospholipids, by *de novo* synthesis, and from the PI anchor of membrane proteins, direct measurement of DAG levels becomes necessary. Excellent methods exist to quantitate *sn*-1,2-DAG mass in crude lipid extracts.[44]

Ability of DAG and/or Phorbol Esters to Mimic Physiological Response

Another necessary requirement to establish a role for protein kinase C in mediating particular cellular responses is to mimic those responses by the addition of exogenous DAGs or phorbol esters. These studies, however, are also subject to limitations in that DAG may have targets other than protein kinase C (such as phospholipase C[45] and phospholipase A$_2$[46]) and other membrane effects (such as membrane fusion[47]), while phorbol esters cause prolonged activation of protein kinase C and its subsequent down regulation with the possible generation of the constitutively active catalytic fragment.[48,49] Therefore, a response mediated by DAG and/or phorbol esters may or may not involve protein kinase C physiologically, but for a response to be mediated by protein kinase C, it must be mimicked by DAG and/or phorbol esters.

[44] J. Priess, C. R. Loomis, W. R. Bishop, R. Stein, J. E. Niedel, and R. M. Bell, *J. Biol. Chem.* **261**, 8597 (1986).
[45] S. L. Hofmann and P. W. Majerus, *J. Biol. Chem.* **257**, 14359 (1982).
[46] R. M. Burch, *FEBS Lett.* **234**, 283 (1988).
[47] D. P. Siegel, J. Banschbach, D. Alford, H. Ellens, L. J. Lis, P. J. Quinn, P. L. Yeagle, and J. Bentz, *Biochemistry* **28**, 3703 (1989).
[48] A. Rodriguez-Pena and E. Rozengurt, *Biochem. Biophys. Res. Commun.* **120**, 1053 (1984).
[49] S. Pontremoli, E. Melloni, M. Michetti, B. Sparatore, F. Salamino, S. Oliviero, and B. L. Horecker, *Proc. Natl. Acad. Sci. U.S.A.* **84**, 3604 (1987).

Down Regulation

Prolonged activation of protein kinase C by phorbol esters results in down regulation of the enzyme.[50] This observation has been employed as a tool to investigate the role of protein kinase in mediating physiologic responses. The loss of cellular response following down regulation of protein kinase provides a strong indication for involvement of protein kinase C. However, it appears that not all protein kinase C activity is subject to down regulation[51] and, therefore, this criterion is not sufficient to rule out a role for protein kinase C if the biological response persists following down regulation of protein kinase C.

Use of Other Inhibitors

Although most other inhibitors of protein kinase C show lack of specificity, the isoquinoline sulfonamide derivative H-7 has found substantial use as a tool to investigate the role of protein kinase C in biological response. H-7 is a competitive inhibitor of ATP[17] and, therefore, inhibits other kinases with variable potencies. Responses inhibited by H-7 but not inhibited by close analogs of H-7 with stronger potency toward other protein kinases are suggested to occur through a protein kinase C-mediated pathway. Because of the different mechanism of action of H-7 and its different specificity, the use of H-7 complements that of sphinosine.

Conclusions

The proper and careful use of sphingosine as a cellular inhibitor of protein kinase C offers important information as to the possible involvement of protein kinase C in a particular cellular response. At present, it appears that any biological response mediated by protein kinase C should be inhibited by sphingosine. Therefore, the demonstration that sphingosine inhibits the response is necessary, but not sufficient, evidence to incriminate protein kinase C. Unfortunately, none of the available tools is sufficient to prove a role for protein kinase C in a particular cellular response. However, a combination of these tools, such as inhibition by sphingosine, overcoming the sphingosine effect by diacylglycerol and/or phorbol esters,

[50] S. Solanki, T. J. Slaga, M. Callaham, and E. Huberman, *Proc. Natl. Acad. Sci. U.S.A.* **78,** 1722 (1981).
[51] K. Kariya, Y. Kawahara, H. Fukuzaki, M. Hagiwara, H. Hidaka, Y. Fukumoto, and Y. Takai, *Biochem. Biophys. Res. Commun.* **161,** 1020 (1989).

mimicking of the response by diacylglycerol and phorbol esters, abolishing the response following down regulation of protein kinase C, and inhibition of the response by H-7 or other additional inhibitors, and demonstration of elevations of diacylglycerol and specific substrate phosphorylation, if taken together, provide powerful evidence for a role for protein kinase C.

[27] Properties and Use of H-Series Compounds as Protein Kinase Inhibitors

By Hiroyoshi Hidaka, Masato Watanabe, and Ryoji Kobayashi

Introduction

Numerous extracellular signals, including a variety of neurotransmitters, growth factors, and hormones, are known to produce many of their diverse physiological effects by regulating the state of phosphorylation of specific phosphoproteins in their target cells.[1-3] Although various aspects of the organizing and functioning of the phosphorylation–dephosphorylation systems have been examined, much remains to be learned of the complex cellular responses in this system. The advent of a new class of effective pharmacological agents is always an event of considerable interest, in particular for new types of antagonists, potentially important groups of compounds that act by specifically blocking one or more of the steps in the phosphorylation–dephosphorylation systems.[4]

Pharmacological agents can be used to determine the physiological significance of the protein phosphorylation systems in various types of cells.[5] Two approaches can be used to study protein phosphorylation systems. First, their effects on *in vitro* protein kinase activities should be examined to yield information on molecular mechanisms and specificity of action of various compounds. Second, these agents should be used in various *in vivo* biological systems, an approach which will provide considerable information on the physiological significance of the protein phosphorylation systems.

[1] A. M. Edelman, D. A. Blumenthal, and E. G. Krebs, *Annu. Rev. Biochem.* **56,** 567 (1987).
[2] T. Hunter and J. A. Cooper, *Cell (Cambridge, Mass.)* **24,** 741 (1981).
[3] H. Schulman, *Adv. Second Messenger Phosphoryl. Res.* **22,** 39 (1988).
[4] H. Hidaka, T. Tanaka, M. Saitoh, and S. Matsushima, *Calcium-Binding Proteins Health Dis.* [*Int. Symp.*], *5th, 1986* p. 170 (1987).
[5] H. Hidaka and T. Tanaka, this series, Vol. 139, p. 153.

Overview of H-Series Protein Kinase Inhibitors

In 1977, the calmodulin antagonist, naphthalene sulfonamide, W-7 [N-(6-aminohexyl)-5-chloro-1-naphthalene sulfonamide], synthesized in our laboratory was first introduced.[6] We proposed that calmodulin antagonists were defined as agents which bind to calmodulin, Ca^{2+} dependently, and inhibit selectively the Ca^{2+}/calmodulin-dependent enzymes.[7] During the synthesizing and selecting of a novel calmodulin inhibitor among the derivatives of naphthalene sulfonamide, we discovered that shorter alkyl chain derivatives of W-7 markedly inhibited protein kinases such as myosin light-chain kinase (MLCK), cAMP-dependent protein kinase (A-kinase) and cGMP-dependent protein kinase (G-kinase), through mechanisms differing from those of W-7.[8] One of these derivatives, 1-(5-chloronaphthalene-1-sulfonyl)-1H-hexahydro-1,4-diazepine (ML-9), proved to be a specific inhibitor of MLCK without interacting with calmodulin.[9,10] ML-9 inhibits MLCK activity competitively with respect to ATP but not myosin light chain or calmodulin, and the K_i value of ML-9 against MLCK is 3.8 μM. Accordingly this compound appears to interact with MLCK at the binding site or near the site of ATP. ML-9 was then used to elucidate the physiological roles of MLCK. In various types of tissue, ML-9 proved to be useful for studying the physiological role of MLCK-dependent phosphorylation.[11]

At higher concentrations, W-7 inhibits Ca^{2+}/phospholipid-dependent protein kinase (C-kinase) activity, in a manner competitive with that of phospholipid. When the naphthalene ring of the naphthalene sulfonamides was replaced by isoquinoline, the derivatives were no longer calmodulin- or phospholipid-interacting agents, but rather compounds that directly suppress protein kinase activities. Among them, 1-(5-isoquinolinesulfonyl)-2-methylpiperazine, referred to as H-7, is a specific inhibitor of C-kinase.[12,13] Inhibition of C-kinase activity by H-7 was competitive with

[6] H. Hidaka, T. Asano, S. Iwadare, I. Matsumoto, T. Totsuka, and N. Aoki, *J. Pharmacol. Exp. Ther.* **207**, 8 (1978).

[7] H. Hidaka, T. Yamaki, M. Naka, T. Tanaka, H. Hayashi, and R. Kobayashi, *Mol. Pharmacol.* **17**, 66 (1980).

[8] M. Inagaki, S. Kawamoto, H. Itoh, M. Saitoh, M. Hagiwara, J. Takahashi, and H. Hidaka, *Mol. Pharmacol.* **29**, 571 (1986).

[9] M. Saitoh, T. Ishikawa, S. Matsushima, M. Naka, and H. Hidaka, *J. Biol. Chem.* **262**, 7796 (1987).

[10] M. Saitoh, M. Naka, and H. Hidaka, *Biochem. Biophys. Res. Commun.* **140**, 280 (1987).

[11] T. Ishikawa, T. Chijiwa, M. Hagiwara, S. Mamiya, M. Saitoh, and H. Hidaka, *Mol. Pharmacol.* **33**, 598 (1988).

[12] H. Hidaka, M. Inagaki, S. Kawamoto, and Y. Sasaki, *Biochemistry* **23**, 5036 (1984).

[13] S. Kawamoto and H. Hidaka, *Biochem. Biophys. Res. Commun.* **125**, 258 (1984).

respect to ATP and noncompetitive with the phosphate acceptor. The K_i value of this compound against C-kinase is 6.0 μM. Thus, it was clearly shown that H-7 directly inhibits C-kinase, and not through interaction with Ca^{2+} or phospholipid. H-7, now in wide use for studies on various biological systems, has significant effects on various functions by inhibiting C-kinase-induced phosphorylation.[14]

On the other hand, a derivative of naphthalene sulfonamide, N-(6-phenylhexyl)-5-chloro-1-naphthalene sulfonamide (SC-9), is a Ca^{2+}-dependent activator of C-kinase.[15] SC-9 and phosphatidyl serine have similar effects on C-kinase activity. Pharmacological activators of C-kinase, such as SC-9, will aid in elucidating the biological role of phosphorylation of this protein in intact cells. Exposure of Swiss 3T3 cells to SC-9 led to increase in hexose uptake, an event also observed when the cells were treated with 12-O-tetradecanoylphorbol-13-acetate (TPA). This activation by SC-9 was inhibited by H-7.[16]

We found that derivatives of H-7 exhibit a selective inhibition toward cyclic nucleotide-dependent protein kinases and that N-[2-(methylamino)ethyl]-5-isoquinoline sulfonamide (H-8) was a potent cyclic nucleotide-dependent protein kinase inhibitor in this series of compounds.[12] The inhibition was freely reversible and of the competitive type, with respect to ATP. We found that H-8 specifically binds to the ATP-binding site of the catalytic subunit with a binding ratio of 1 : 1 and that H-8 has unique features that differ from the ATP analogs reported by Flockhart et al.[17] or others, in the following points: (1) H-8, among other protein kinases, specifically inhibits cyclic nucleotide-dependent protein kinase; (2) the binding constant of H-8 to the enzyme was much lower than that of ATP; (3) the binding of H-8 to the enzyme was independent of magnesium ion; and (4) the binding subsite of H-8 at the active site of the enzyme differed slightly from that of the ATP.[18]

Newly Synthesized H-Series Protein Kinase Inhibitors

In addition to these pharmacological probes, we have synthesized additional compounds that are specific inhibitors of casein kinase I (CKI-7), A-kinase (H-89), and Ca^{2+}/CaM kinase II (KN-62). The potent selective

[14] M. Inagaki, S. Kawamoto, and H. Hidaka, J. Biol. Chem. 259, 14321 (1984).
[15] M. Ito, T. Tanaka, M. Inagaki, K. Nakanishi, and H. Hidaka, Biochemistry 25, 4179 (1986).
[16] H. Nishino, K. Kitagawa, A. Iwashima, M. Ito, T. Tanaka, and H. Hidaka, Biochim. Biophys. Acta 889, 236 (1986).
[17] D. A. Flockhart, W. Freist, J. Hoppe, T. M. Lincoln, and J. D. Corbin, Eur. J. Biochem. 140, 289 (1984).
[18] M. Hagiwara, M. Inagaki, and H. Hidaka, Mol. Pharmacol. 31, 523 (1987).

inhibitors we synthesized will aid in studies on the cellular response elements in diverse cellular functions and in the investigation of the physiological significance and molecular mechanisms of intracellular messenger systems. Structures and characteristics of the aforementioned are summarized in Fig. 1.

Molecular Mechanisms of Inhibitory Effects of KN-62 on Ca^{2+}/CaM Kinase II

KN-62   [1-(N,O-bis-1,5-isoquinolinesulfonyl)-N-methyl-L-tyrosyl-4-phenylpiperazine] affects the interaction between calmodulin and Ca^{2+}/CaM kinase II following inhibition of this kinase activity by directly binding to the calmodulin-binding site of the enzyme, but does not affect the calmodulin-independent activity of the already autophosphorylated (activated) enzyme.[19] Figure 2 shows the effect of KN-62 on the activities of Ca^{2+}/CaM kinase II, MLCK, A-kinase, and C-kinase. More than 80% of Ca^{2+}/CaM kinase II activity was inhibited by adding 10^{-6} M of KN-62; however, the activities of the other kinases were affected only slightly in the presence of even higher concentrations of KN-62. Thus, this newly synthesized compound seems to be a selective and potent inhibitor of Ca^{2+}/CaM kinase II. To elucidate mechanisms involved in the inhibition of this kinase activity, KN-62 was tested for its ability to compete with Ca^{2+}-calmodulin or ATP binding to the enzyme. Double-reciprocal analyses revealed that inhibition of Ca^{2+}/CaM kinase II by KN-62 was competitive with calmodulin and noncompetitive with respect to ATP (Fig. 3). The effect of KN-62 on the activity of autophosphorylated Ca^{2+}/CaM kinase II was also examined. We found an approximate 75% decrease in the exogenous substrate phosphorylation of Ca^{2+}/CaM kinase II when KN-62 was added before the autophosphorylation. Nevertheless, Ca^{2+}/CaM kinase II activity seen in the addition of KN-62 after autophosphorylation was essentially unchanged. KN-62 had no effect on the CaM-independent activity of the enzyme (Fig. 4).

To demonstrate that KN-62 binds directly to Ca^{2+}/CaM kinase II, the mixture of the enzyme and calmodulin in the presence of EGTA was applied to a KN-62-coupled Sepharose 4B column and a pass-through fraction and a bound fraction were analyzed by sodium dodecyl sulfate-polyacrylamide gel electrophoresis (SDS–PAGE). Calmodulin was eluted in void fractions, but Ca^{2+}/CaM kinase II bound to the column and was eluted with the buffer containing 8 M urea. This suggests that KN-62 binds

---

[19] H. Tokumitsu, T. Chijiwa, M. Hagiwara, A. Mizutani, M. Terasawa, and H. Hidaka, *J. Biol. Chem.* **265,** 4315 (1990).

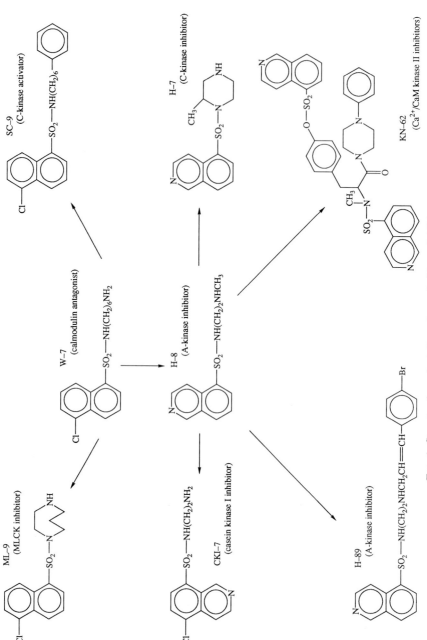

FIG. 1. Structural manipulation of protein kinase inhibitors.

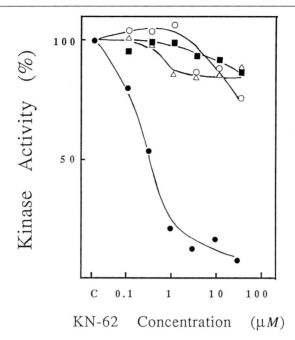

FIG. 2. Effect of KN-62 on various protein kinase activities. (■), MLCK; (△), C-kinase; (○), A-kinase; (●), Ca^{2+}/CaM kinase II.

directly to Ca^{2+}/CaM kinase II and is not a calmodulin antagonist, as is W-7.

Effects of KN-62 on Protein Phosphorylation in Intact Cells

The physiological role of Ca^{2+}/CaM kinase II was postulated to prolong effects triggered by transient Ca^{2+} signals[20–22] and to play a role in the long-term modulation of synaptic transmission.[23] Treatment of PC12 cells with nerve growth factor (NGF) and other extracellular signals was found to regulate the phosphorylation of a number of different substrate proteins.[24–26] The PC12 cell line therefore is an appropriate system for observ-

[20] S. G. Miller and M. B. Kennedy, Cell (Cambridge, Mass.) 44, 861 (1986).
[21] Y. Lai, A. C. Nairn, and P. Greengard, Proc. Natl. Acad. Sci. U.S.A. 83, 4253 (1987).
[22] C. M. Schworer, R. J. Colbran, and T. R. Soderling, J. Biol. Chem. 261, 8581 (1986).
[23] A. C. Nairn, H. C. Hemmings, and P. Greengard, Annu. Rev. Biochem. 54, 931 (1985).
[24] S. Halegoua and J. Patrick, Cell (Cambridge, Mass.) 22, 571 (1980).
[25] L. A. Greene, R. K. H. Liem, and M. L. Shelanski, J. Cell Biol. 96, 76 (1983).
[26] A. C. Nairn, R. A. Nichols, M. J. Brady, and H. C. Palfrey, J. Biol. Chem. 262, 14265 (1987).

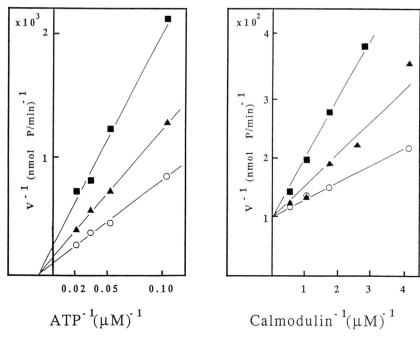

FIG. 3. Kinetic analysis of inhibition by KN-62 of Ca^{2+}/CaM kinase II activity. (○), In the absence of KN-62; (▲), 0.5 μM of KN-62; (■), 2.5 μM of KN-62.

ing molecular events related to Ca^{2+}/CaM kinase II and for determining the relation between phosphorylation and cell function. PC12D cells were labeled with ortho[^{32}P]phosphate and stimulated by A-23187 in the presence and absence of KN-62. Phosphoproteins immunoprecipitated with

TABLE I
SPECIFICITY OF H-89 FOR VARIOUS PROTEIN KINASES

	K_i (μM)		
Kinase	H-8	H-89	H-85[a]
A-kinase	1.2	0.04	>100
G-kinase	0.48	0.34	>100
C-kinase	15	14	>100
MLCK	68	39	28
Ca^{2+}/CaM kinase II	—	11	47
Casein kinase I	133	34	>100
Casein kinase II	950	124	>100

[a] These values represent the IC_{50}.

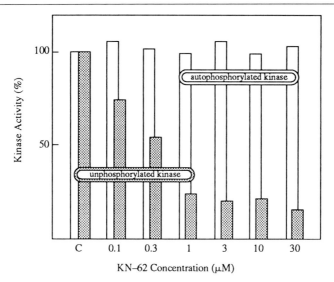

FIG. 4. Effect of KN-62 on autophosphorylated Ca^{2+}/CaM kinase II.

anti-Ca^{2+}/CaM kinase II antibodies were resolved by SDS–PAGE and visualized by autoradiography. Autophosphorylation of the immunoprecipitated 53-kDa subunit of Ca^{2+}/CaM kinase II induced by A-23187 was inhibited by treatment of KN-62 dose dependently. These results suggested that KN-62 is cell permeable and blocks the Ca^{2+}/CaM kinase II activity in PC12D cells.

Molecular Mechanisms of Inhibitory Effects of H-89 on Various Protein Kinase Activities

The effects of this newly synthesized isoquinoline sulfonamide on A-kinase, G-kinase, C-kinase, MLCK, Ca^{2+}/CaM kinase II, and casein kinases I and II were investigated. K_i values for various protein kinases are summarized in Table I. H-89 {N-[2-(p-bromocinnamylamino)ethyl]-5-isoquinoline sulfonamide} proved to be the most potent and selective A-kinase inhibitor among the isoquinoline sulfonamide compounds tested and the K_i value for A-kinase was 0.04 μM.[27] The K_i value of this drug for G-kinase was about nine times higher (0.34 μM), suggesting that H-89 was

[27] T. Chijiwa, A. Mishima, M. Hagiwara, M. Sano, K. Hayashi, T. Inoue, K. Naito, T. Toshioka, and H. Hidaka, *J. Biol. Chem.* **265,** 5267 (1990).

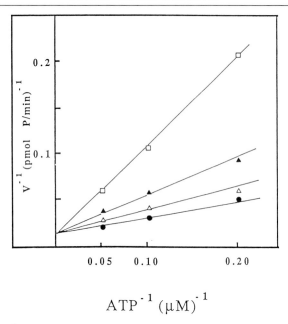

FIG. 5. Kinetic analysis of inhibition by H-89 of A-kinase. ●, In the absence of H-89; △, 0.01 μM of H-89; ▲, 0.03 μM of H-89; □, 0.1 μM of H-89.

a specific A-kinase inhibitor. H-89 was shown to cause a more selective inhibition of A-kinase than G-kinase, and have much weaker effects on other kinases (at least 200 to 300-fold less potent). Inhibition patterns of A-kinase activity by H-89 were then analyzed by double-reciprocal plots (Fig. 5). H-89 inhibited the kinase activity competitively with ATP.

Pharmacological Action of H-89 *in vivo*

To investigate whether this inhibitor could serve as a pharmacological tool for examining viable cells, we studied the effects of H-89 on forskolin- and NGF-induced protein phosphorylation in PC12D cells. Pretreatment of the cells with H-89 markedly inhibited the forskolin-induced protein phosphorylation, in a dose-dependent manner. The inhibition of NGF-induced protein phosphorylation was not observed in the PC12D cells pretreated with H-89. These results suggest that H-89 is a useful compound with respect to the selective inhibition of A-kinase *in vivo*. H-89 did not directly influence activities of cAMP-synthesizing (adenylate cyclase) and -metabolizing (phosphodiesterase) enzymes *in vitro*. Thus, H-89 may act directly on A-kinase in the PC12D cells.

To investigate the role of A-kinase differentiation of PC12D cells, the effect of H-89 on forskolin-, NGF-, and dibutyryl-cAMP-induced neurite outgrowth was investigated using phase-contrast microscopy. When the cells were pretreated with H-89 (20 μM) for 30 min before the addition of forskolin, H-89 (5–30 μM) significantly inhibited the forskolin-induced neurite outgrowth. In addition, H-89 inhibited dibutyryl-cAMP-induced neurite outgrowth dose dependently. However, pretreatment with H-89 exerted no inhibitory action on the NGF-induced neurite outgrowth of PC12D cells. It would thus appear that A-kinase acts as a mediator of forskolin- or dibutyryl-cAMP-induced neurite outgrowth, but not of NGF-induced neurite outgrowth. When the cells were treated with ML-9, H-7, or KN-62, none inhibited forskolin-induced neurite outgrowth of PC12D cells.

H-85   (N-[2-N-formyl-p-chlorocinnamylamino)ethyl]-5-isoquinoline sulfonamide), a derivative of H-89, was chosen as a control agent for H-89. H-85 was not selective in producing inhibition of A-kinase but produced a similar weak inhibition of all the tested kinases, and hence this compound can be used as a negative control for H-89 (Table I).

Newly Synthesized Selective Casein Kinase I Inhibitor CKI-7

Casein kinase I is a cyclic nucleotide-independent protein kinase and has been highly purified from various tissues, including calf thymus,[28] rabbit reticulocytes,[29] liver,[30] and skeletal muscle.[31] The widespread distribution of the enzyme suggests its importance in cellular function, although the exact details remain to be clarified. H-9 has weak effects on casein kinases. When the 5-aminoethylsulfonamide chain of H-9[32] was moved to position 8 on the aromatic ring, the derivative, CKI-6 [N-(2-aminoethyl)-isoquinoline 8-sulfonamide], produced a more potent inhibition of casein kinase activity than did H-9.

CKI-7   [N-(2-aminoethyl)-5-chloroisoquinoline 8-sulfonamide, the chlorinated derivative of CKI-6] revealed a potent inhibition of casein kinase I with an IC_{50} value of 9.5 μM for casein kinase I and 90 μM for casein kinase II.[33] CKI-7 at a concentration up to 1000 μM caused only a weak inhibition of C-kinase. IC_{50} values of CKI-7 were 550 μM for A-kinase and 195 μM for Ca^{2+}/CaM kinase II activity. Kinetic analysis

[28] M. E. Dahmus, J. Biol. Chem. 256, 3319 (1981).
[29] G. M. Hathaway and J. A. Traugh, J. Biol. Chem. 254, 762 (1979).
[30] Z. Ahmad, M. Camici, A. A. DePaoli-Roach, and P. J. Roach, J. Biol. Chem. 259, 3420 (1984).
[31] E. Itarte and K.-P. Huang, J. Biol. Chem. 254, 4052 (1979).
[32] M. Inagaki, M. Watanabe, and H. Hidaka, J. Biol. Chem. 260, 2922 (1985).
[33] T. Chijiwa, M. Hagiwara, and H. Hidaka, J. Biol. Chem. 264, 4924 (1989).

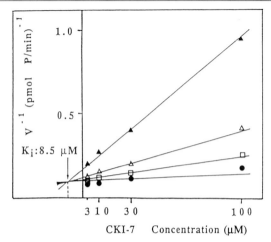

FIG. 6. Kinetic analysis of inhibition by CKI-7 of casein kinase I. ●, 5 μM of ATP; □, 10 μM of ATP; △, 20 μM of ATP; ▲, 30 μM of ATP.

revealed that CKI-7 inhibited casein kinase I competitively with respect to ATP, and that the K_i value was 8.5 μM for the enzyme (Fig. 6).

Affinity Chromatography Using CKI Compounds as Ligands

CKI compounds were used as affinity ligands to purify casein kinase I. Crude casein kinase I fraction was prepared from bovine testis using a modification of the method of Hathaway and Traugh.[29] The final step of the purification of casein kinase I was performed by CKI compound-coupled affinity chromatography. We made two kinds of affinity column by using either CKI-7 or CKI-8 [1-(5-chloroisoquinoline 8-sulfonamide] as affinity ligands. To compare the properties of two affinity columns, partially purified casein kinase I and II were applied to these affinity columns. The CKI-7 affinity column absorbed casein kinase I only. Casein kinase II did not bind to this affinity column. The CKI-8 affinity column absorbed both casein kinase I and II. These results suggested that CKI-7 has a specific affinity for casein kinase I. CKI-8 has about 10 times weaker inhibition for casein kinase I, as compared with CKI-7. However, the recovery of casein kinase I from the CKI-7 affinity column is lower than the CKI-8 affinity column, suggesting that casein kinase I was bound tightly to the column. Therefore we chose the CKI-8 affinity column as the final step of the purification of casein kinase I. Casein kinase I eluted from the affinity column with between 0.53 and 0.8 M L-arginine. Eighteenfold purification was achieved only by this step. Analysis of the purified casein kinase I by SDS–PAGE revealed a single protein band with a molecular weight of 37,000. The K_m value of the purified enzyme was determined as 10 μM for ATP. The purified casein kinase I can utilize ATP but not GTP as the phosphate donor and was not inhibited by heparin, which inhibited casein kinase II activity by 70%. Although several casein kinase II inhibitors has been used to investigate the physiological role of the enzyme,[29,34] a selective casein kinase I inhibitor has not been available. CKI-7 and CKI-8 should find a useful place among the armamentarium needed to determine the physiological role and distribution of casein kinase I in different tissues.

Conclusion

There is now convincing evidence that protein phosphorylation is a common cellular mechanism of fundamental importance in biological regulation. The large number of cellular mechanisms involving protein phosphorylation have been determined; however, interrelationships among each protein phosphorylation are complex and uncertainties concerning the complex cellular responses in this system remain. It is difficult to evaluate which protein phosphorylation system is mainly responsible for cell function. The focus of this chapter has been on H-series protein phosphorylation inhibitors. To clarify the physiological significance and molecular mechanisms of the protein phosphorylation systems, the potent and selective inhibitors which we synthesized will be useful as pharmacological probes. The H-series compounds, W-7, ML-9, SC-9, H-8, and H-7, can be obtained from Seikagaku America, Inc. (Rockville, MD).

[34] G. M. Hathaway and J. A. Traugh, *Curr. Top. Cell. Regul.* **21,** 101 (1982).

[28] Use and Specificity of Staurosporine, UCN-01, and Calphostin C as Protein Kinase Inhibitors

By TATSUYA TAMAOKI

Protein kinases are largely classified into protein-serine/threonine kinases and protein-tyrosine kinases.[1] Many protein kinases have been identified as protein-serine/threonine kinases; for example, cAMP-dependent, cGMP-dependent, Ca^{2+}/phospholipid-dependent protein kinases, and oncogenic protein kinases such as the mos and raf proteins. Among the protein-tyrosine kinases are receptors for growth factors and hormones and viral oncogene products such as $p60^{v-src}$ and $p160^{v-abl}$. It is noteworthy that all protein kinases have striking sequence homology in their catalytic domain.

The Ca^{2+}/phospholipid-dependent protein kinase (protein kinase C) is a protein-serine/threonine kinase involved in the regulation of many cellular processes, including cellular growth, differentiation, and tumor promotion.[2] The primary structure of protein kinase C has two functional domains, the catalytic domain and the regulatory domain.[3] The carboxyl-terminal region of protein kinase C is the catalytic domain, and the amino-terminal region has cysteine-rich repeats and is presumed to be the regulatory domain, which has phospholipid and diacylglycerol/phorbol ester binding regions.

In order to elucidate the physiological role of protein kinases, potent and specific inhibitors would be of great value.

In the course of screening for protein kinase C inhibitors from microbial sources, we found staurosporine, which is the most potent general inhibitor of protein kinases, UCN-01, which selectively inhibits protein kinase C, and calphostin C, which is a specific inhibitor of protein kinase C. The present chapter describes the properties, inhibition, and biological activities of staurosporine, UCN-01, and calphostin C.

[1] T. Hunter, *Cell* (*Cambridge, Mass.*) **50**, 823 (1987).
[2] Y. Nishizuka, *Science* **233**, 305 (1986).
[3] Y. Nishizuka, *Nature* (*London*) **334**, 661 (1988).

Assay of Protein Kinases

Inhibitory activities of staurosporine, UCN-01, and calphostin C were determined against protein kinase C, cAMP-dependent protein kinase, and the protein-tyrosine kinase, p60$^{v\text{-}src}$.

Protein Kinase C Assay

Protein kinase C is partially purified from rat brain using DE-52 (Whatman, Clifton, NJ) column chromatography by the method of Kikkawa *et al.*[4] Protein kinase C activity is assayed in a reaction mixture (0.25 ml) containing 20 mM Tris-HCl, pH 7.5, 10 mM magnesium acetate, 50 μg histone IIIS (Sigma, St. Louis, MO), 20 μg phosphatidylserine (Serdary Research Laboratory), 0.89 μg diolein (Nakarai Chemical), 5 × 10^{-5} M calcium chloride, 5 μM [γ-^{32}P]ATP (Amersham, Arlington Heights, IL) (10^5 cpm/nmol), 5 μg partially purified enzyme, and 10 μl inhibitor solution, according to the method of Kikkawa *et al.*[4]

The activity of the proteolytically generated catalytic domain of protein kinase C is determined as follows: After protein kinase C is treated with 6 μg/ml (final) of Ca^{2+}-activated neutral protease (calpain) for 30 min at 30° in the presence of 80 mM Tris-HCl, pH 7.5, 43 mM 2-mercaptoethanol, and 3 mM calcium chloride, the activity of calpain-treated protein kinase C (50 μl) is assayed in the reaction mixture (0.25 ml) containing 20 mM Tris-HCl, pH 7.5, 10 mM magnesium acetate, 50 μg histone IIIS, 1 mM EGTA, 5 μM [γ-^{32}P]ATP (10^5 cpm/nmol), and 10 μl inhibitor solution by conventional method.

cAMP-Dependent Protein Kinase (PKA) Assay

PKA is partially purified from bovine heart with ammonium sulfate precipitation, and its activity is assayed by the method of Kuo and Greengard.[5] The reaction mixture (0.25 ml) contains 20 mM Tris-HCl, pH 6.8, 10 mM magnesium acetate, 1 μM cAMP, 100 μg histone IIS (Sigma), 5 μM [γ-^{32}P]ATP (10^5 cpm/nmol), and 10 μl inhibitor solution.

Protein-Tyrosine Kinase (PTK) Assay of p60$^{v\text{-}src}$

For the measurement of protein kinase activity of p60$^{v\text{-}src}$, an autophosphorylating reaction of p60$^{v\text{-}src}$ in the immunocomplex is used. Secondary cultures of chicken embryo fibroblasts (CEF) are infected with wild-type Rous sarcoma virus (SR-A) and grown in Dulbecco's modified Eagle's

[4] U. Kikkawa, Y. Takai, R. Minamikuchi, S. Inohara, and Y. Nishizuka, *J. Biol. Chem.* **257**, 13341 (1982).
[5] J. F. Kuo and P. Greengard, *Proc. Natl. Acad. Sci. U.S.A.* **64**, 1349 (1969).

medium (DMEM) supplemented with 5% (v/v) calf serum and 10% (w/v) tryptose phosphate broth. Fully transformed cells are lysed in RIPA buffer [50 mM Tris-HCl, pH 7.2, 150 mM NaCl, 1% (v/v) Triton X-100, 1% (v/v) sodium deoxycholate, 0.1% (w/v) sodium dodecyl sulfate (SDS), and 1 mM EDTA]. Anti-p60 serum (4 μl) is added to 80 μl of cell lysate and kept on ice for 1 hr. The resultant immunoprecipitates are washed five times with RIPA buffer and 40 mM Tris-HCl, pH 7.2. Washed immunoprecipitates are resuspended in 30 μl of kinase assay buffer containaing 20 mM Tris-HCl, pH 7.2, 5 mM magnesium chloride, 10 μCi [γ-^{32}P]ATP (3000 Ci/mmol), and 0.5% (v/v) DMSO, with inhibitor. After a 30-min incubation at 20°, the reaction is terminated by adding 1 ml RIPA buffer (4°), and analyzed by 10% (w/v) SDS/polyacrylamide gel as described by Iba *et al.*[6]

Phorbol/Ester Binding Assay

The binding of [^{3}H]phorbol dibutyrate (PDBu) to protein kinase C is determined by the method of Miyake *et al.*[7] Reaction mixture (0.2 ml) contains 4 μmol Tris/malate, pH 6.8, 20 μmol KCl, 30 nmol calcium chloride, 20 μg phosphatidylserine, 5 μg partially purified protein kinase C, 0.5% (final) DMSO, 10 pmol [^{3}H]PDBu (1–3 × 10^4 cpm/pmol), and 10 μl inhibitor solution. After incubation of the reaction mixture for 20 min at 30°, the reaction is terminated by 3 ml of ice-cold DMSO (0.5%). This mixture is collected on a polyethyleneimine-treated Whatman (Clifton, NJ) GF/B glass filter and washed with 6 ml of 0.5% DMSO. The radioactivity on the filter is determined.

Antiproliferative Activity

Human cervical cancer HeLa S3 cells are cultured in DMEM supplemented with 10% fetal bovine serum (FBS). Human breast cancer MCF-7 cells are cultured in RPMI 1640 supplemented with 10% (v/v) FBS, 10 μg/ml insulin, and 10^{-8} M estradiol. For determination of cytotoxicity, the cells are preincubated for 24 hr at 37° in 96-well plastic plates and then cultured in the presence of different dilutions of inhibitors for 72 hr at 37°. Thereafter the concentration of inhibitors required for 50% inhibition of cell growth is determined by the uptake of neutral red dye by the modified method of Gomi *et al.*[8]

---

[6] H. Iba, F. R. Cross, E. A. Gaber, and H. Hanafusa, *Mol. Cell. Biol.* **5**, 1058 (1985).
[7] R. Miyake, Y. Tanaka, T. Tsuda, K. Kaibuchi, U. Kikkawa, and Y. Nishizuka, *Biochem. Biophys. Res. Commun.* **121**, 649 (1984).
[8] K. Gomi, S. Akinaga, T. Oka, and M. Morimoto, *Cancer Res.* **46**, 6211 (1986).

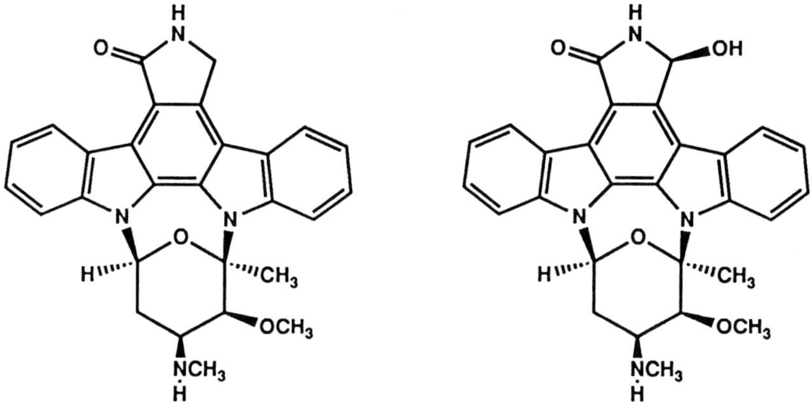

Staurosporine   **UCN-01**

FIG. 1. Chemical structures of staurosporine and UCN-01.

Staurosporine and UCN-01

Properties and Structures

Staurosporine[8a] was first reported to be a microbial alkaloid having an indolo[2,3-a]carbazole chromophore produced by *Streptomyces* species.[9] (Fig. 1) The molecular formula and molecular weight of staurosporine are $C_{28}H_{26}N_4O_3$ and 466, respectively. UCN-01 was found in the culture broth of *Streptomyces* sp. No. 126, which produces staurosporine.[10] The structure of UCN-01 differs from staurosporine in that the C-7 carbon bears a hydroxyl group (Fig. 1); its molecular formula is $C_{28}H_{26}N_4O_4$, and its molecular weight is 482. Both compounds show low solubility in water and various other solvents. Therefore, it is best to dissolve them in dimethyl sulfoxide (DMSO) or dimethylformamide (DMF). DMSO and DMF solutions of these compounds are stable at 4° in the dark, and should be diluted just before use.

[8a] Staurosporine and calphostin C are commercially available from Kamiya Biomedical Co. (Thousand Oaks, CA) and Kyowa Medex Co., Ltd., Japan.

[9] S. Omura, Y. Iwai, A. Hirao, A. Nakagawa, J. Awaya, H. Tsuchiya, Y. Takahashi, and R. Masuma, *J. Antibiot.* **30**, 275 (1977).

[10] I. Takahashi, E. Kobayashi, K. Asano, I. Kawamoto, T. Tamaoki, and H. Nakano, *J. Antibiot.* **42**, 564 (1989).

TABLE I
INHIBITORY ACTIVITIES OF PROTEIN KINASE INHIBITORS

| | IC$_{50}$ (μM) | | | Cytotoxicity (μM)[a] | |
	Protein kinase C	cAMP-dependent protein kinase	p60$^{v\text{-}src}$ tyrosine kinase	HeLa S3	MCF-7
Staurosporine	0.0027	0.0082	0.0064	<0.003	0.09
UCN-01	0.0041	0.042	0.045	0.23	0.31
Calphostin C	0.05	>50	>50	0.23	0.18

[a] Growing cells were incubated with protein kinase inhibitors for 72 hr.

Inhibition of Protein Kinases

Staurosporine is the most potent inhibitor of protein kinases *in vitro* to date, with an IC$_{50}$ value of 2.7 nM for protein kinase C, 8.2 nM for PKA, 6.4 nM for PTK of p60$^{v\text{-}src}$ (Table I), and 630 nM for PTK of epidermal growth factor (EGF) receptor.[11–13] Staurosporine inhibits the proteolytically generated catalytic domain of protein kinase C, and has no effect on the binding of [^{3}H]PDBu to protein kinase C, indicating that it targets the catalytic domain of protein kinase C.[11] The limited selectivity of staurosporine for different protein kinases is attributable to its interaction with the essential region of the catalytic domain of protein kinases that share a homologous region. Therefore, many protein kinases other than protein kinase C, e.g., PKA and PTK of p60$^{v\text{-}src}$, are strongly inhibited by staurosporine.

In contrast, UCN-01 shows selectivity for protein kinase C inhibition *in vitro* (IC$_{50}$ values: 4.1 nM for protein kinase C, 42 nM for PKA, 45 nM for PTK of p60$^{v\text{-}src}$), as shown in Table I.[14] It is interesting how the hydroxyl group at C-7 contributes to the selectivity for protein kinase C inhibition.

Biological Activities

Staurosporine was first reported to show antifungal and hypotensive activities by Omura *et al.*[9] We found that staurosporine shows a strong cytotoxic effect on the growth of various mammalian cells (IC$_{50}$ values:

[11] T. Tamaoki, H. Nomoto, I. Takahashi, Y. Kato, M. Morimoto, and F. Tomita, *Biochem. Biophys. Res. Commun.* **135**, 397 (1986).

[12] H. Nakano, E. Kobayashi, I. Takahashi, T. Tamaoki, Y. Kuzuu, and H. Iba, *J. Antibiot.* **40**, 706 (1987).

[13] Y. Fujita-Yamaguchi and S. Kathuria, *Biochem. Biophys. Res. Commun.* **157**, 955 (1988).

[14] I. Takahashi, E. Kobayashi, K. Asano, M. Yoshida, and H. Nakano, *J. Antibiot.* **40**, 1782 (1987).

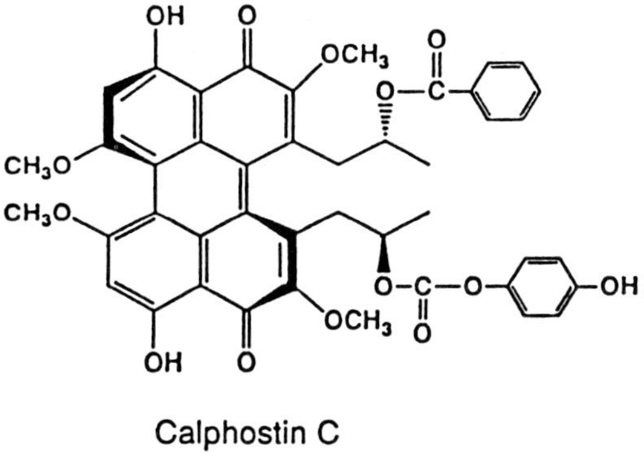

Calphostin C

FIG. 2. Chemical structure of calphostin C.

<3 nM for HeLa S3 cells, 90 nM for MCF-7 cells). In addition, the various biological effects of staurosporine have been reported, for example induction of differentiation, [15,16] and effects on the various functions of platelet and smooth muscle cells.[17,18] UCN-01 also shows strong cytotoxic effects on the growth of mammalian cells (IC$_{50}$ values: 230 nM for HeLa S3 cells, 310 nM for MCF-7 cells). Interestingly, UCN-01 shows antitumor activity *in vivo* against the P388 murine leukemia model and other models, including cells transformed with oncogenes, while staurosporine does not show antitumor activity against any *in vivo* models so far tested.[19] Furthermore, staurosporine and UCN-01 do not interact with DNA *in vitro*. Thus, the biological effects of staurosporine and UCN-01 could be due to the inhibition of protein kinases.

Calphostin C

Property and Structure

Calphostin homologs were newly isolated from the culture broth of the fungus *Cladosporium cladosporioides*.[20] Among five calphostin homologs, calphostin C[8a] has the most potent biological activities. As shown in Fig. 2, calphostin C has a unique structure, which includes a 3,10-perylenequi-

[15] H. Morioka, M. Ishihara, H. Shibai, and T. Suzuki, *Agric. Biol. Chem.* **49**, 1959 (1985).
[16] T. Sato, A. I. Tauber, A. Y. Jeng, S. H. Yusupa, and P. M. Blumberg, *Cancer Res.* **48**, 4646 (1988).
[17] S. P. Watson, J. McNally, L. J. Shipman, and P. P. Godfrey, *Biochem. J.* **249**, 345 (1988).
[18] I. Laher and J. A. Bevan, *Biochem. Biophys. Res. Commun.* **158**, 58 (1989).
[19] S. Akinaga, K. Gomi, M. Morimoto, T. Tamaoki, T. Hirata, and M. Okabe, in preparation.
[20] E. Kobayashi, K. Ando, H. Nakano, T. Iida, H. Ohno, M. Morimoto, and T. Tamaoki, *J. Antibiot.* **42**, 1470 (1989).

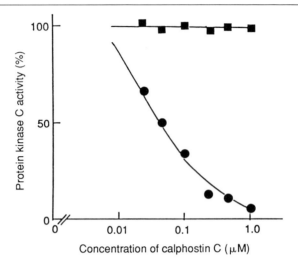

FIG. 3. Inhibition of protein kinase C and protein kinase C catalytic domain by calphostin C. The definition of percentage activity was as follows: (protein kinase C activity in the presence of calphostin C/protein kinase C activity in the absence of calphostin C) × 100. Inhibition of protein kinase C (●) and protein kinase C catalytic domain (■).

none skeleton. Physicochemical properties of calphostin C are as follows: appearance, dark red crystal; molecular formula, $C_{44}H_{38}O_{14}$; molecular weight, 790. It is poorly soluble in water and most solvents, except for DMSO.[19] DMSO solution is stable at 4° in the dark, and should be diluted just before use.

Inhibition of Protein Kinases

Calphostin C inhibits protein kinase C with an IC_{50} value 0.05 μM, but does not inhibit PKA and PTK of p60[v-src] even at 50 μM (Table I), indicating that it is a potent and specific inhibitor of protein kinase C.[21] In addition, calphostin C does not inhibit the proteolytically generated catalytic domain of protein kinase C (Fig. 3), but does inhibit [³H]PDBu binding to protein kinase C (Table II), indicating that calphostin C interacts with the regulatory domain of protein kinase C, which is distinguished from those of other protein kinases, presumably with its diacylglycerol/phorbol ester binding region. This is supported by the fact that inhibitory activity of calphostin C is not affected by the concentration of Ca^{2+} and phospholipid, both of which interact with the regulatory domain.

[21] E. Kobayashi, H. Nakano, M. Morimoto, and T. Tamaoki, *Biochem. Biophys. Res. Commun.* **159**, 548 (1989).

TABLE II
INHIBITION OF [³H]PDBu BINDING TO PROTEIN KINASE C
BY CALPHOSTIN C

Calphostin C (μM)	Inhibition of [³H]PDBu binding[a]
0	0
0.03	4.6
0.1	47.1
0.5	69.5
1.0	91.3

[a] Percentage inhibition = [1 − ([³H]PDBu binding in the presence of drugs/[³H]PDBu binding in the absence of drugs)] × 100.

Biological Effects

Calphostin C has a cytotoxic effect on the growth of mammalian cells (IC_{50} values: 0.23 μM for HeLa S3 cells, 0.18 μM for MCF-7 cells). Calphostin C does not interact with DNA *in vitro*, and cytotoxicities of calphostin homologs are well correlated to their inhibitory activities against protein kinase C, suggesting that the inhibition of protein kinase C contributes to the cytotoxicity of calphostin C.

Acknowledgments

I would like to thank my many scientific colleagues in Tokyo Research Laboratories and Pharmaceutical Research Laboratories of Kywa Hakko Company, in particular Dr. Hirofumi Nakano for his support and helpful discussion, with special thanks to my collaborators, Dr. Hideo Iba, University of Tokyo, and Dr. Koichi Suzuki, Tokyo Metropolitan Institute of Medical Science.

[29] Inhibition of Protein-Tyrosine Kinases by Tyrphostins

By ALEXANDER LEVITZKI, AVIV GAZIT, NIR OSHEROV,
ISRAEL POSNER, and CHAIM GILON

Introduction

Many of the known protein tyrosine kinases (PTKs) catalyze the phosphorylation of tyrosine residues on exogenous substrates as well as the autophosphorylation of tyrosine residues. In order to test the ability of PTKs to phosphorylate an exogenous substrate, one chooses as substrates synthetic copolymers of amino acids which contain tyrosine residues, but are devoid of serine and threonine. Once the polymers are examined at three arbitrary concentrations the concentration dependence of substrate phosphorylation is examined in detail. Thus, for the epidermal growth factor (EGF) receptor kinase,[1-3] the insulin receptor kinase,[4] and the abl kinase,[4] the concentration dependence was found to be Michaelian (noncooperative). The kinase used can either be in the form of free purified enzyme, partially purified enzyme, or as a fresh immunoprecipitate isolated from broken cells. Indeed, we have shown that immunoprecipitates of abl kinases can be used for this purpose. Protein-tyrosine kinases can be specifically inhibited both *in vivo* and *in vitro*[1-9] by synthetic blockers which we refer to as "tyrphostins." In this chapter, we shall describe the inhibition of *in vitro* phosphorylation and the antiproliferative activity of these compounds.

Synthesis of Tyrphostins

A large number of tyrphostins have been synthesized so far. In this chapter we shall focus our attention on a selected number of representative PTK inhibitors which were found to be potent blockers of EGF receptor kinase *in vitro* and inhibitors of EGF-dependent cell proliferation.

Five of the inhibitors shown in Table I (**1, 2, 5, 6,** and **7**) are prepared by straightforward Knoevenagel condensation of benzaldehyde derivatives with malononitrile, cyanothioacetamide, malononitrile dimer, cyanoacetanilide, and benzoyl acetonitrile, according to the following general synthetic scheme (Scheme 1):

Compounds **3** and **4** are prepared according to the synthetic schemes shown below. Most compounds are purified by recrystallization and some by chromatography. All compounds are pure and have the appropriate

[1] P. Yaish, A. Gazit, C. Gilon, and A. Levitzki, *Science* **242**, 933 (1988).

[2] A. Gazit, P. Yaish, C. Gilon, and A. Levitzki, *J. Med. Chem.* **32**, 2344 (1989).

[3] A. Gazit, N. Osherov, I. Posner, P. Yaish, E. Pradosu, C. Gilon, and Z. Levitzki, *J. Med. Chem.,* in press. (1991).

[4] Y. Schechter, P. Yaish, M. Chorev, C. Gilon, S. Braun, and A. Levitzki, *EMBO J.* **8**(6), 1671 (1989).

[5] M. Anafi, A. Gazit, C. Gilon, Y. Ben-Neria, and A. Levitzki, in preparation.

[6] R. M. Lyall, A. Zilberstein, A. Gazit, C. Gilon, A. Levitzki, and J. Schlessinger, *J. Biol. Chem.* **264**, 14503 (1989).

[7] C. Roifman, G. R. Mills, K. Chin, A. Gazit, C. Gilon, and A. Levitzki, *J. Immunol.,* in press. (1990).

[8] A. Dvir, Y. Milner, O. Chomsky, C. Gilon, A. Gazit, and A. Levitzki, *J. Cell Biol.,* in press. (1991).

[9] G. Bilder, J. A. Krawiec, A. Gazit, C. Gilon, M. McVety, R. Lyall, A. Zilberstein, A. Levitzki, M. H. Penmore, and A. B. Schreiber, *Am. J. Physiol.,* in press. (1991).

TABLE I
STRUCTURES AND BIOLOGICAL ACTIVITIES OF TYRPHOSTINS

No. (AG)	Structure	Molecular formula	M_r	Phosphorylation of poly(GAT)		Autophosphorylation [IC$_{50}$ (μM)]	Inhibition of cell growth [IC$_{50}$ (μM)]	
				IC$_{50}$ (μM)	K_i		EGF	Serum
1 (18)	*(structure)*	C$_{10}$H$_6$N$_2$O$_2$	186	35	10	31	30	60
2 (213)	*(structure)*	C$_{10}$H$_6$SN$_2$O$_2$	220	1.6	0.8	7.5	10	>50
3 (308)	*(structure)*	C$_{12}$H$_8$N$_2$O$_2$	212	0.5	ND^a	30	ND	30
4 (336)	*(structure)*	C$_{11}$H$_5$N$_3$O$_3$	227	2.3	ND	ND	>24^b	>24^b
5 (293)	*(structure)*	C$_{13}$H$_7$BrN$_4$O$_2$	331	0.5	ND	3.23	ND	ND
6 (494)	*(structure)*	C$_{16}$H$_{12}$N$_2$O$_3$	280	0.7	ND	1.2	6	24
7 (473)	*(structure)*	C$_{16}$H$_{11}$NO$_3$	265	1.0	ND	1.7	20	>20

^{a} ND, Not determined.
^{b} Unstable in neutral solutions. Therefore there is a large discrepancy between results *in vitro* and inhibitory action on cells.

elemental analysis, mass spectroscopy (MS), and nuclear magnetic resonance (NMR) spectra. All inhibitors (except **4**) are stable compounds.

Synthetic Procedures

For compound structures please see Table I.

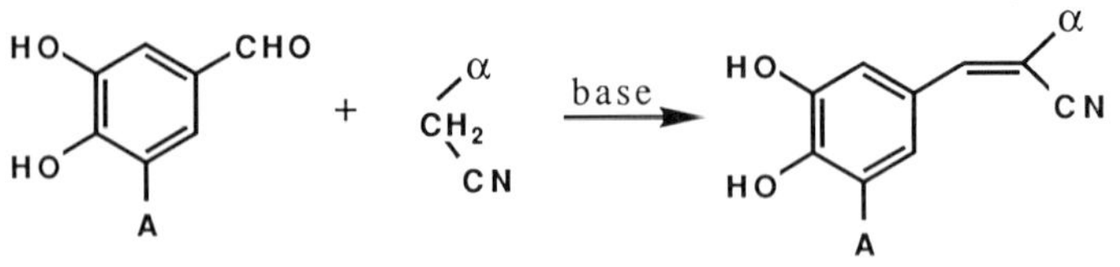

SCHEME 1. General synthetic scheme of tyrphostins.

Compound No.	α	A
1	$-CN$	H
2	$\overset{\overset{S}{\|}}{-C-NH_2}$	H
5	$\underset{\diagdown CN}{\overset{H_2N \quad CN}{>\!=\!<}}$	Br
6	$\overset{\overset{O}{\|}}{-C-NH-C_6H_5}$	H
7	$\overset{\overset{O}{\|}}{-C-C_6H_5}$	H

3,4-Dihydroxybenzylidene malononitrile (1)

To 11 g (80 mmol) of 3,4-dihydroxybenzaldehyde and 5.5 g (83 mmol) of malononitrile in 60 ml of ethanol is added 10 drops of piperidine and the reaction is refluxed for 0.5 hr. Water is added and the solid is filtered and dried to give 12.7 g (86% yield) of yellow solid: mp 225° (Ref. 2, mp 225°).

MS: m/e 186 (M$^+$, 100%), 159 (M $-$ HCN, 10%), 158 (M $-$ CO, 59%), 157 (M $-$ HCO, 26%), 130 (M $-$ 2CO, 90%), 129 (54%), 113 (21%), 103 (M $-$ 2CO $-$ HCN, 74%); NMR (acetone-$d_6\delta$): 7.03 (1H, d, $J_{5,6}$ = 8.4 Hz, H$_5$), 7.41 (1H, dd, $J_{5,6}$ = 8.4 Hz, $J_{2,6}$ = 2.1 Hz, H$_6$), 7.68 (1H, d, $J_{2,6}$ = 2.1 Hz, H$_2$), 8.04 (1H, s, vinylic proton); ^{13}C NMR (acetone-d_6): 77.4 (C$_\alpha$), 114.7, 115.6 (CN), 116.9 (C$_5$), 125.0 (C$_1$), 130.0 (C$_2$), 146.6 (C$_6$), 153.3 (C$_4$), 160.7 (C$_\beta$).

(3,4-Dihydroxybenzylidene)cyanothioacetamide (2)

To 0.83 g (6 mmol) of 3,4-dihydroxybenzaldehyde and 0.7 g (7 mmol) of cyanothioacetamide in 30 ml ethanol is added four drops of piperidine. The reaction mixture is refluxed 1 hr and then cooled and poured into ice water. The solid is collected by filtration and dried to give 0.57 g (41% yield) of orange solid: mp 213°.

MS: m/e 220 (M^+, 93%), 219 (M − 1, 100%), 203 (M − OH, 26%), 186 (M − 2OH, 24%), 123 (33%), 110 (30%), 100 (43%).

Synthesis of Tyrphostin 3

This compound (Table I) is synthesized according to Scheme 2:

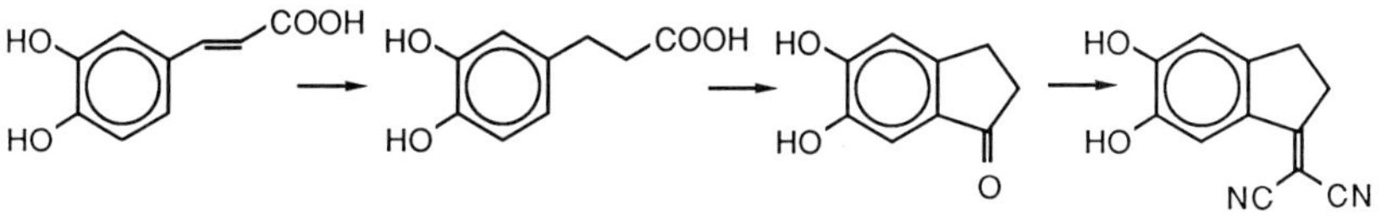

SCHEME 2. Synthesis of 5,6-dihydroxy-1-indanylidene malononitrile (3).

3,4-Dihydroxyphenylpropionic acid (**3a**): Caffeic acid, 1.7 g, in 60 ml ethanol with 0.2 g 5% Pd/C is hydrogenated in a Paar apparatus for 20 hr. Filtering and solvent evaporation give 1.37 g of white solid [81% yield, mp 125° (Aldrich—dihydrocaffeic acid, mp 138°].

NMR (acetone $d_6\delta$): 7.80 (1H, br. s, COOH), 6.76 (1H, d, J = 2.0 Hz, H_2), 6.73 (1H, d, J = 8.0 Hz, H_5), 6.58 (1H, dd, J = 8.0, 2.0 Hz, H_6), 2.77, 2.56 (4H, AA′BB′ multi.).

5,6-Dihydroxy 1-indanone (**3b**): The above acid (**3a**), 2 g, in 20 ml neat liquefied HF is stirred 24 hr in a KelF system at room temperature. HF is evaporated and the gray solid extracted with ethyl acetate and chromatographed on silica gel to give 0.26 g light brown–white solid (14% yield)

NMR (acetone $d_6\delta$): 7.06 (1H, s, H_7), 6.99 (1H, s, H_4), 2.93, 2.54 (4H, AA′BB′ mult). *Note:* Attempts to prepare **3b** with trifluoroacetic acid (TFA), 24 hr at room temperature, gave only starting material (**3a**). With PPA, 5 hr at 100°, a mixture of **3b** and an unidentified produce was obtained.

5,6-Dihydroxy-1-indanylidene malononitrile (**3**): To 0.25 g (1.5 mmol) **3b** in 30 ml ethanol and 0.2 g (3 mmol) malononitrile is added 80 mg β-alanine and the reaction is refluxed 20 hr. Evaporation to dryness gives a yellow-brown solid.

Purification: R_f (EtOH: CH_2Cl_2, 1 : 9) = 0.5, 0.45. [On standing overnight the yellow thin-layer chromatography (TLC) spots turn brown.]

High-performance liquid chromatography (HPLC): RP-8 (25 × 4 cm, 1 mm, 5 μ), F = 1.0 ml/min, UV = 254 nm, 70 : 30 CH_3OH : H_2O, R_f = 1.88 min for **3b**, 3.56 min for **3**. Separation on a preparative HPLC column (RP-18, 50 g of Lichroprep 25 to 35-μ step gradient from 15 : 85 CH_3OH : H_2O to 45 : 55 CH_3OH : H_2O + 0.1% TFA, UV = 254 nm, F = 9 ml/min) gives 40 mg of yellow solid (R_f = 2.5 hr).

NMR (acetone $d_6\delta$): 7.76 (1H, *s*, H$_7$), 7.02 (1H, *s*, H$_4$), 3.20, 3.11 (4H, AA'BB' mult.)

Synthesis of Compound 4

Compound **4** is synthesized according to Scheme 3:

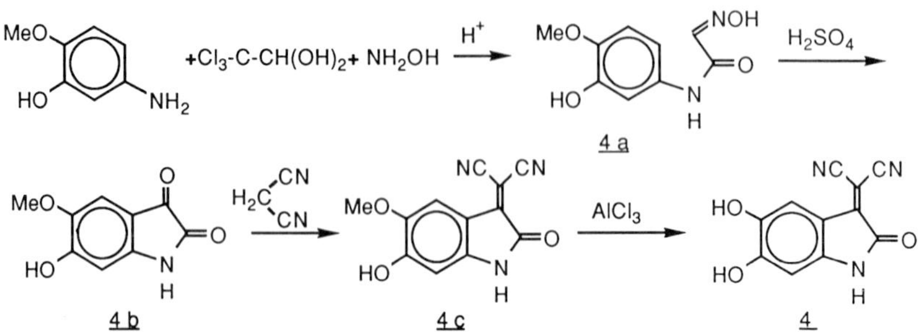

SCHEME 3. Synthesis of 5,6-dihydroxy isatinylidene malononitrile (**4**).

3-OH-4-OCH$_3$ isonitroso acetanilide (**4a**)

To 6.6 g (39 mmol) chloral hydrate and 30 g Na$_2$SO$_4$ in 100 ml water is added 5 g (36 mmol) of 3-OH-4-OCH$_3$ aniline in 20 ml H$_2$O and 3 ml HCl conc., followed by 7.2 g NH$_2$OH · HCl. The reaction is heated 0.25 hr, cooled with ice, and filtered to give 1.5 (20% yield) of black solid 3-OH-4-OCH$_3$ isonitroso acetanilide (**4a**).

NMR (acetone d$_6\delta$): 7.51 (1H, *s*, CH=NOH), 7.36 (1H, *d*, $J_{2.6}$ = 2.5 Hz, H$_2$), 7.17 (1H, *dd*, $J_{8.7}$ = 2.5 Hz, H$_6$), 6.88 (1H, *d*, $J_{5.6}$ = 8.7 Hz, H$_5$), 3.81 (3H, *s*, OCH$_3$).

5-Methoxy-6-OH isatin (**4b**):

Compound **4a** (1.5 g) in 8 ml H$_2$SO$_4$ conc. + 2 ml H$_2$O is heated 10 min, with internal temperature of 80°. The mixture is decomposed on crushed ice and extracted with EtOAc to give 80 mg of red–brown solid (6% yield, mp 265°). NMR (acetone-d$_6\delta$): 7.09 (1H, *s*, H$_4$), 6.50 (1H, *s*, H$_7$), 3.87 (3H, *s*, OCH$_3$).

5-Methoxy-6-hydroxy isatinylidene malononitrile (**4c**):

To 80 mg (0.4 mmol) isatin (**4b**) is added 100 mg (1.5 mmol) malononitrile and the reaction mixture refluxed 6 hr, until TLC shows the reaction is finished. Evaporation gives a violet solid, which is purified on silica gel (5:95, CH$_3$OH : CHCl$_3$). The dark green band is collected to give 80 mg (80% yield) of violet solid, mp 300°. Compound **4c** is stable on standing but in basic water (green solution) it decomposes in 5 min. On standing several hours on a TLC plate its color is changed (R_f CH$_2$Cl$_2$ = 0.5, green–gray spot). NMR (acetone d$_6\delta$): 7.51 (1H, *s*, H$_4$), 6.54 (1H, *s*, H$_7$), 3.87 (3H, *s*,

OCH_3); MS: m/e 241 (M$^+$, 66%), 227 (12%), 226 (M—CH$_3$, 100%), 198 (M—CONH, 28%).

5,6-Dihydroxy Isatinylidene Malononitrile (4)

To 60 mg (0.3 mmol) compound **4c** in 10 ml trichloroethylene is added 110 mg AlCl$_3$ and seven drops pyridine. The suspension is refluxed for 1.5 hr, filtered, and the solid added to water–HCl and extracted with EtOAc. Evaporation gives 15 mg (27% yield) light blue solid (R_f CH$_2$Cl$_2$ = 0.3, green spot).

NMR (acetone d$_6$δ): 7.47 (1H, s, H$_4$), 6.54 (1H, s, H$_7$).

3,4-Dihydroxy-5-bromo(β-amino-γ-cyano)cinnamylidene Malononitrile (5)

To 0.38 g (1.7 mmol) 3,4-dihydroxy-5-bromobenzaldehyde and 0.26 g (2 mmol) malononitrile dimer in 20 ml ethanol is added 50 mg β-alanine. The reaction is refluxed 1 hr. Water is added and the reaction extracted with EtOAc. Evaporation gives 0.45 g (77% yield) orange solid, mp 241°.

NMR (acetone d$_6$δ): 7.86 (1H, s, vinyl), 7.71 (1H, d, J = 2.1 Hz, H$_6$), 7.67 (1H, d, J = 2.1 Hz, H$_2$).

Cyanoacetanilide (6a)

To 5.5 g (55 mmol) methylcyano acetate is added 5.9 g (65 mmol) aniline. The reaction is heated 14 hr at 120°, without condensor. Cooling gives a dark solid which is recrystallized from ethanol to give 1.5 g white solid (17% yield, mp 193°).

NMR (acetone d$_6$δ): 3.56 (2H, s, CH$_2$CN), 7.50–7.20 (5H, m, aromatic).

3,4-Dihydroxy(α-benzamido)cis-cinnamonitrile (6)

To 0.55 g (4 mmol) of 3,4-dihydroxybenzaldehyde and 0.75 g (4.7 mmol) of **6a** in 30 ml ethanol is added two drops piperidine. The reaction mixture is refluxed 4 hr, cooled, filtered, and washed with cold ethanol to give 0.77 g (69% yield) fluffy, golden yellow solid (mp 258°).

NMR (acetone d$_6$δ): 8.12 (1H, s, vinyl), 7.76 (1H, d, J = 1.4 Hz, H$_2$), 7.41 (1H, dd, J = 8.0, 1.4 Hz, H$_6$), 6.98 (1H, d, J = 8.0 Hz, H$_5$), 9.15 (1H, br. s, NH), 7.70–7.10 (5H, m, aromatic); MS: m/e 280 (M$^+$, 42), 202 (18), 188 (M − NHPh, 35) 164 (13), 114 (18), 93 (100), 91 (26).

3,4-Dihydroxy(α-benzoyl)cis-cinnamonitrile (7)

To 0.7 g (5 mmol) 3,4-dihydroxybenzaldehyde and 0.75 g (5.2 mmol) benzoyl acetonitrile in 30 ml ethanol is added five drops piperidine. The reaction mixture is refluxed 5 hr, water and 2 ml conc. HCl are added, and the solution extracted with EtOAc. Evaporation gives 1.1 g (82% yield) yellow solid, mp 180°.

MS: m/e 265 (M$^+$, 5%), 105 (PhCO$^+$, 100) 77 (67). NMR (acetone d$_6\delta$): 7.97 (1H, s, vinyl), 7.81 (1H, d, H$_2$), 7.47 (1H, dd, J = 8.4 Hz, H$_6$), 7.03 (1H, d, J = 8.4 Hz, H$_5$), 7.86–7.50 (5H, m, aromatic).

Assay of Tyrphostins

In this chapter we concentrate on the assay of tyrphostins as inhibitors of EGF receptor kinase activity. The protocols described are applicable to other protein-tyrosine kinases and have already been applied to the insulin receptor kinase (IRK)[1,3] and more recently to human p210$^{bcr-abl}$, p185$^{bcr-abl}$, p140^{c-abl}, and mouse p140^{c-abl}, and p160$^{gag-abl}$, immunoprecipitates using specific antibodies.[5] The quality of the results obtained with immunoprecipitates is similar to that obtained with systems of purified receptors.

In all cases one can investigate the ability of tyrphostins to inhibit kinase activity against different exogenous substrates as well as their potency in inhibiting PTK autophosphorylation.

Assay of EGF Receptor Tyrosine Kinase Activity Using an Exogenous Substrate

Materials

Culture plates with 96 conical wells (not sterile): Two plates, a preparatory plate and a reaction plate, are needed for each experiment
Multipipettor for eight tips graduated to deliver from 6 to 50 μl
Chromatography filter paper (No. 3MM; Whatman, Clifton, NJ) in rolls 3 cm wide
Paper clamps: Twelve, each 3 in. wide

Reagents

Tris-MES (TM), pH 7.6: 250 and 50 mM
Triton X-100 (0.1%, w/v): 150 mM NaCl: 10% (v/v) glycerol (v/v/v) in 50 mM TM (TM-TNG)
Mg(Ac)$_2$, 480 mM
Poly(Glu$_6$Ala$_3$Tyr), 3.2 mM (GAT; Sigma, St. Louis, MO): ε_{240} = 11,000 M^{-1} cm^{-1}, ε_{295} = 2900 M^{-1} cm^{-1} calculated per unit Glu$_6$Ala$_3$Tyr
[γ-^{32}P]ATP
EGF, 10^{-5} M
TCA, 10% in sodium pyrophosphate

Reaction cocktail: Prepared by mixing equal volumes of Mg(Ac)$_2$, 250 mM TM, and 2.0 mM ATP; [γ-^{32}P]ATP is added to a concentration of 1–2 μCi/15 μl reaction cocktail

Procedure

1. The filter paper is cut with scissors into sections 2.0 cm long and these are cut almost all the way across into eight strips, by seven double cuts to create narrow spaces between strips. Each filter paper is then held by a paper clamp from the cut end and the strips are numbered. It is convenient to arrange the paper clamps in numerical order in rows of four (lowest numbers at bottom left, highest numbers at top right) near the water bath.

2. GAT is added in increasing concentrations (0.0 to 3.2 mM) to a row of 8 wells across (left to right) the preparatory plate and the 5-μl aliquots of GAT are transferred into rows of wells down the reaction plate with the aid of a multipipettor (final GAT concentrations will be 0.0 to 400 μM, eight concentrations).

3. An inhibitor is dissolved in dimethylsulfoxide (DMSO) and then diluted with ethanol : water (1 : 1, v/v); further dilutions are in DMSO : ethanol : water (10 : 45 : 45, v/v/v). Six inhibitor concentrations are prepared. The inhibitors are added into a row of six wells *down* the preparatory plate to eight positions across. The inhibitors are then transferred to the reaction plate in 5-μl aliquots into series of six wells from left to right (eight positions). The final inhibitor concentrations will be one-eighth the respective concentration in the preparatory plate. In this fashion six inhibitor concentrations are used for each of eight concentrations of the substrate.

4. Aliquots of the reaction cocktail are pipetted into a row of eight wells across the preparatory plate. Next, 15 μl cocktail is transferred to rows of wells down the reaction plate.

5. EGF receptor is mixed with 3 vol TM-TNG and 2 vol 10^{-5} M EGF diluted 1 : 4 with 50 mM TM (for controls the receptor is mixed with 3 vol TM-TNG and 2 vol 50 mM TM). The receptor preparations are incubated at 4° for 20 min and then aliquots are pipetted into a row of eight wells across the preparatory plate. Both the preparatory and the reaction plates are now incubated in a water bath with a low water level at 22° for 3 min.

6. The reaction is started by the addition of 15-μl aliquots of receptor with the aid of the multipipettor into rows of eight wells down the reaction plate at 10-sec intervals. At the end of 4 min of incubation, aliquots of 20 μl are taken with the aid of the multipipettor from rows of eight wells in the order in which the receptor was added and transferred to respective paper strips held in a paper clamp. The strips are immediately released

into a beaker containing 10% TCA and pyrophosphate. The paper strips are washed three times with TCA–pyrophosphate at 30-min intervals and then overnight by gentle shaking. The strips are washed once with ethanol and then with acetone, dried, and radioactivity is determined by Cerenkov counting.

The above arrangement allows for an experiment in which the substrate is varied from left to right using eight different concentrations and the inhibitor is varied from top to bottom using six different concentrations; the whole experiment is done in duplicate. Dixon plots[1] can be readily constructed and IC_{50} values may be calculated at any desired substrate concentration.

For a kinetic study in which ATP is varied at different fixed concentrations of inhibitor, the setup is similar to the above with the following exceptions: a reaction cocktail is prepared by mixing equal volumes of $Mg(Ac)_2$, 250 mM TM, 3.2 mM GAT, and [γ-^{32}P]ATP. Various dilutions of ATP (ranging from 0.0 to 2.0 mM) are prepared and an aliquot of each dilution is mixed with three times its volume of reaction cocktail, yielding reaction cocktails with eight different concentrations of ATP. These are added to a row of wells in the preparatory plate and, following the transfer of the inhibitor to the reaction plate as above, 20 μl aliquots of the reaction cocktails are transferred to rows of wells down the reaction plate. Final ATP concentrations in the reaction plate will vary from 0.0 to 250 μM. Steps 5 and 6 above are now followed.

For a kinetic study where GAT is varied at different fixed concentrations of ATP or vice versa, the scheme described for Dixon plots may be followed with ATP substituting for the inhibitor (i.e., 5 μl GAT at varying concentrations will be added down the reaction plate and 5 μl of varying ATP concentrations will be added from left to right in the reaction plate). A reaction cocktail is prepared by mixing equal volumes of 250 mM TM and $Mg(Ac)_2$ and adding a proper amount of [γ-^{32}P]ATP; 15 μl cocktail/well is used and the reaction is carried out by following steps 5 and 6 above.

For the determination of IC_{50} values the following procedure may be followed:

1. Six inhibitor concentrations are prepared and are added to six wells down the preparatory plate and 5-μl aliquots are then transferred to two series of wells down the reaction plate.

2. The cocktail is prepared by mixing equal volumes of 250 mM TM, $Mg(AC)_2$, 2 mM ATP, 0.8 mM GAT, and [γ-^{32}P]ATP to a final concentration of 1–2 μCi/20 μl cocktail. Fractions of the reaction cocktail are

placed in two rows of wells in the preparatory plate and 20-μl aliquots are transferred down the reaction plate.

3. The receptor is prepared and the reaction is started and terminated as above.

This arrangement allows for the simultaneous determination of IC$_{50}$ values for eight different inhibitors in duplicate.

Note: It is convenient to outline areas in the preparatory plate with colored markers for easy identification of samples during transfer to the reaction plate.

Inhibition of EGF Receptor Autophosphorylation in Membranes of Cells Overexpressing EGF Receptor

In this section we refer specifically to NIH 3T3 cells transfected with EGF receptor. We describe (1) the growth of appropriate cells which possess tyrosine kinase, (2) the preparation of crude membrane extracts, and (3) the inhibition of autophosphorylation by various tyrphostins.

Cell Cultures

DHER 14 cell line: These are NIH 3T3 cells that overexpress EGFR kinase activity.[6] The protocols described are valid for any cell line for any kinase tested.

Frozen cells in an ampoule in liquid nitrogen containing 1×10^6 cells in 2 ml Dulbecco's modified Eagle's medium (DMEM) : 10% calf serum (CS) : 10% DMSO are thawed rapidly at 37° and poured into an 85 cm² flask containing DMEM : 10% CS plus antibiotics. On the following day, when the cells cover about 80% of the flask bottom, the cells are released with trypsin. After the addition of DMEM : CS and low-speed centrifugation, the cell pellet is dispersed with the aid of a micropipettor, the cells are divided into two 175-cm² flasks, and about 100 ml medium is added per flask. When the cells reach about 80% confluence, they are treated as above and divided into five flasks. Then the process is repeated and cells are divided into 20 flasks.

Preparation of Crude Membrane Extracts

1. The 20 flasks are removed from the incubator to a work bench, media are decanted, and the cells are rinsed once with phosphate-buffered saline (PBS) (20 ml/flask) at room temperature. PBS containing 1 mM ethylenediaminetetraacetic acid (EDTA) is added (20 ml/flask), the flasks are placed on a slowly moving platform to aid cell release, and, after about 1 hr, the cells are collected in a 1-liter beaker. Next, 0.5-ml aliquots are taken for counting in a Coulter counter. The total number of cells should

be in the neighborhood of 30–40 $\times$ 10^6 cells. Densely grown cells usually amount to double this number, but may result in a decrease in the total number of receptors. The cells are now collected by centrifugation for 10 min at 200 g, dispersed with the aid of a multipipettor, and pooled into one tube.

2. Cold lysis buffer [10.0 mM N-2-hydroxyethylpiperazine-N'-2-ethanesulfonic acid (HEPES), pH 7.6, 10 μM NaCl, 2 μM EDTA, 1 μM phenylmethylsulfonyl fluoride (PMSF), 10 μg/ml aprotenin, and 10 μg/ml leupeptin] is added to the final cell pellet (1 ml lysis buffer/1 $\times$ 10^6 cells) and the cells are extracted by rapid vortexing. The cell extract is kept at 4°, the temperature of all subsequent steps. The lysate is then homogenized immediately in a Dounce homogenizer using a tightly fitting pestle (10 up-and-down strokes/fill). Nuclei and cell debris are removed by centrifugation for 10 min at 1000 g. The membranes are removed by ultracentrifuging for 30 min at 1 $\times$ 10^5 g. The membrane pellet is suspended in about 1.5 ml HNG buffer (50 mM HEPES, pH 7.6, 125 mM NaCl, and 10% glycerol). The above procedure should yield around 2 mg/ml protein.

Determination of protein concentration can be performed by the method of Peterson[11] or by the DNFB procedure.[12] The method of Lowry et al.[13] is adversely affected by HEPES and cannot be applied. The membrane extract is divided into 20-μl aliquots which are immediately frozen in liquid nitrogen and stored at −75°. The tyrosine kinase activity of membranes thus kept is stable for several months.

Inhibition of Autophosphorylation

Preparation of tyrphostin solutions: Tyrphostin (2–3 mg) is accurately weighed and dissolved in DMSO by vortexing to bring the final inhibitor concentration to 10 mM. If the tyrphostin does not enter solution, it should be sonicated for 5 min. An aliquot of the 10 mM tyrphostin solution is diluted 1 : 10 (v/v) with 50 mM HEPES, pH 7.6, containing 125 mM NaCl (HN) and 50% ethanol by volume. Vortexing, if necessary, will ensure the solubility of the tyrphostin. Further dilutions of the inhibitor are performed with DMSO : ethanol : HN buffer (50 mM HEPES pH 7.6, 125 mM NaCl) (1.0 : 4.5 : 4.5, v/v/v). Since the final dilution into the reaction mixture is sixfold, the final concentrations of ethanol and DMSO will be 7.5 and

[10] J. B. Stanley, R. Gorczynski, C.-K. Huang, J. Love, and G. B. Mills, *J. Immunol.* **145**, 2189 (1990).

[11] G. L. Peterson, *Anal. Biochem.* **83**, 346 (1977).

[12] R. M. Schultz, J. D. Bleil, and P. M. Wasserman, *Anal. Biochem.* **91**, 354 (1978).

[13] O. H. Lowry, N. J. Rosenbrough, A. L. Farr, and R. J. Randall, *J. Biol. Chem.* **193**, 265 (1951).

1.67%, respectively. At these concentrations of solvents about 80% of the EGFR autophosphorylation activity is retained. Independent experiments demonstrate that the parameters characterizing the inhibitors are unchanged as compared to those determined in solutions devoid of ethanol and DMSO. The above solvent mixture enables the solubilization of even the highly hydrophobic tyrphostins. The inhibitors should be used as soon as possible. Hygroscopic tyrphostins should be desiccated overnight *in vacuo* over Drierite.

Assay of autophosphorylation: 1. Make up a reaction cocktail containing 50 mM HEPES, pH 7.6, 125 mM NaCl, 24 mM Mg(Ac)$_2$, 4.0 mM MnCl$_2$, 2 mM NaVO$_3$, 2.0 μM ATP, and [^{32}P]ATP to yield 2–3 $\times$ 10^6 cpm/ 12 μl, i.e., the volume of reaction cocktail needed per assay.

2. An aliquot of membrane extract is diluted with HN buffer (50 mM HEPES, pH 7.6, 125 mM NaCl) which contains EGF (final concentration 200 nM) to yield about 1 μg membrane protein/8 μl (i.e., per assay; see below). Samples are left at 4° for 10 min to allow for receptor activation by EGF.

3. To series of wells are pipetted aliquots of 4 μl inhibitor of varying concentrations (or solvent containing no inhibitor for control) and 12 μl reaction cocktail. The reaction is started by the addition of 8 μl diluted membrane extract in a total volume of 24 μl. To make sure that the receptor is activated by EGF, membrane extracts are diluted with HN buffer, which contains no EGF, and these control samples are run in parallel with the activated samples. After 30 sec at 4°, the reaction is stopped by the addition of 8 μl SDS–PAGE sample buffer (4× concentration). The samples are heated for 5 min at 95° and subjected to electrophoresis in 1.5-mm minigels with 14 wells. The gels are fixed in 40% methanol : 15% acetic acid for 30 min and are washed four times at 45-min intervals with 20% methanol : 7% acetic acid and then overnight in 20% methanol : 7% acetic acid to obtain clear autoradiograms.

4. The gels are dried and laid over Kodak (Rochester, NY) X-OR films for about 2 hr at −70°. Under these conditions, good autoradiograms will be obtained (it is important not to overexpose, as this will result in loss of linearity). The exposed films are then scanned in a densitometer at 525 nm.

Inhibition of EGF Receptor Kinase Using EGF Receptor Immunoprecipitates

In this section we describe the assay of EGF receptor autophosphorylation following immunoprecipitation with monoclonal antibody 108 (Mab 108)[5] and its inhibition by tyrphostins. The results obtained with immunoprecipitates were found to be similar to those obtained with membrane extracts of cells overexpressing the EGFR. Quantities cited are for 20 reaction mixtures.

Binding of MAb 108 to Protein A–Sepharose. Dry protein A–Sepharose (30 mg; Pharmacia, Piscataway, NJ) is washed three times with HTNG [50 mM HEPES, pH 7.6, 125 mM NaCl, 0.1% (w/v) Triton X-100, 10% (v/v) glycerol] on a rotatory shaker at room temperature. MAb 108 (100 μl) is added and rotation is contuinued for an additional 25-min period at 4°. The gel is washed three times with 0.5 ml HTNG and is finally suspended in 200 μl HTNG.

Preparation of Membrane Lysates. To 100 μl crude membrane extract of DHER14 cells (about 250 μg protein) is added 900 μl HTNG containing 1% Triton X-100. After 10 min of end-over-end rotation, the tube is spun for 10 min at 1×10^5 g at 4°. The supernatant containing the solubilized membranes is mixed with 100 μl EGF (10 μM) and is incubated at 10 min at 4°.

Immunoprecipitation. The EGF–membrane extract mixture is added to the MAb 108–protein A– Sepharose and the tube is rotated end over end for 1 hr at 4°. The gel is washed four times with HTNG (0.1% Triton X-100) and is suspended in about 360 μl HTNG to a final volume of around 400 μl.

Inhibition by Tyrphostins. Each assay is performed as follows: Aliquots of 3 μl tyrphostin of varying concentrations (dissolved in 10% DMSO : 45% ethanol : 45% HN buffer, v/v/v) and 10 μl reaction cocktail are added to series of tubes and the autophosphorylation reaction is started by the addition of 17 μl immobilized receptor. After 30 min at 4°, 10 μl SDS–PAGE sample buffer is added to stop the reaction. The following steps are as described in the previous section.

Action of Tyrphostins on Cultured Cells

When tyrphostins are added to tissue culture it takes approximately 20 min to 2 hr for most of them to penetrate cells (Ref. 6, and unpublished experiments). The biochemical half-life of some tyrphostins was measured and found to be around 16 to 24 hr. During that period, some 50% of the compound is degraded. It is therefore recommended that the tissue culture medium be replaced every 24 to 48 hr if one wants to determine the effect of tyrphostins on cells grown in culture. Using this practice we[1,6] and others (Ref. 10 and unpublished personal communications) could examine in detail the effect of tyrphostins on biological systems and correlate their antiproliferative activity with their tyrosine kinase inhibitory activity. The effect of tyrphostins on the EGF-dependent growth of tissue culture cells,[1,8] human and guinea pig keratinocytes,[8] and platelet-derived growth

TABLE II
DIFFERENTIAL ACTIVITY OF TYRPHOSTINS ON VARIOUS TYROSINE KINASES

No.	Structure	EGFR[a]	InsR[b]	PDGFR[c]	p210[bcr-abl d]
		IC_{50} (μM)			
1		35	4000	20–30	75
2		460	ND[e]	0.5	ND
3		2.4	640	2–5	5.9
4		820	ND	20[f]	ND

[a] Measured as inhibitors of poly(GAT) phosphorylation by EGF receptor, described in Ref. 1.

[b] Measured as inhibitors of poly(GAT) phosphorylation by insulin receptor, described in Ref. 1.

[c] Measured as blockers of PDGF-dependent phosphorylation of PLCγ and other intracellular substrates in intact rabbit vascular smooth muscle cells (Ref. 14).

[d] Anafi et al., unpublished results.

[e] ND, Not determined.

[f] Measured as blockers of PDGF-dependent phosphorylation of endogenous substrates in human bone marrow cells (Ref. 15).

factor (PDGF)-stimulated growth of rabbit vascular smooth muscle cells[14] was studied in detail using this protocol.

Differential Activity of Tyrphostins

Tyrphostins possess either broad specificities or are more restricted in specificity, as can be seen from Table II.

[14] G. Bilder, J. A. Krawiec, A. Gazit, C. Gilon, K. McVety, R. Lyall, A. Zilberstein, A. Levitski, M. H. Penire, and A. B. Schreiber, Am. J. Physiol., in press. (1990).

[15] M. C. Bryckaert, A. Eldor, A. Gazit, N. Osherov, C. Gilon, M. Fonteray, A. Levitzki, and G. Tabelem, submitted for publication and unpublished experiments.

[30] Use and Specificity of Genistein as Inhibitor of Protein-Tyrosine Kinases

By Tetsu Akiyama and Hiroshi Ogawara

Genistein [5,7-dihydroxy-3-(4-hydroxyphenyl)-4H-1-benzopyran-4-one; 4′,5,7-trihydroxyisoflavone] (Fig. 1) was isolated as an inhibitor of tyrosine phosphorylation from a culture broth of *Pseudomonas*.[1] This compound has been shown to inhibit the activity of tyrosine kinases such as the epidermal growth factor (EGF) receptor and pp60[src], but scarcely to inhibit the activity of serine and threonine kinases such as cAMP-dependent protein kinase.[2] Therefore, genistein could be a good tool to elucidate cellular processes that are mediated by tyrosine phosphorylation, in particular the molecular cascades involved in cellular proliferation, transformation, and differentiation.

Physical and Chemical Properties of Genistein

Isolation of Genistein. Genistein can be purified from the hydrolysate of soybean meal.[3] We previously purified genistein from a fermentation broth of *Pseudomonas* by chromatography on silica gel and preparative silica gel thin-layer chromatography.[1] *Streptomyces xanthophaeus* is also reported to produce genistein.[4]

Synthesis of Genistein. Genistein can be synthesized from anhydrous phloroglucinol and *p*-methoxyphenylacetonitrile.[5,6] Genistein is commercially available from Funakoshi (Tokyo, Japan) and ICN (Costa Mesa, CA).

Physical Properties. The molecular weight of genistein is 270.23. The ultraviolet absorption spectrum has a peak at 262.5 nm with a molar extinction coefficient of 138. The melting point is 297–298° with decomposition.

---

[1] H. Ogawara, T. Akiyama, J. Ishida, S. Watanabe, and K. Suzuki, *J. Antibiot.* **39**, 606 (1986).

[2] T. Akiyama, J. Ishida, S. Nakagawa, H. Ogawara, S. Watanabe, N. Itoh, M. Shibuya, and Y. Fukami, *J. Biol. Chem.* **262**, 5592 (1987).

[3] A. C. Eldridge, *J. Chromatogr.* **234**, 494 (1982).

[4] T. Hazato, H. Naganawa, M. Kumagai, T. Aoyagi, and H. Umezawa, *J. Antibiot.* **32**, 217 (1979).

[5] R. L. Shriner and C. J. Hull, *J. Org. Chem.* **10**, 288 (1945).

[6] W. Baker, J. Chadderton, J. B. Harborne, and W. D. Ollis, *J. Chem. Soc. (London)* p. 1852 (1953).

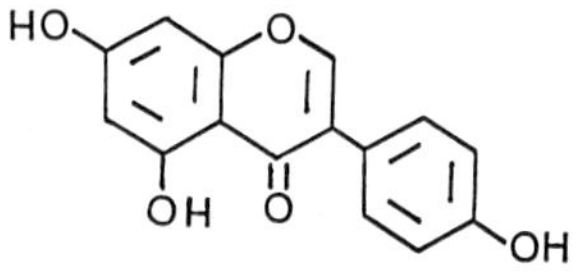

FIG. 1. Structure of genistein.

Chemical Properties. Genistein is soluble in the usual organic solvents and diluted alkali, but poorly soluble in water (less than 200 μg/ml). Alkaline hydrolysis yields phloroglucinol and *p*-hydroxyphenylacetic acid.

Inhibition of Tyrosine Kinase Activity *in Vitro*

Genistein shows inhibitory activity against tyrosine kinases such as the EGF receptor, pp60^{src}, and pp110$^{gag-fes}$ *in vitro*.[2] When the inhibitory activity against the autophosphorylation activity of the EGF receptor is examined by using an A431 cell membrane fraction, genistein exhibits the half-maximal effect at 2.6 μM. The half-maximal effect against phosphorylation of an exogenous substrate protein, histone H2B, is 20.4 μM, which is considerably higher than that observed for autophosphorylation. Phosphorylation of IgG and casein by pp60^{src} is inhibited by genistein at the half-maximal doses of 29.6 and 25.9 μM, respectively. The half-maximal effect against autophosphorylation of pp110$^{gag-fes}$ is 24.1 μM (Table I).

By contrast, genistein exhibits a weak inhibitory effect against protein-serine/threonine kinases (Table I). Genistein scarcely inhibits cAMP-dependent protein kinase activity, even at 370 μM. Activity of protein kinase C is inhibited approximately 40% by 370 μM genistein.

Mechanism of Inhibition

Genistein is a competitive inhibitor with respect to ATP (K_i 13.7 μM). Since the reaction mechanism of the EGF receptor kinase is a sequential ordered Bi–Bi reaction with a peptide as the first substrate and ATP as the second, genistein could be expected to act as an uncompetitive inhibitor with respect to a phosphate acceptor, i.e., genistein could bind to the enzyme only after a phosphate acceptor combines. However, genistein is a noncompetitive inhibitor with respect to a phosphate acceptor. Because genistein bears no structural relationship to ATP, inhibition of the EGF receptor kinase activity by genistein may not be due to true competition for exactly the same site as that utilized by ATP. Thus it would be possible that genistein binds to the EGF receptor in multiple places in the reaction

TABLE I
EFFECT OF GENISTEIN ON ENZYME ACTIVITIES[a]

Enzymes	IC$_{50}$ (μM)	
	Genistein	Quercetin
EGF receptor	22.2[b]	26.5[b]
pp60$^{v\text{-}src}$	25.9[b]	26.5[b]
	29.6[c]	—
pp110$^{gag\text{-}fes}$	24.1[d]	—
cAMP-dependent protein kinase	>370	>331
Protein kinase C	>370	82.8
Phosphorylase kinase	>370	16.6
5'-Nucleotidase	>370	99.3
Phosphodiesterase	>370	—

[a] Enzyme activities were measured at 30° for 3 min as described in the text in the absence or presence of various concentrations of genistein. The concentration for 50% inhibition was determined from the inhibition curve, taking probit of inhibition percentage at varied concentrations of genistein on the ordinate and the logarithm of concentrations of genistein on the abscissa. From Akiyama *et al.*[2]
[b] Purified EGF receptor (0.1 μg) and pp60$^{v\text{-}src}$ (0.05 μg) were used to assay the phosphorylation of exogenous substrates (histone H2B) (25 μg) and casein (25 μg), respectively.
[c] Phosphorylation of bound IgG by immunoprecipitated pp60$^{v\text{-}src}$ was assayed.
[d] Autophosphorylation of immunoprecipitated p110$^{gag\text{-}fes}$ was assayed.

pathway and consequently appears noncompetitive with respect to a phosphate acceptor.

Phosphorylation Assays

The tyrosine kinase reaction is performed in a final volume of 50 μl containing 20 mM HEPES–NaOH, pH 7.2, 10 mM MgCl$_2$, 3 mM MnCl$_2$, 1 mM dithiothreitol (DTT), 100 mM sodium vanadate, 10 μM [γ-^{32}P]ATP (4 mCi/μmol), an enzyme, and genistein, which is dissolved in dimethyl sulfoxide (DMSO). A membrane fraction, an immunoprecipitated enzyme, or a purified enzyme can be used as a kinase source. When the immunoprecipitated kinase or purified growth factor receptor is used as a kinase source, 0.05% Triton X-100 or Nonidet P-40 (NP-40) is included in the reaction mixture. For the assay of growth factor receptor-associated tyrosine kinase activity, the reaction mixture is preincubated for 10–30 min in the presence or absence of the corresponding ligand. When the kinase activity against exogenous substrates is assayed, histone H2B (25 μg), casein, enolase, or src peptide (final concentration, 2 mM), a tyrosine-containing peptide closely resembling the autophosphorylation site in pp60^{src}, is included in the reaction mixture. The reaction is continued for 1–10 min at 4° and terminated by the addition of Laemmli sodium dodecyl sulfate (SDS) sample buffer. The samples are analyzed by SDS–polyacrylamide gel electrophoresis followed by autoradiography. The bands of the autophosphorylated tyrosine kinase or the exogenous substrate are excised from the gels and the radioactivity is counted with a liquid scintillation counter. Phosphorylation of the src peptide is assayed by binding the phosphopeptide to phosphocellulose paper as described previously.[7]

Inhibition of Tyrosine Phosphorylation in Vivo

Genistein inhibits tyrosine phosphorylation not only in vitro but also in intact cells.[2] When genistein is added to the culture of A431 cells, it inhibits the EGF-stimulated increase in cellular phosphotyrosine level as well as the increase in the phosphorylation of the EGF receptor. The half-maximal effect on the phosphorylation of the EGF receptor in vivo is observed at 111 μM. It should be noted that genistein inhibits not only tyrosine phosphorylation of the EGF receptor but also EGF-stimulated serine and threonine phosphorylation of the receptor. This may result from direct inhibition of serine and threonine kinase activity, which is responsible for phosphorylation of the EGF receptor. Alternatively, inhibition of the EGF receptor kinase activity may block a putative cellular pathway regulating serine and threonine kinase activity that phosphorylates the EGF receptor.

Assay Methods

A431 cells are labeled for 6 hr in phosphate-free Dulbecco's modified Eagle's medium supplemented with 2–4% (v/v) fetal calf serum and 1 mCi/ml of [^{32}P]phosphate, incubated with genistein (dissolved in dimethyl sulfoxide) for 30 min, and then incubated with EGF (100 μg/ml) for the last 3–15 min. The labeled cells are lysed in RIPA buffer (40 mM HEPES–NaOH, pH 7.2, 1% NP-40 or Triton X-100, 0.5% sodium deoxycholate, 150 mM NaCl, 10 mM sodium pyrophosphate, 100 mM sodium fluoride, 4 mM EDTA, 2 mM sodium orthovanadate, 1 mM phenylmethylsulfonyl fluoride, 100 units (U)/ml aprotinin, and 5 μg/ml leupeptin), and the lysate is incubated with anti-EGF receptor antibodies. The immunocomplexes are adsorbed to protein A-Sepharose 4B and washed exten-

[7] L. J. Pike, B. Gallis, J. E. Casnellie, P. Bornstein, and E. G. Krebs, Proc. Natl. Acad. Sci. U.S.A. 79, 1443 (1982).

sively with RIPA buffer. The immunoprecipitates are analyzed on SDS–polyacrylamide gel electrophoresis followed by autoradiography.

The phosphorylated EGF receptor separated by SDS gel electrophoresis is eluted from the gel and subjected to acid hydrolysis in 6 N HCl for 1.5 hr at 110°. The phosphoamino acids are resolved by two-dimensional electrophoresis at pH 1.9 followed by electrophoresis at pH 3.5.[8]

For the estimation of the level of phosphotyrosine in total cellular protein, cell protein is extracted with phenol from the detergent lysate of [^{32}P]phosphate-labeled A431 cells and then precipitated with 20% trichloroacetic acid (TCA). The cell protein thus obtained is acid hydrolyzed and subjected to two-dimensional thin-layer electrophoresis.[8]

Inhibition of Phosphatidylinositol Turnover

Several growth factors, including EGF and platelet-derived growth factor (PDGF), are known to stimulate phosphatidylinositol turnover in certain cells.[9] More recently it has been shown that EGF and PDGF induce tyrosine phosphorylation of phospholipase C-γ.[10–12] This finding suggests the possibility that tyrosine phosphorylation of phospholipase C-γ by EGF and PDGF receptors leads to its activation, and a consequent increase in phosphatidyl turnover.[12] Consistent with this concept, genistein shows inhibitory activity against EGF- or PDGF-induced hydrolysis of phosphatidylinositol 4,5-bisphosphate (PIP$_2$) and the generation of inositol 1,4,5-trisphosphate (IP$_3$) and diacylglycerol in A431 or BALB 3T3 cells, respectively.[13] However, genistein also inhibits bombesin-induced phosphatidylinositol turnover. Since bombesin does not induce tyrosine phosphorylation of phospholipase C-γ,[10] genistein may also exert its effect by inhibiting other steps in the cascade leading to stimulation of phosphatidyl turnover. This notion is further supported by the fact[13] that genistein inhibits phosphatidylinositol turnover induced by the treatment of cells with AlF$_4$, a potent G protein activator. Genistein also inhibits incorporation of phosphate into phosphatidylinositol, phosphatidylinositol 4-monophosphate, phosphatidylinositol 4,5-bisphosphate, phosphatidylcholine,

[8] J. A. Cooper, B. M. Sefton, and T. Hunter, this series, Vol. 99, p. 387.
[9] M. J. Berridge, *Annu. Rev. Biochem.* **54**, 205 (1985).
[10] M. I. Wahl, S. Nishibe, P.-G. Suh, S. G. Rhee, and G. Carpenter, *Proc. Natl. Acad. Sci. U.S.A.* **86**, 1568 (1989).
[11] B. Margolis, S. G. Rhee, S. Felder, M. Mervic, R. Lyall, A. Levitzki, A. Ullrich, A. Zilberstein, and J. Schlessinger, *Cell (Cambridge, Mass.)* **57**, 1101 (1989).
[12] J. Meisenhelder, P.-S. Suh, S. G. Rhee, and T. Hunter, *Cell (Cambridge, Mass.)* **57**, 1109 (1989).
[13] T. Akiyama and H. Ogawara, unpublished observation.

phosphatidylethanolamine, and phosphatidic acid *in vivo*. Genistein does not show inhibitory activity against phospholipase C, phosphatidylinositol kinase, and diacylglycerol kinase *in vitro*.

Genistein has been shown to prevent T cell receptor-CD3-mediated phospholipase C activation, interleukin 2 (IL-2) receptor expression, T cell proliferation at doses that inhibit the activity of total membrane-bound T cell tyrosine kinases and pp56*lck* (ID_{50} 111 μM), and tyrosine phosphorylation of the T cell receptor ζ subunit.[14] Another tyrosine kinase inhibitor, herbimycin A, whose structure is not related to genistein, is also reported to inhibit T cell receptor-mediated phosphoinositide hydrolysis.[15] These findings suggest that tyrosine phosphorylation is required for phospholipase C activation and T cell receptor-mediated signal transduction. In this context, taking advantage of the reversibility of the genistein action, Hill *et al*.[16] studied the exact relationship between PDGF-induced activation of phospholipase C and induction of DNA synthesis in C3H/10T1/2 mouse fibroblasts. Genistein (100–200 μM) inhibits PDGF-induced DNA synthesis as well as PDGF-induced activation of phospholipase C and subsequent accumulation of inositol phosphate, increase in intracellular Ca^{2+}, and protein kinase C-dependent 80-kDa protein phosphorylation. However, removal of genistein results in DNA synthesis without the occurrence of phospholipase C activation. From these findings, Hill *et al*. concluded that PDGF-induced activation of phospholipase C is not necessary for induction of DNA synthesis.

Effect of Genistein on Growth of Cultured Cells

Genistein exhibits inhibitory activity against proliferation of cultured cells. The half-maximal effect of genistein against the growth of Rous sarcoma virus-transformed rat 3Y1 cells is observed at about 25.9 μM.[17] Okura *et al*.[18] reported that genistein selectively inhibits the growth of [Val^{12}]H-*ras*-transformed cells but not normal NIH 3T3 cells at 37.0 μM when it is present in the culture longer than 10 days. Linassier *et al*.[19]

[14] T. Mustelin, K. M. Coggeshall, N. Isakov, and A. Altman, *Science* **247**, 1584 (1990).

[15] C. H. June, M. C. Fletcher, J. A. Ledbetter, G. L. Schieven, J. N. Siegel, A. F. Phillips, and L. E. Samelson, *Proc. Natl. Acad. Sci. U.S.A.* **87**, 7722 (1990).

[16] T. D. Hill, N. M. Dean, L. J. Mordan, A. F. Lau, M. Y. Kanemitsu, and A. L. Boynton, *Science* **248**, 1660 (1990).

[17] H. Ogawara, T. Akiyama, S. Watanabe, N. Ito, M. Kobori, and Y. Seoda, *J. Antibiot.* **42**, 340 (1989).

[18] A. Okura, H. Arakawa, H. Oka, T. Yoshinari, and Y. Monden, *Biochem. Biophys. Res. Commun.* **157**, 183 (1988).

[19] C. Linassier, M. Pierre, J.-B. L. Pecq, and J. Pierre, *Biochem. Pharmacol.* **39**, 187 (1990).

showed that genistein exhibits a cytostatic effect against NIH 3T3 cells at low concentrations (less than 37.0 μM) and a cytotoxic effect at higher concentrations (more than 37.0 μM). They also reported that genistein blocks the mitogenic effect mediated by EGF on NIH 3T3 cells (ID_{50} 12.2 μM), by insulin (ID_{50} 19.6 μM), or by thrombin (ID_{50} 20.4 μM).

Induction of Cell Differentiation

Genistein is able to induce differentiation of several cell lines. When genistein (3.7–11.1 μM) is added to the culture of human neuroblastoma SK-N-DZ cells, morphology of this cell line changes to that of neuron-like cells and enhancement of synthesis of neurofilament is induced.[13] Genistein is reported to induce terminal erythroid differentiation of mouse erythroleukemia (MEL) cells in a synergistic manner with an agent that blocks DNA replication, such as mitomycin C (MMC).[20] According to Watanabe et al.,[20] maximum erythroid induction (more than 70% induction) is obtained with 3.7 μM genistein and 0.3 μM MMC in the culture medium, while erythroid induction by genistein alone requires much higher genistein concentrations (more than 37.0 μM). These findings suggest that the inhibition of tyrosine phosphorylation of cellular proteins is involved in cell differentiation. This notion is further supported by the fact that other inhibitors of tyrosine kinases, herbimycin A and ST638 (α-cyano-3-ethoxy-4-hydroxy-5-phenylthiomethylcinnamamide), whose structure is not related to genistein, also possess differentiation-inducing activity.[13,21]

On the other hand, genistein is reported to inhibit keratinocyte differentiation[22]; treatment of primary mouse keratinocytes with genistein (111–222 μM) prevents induction of tyrosine phosphorylation by Ca^{2+} and TPA and interferes with the differentiative effects of these agents. These findings suggest that tyrosine phosphorylation plays a crucial role in keratinocyte differentiation.

Other Activities

Genistein is reported to inhibit the activity of several enzymes other than tyrosine kinases. Hazato et al.[4] reported that genistein inhibits the activity of β-galactosidase. Okura et al.[18] showed that genistein inhibits the activity of DNA topoisomerases II and I; genistein increases the enzyme–DNA complex in L1210 cells at 3.7 μM and interferes with

[20] T. Watanabe, T. Shiraishi, H. Sasaki, and M. Oishi, *Exp. Cell Res.* **183**, 335 (1989).
[21] K. Kondo, T. Watanabe, H. Sasaki, Y. Uehara, and M. Oishi, *J. Cell Biol.* **109**, 285 (1989).
[22] E. Filvaroff, D. F. Stern, and G. P. Dotto, *Mol. Cell. Biol.* **10**, 1164 (1990).

TABLE II
EFFECT OF FLAVONOIDS ON PROTEIN-TYROSINE KINASE ACTIVITY OF EGF RECEPTOR[a]

| Compound | Position | | | | | IC$_{50}$ (μg/ml) | |
	2	5	7	3'	4'	PKI	RSV-3Y1
Genistein	—	OH	OH	—	OH	0.7	7.0
Prunetin	—	OH	OCH$_3$	—	OH	4.2	25.0
Daidzein	—	—	OH	—	OH	>100	25.0
Biochanin A	—	OH	OH	—	OCH$_3$	26.0	18.0
Genistin	—	OH	Glucose	—	OH	>100	>100

| Compound | Position | | | | | IC$_{50}$ (μg/ml) | |
	2	5	7	3'	4'	PKI	RSV-3Y1
Apigenin	—	OH	OH	—	OH	25.0	11.0
Acacetin	—	OH	OH	—	OCH$_3$	40.0	24.0
Flavone	—	—	—	—	—	50.0	7.0
Kaempferol	OH	OH	OH	—	OH	3.2	14.0
Quercetin	OH	OH	OH	OH	OH	5.0	12.0

[a] Protein kinase inhibition (PKI). RSV, Rous sarcoma virus.

pBR322 DNA relaxation by the enzyme. Genistein is also reported to inhibit S6 kinase activity.[19] Interestingly, isoflavones, including genistein, show estrogenic activity in mammals and birds.[23]

Activity of Other Isoflavones and Related Compounds

Several genistein-related compounds also exhibit an inhibitory effect against the activity of tyrosine kinases.[2] In Table II, the effects of other isoflavones and flavones on EGF receptor kinase activity are shown.

[23] R. B. Bradbury and D. E. White, *J. Chem. Soc.* (*London*) p. 3447 (1951).

Compounds listed in Table II are commercially available; acacetin, prunetin, genistin, and daidzein can be obtained from Funakoshi. Biochanin A can be purchased from Aldrich (Milwaukee, WI), flavone, kaempferol, and apigenin from Sigma (St. Louis, MO), and quercetin from Nakarai (Japan).

Among isoflavones, prunetin, which has a methyl group in position 7, exhibits rather strong activity, while biochanin A, which has a methoxy group at the 4' position, shows lower inhibitory activity. Daidzein, which has no hydroxyl group in position 5, does not show any inhibitory activity at 393 μM. Genistin, which has a glucose residue at position 7, also does not show inhibitory activity. Flavones such as apigenin, acacetin, and flavone exhibit only low activity. Flavonols such as kaempferol and quercetin are highly active but less specific than genistein for tyrosine kinases. Thus a hydroxyl group at position 5 seems to be essential for isoflavones and flavones to express inhibitory activity. Hydroxyl groups at positions 7 and 4' are also necessary for full expression of the activity.

The effect of these compounds on the growth of Rous sarcoma virus (RSV)-transformed cells is also shown in Table II.[17] With the exception of genistin, all the isoflavonoids and flavonoids listed in Table II exhibit a fairly inhibitory activity on the proliferation of RSV-3Y1 cells.

[31] Use and Selectivity of Herbimycin A as Inhibitor of Protein-Tyrosine Kinases

By YOSHIMASA UEHARA and HIDESUKE FUKAZAWA

Protein-tyrosine phosphorylation is one of the basic mechanisms of signal transduction for cell growth and differentiation. Specific inhibitors of protein tyrosine kinases, therefore, may provide useful means to examine the role of tyrosine phosphorylation in a variety of cellular events. In a search for agents that reverse Rous sarcoma virus (RSV) transformation, we found that herbimycin A induced inactivation of v-src tyrosine kinase and reduced cellular phosphotyrosine content in RSV-transformed cells. Because herbimycin A was also effective in reversing transformation by other tyrosine kinase oncogenes, the use of this antibiotic should be valuable in examining the role of tyrosine phosphorylation in cell transformation, growth, and differentiation. Here, we describe the use and selectivity of herbimycin A as an inhibitor of protein-tyrosine kinases.

Herbimycin A as Inhibitor of src Oncogene Action

A fermentation broth of *Streptomyces* (MH237-CF8) isolated at the Institute of Microbial Chemistry was found to reverse RSV transformation,[1] and the active compound was identified as the benzenoid ansamycin antibiotic herbimycin A, which had previously been isolated for its herbicidal activity.[2] Among benzenoid ansamycins, macbecin and geldanamycin were also active, but ansamitocin, an inhibitor of microtubules, and naphthalenoid ansamycins, such as streptovaricin and rifamycins (which inhibit reverse transcriptase), did not reverse *src* transformation.[3] These studies indicated that the newly found activity might be unique to ansamycins that contain a benzoquinone moiety. The possible involvement of this moiety for its interaction with src kinase will be discussed later.

The site of action of herbimycin A was thought to be the *src* gene product, $p60^{v-src}$, or to be near the very upstream site of the process in which the transduction of the tyrosine phosphorylation signal occurs for transformation. Herbimycin A reversed various transformed phenotypes, including rounded cell morphology, increased glucose uptake, lowered serum requirement, and anchorage independence for cell growth of rat NRK cells expressing v-src oncogene.[4] The immune complex prepared by mixing herbimycin-treated cell extracts with antibody against $p60^{src}$ was found to have reduced phosphorylation activity on the complex itself or on casein as an exogenous substrate *in vitro*.[3,5] Because the reduced kinase activity was not due to a decrease in the amount of $p60^{v-src}$ protein, it appeared that herbimycin A induced inactivation of $p60^{v-src}$, thereby reversing various transformed phenotypes.

Screening Method

A rat kidney cell line infected with *ts25*, a T-class mutant of RSV Prague strain (src^{ts}/NRK) isolated by Chen et al.,[6] was obtained from Dr. M. Yoshida (University of Tokyo, Japan) and used to screen active

[1] Y. Uehara, M. Hori, T. Takeuchi, and H. Umezawa, *Jpn. J. Cancer. Res.* **76,** 672 (1985).
[2] S. Ōmura, Y. Iwai, Y. Takahashi, N. Sadakane, A. Nakagawa, H. Ōiwa, Y. Hasegawa, and T. Ikai, *J. Antibiot.* **32,** 255 (1979).
[3] Y. Uehara, M. Hori, T. Takeuchi, and H. Umezawa, *Mol. Cell. Biol.* **6,** 2198 (1986).
[4] Y. Murakami, S. Mizuno, M. Hori, and Y. Uehara, *Cancer Res.* **48,** 1587 (1988).
[5] Y. Uehara, Y. Murakami, Y. Sugimoto, and S. Mizuno, *Cancer Res.* **49,** 780 (1989).
[6] Y. C. Chen, M. J. Hayman, and P. K. Vogt, *Cell (Cambridge, Mass.)* **11,** 513 (1977).

compounds that reversed transformed morphology. When src^{ts}/NRK cells are grown at a permissive temperature (33°) for transformation, they are small, densely packed, and spindle shaped. At a nonpermissive temperature (39°), however, the cells flatten out and are normal (Fig. 1). These morphological changes are reversible in either direction within 16 to 20 hr after the temperature shift. When a fermentation broth of *Streptomyces* MH237-CF8 (0.5%, v/v) is added to a 33° culture, the transformed cell morphology changes to one indistinguishable from that of the cells grown at 39°. The active compound in the fermentation broth was purified and identified as herbimycin A (Fig. 1). When herbimycin A was added to a 33° culture, morphological reversion occurred within 16 to 20 hr and lasted as long as several days at concentrations ranging from 0.1 to 1.0 μg/ml. At these herbimycin A concentrations cell growth was inhibited from 10 to 90%. Removal of the antibiotic allowed the cells to revert to the transformed morphology within 24 hr, indicating that the action of herbimycin A to the cells is reversible. At 0.5 μg herbimycin A/ml, all cells were converted to the normal morphology (Fig. 1) and the growth rate was similar to that of cells grown at 39°, suggesting that the antibiotic action is selective for events that are involved in transformation. Other inhibitors, such as quercetin, genistein, erbstatin, and staurosporine, did not exert as much morphological reversion activity as herbimycin A in src^{ts}/NRK cells.

Method for Use of Herbimycin A

Because of the restricted water solubility of herbimycin A, preparation of concentrated stock solutions can best be done in an organic solvent such as dimethyl sulfoxide (DMSO) or methanol. We generally store herbimycin A as a 1 mg/ml solution in DMSO, dilute it with the culture medium just prior to use, and add a small volume of the solution to the medium to observe its effect on cells. Stock solutions of herbimycin A in DMSO are relatively stable, and can be stored at −20° for several months. This compound reverses the morphology of many, if not all, cells transformed by *src* family oncogenes, and also exhibits various effects on cellular events involving cytoplasmic protein-tyrosine kinases (see below). The concentration required to reverse morphology varies among cells. The beginning of morphological changes can usually be observed within several hours, and is preceded by a reduction in tyrosine kinase activity.

Inactivation of p60^{v-src} Kinase by Herbimycin A *in Vitro*

Initial *in vitro* studies did not show any direct effect of herbimycin A on p60^{v-src}, and the antibiotic was thought to reduce the kinase activity in cells through an indirect mechanism. We later found that herbimycin A

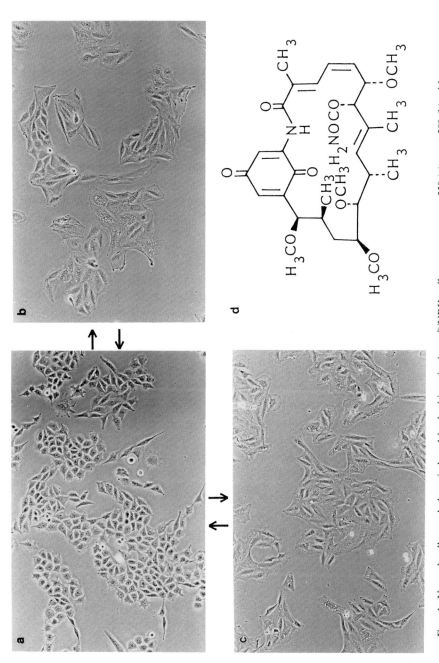

FIG. 1. Normal cell morphology induced by herbimycin A. src^{ts}/NRK cells were grown at 33° (a) or at 39° (b) without any test materials. Cells were grown at 33° with 0.5 µg/ml of herbimycin A for 24 hr (c). (d) Structure of herbimycin A.

was inactivated by sulfhydryl compounds that were included in the *in vitro* kinase reaction mixture, and demonstrated that, in the absence of such compounds, herbimycin A irreversibly inhibited p60$^{v\text{-}src}$, possibly by binding to reactive SH group(s) of the kinase. The method by which we observed the direct effect of herbimycin A on p60$^{v\text{-}src}$ kinase *in vitro*[7] is described in detail.

Method

Rous sarcoma virus (RSV)-infected cells are collected, washed, and lysed with 20 mM HEPES, pH 7.4, 1 mM EDTA, 0.1 mM Na$_3$VO$_4$, 150 mM NaCl, and 1% (v/v) Triton X-100, containing 25 μg/ml each of protease inhibitors phenylmethylsulfonyl fluoride (PMSF), antipain, leupeptin, and pepstatin A, in a microcentrifuge tube. After 10 min on ice, the lysate is centrifuged at 15,000 g for 30 min at 4°. The supernatant is collected and the protein concentration adjusted to 1 mg/ml. Then 1 ml of the extract is incubated with 10 μl of monoclonal antibody 327 (purchased from Oncogene Science, Inc., Manhasset, NY) for 1 hr on ice. The antigen–antibody complex is collected in tubes onto formalin-fixed *Staphylococcus aureus* (200 μl of 10% suspension) with rabbit anti-mouse IgG as a second antibody. After a 1-hr incubation on ice, the immune complex is pelleted by centrifugation at 15,000 g for 30 sec at 4°, washed three times with the lysis buffer without protease inhibitors, twice the STE buffer (150 mM NaCl, 10 mM Tris hydrochloride, pH 7.2, 1 mM EDTA), suspended in 900 μl of 20 mM HEPES, pH 7.4, and then divided into 90-μl aliquots. Ten microliters of herbimycin A, dissolved in DMSO at various concentrations, is added to the aliquots and the mixtures are incubated at 25° for 30 min. The immune complex is pelleted, washed with 20 mM HEPES, pH 7.4, and then assayed for p60$^{v\text{-}src}$ activity. The assay mixture in 20 μl contains 20 mM HEPES, pH 7.4, 5 mM MgCl$_2$, 1 mM dithiothreitol (DTT), and 10 μM [γ-^{32}P]ATP (10 μCi) either in the absence or presence of 0.5 mg/ml α-casein. After incubation at 25° for 20 min, the reaction is stopped by the addition of 10 μl of 3× concentrated electrophoresis sample buffer. The radiolabeled proteins are analyzed using sodium dodecyl sulfate (SDS)-polyacrylamide gel electrophoresis and visualized by autoradiography.

Comments

The IC$_{50}$ under the condition described above is about 5 μg/ml. At similar concentrations, herbimycin A also inactivates other cytoplasmic tyrosine kinases such as p120$^{v\text{-}abl}$, p130$^{v\text{-}fps}$, p210$^{bcr\text{-}abl}$, and an activated

[7] Y. Uehara, H. Fukazawa, Y. Murakami, and S. Mizuno, *Biochem. Biophys. Res. Commun.* **163,** 803 (1989).

form of p60$^{c\text{-}src}$ of human colon cancer cells. Irreversible inhibition of p60$^{v\text{-}src}$ kinase by herbimycin A does not occur if an immune complex prepared with monoclonal antibody 327 is incubated in an ice bath. The degree of inhibition is dependent on the temperature and time of treatment. The inactivation reaction proceeds at a faster rate at 37° than at 25°, but incubation at higher temperature alone results in considerable loss of p60$^{v\text{-}src}$ kinase activity. If the immune complex is prepared with tumor-bearing rabbit (TBR) serum, herbimycin A inactivates p60$^{v\text{-}src}$ even at 0°. Optimum incubation conditions for inactivation should be determined for each antibody.

Herbimycin A is readily inactivated by sulfhydryl compounds such as 2-mercaptoethanol, dithiothreitol, cysteine, and the reduced form of glutathione. Contact with such compounds abolishes the ability of herbimycin A to block p60$^{v\text{-}src}$ kinase *in vitro* as well as the ability to reverse morphology of RSV-transformed cells. It should be noted that many reagents, [^{35}S]methionine for example, contain sulfhydryl compounds as stabilizers. The inactivation probably occurs through conjugation between highly polarized double bonds in the benzoquinone moiety of herbimycin A and the highly reactive SH group of thiols.[7] We speculate that the antibiotic binds irreversibly to SH group(s) of p60$^{v\text{-}src}$ in the same manner.

The existence of an SH group in p60$^{v\text{-}src}$ essential for its kinase activity is yet to be elucidated. We examined effects of various SH reagents in the *in vitro* kinase assay, and observed significant variation in the degree of p60$^{v\text{-}src}$ inactivation. Iodoacetamide, for example, did not reduce p60$^{v\text{-}src}$ kinase activity even at a concentration as high as 100 mM. However, pretreatment of p60$^{v\text{-}src}$ immune complex with iodoacetamide protected the kinase from inactivation by herbimycin A and other p60$^{v\text{-}src}$-inactivating SH reagents.[7a] We speculate from these results that the active SH group(s) to which herbimycin binds is not essential for kinase activity, but is positioned in the vicinity of the active center. Binding of bulky compounds to such SH group(s) should lower the kinase activity, possibly by inhibiting the access of ATP molecule.

Selectivity of Herbimycin A

A number of oncogenes encode protein-tyrosine kinases whose catalytic domains show striking homology, and we found it interesting to test whether herbimycin A was inhibitory against other tyrosine kinase oncogene products. We examined the effectiveness of the antibiotic in reversing the morphology of chicken and mammalian cells transformed

---

[7a] H. Fukazawa, S. Mizuno, and Y. Uehara, *Biochem. Biophys. Res. Commun.* **173**, 276 (1990).

by various oncogenes and found that it was effective against the cells transformed by tyrosine kinase oncogenes *src, yes, fps, ros, abl,* and *erbB,* but that it did not reverse the transformed morphologies induced by oncogenes *ras, myc,* and the serine/threonine kinase *raf.*[8] The reverse transformation was accompanied by significant decreases in phosphotyrosine content of the total cellular proteins and p36 protein, which is one of the cellular targets of tyrosine kinase oncogene products. The results provided supporting evidence of the selective inhibition by herbimycin A of tyrosine kinase oncogene products, although an examination of the effect on these kinases *in vitro* will be necessary for conclusive results.

Another interesting question with regard to selectivity is whether herbimycin A is effective against other types of protein kinases, such as cyclic AMP-dependent kinase and protein kinase C. Herbimycin A showed no inhibitory effect on the endogenous protein phosphorylation that was stimulated by the addition of cAMP *in vitro* (unpublished results). In agreement with this, the serine at position 17 of p60^{v-src}, which is phosphorylated by cAMP-dependent protein kinase,[9] was not reduced as compared with the tyrosine-416 autophosphorylation in herbimycin-treated cells, suggesting that herbimycin A does not inhibit the protein kinase A activity, at least the one that is involved in the phosphorylation of p60^{v-src}.

Herbimycin A also showed no inhibitory effect on the TPA (12-*O*-tetradecanoyl-13-acetate) stimulated phosphorylation of the heat-stable p80 protein in an *in vitro* kinase assay (unpublished results), suggesting that herbimycin A does not inhibit protein kinase C. The effect of the antibiotic on the activity of protein kinase C was also estimated, although indirectly, using HL-60 cells after the treatment of retinoic acid. In these differentiation-induced cells, increased generation of superoxide anion mediated by NADPH oxidase could be observed by treatment with TPA, which activates protein kinase C. Herbimycin A showed no inhibitory effect on the TPA-induced superoxide anion generation, suggesting that protein kinase C, at least the one that is involved in the activation of NADPH oxidase, was not likely to be inhibited by herbimycin A (H. Henmi, personal communication).

Use of Herbimycin A as Inhibitor of Tyrosine Kinases

Since herbimycin A appears to exhibit specific inhibitory activity against tyrosine kinases in cells, the use of this antibiotic may be beneficial in determining whether tyrosine phosphorylation is involved in the mechanisms for cell transformation, growth, and differentiation. For example,

[8] Y. Uehara, Y. Murakami, S. Mizuno, and S. Kawai, *Virology* **164**, 294 (1988).
[9] M. S. Collett, E. Erikson, and R. L. Erikson, *J. Virol.* **29**, 770 (1979).

herbimycin A was used to determine whether the cell transformation is dependent on the functional expression of p60$^{v\text{-}src}$ tyrosine kinase. Young et al.[10] isolated a host–range mutant of RSV that induced a temperature-dependent transformation in Rat-3 cells, but not in chicken cells: it transforms rat cells at 34° but not at 39.5°, while chicken cells are transformed at both 36° and 41.5°. They isolated retransformants that grew in soft agar suspension at 39.5° from the mutant RSV-infected rat cells. These retransformants showed an unchanged level of p60$^{v\text{-}src}$ at 39.5°, and the rescued virus induced a temperature-dependent transformation when reintroduced into rat cells. They treated the retransformants with herbimycin A and observed that the cells reverted to normal, indicating that the transformation was dependent on continued exprerssion of p60$^{v\text{-}src}$. Thus, they concluded that the retransformants were pseudorevertants and that the temperature dependence of transformation by the mutant RSV is affected by the host cell environment. The use of herbimycin A and the host-dependent mutants of RSV may provide information regarding host factors that are involved in transformation.

Honma et al. examined the effect of herbimycin A on the differentiation of the human chronic myelogenous leukemia cell line K562,[11] which is known to express a structurally altered c-abl proto-oncogene product with increased tyrosine kinase activity. They found that herbimycin A induced erythroid differentiation concomitant with rapid reduction of tyrosine phosphorylation of c-abl gene product, suggesting a possibility that protein-tyrosine phosphorylation may influence the differentiation of K562 cells. With respect to inhibition of cell proliferation, they also found that K562 cells were the most sensitive to herbimycin A among the several human cell lines they tested, suggesting the involvement of tyrosine kinase activity in the growth control mechanism.

Kondo et al.[12] found that herbimycin A was an effective inducing agent capable of triggering differentiation in two typical mouse in vitro differentiation systems that have been considered to be quite different in their mechanism of induction: endoderm differentiation of embryonal carcinoma (F9) cells and terminal erythroid differentiation of erythroleukemia (MEL) cells. MEL cells are induced to differentiate into erythroid cells by a number of compounds, including DMSO, hexamethylenebisacetamide (HMBA), and butyric acid, but none of these compounds induces F9 differentiation. In contrast, retinoic acid, which induces F9 cells, has

[10] J. C. Young, E. Liebel, and G. S. Martin, Virology 166, 561 (1988).
[11] Y. Honma, J. Okabe-Kado, M. Hozumi, Y. Uehara, and S. Mizuno, Cancer Res. 49, 331 (1989).
[12] K. Kondo, T. Watanabe, H. Sasaki, Y. Uehara, and M. Oishi, J. Cell Biol. 109, 285 (1989).

no effect on MEL cell differentiation. Thus, they suggest that there is a common step that is sensitive to herbimycin A in the intracellular differentiation cascade, directly or indirectly associated with phosphorylation at specific tyrosine residues of cellular proteins.

Herbimycin A has been found to inhibit angiogenesis in chick embryo chorioallantoic membrane and to suppress capillary growth induced by tumor angiogenesis factor in rabbit cornea, and was recognized as a potent inhibitor of angiogenesis.[13,14] In an *in vitro* model of angiogenesis, in which capillary endothelial cells can be induced, it was observed that sodium vanadate, an inhibitor of phosphotyrosine phosphatase, was an effective inducer, suggesting the possible involvement of tyrosine phosphorylation.[15] Thus, it might be speculated that herbimycin A induced antiangiogenic activity by selectively reducing the activity of certain tyrosine kinases, but precise mechanisms remain to be elucidated. Although we have described examples of the usefulness of herbimycin A in a study of the possible involvement of protein-tyrosine kinases in a variety of cells, few, if any, inhibitors will block one enzyme without affecting others. Therefore, it should be added that cellular events caused by the "specific" inhibitor herbimycin A should be cautiously interpreted.

Although oncogenes are classified into several groups according to structural and functional similarities, the tyrosine-specific protein kinase activity is known to be associated with the largest family of oncogenes. Nearly 50 oncogenes have been identified to date as viral and cellular transforming genes and new oncogenes are still being isolated. Before cloning, sequencing, and searching for homology, one might be able to predict to which group the particular activated oncogene belongs if specific inhibitors for each set of the oncogene products were available. Herbimycin A, therefore, may be the first inhibitor to be used for the purpose of classifying newly found oncogenes because of its selectivity toward the tyrosine kinase oncogenes.

Oncogene activation has been implicated as a cause of some human cancers. We consider that specific inhibitors of oncogene functions may also be used to determine whether particular activated oncogenes are crucially involved in the malignant aspect of human tumors. For instance, in the case of the *src* oncogene, increased levels of $p60^{c-src}$ tyrosine kinase

---

[13] T. Yamashita, M. Sakai, Y. Kawai, M. Aono, and K. Takahashi, *J. Antibiot.* **42**, 1015 (1989).

[14] T. Oikawa, K. Hirotani, M. Shimamura, H. Ashino-Fuse, and T. Iwaguchi, *J. Antibiot.* **42**, 1202 (1989).

[15] R. Montesano, M. S. Pepper, D. Belin, J. D. Vassali, and L. Orci, *J. Cell. Pysiol.* **134**, 460 (1988).

activity have been reported in human colon carcinomas, neuroblastomas, and other tumors,[16] although the significance of such *src* activation remains unclear. To determine if these highly expressed oncogenes are involved in tumorigenicity, an examination using herbimycin A to study phenotypical reversion will provide us with valuable information.

[16] N. Rosen, J. B. Bolen, A. M. Schwartz, P. Cohen, V. DeSeau, and M. A. Israel, *J. Biol. Chem.* **261**, 13754 (1986).

[32] Use of Erbstatin as Protein-Tyrosine Kinase Inhibitor

By KAZUO UMEZAWA and MASAYA IMOTO

Microbial Secondary Metabolites

Streptomyces and other microorganisms produce antibiotics, anticancer agents, and enzyme inhibitors as secondary metabolites. It is possible that microorganisms produce antibiotics to suppress the growth of their competitors. However, most enzyme inhibitors produced by microorganisms have weak or no antimicrobial activity. The discovery of enzyme inhibitors indicates that microorganisms produce compounds having various structures that may play no role in the growth of the microorganisms themselves. Thus, microorganisms are a treasury of organic compounds that have various structures and biological activities. Isolation of specific bioactive products in culture broths depends on the assay system. Since the oncogene theory has been extensively developed, we have screened microbial secondary metabolites for oncogene function inhibitors. Erbstatin (Fig. 1) was isolated from a culture filtrate of *Streptomyces* as a specific inhibitor of protein-tyrosine kinase activity.

Isolation of Erbstatin

Erbstatin is produced by *Streptomyces* MH435-hF3. For isolation of erbstatin,[1] the producing strain is cultured in a 500-ml flask containing 110 ml of medium, consisting of 3% (v/v) glycerol, 2% (w/v) fish meal, and 0.2% (w/v) $CaCO_3$, pH 7.4 (before sterilization), on a rotary shaker at 27° for 48 hr, and 3.0 ml of the cultured broth is inoculated into a 500-ml flask containing 110 ml of the medium described above. The fermentation was

[1] H. Umezawa, M. Imoto, T. Sawa, K. Isshiki, N. Matsuda, T. Uchida, H. Iinuma, M. Hamada, and T. Takeuchi, *J. Antibiot.* **39**, 170 (1986).

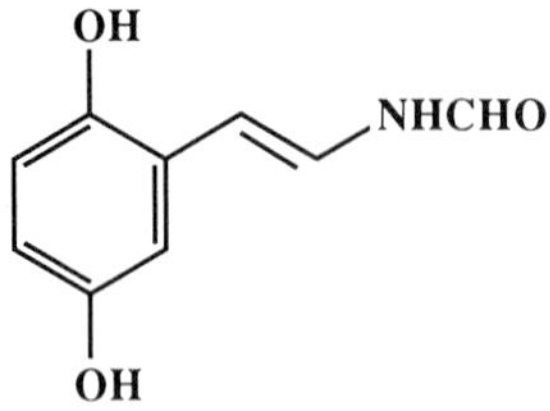

OH

NHCHO

OH

FIG. 1. Structure of erbstatin.

carried out at 27° for 4 days. Erbstatin in culture filtrates (5 liters) is extracted with 5 liters of butyl acetate, and the extract is concentrated under reduced pressure to give a yellowish powder (370 mg). The dried material is dissolved in $CHCl_3$ and subjected to silicic acid column chromatography. After washing with $CHCl_3$–methanol (100 : 2), the active fraction is eluted with $CHCl_3$–methanol (100 : 5) and concentrated *in vacuo* to give a yellow powder (110 mg). It is dissolved in methanol–$CHCl_3$ (1 : 1) and kept at 4° to yield 60 mg of crude crystals of erbstatin. It is recrystallized from methanol–$CHCl_3$, mp 156–157°. It is soluble in dimethyl sulfoxide (DMSO), methanol, ethanol, and acetone, slightly soluble in $CHCl_3$, and ethyl acetate, hardly soluble in H_2O and *n*-hexane. Erbstatin has been chemically synthesized in several laboratories with various synthetic routes.[2-7] Erbstatin is not yet commercially available.

Inhibition of Protein-Tyrosine Kinase *in Vitro*

Erbstatin inhibits epidermal growth factor (EGF) receptor-associated tyrosine kinase with an IC_{50} of 0.15 μg/ml. Protein-tyrosine kinase activity is measured by a modification of the method described by Carpenter *et al.*[8] The reaction mixture contains $MnCl_2$ (1 mM), bovine serum albumin (BSA, 0.125 mg/ml), mouse EGF (100 ng/ml; Collaborative Research Inc., Bedford, MA), the membrane fraction of A431 cells (40 μg protein/assay, prepared by the method described by Thom *et al.*[9]), tridecapeptide (1 mg/ ml; RR-SRC, Peptide Institute, Osaka, Japan), and $[\gamma\text{-}^{32}P]ATP$ (12 Ci/

[2] D. G. Hangauer, *Tetrahedron Lett.* **27**, 5799 (1986).
[3] W. K. Anderson, T. T. Dabrah, and D. M. Houston, *J. Org. Chem.* **52**, 2945 (1987).
[4] K. Isshiki, M. Imoto, T. Takeuchi, H. Umezawa, T. Tsuchida, T. Yoshioka, and K. Tatsuta, *J. Antibiot.* **40**, 1207 (1987).
[5] R. L. Dow and M. J. Flynn, *Tetrahedron Lett.* **28**, 2217 (1987).
[6] J. Kleinschroth and J. Hartenstein, *Synthesis* p. 970 (1988).
[7] M. N. Deshmukh and S. V. Joshi, *Synth. Commun.* **18**, 1483 (1988).
[8] G. Carpenter, J. R. Lloydking, and S. Cohen, *Nature* (*London*) **276**, 409 (1978).
[9] D. Thom, A. J. Powell, C. W. Lloyd, and D. A. Rees, *Biochem. J.* **168**, 187 (1977).

mmol, 1 μM) in a final volume of 60 μl HEPES buffer (20 mM, pH 7.2). Erbstatin is dissolved in DMSO (10 mg/ml) and diluted with water. The EGF receptor is first incubated with EGF and erbstatin at 0° for 10 min before assay of kinase activity. The kinase reactions are initiated by the addition of the peptide and [γ-^{32}P]ATP, and the reaction mixture is incubated for 30 min at 0°. The reactions are terminated by addition of 25 μl of 10% (v/v) trichloroacetic acid (TCA) and 6 μl BSA (10 mg/ml). Precipitated proteins are removed by centrifugation and 45-μl aliquots of the supernatants are spotted on Whatman (Clifton, NJ) P81 phosphocellulose papers (2 × 2 cm), which are immediately immersed in 30% (v/v) acetic acid at 25°. The papers are washed three times for 15 min in 15% acetic acid, for 5 min in Me$_2$CO, and then dried. The radioactivity is counted with a liquid scintillation counter.

Erbstatin does not inhibit purified rabbit brain protein kinase C at 100 μg/ml (H. Hidaka, Nagoya University, personal communication, 1990), and weakly inhibits protein kinase A[10] and phosphatidylinositol kinase[11] with IC$_{50}$ values of 100 and 25 μg/ml, respectively. Erbstatin does not inhibit tyrosine hydroxylase (monooxygenase) activity (Prof. T. Nagatsu, Nagoya University, Japan, personal communication).

The inhibitory pattern in the Lineweaver–Burk plot of erbstatin vs peptide is a typical competitive inhibition, while the inhibitory pattern of erbstatin vs ATP is noncompetitive.[12] From the Dixon plot analysis the K_i value was found to be about 5.58 μM. Thus, erbstatin competes with the peptide substrate, and the mechanism of inhibition is clearly different from that of orobol[12] and genistein,[13] which compete with ATP.

Use of Erbstatin in Cell Culture

Phosphorylation of the 170K EGF receptor is stimulated by addition of EGF for 15 min to the monolayer culture of A431 cells, and addition of erbstatin to A 431 cells induces dose-dependent inhibition of EGF-stimulated receptor autophosphorylation detected by anti-EGF receptor antibody.[13] The half-maximal effect is observed at about 15 μg/ml of erbstatin, and 50 μg/ml of the drug inhibits the EGF-stimulated phosphorylation completely. The turnover of the EGF receptor during this period as

[10] C. S. Rubin, J. Erlichman, and O. M. Rosen, *J. Biol. Chem.* **247**, 36 (1972).
[11] M. Imoto, K. Umezawa, T. Sawa, T. Takeuchi, and H. Umezawa, *Biochem. Int.* **15**, 989 (1987).
[12] M. Imoto, K. Umezawa, K. Isshiki, S. Kunimoto, T. Sawa, T. Takeuchi, and H. Umezawa, *J. Antibiot.* **40**, 1471 (1987).
[13] H. Ogawara, T. Akiyama, J. Ishida, S. Watanabe, and K. Suzuki, *J. Antibiot.* **39**, 606 (1986).

determined by [^{35}S] methionine incorporation is not influenced by erbstatin at 50 μg/ml.

Erbstatin also blocks autophosphorylation of *erbB-2* product, which is detected by antiphosphotyrosine antibody. For this assay 5×10^5 activated *erbB-2*-transfected NIH 3T3 cells (kindly supplied by Dr. T. Akiyama, Osaka University, Japan) plated in 35-mm dishes are grown for 16 hr before use. The cells are washed twice with Ca^{2+},Mg^{2+}-free phosphate-buffered saline (PBS$^-$) and they are incubated for 15 min in serum-free Dulbecco's modified Eagle's medium (DMEM) with or without 50 μg/ml erbstatin dissolved in DMSO at 10 mg/ml. Then, the cells are washed twice with cold PBS$^-$ and extracted with 200 μl RIPA buffer [25 mM HEPES, 1.5% (v/v) Triton X-100, 1% (v/v) sodium deoxycholate, 0.1% (w/v) sodium dodecyl sulfate, 0.5 M NaCl, 5 mM EDTA, 50 mM NaF, 100 μM sodium vanadate, 1 mM phenylmethylsulfonyl fluoride, 1 mg/ml of leupeptin, pH 7.8] for 30 min at 4°. The cell lysate is centrifuged at 15,000 g for 5 min. A portion of the cell extract (100 μl) is mixed with 100 μl of electrophoresis loading buffer [42 mM Tris-HCl, 10% (v/v) glycerol, 2.3% (w/v) SDS, 5% (v/v) 2-mercaptoethanol, 0.002% (v/v) Bromphenol Blue, pH 6.3], boiled for 5 min, then centrifuged at 15,000 g for 2 min. The supernatant is electrophoresed on SDS-polyacrylamide gel. Using Western blotting, the proteins are separated on a nitrocellulose membrane and anti-phosphotyrosine antibody added. The tyrosine-phosphorylated proteins are visualized by the biotin–avidin method.[14]

When erbstatin is added to cultured Rous sarcoma virus-infected rat kidney cells for 3 hr with labeled phosphate, it also inhibits autophosphorylation of the p60^{src} at 25–50 μg/ml.[11] Using A431 cells no significant decrease in EGF binding is observed with erbstatin up to 100 μg/ml. On the other hand, erbstatin at a concentration of 25–50 μg/ml inhibits internalization of EGF receptor.[11] The half-maximal effect is observed at about 30 μg/ml.

As shown in Fig. 2, erbstatin delays the S-phase induction by EGF for about 3 hr in quiescent normal kidney (NRK) cells, without affecting the total amount of DNA synthesis.[15] The reversible effect of erbstatin is probably due to the inactivation of erbstatin. We found that 90% of the labeled erbstatin is degraded within 2 hr in the culture medium of A431 cells. For the S-phase induction assay, NRK cells (4×10^5) are plated in a 16-mm glassware dish and cultured in 1 ml of DMEM containing 10%

[14] M. S. Blake, K. H. Johnson, G. J. Russel, and E. C. Gotschli, *Anal. Biochem.* **136,** 175 (1984).
[15] K. Umezawa, T. Hori, H. Tajima, M. Imoto, K. Isshiki, and T. Takeuchi, *FEBS Lett.* **260,** 198 (1990).

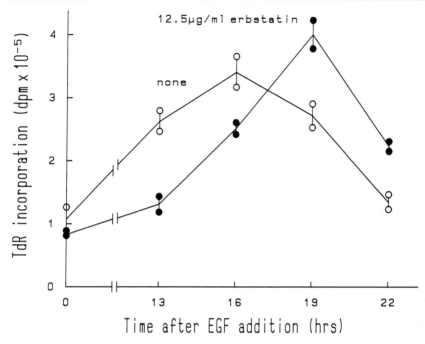

FIG. 2. Inhibition of EGF-induced DNA synthesis by erbstatin. Quiescent NRK cells were incubated with EGF (○) or EGF and 12.5 μg/ml erbstatin (●). Details are described in the text. TdR, deoxythymidine.

calf serum for 60 hr. During this period the cells become quiescent by density-dependent inhibition. Then the cells are washed twice with 0.5 ml of serum-free DMEM and given 1 ml of serum-free DMEM containing 12.5 μg/ml erbstatin and/or 100 ng/ml EGF. After incubation for indicated periods, the cells are washed twice with 0.5 ml Dulbecco's phosphate-buffered saline, followed by the addition of 1 ml serum-free DMEM containing 1 μCi/ml [³H]thymidine. After 1 hr the cells are washed twice with 0.5 ml Dulbecco's phosphate-buffered saline, and 0.5 ml cold 10% trichloroacetic acid (TCA) is added. After incubation for 10 min at 4°, the cells are further washed twice with 0.5 ml 10% TCA. Then they are solubilized by 0.5 ml of 0.5 N NaOH and an aliquot is counted for radioactivity in a liquid scintillation counter. Erbstatin does not inhibit S-phase induction at 3 μg/ml, and at 50 μg/ml suppresses the total amount of DNA synthesis.

Erbstatin inhibited growth of the cells with IC$_{50}$ values of 3.6, 4.1, and 3.1 μg/ml for A431, NRK, and NIH 3T3 cells, respectively. In cell culture,

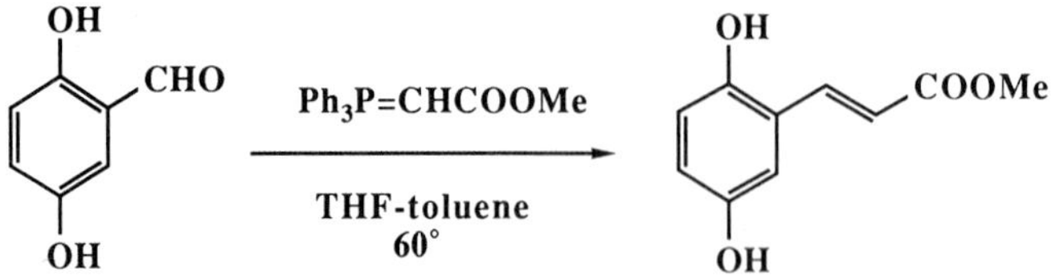

Methyl 2,5-dihydroxycinnamate

FIG. 3. Synthesis of methyl 2,5-dihydroxycinnamate.

it can be used at 1–50 μg/ml, depending on the incubation period and cell lines.

Erbstatin Analog: Methyl 2,5-Dihydroxycinnamate

Erbstatin is easily inactivated in calf serum,[16] and more stable analogs of erbstatin have been looked for. Previously, we synthesized various erbstatin analogs and tested them for their inhibitory activity against tyrosine kinase.[17] Among them only 2,5-dihydroxycinnamic acid, which has a carboxylic acid moiety instead of the N-formyl moiety, would be more stable. This analog inhibits tyrosine kinase *in vitro,* but it was not expected to be active *in vivo,* because, being a polar molecule, it would penetrate the cells poorly. Therefore we synthesized methyl 2,5-dihydroxycinnamate to reduce the polarity of the molecule.[15]

Synthesis of methyl 2,5-dihydroxycinnamate is shown in Fig. 3. 2,5-Dihydroxybenzaldehyde (4.3 g) and methyl(triphenylphosphoranilidene)acetate (11.6 g) are mixed in 10 ml tetrahydrofuran (THF) and 120 ml toluene and stirred for 30 min at 60°. Then the mixture is evaporated and applied to a silica gel column (305 g, CHCl$_3$–methanol, 20 : 1) to give 5.7 g of yellow flakes. The crude product is further purified by LH-20 column chromatography with methanol to give 4.86 g of methyl 2,5-dihydroxycinnamate as a yellow powder.

Methyl 2,5-dihydroxycinnamate is about four times more stable than erbstatin in calf serum. Erbstatin disappears completely in 30 min, whereas methyl 2,5-dihydroxycinnamate can be recovered in appreciable amount even after a 60-min incubation.

Methyl 2,5-dihydroxycinnamate inhibits EGF receptor tyrosine kinase

[16] M. Imoto, K. Umezawa, K. Komuro, T. Sawa, T. Takeuchi, and H. Umezawa, *Gann* **78,** 329 (1987).
[17] K. Isshiki, M. Imoto, T. Sawa, K. Umezawa, T. Takeuchi, and H. Umezawa, *J. Antibiot.* **40,** 1209 (1987).

in vitro with an IC_{50} of 0.15 μg/ml, which is equivalent to the activity of erbstatin. Lineweaver–Burk plots show that it inhibits EGF receptor kinase competitively with the peptide and noncompetitively with ATP, as is the case for erbstatin.[12] The methyl ester derivative inhibits EGF receptor autophosphorylation in cultured A431 cells at 25–100 μg/ml. This *in situ* activity of methyl 2,5-dihydroxycinnamate is slightly weaker than that of erbstatin.[11] The decrease in autophosphorylation is not due to the inhibition of EGF binding, since the inhibitor does not affect EGF receptor binding in A431 cells at 100 μg/ml. Methyl 2,5-dihydroxycinnamate at 12.5 μg/ml delays the S-phase induction by epidermal growth factor for about 10 hr in quiescent normal rat kidney cells, without affecting the total amount of DNA synthesis. It is not effective at 3 μg/ml, and at 50 μg/ml it reduces the total amount of DNA synthesis. The effect of methyl 2,5-dihydroxycinnamate is more prominent than that of erbstatin, possibly because of its longer lifetime. Methyl 2,5-dihydroxycinnamate can be dissolved in DMSO at 10 mg/ml and diluted with water or medium.

We have developed 5'-*O*-methylerbstatin[17,18] as an inactive analog of erbstatin. It does not inhibit EGF receptor tyrosine kinase at 30 μg/ml and is more stable than erbstatin in calf serum. Tyrphostins are potent and specific tyrosine kinase inhibitors derived from erbstatin and they are discussed in [29] in this volume.

[18] K. Umezawa, K. Tanaka, T. Hori, S. Abe, R. Sekizawa, and M. Imoto, *FEBS Lett.* **279,** 132 (1991).

Section III

Protein Phosphatases

[33] Classification of Protein-Serine/Threonine Phosphatases: Identification and Quantitation in Cell Extracts

By PHILIP COHEN

Several years ago it became clear that most, if not all, of the protein phosphatases (PP) that dephosphorylated the regulatory enzymes controlling metabolism in skeletal muscle and liver were explained by four types of catalytic subunit.[1-3] In view of their broad and overlapping substrate specificities, a classification procedure was therefore introduced that could be used to distinguish protein phosphatases by activity measurements, even in crude tissue extracts.[1,2,4] The enzymes were divided into two groups, type 1 and type 2, depending on whether they dephosphorylated the β subunit of phosphorylase kinase specifically and were inhibited by nanomolar concentrations of two small heat- and acid-stable proteins, termed inhibitor 1 (I-1) and inhibitor 2 (I-2) (type 1, PP1), or whether they dephosphorylated the α subunit of phosphorylase kinase preferentially and were insensitive to inhibitors 1 and 2 (type 2, PP2). The type 2 phosphatases could, in turn, be subdivided into three distinct enzymes, PP2A, PP2B, and PP2C, in a number of ways, but most simply by their dependence on divalent cations. PP2B and PP2C had absolute requirements for Ca^{2+} and Mg^{2+}, respectively, while PP2A (like PP1) was at least partially active toward most substrates in the absence of divalent cations.[1,2] Using these criteria, it was found that the four types of catalytic subunit were present in all mammalian tissues examined, and that their levels were as high, if not higher, in other tissues, such as the brain.[4] These observations, in conjunction with their broad substrate specificities *in vitro,* suggested that type 1 and type 2 protein phosphatases were involved in the regulation of many cellular processes, a prediction that has been borne out by many findings.[5-7]

Other useful criteria for distinguishing type 1 and type 2 protein phos-

[1] T. S. Ingebritsen and P. Cohen, *Eur. J. Biochem.* **132**, 255 (1983).
[2] T. S. Ingebritsen and P. Cohen, *Science* **221**, 331 (1983).
[3] P. Cohen, *Eur. J. Biochem.* **151**, 439 (1985).
[4] T. S. Ingebritsen, A. A. Stewart, and P. Cohen, *Eur. J. Biochem.* **132**, 297 (1983).
[5] P. Cohen, *Annu. Rev. Biochem.* **58**, 453 (1989).
[6] P. Cohen and P. T. W. Cohen, *J. Biol. Chem.* **264**, 21435 (1989).
[7] P. Cohen, C. F. B. Holmes, and Y. Tsukitani, *Trends Biochem. Sci.* **15**, 98 (1990).

METHODS IN ENZYMOLOGY, VOL. 201

phatases are summarized in Table I.[8-30] Of these, the most important is the effect of okadaic acid, a complex fatty acid polyketal first extracted from the marine sponge *Halichondria okadaii*. Produced by certain types of marine plankton, okadaic acid accumulates in the digestive glands of sponges and mollusks. It is one of the toxins responsible for diarrhetic seafood poisoning and a potent tumor promoter. Okadaic acid was shown to be a potent and specific inhibitor of PP1 and PP2A by Takai and co-workers[10,31] and, as detailed below, it has improved the procedures for identifying and quantitating PP1, PP2A, and PP2C in eukaryotic cell extracts when used in combination with either inhibitor 1 or inhibitor 2. Okadaic acid can also be used in intact cells to identify physiological substrates of PP1 and PP2A and to reveal novel intracellular processes that are controlled by phosphorylation as described elsewhere in this volume.[32] For a more detailed review of this topic readers are referred to Cohen *et al.*[7]

[8] A. A. Stewart, T. S. Ingebritsen, A. Manalan, C. B. Klee, and P. Cohen, *FEBS Lett.* **137**, 80 (1982).

[9] A. A. Stewart, T. S. Ingebritsen, and P. Cohen, *Eur. J. Biochem.* **132**, 289 (1983).

[10] C. Bialojan and A. Takai, *Biochem. J.* **256**, 283 (1988).

[11] P. Cohen, S. Klumpp, and D. L. Schelling, *FEBS Lett.* **250**, 596 (1989).

[12] S. Jakes and K. K. Schlender, *Biochim. Biophys. Acta* **967**, 11 (1988).

[13] P. Agostinis, J. Goris, E. Waelkens, L. A. Pinna, F. Marchiori, and W. Merlevede, *J. Biol. Chem.* **262**, 1060 (1987).

[14] A. Burchell and P. Cohen, *Abstr. 11th FEBS Meet.* **A1-4**, 003 (1977).

[15] D. Gratecos, T. C. Detwiler, S. Hurd, and E. H. Fischer, *Biochemistry* **16**, 4812 (1977).

[16] J. Di Salvo, E. Waelkens, D. Gifford, J. Goris, and W. Merlevede, *Biochem. Biophys. Res. Commun.* **117**, 493 (1983).

[17] R. L. Mellgren and K. K. Schlender, *Biochem. Biophys. Res. Commun.* **117**, 501 (1983).

[18] S. Pelech and P. Cohen, *Eur. J. Biochem.* **148**, 245 (1985).

[19] P. Gergely, F. Erdodi, and G. Bot, *FEBS Lett.* **169**, 45 (1984).

[20] F. Erdodi, C. Csortos, G. Bot, and P. Gergely, *Biochim. Biophys. Acta* **827**, 23 (1985).

[21] H. Usui, M. Imazu, K. Maeta, H. Tsukamoto, K. Azume, and M. Takeda, *J. Biol. Chem.* **263**, 3752 (1988).

[22] J. Goris and W. Merlevede, *Biochem. J.* **254**, 501 (1988).

[23] F. Erdodi, C. Csortos, G. Bot, and P. Gergely, *Biochim. Biophys. Res. Commun.* **128**, 705 (1985).

[24] P. Strålfors, A. Hiraga, and P. Cohen, *Eur. J. Biochem.* **149**, 295 (1985).

[25] S. Alemany, S. Pelech, C. H. Brierley, and P. Cohen, *Eur. J. Biochem.* **156**, 101 (1986).

[26] A. A. K. Chisholm and P. Cohen, *Biochim. Biophys. Acta* **968**, 392 (1988).

[27] M. D. Pato and E. Kerc, *J. Biol. Chem.* **260**, 12359 (1985).

[28] D. L. Brautigan, P. A. Gruppuso, and M. Mumby, *J. Biol. Chem.* **251**, 14924 (1986).

[29] M. Bollen, J. R. Vandenheede, J. Goris, and W. Stalmans, *Biochim. Biophys. Acta* **969**, 66 (1988).

[30] H. Olssen and P. Belfrage, *Eur. J. Biochem.* **168**, 399 (1987).

[31] A. Takai, C. Bialojan, M. Troschka, and J. C. Ruegg, *FEBS Lett.* **217**, 81 (1987).

[32] D. G. Hardie, T. J. Haystead, and A. T. R. Sim, this volume [40].

TABLE I

METHODS FOR DISTINGUISHING PROTEIN PHOSPHATASES

Method	PP1	PP2A	PP2B	PP2C	Refs.
Preference for α or β subunit of phosphorylase kinase	β subunit	α subunit	α subunit	α subunit	1,2
Inhibition by I-1 and I-2	Yes[a]	No[b]	No	No	1,2
Absolute requirement for divalent cations	No[c]	No[c]	Yes (Ca^{2+})	Yes (Mg^{2+})	1,2
Stimulation by calmodulin	No	No	Yes	No	8,9
Inhibition by trifluoperazine	No	No	Yes	No	9
Inhibition by okadaic acid	Potent[d]	Very potent[d]	Weak[d]	No	10,11
Phosphorylase phosphatase activity	High	High	Very low	Very low	1
Activity toward histone H1 phosphorylated by protein kinase C	Very low	Very high			12
Activity toward casein phosphorylated by A-kinase	Very low	High			13
Effect of highly basic proteins (H1) on phosphorylase phosphatase activity	Inhibition[e]	Activation[f]			14–18
Effect of heparin on phosphorylase phosphatase activity	Inhibition[e,g]	No effect or[h] activation			18–21
Effect of p-nitrophenyl phosphate on phosphorylase phosphatase activity	Activation	Inhibition			22
Binding to heparin-Sepharose at 0.1 M NaCl	Retained	Excluded			23

[a] The catalytic subunit of PP1 is inhibited instantaneously by I-1 or I-2, but inhibition of some high-molecular-mass forms of PP-1 is time dependent,[24–26] while other forms may be resistant to I-2.[27]

[b] The catalytic subunit of PP2A was reported to be inhibited by I-2 at extremely high (micromolar) concentrations,[28] but other investigators could not confirm this result.[29]

[c] Dephosphorylation of some substrates (e.g., I-1) is strongly stimulated by Mn^{2+}.[4]

[d] PP1, PP2A, and PP2B are inhibited with IC$_{50}$ values of about 10 nM, 0.1 nM, and 5 μM, respectively, under normal assay conditions (Fig. 1).

[e] Inhibition by highly basic proteins (histone H1, protamine, polylysine) and heparin is observed only with phosphorylase a as substrate. With most other substrates (e.g., glycogen synthase, pyruvate kinase, acetyl-CoA carboxylase, hormone-sensitive lipase) basic proteins and heparin are activators rather than inhibitors of PP1.[18,30]

[f] PP2A is activated at low (submicromolar) concentrations of basic proteins, but inhibited at higher concentrations.[16,17] Optimal differentiation between PP1 and PP2A is observed at about 300 μg/ml protamine.[18]

[g] Heparin binds to many proteins and the concentration required for 50% inhibition depends on the protein concentration in the assay; 90% inhibition is observed with 150 μg/ml heparin at 1 mg/ml serum albumin.[18]

[h] Whether any activation is observed depends on the form of PP2A, and the presence or absence of Mn^{2+}.[18,21]

Procedure for Identifying and Quantitating Protein Phosphatases 1, 2A, and 2C in Cell Extracts

Principle

The procedures for assaying PP1, PP2A, PP2B, and PP2C were described in volume 159 of this series[33-36] and readers are referred to these four chapters for the detailed protocols. PP1 and PP2A are the only enzymes with significant activity toward glycogen phosphorylase and are therefore generally assayed by measuring the dephosphorylation of this substrate (although others can also be used). Inhibitor 1 (for PP2B) and casein (for PP2C) are the preferred substrates for the other two phosphatases. One unit of activity is that amount which catalyzes the dephosphorylation of 1 μmol of substrate in 1 min.

In order to measure PP1 and PP2A, assays are performed in the absence of divalent cations, and cell extracts are diluted to obtain a final phosphorylase phosphatase concentration of <0.1 mU/ml in the assays. About a 600-fold final dilution is generally required when extracts are prepared by homogenizing mammalian tissues in 3 vol of buffer. Under these conditions 1 nM okadaic acid inhibits PP2A completely (IC$_{50}$ ~0.1 nM, Fig. 1). PP1 is unaffected by 1 nM okadaic acid, but completely inhibited at 5 μM okadaic acid (IC$_{50}$ ~10–15 nM, Fig. 1). Assays are therefore performed in the presence of 1 nM okadaic acid (to block PP2A), 0.2 μM I-1 or I-2 (to block PP1), 0.2 μM I-1 or I-2 plus 1 nM okadaic acid (to block PP1 and PP2A), and 5 μM okadaic acid (to block PP1 and PP2A). PP1 is the activity that is sensitive to I-1 or I-2, or the activity at 5 μM okadaic acid subtracted from the activity measured at 1 nM okadaic acid. PP2A is the activity that is unaffected by I-1 or I-2, or the amount that is blocked by 1 nM okadaic acid. The values obtained by either procedure agree to ±10%.

In order to measure PP2C, assays are carried out in the presence of 5 μM okadaic acid (to block PP1 and PP2A) and 0.1 mM EGTA (to block PP2B) and in the absence and presence of Mg^{2+} (20 mM). PP2C is the activity in the absence of Mg^{2+} (which is negligible) subtracted from the activity measured in the presence of Mg^{2+}. For measurement of PP2C activity with casein as substrate, a twofold lower dilution of the cell

[33] P. Cohen, S. Alemany, B. A. Hemmings, T. J. Resink, P. Strålfors, and H. Y. L. Tung, this series, Vol. 159, p. 390.
[34] A. A. Stewart and P. Cohen, this series, Vol. 159, p. 409.
[35] C. H. McGowan and P. Cohen, this series, Vol. 159, p. 416.
[36] P. Cohen, J. G. Foulkes, G. A. Nimmo, and N. K. Tonks, this series, Vol. 159, p. 426.

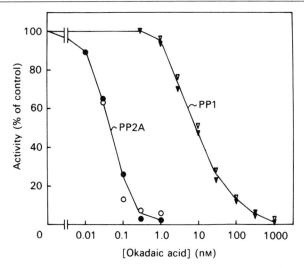

FIG. 1. Effect of okadaic acid in plant and mammalian type 1 and type 2A protein phosphatases. The cytosol of *Brassica napus* was assayed at 150-fold dilution with ^{32}P-labeled phosphorylase as substrate. PP2A was measured after preincubation for 15 min with 100 nM inhibitor 1 (O). PP1 was determined in the absence of inhibitor 1 (∇) and the graph plotted after subtracting activity measured at 0.3 nM okadaic acid to correct for PP2A activity. The closed triangles and circles show experiments with the homogeneous catalytic subunits of rabbit skeletal muscle PP1 and PP2A, respectively. (Reproduced from Cohen *et al.*[38])

extracts is generally required than is employed for the assay of PP1 and PP2A.

Comments on Procedure

The concentration of okadaic acid required for inhibition of PP2A is similar to that of the phosphatase in the assays, and the IC$_{50}$ therefore increases with increasing PP2A concentration (Fig. 2). In contrast, at dilutions used in assays inhibition of PP1 is independent of phosphatase concentration (Fig. 2). In order to use okadaic acid as a specific inhibitor of PP2A it is therefore essential that extracts be diluted to give a phosphorylase phosphatase concentration in the assays of <0.1 mU/ml. This high dilution also minimizes the effects of inhibitory substances that are present in concentrated tissue extracts. Okadaic acid inhibits PP1 and PP2A instantaneously (and reversibly), and the free catalytic subunits are inhibited in a manner identical to the inhibition of native forms, in which the catalytic subunits are complexed to other proteins.[11] A report that the

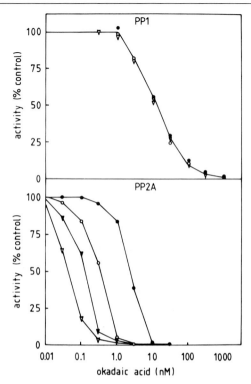

FIG. 2. Influence of enzyme concentration on inhibition of the purified catalytic subunits of rabbit skeletal muscle PP1 and PP2A by okadaic acid. Assays were carried out with glycogen phosphorylase as substrate. Phosphatase concentrations in the assays were 0.03 (∇), 0.1 (▼), 0.3 (○), and 3.0 (●) mU/ml. (Reproduced from Cohen et al.[11])

PP2A catalytic subunit was more sensitive to okadaic acid than the native forms[10,37] is explained by the failure to carry out experiments at identical protein phosphatase concentrations (see Fig. 2).

The sensitivity of PP1 and PP2A to okadaic acid, and that of PP1 to mammalian inhibitors 1 and 2, have been remarkably conserved during evolution and are essentially identical in organisms as diverse as mammals, yeast,[38] and higher plants[39] (Fig. 1) and with a variety of substrates[11] (Fig.

[37] J. Hescheler, G. Mieskes, J. C. Ruegg, A. Takai, and W. Trautwein, Pfluegers Arch. 412, 248 (1988).
[38] P. Cohen, D. L. Schelling, and M. J. Stark, FEBS Lett. 250, 601 (1989).
[39] C. MacKintosh and P. Cohen, Biochem. J. 262, 335 (1989).

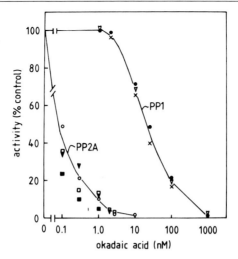

FIG. 3. Effect of okadaic acid on PP1 and PP2A activities in tissue extracts. Assays were performed at a 600-fold dilution of the extracts in the presence of 1 nM okadaic acid (PP1) or after preincubation for 15 min with 0.2 μM inhibitor-2 (PP2A). Phosphorylase phosphatase activity was measured in extracts of bovine adrenal medulla (○), rat liver (▽, PP1; ▼, PP2A), and rabbit brain (□); phosphorylase kinase phosphatase in an extract of rabbit skeletal muscle (×), glycogen synthase phosphatase with rat liver glycogen particles (●), and casein phosphatase in an extract of rabbit brain cortex (■). (Reproduced from Cohen et al.[11])

3). The procedure described above is therefore applicable to virtually all eukaryotic cells. The only exception, so far, is *Paramecium*, the PP2A-like activity in this ciliated protozoan being unaffected by okadaic acid, even at 30 μM.[40]

The phosphorylase phosphatase activity, which is unaffected by 1 nM okadaic acid, can be inhibited >90% by inhibitor 1 or inhibitor 2 in mammalian tissues (Fig. 4), yeast,[38] and higher plants,[39] indicating that the native forms of PP1 present in these extracts can be inhibited almost completely by the inhibitor proteins. However, the diluted extracts must be preincubated with inhibitor 1 or inhibitor 2 for 15 min prior to assay, in order to achieve complete inhibition.[24-26] In contrast, the PP1 catalytic subunit is inhibited almost instantaneously by inhibitors 1 and 2.[24] A PP1-like activity in avian smooth muscle with high myosin phosphatase activity is particularly resistant to inhibitor 2,[27] although its sensitivity to okadaic acid is similar to that of other forms of PP1.[41] A fragment of inhibitor 1 comprising residues 9 to 41 of inhibitor 1 is fully active, provided threonine-

[40] S. Klumpp, P. Cohen, and J. E. Schultz, *EMBO J.* **9,** 685 (1990).
[41] M. M. Sola, and P. Cohen, unpublished observations.

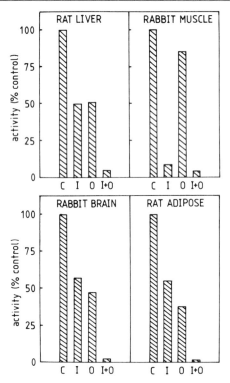

FIG. 4. Effect of okadaic acid and inhibitor 2 on phosphorylase phosphatase (PhP) activity in tissue extracts. Assays were performed in the absence of divalent cations at 600-fold dilution (muscle, liver, brain) or 60-fold dilution (adipocyte) in the presence of 2 nM okadaic acid (O), after preincubation for 15 min with 0.2 μM inhibitor 2 (I), 15-min preincubation with inhibitor 2 plus okadaic acid (I + O), or in the absence of okadaic acid and inhibitor 2 (C, control). (Reproduced from Cohen et al.[11])

35 is phosphorylated or thiophosphorylated.[42] Synthesis of this fragment is the simplest way of obtaining large amounts of active inhibitor 1.

In addition to mammalian tissues, PP2B has been detected in a variety of invertebrates[43] and in yeast, but not in higher plants.[39] PP2C has been found in yeast[38] and in higher plants,[44] while *Escherichia coli* extracts have no detectable activity toward the major substrates (glycogen phosphorylase, phosphorylase kinase, or casein) used to assay PP1, 2A, 2B, and 2C.[45]

[42] C. F. B. Holmes, in preparation.
[43] C. B. Klee and P. Cohen, *Mol. Aspects Cell. Regul.* **6**, 225 (1988).
[44] C. MacKintosh, J. Coggins, and P. Cohen, *Biochem. J.* **273**, 733 (1991).
[45] P. Cohen and P. T. W. Cohen, *Biochem. J.* **260**, 931 (1989).

The procedure described above measures the active forms of each protein phosphatase. However, it should be noted that inactive forms of PP1 and PP2A are also present in mammalian tissue extracts.

Classification of Protein-Serine/Threonine Phosphatases: An Update

Although the current classification of protein-serine/threonine phosphatases is now widely accepted and understood, some modification to the nomenclature of the mammalian enzymes is needed as a result of several findings that have been made since the introduction of recombinant DNA technology to this area,[6] which is discussed in [34] in this volume.[46]

cDNA cloning has revealed that at least two isoforms of PP1, PP2A and PP2B, present in mammalian tissues are the products of separate genes. These isoforms, which have very similar amino acid sequences, are termed PP1α and PP1β (85% identity), PP2Aα and PP2Aβ (97% identity), and PP2Bα and PP2Bβ (89% identity). It is suggested that additional isoforms be termed γ, δ, etc., as they are isolated. For consistency, the two isoforms of PP2C in rabbit tissues that have been resolved by anion-exchange chromatography and were previously termed $2C_1$ and $2C_2$,[6,47] are renamed PP2Cα and PP2Cβ.

The native forms of PP1 and PP2A, in which the catalytic subunit is complexed to other subunits, are denoted by different subscripts. For example, the native forms of PP1 present in skeletal muscle have been termed PP1$_G$ (glycogen associated), PP1$_M$ (myofibrillar), and PP1$_I$ (inactive) to indicate their subcellular location or activity state. The native forms of PP2A in skeletal muscle and liver have been termed PP2A$_0$, PP2A$_1$, and PP2A$_2$ by the order in which they are eluted from DEAE-cellulose, the subscript 2A$_0$ denoting a form with almost "zero" activity in the standard assay.[5,6]

The catalytic domains of PP1 and PP2A are about 50% identical in amino acid sequence. A potential problem with the current nomenclature has therefore arisen with the identification by cDNA cloning of additional protein phosphatase catalytic subunits that are structurally related to PP1 and PP2A. For example, PPY (isolated from a Drosophila cDNA library) is more similar to PP1 (~70% identity) than PP2A (~40% identity), while PPX (isolated from a liver library) is more closely related to PP2A (65%

[46] P. T. W. Cohen, this volume [34].
[47] C. H. McGowan and P. Cohen, *Eur. J. Biochem.* **166,** 713 (1987).

identity) than PP1 (~40% identity).[6,46] It is clearly essential to study the enzymatic properties of these novel enzymes, either by isolating them or expressing the cDNA clones, in order to determine whether they can be classified as type 1 or type 2A, or whether their properties are so distinctive that additions to the current nomenclature are required.

[34] Cloning of Protein-Serine/Threonine Phosphatases

By PATRICIA T. W. COHEN

Four major types of protein-serine/threonine phosphatase (PP) catalytic subunit (PP1, PP2A, PP2B, and PP2C) have been identified in mammalian cells on the basis of their enzymatic properties.[1] Over the past few years cDNAs encoding each of these enzymes have been cloned by constructing suitable oligonucleotides to peptide sequences derived from the purified phosphatases, and employing them in hybridization studies to isolate appropriate clones from bacteriophage cDNA libraries. Since these approaches have been used to clone many other proteins, they will be described fairly briefly in this chapter with emphasis on aspects of the methodology that the author has found particularly useful or are not documented clearly elsewhere. Subsequent library screening with the cDNAs encoding PP1 and PP2A led to the isolation of isoforms, homologs from different species, and cDNAs encoding putative protein phosphatases that had not been detected previously by the techniques of protein chemistry and enzymology. Procedures for detecting such novel enzymes are also discussed in this chapter, because additional phosphatases related to PP1 and PP2A are likely to be present in eukaryotic cells.

Procedures for Cloning of PP1 and PP2A

Isolation of Peptides for Sequence Analysis

PP1 and PP2A are low-abundance proteins composing <0.1% of the protein in any cell. They have been purified as either the free catalytic subunits, or as the native holoenzymes in which the catalytic subunits are complexed to other (regulatory) polypeptides. In the latter case, or if the

---

[1] P. Cohen and P. T. W. Cohen, J. Biol. Chem. **264**, 21435 (1989).

preparations are impure, the catalytic subunits are first resolved from other proteins by sodium dodecyl sulfate (SDS)–polyacrylamide gel electrophoresis and a piece of the gel containing this component is digested with a proteinase. The procedure described below is used for the glycogen-bound form of PP1.[2]

1. Glycogen-bound PP1 from rabbit skeletal muscle (20 nmol) is subjected to SDS–polyacrylamide gel electrophoresis to separate the catalytic (C) subunit (37 kDa) from the much larger glycogen-binding G subunit. The gel is stained with Coomassie Blue, destained in 43% methanol/7% acetic acid, washed with water, and the band containing the C subunit is excised and broken into small pieces.

2. The gel pieces are dried, suspended in 2.5 ml of 0.2 M N-methylmorpholine acetate, pH 8.1 and, after addition of trypsin or chymotrypsin (5 μg), the suspension is incubated at 37°. Two further additions of proteinase (5 μg) are made after 3 and 16 hr, and after 24 hr the supernatant is removed and the gel washed with methanol (2 ml).

3. The methanol extract and the first supernatant are combined, dried, and redissolved in 1.0 ml of 1.0% (v/v) trifluoroacetic acid, centrifuged for 15 min at 13,000 g at room temperature, and the supernatant containing the peptides is removed.

4. The peptides are chromographed on a Vydac C_{18} column (Separations Group, Hesperia, CA) and sequenced.

Construction of Oligonucleotide Probes

Short oligonucleotides (14–20 bases) comprising all possible sequences coding for small peptides have been successful for isolating cDNAs. However, low signal-to-noise ratios during low-temperature hybridization and filter washing, and the fact that the correct matching sequence may constitute a very small proportion of the probe mixture, can cause problems. Longer oligonucleotides (21–60 bases) are therefore used to clone PP1 and PP2A, based on the optimum codon usage for a given mammalian species[3,4] and the incorporation of up to four mixed sites. Since pairing of G with T does not disrupt duplex formation, placing G in the probe sequence when the target sequence is C or T is advantageous.[5] In addition inosine (I), which occurs naturally in the wobble position of the anticodon of some tRNAs, where it pairs with A, C, and U, can be valuable for pairing with

[2] N. Berndt, D. G. Campbell, F. B. Caudwell, P. Cohen, E. F. da Cruz e Silva, O. B. da Cruz e Silva, and P. T. W. Cohen, *FEBS Lett.* **223,** 340 (1987).
[3] R. Lathe, *J. Mol. Biol.* **183,** 1 (1985).
[4] T. Maruyama, *Nucleic Acids Res.* **14,** Suppl., r151 (1986).
[5] M. Jaye, H. de la Salle, F. Schamber, A. Balland, V. Kohli, A. Findeli, P. Tolstoshev, and J.-P. Lecocq, *Nucleic Acids Res.* **11,** 2325 (1983).

any of the four bases at the target site, since it stabilizes the DNA structure when pairing with A and C, and may not disrupt the duplex when pairing with G and T.[6,7] However, the stability of a G : I pair is not entirely clear, and may depend on the surrounding sequence. Insertion of a C/I mixed site in the probe may therefore be preferable for pairing with all four bases.[2,8] The GC content of the oligonucleotides is 45–60% and self-complementarity is avoided. Oligonucleotides are synthesized by the phosphoramidite method on an Applied Biosystems (Warrington, UK) 381A DNA synthesizer and purified by chromatography on a Vydac C_{18} column. Alternatively, they are made 0.4 M in sodium acetate, pH 5.0, and precipitated with 2.5 vol ethanol at $-20°$. After centrifugation for 15 min at 15,000 g, the supernatant is discarded and the precipitate washed with 70% ethanol and redissolved in 20 mM Tris-HCl, pH 8.0, 0.1 mM EDTA. One A_{260} unit is assumed to be 25 μg/ml for oligonucleotides purified by chromatography on the C_{18} column, but 12.5 μg/ml for ethanol-precipitated oligonucleotides which are contaminated by other A_{260} absorbing material. Labeling at the 5'-OH end, carried out using T4 polynucleotide kinase with equimolar proportions of oligonucleotide and [γ-^{32}P]ATP ($>$3000 Ci/mmol) in a standard reaction mixture,[9] proceeds virtually to completion, and removal of unincorporated ATP is unnecessary.

Screening of Bacteriophage Libraries

Libraries in bacteriophage are utilized in order that large numbers of cDNA clones (10^5–10^6 pfu) can be rapidly screened. Since peptide sequences of PP1 and PP2A had been obtained from the rabbit skeletal muscle enzymes, cDNA libraries are constructed in phage λgt10[10] from rabbit skeletal muscle and liver RNA using cDNA synthesized by the method of Gubler and Hoffman,[11] or purchased from Clontech (Palo Alto, CA). Libraries are then screened by hybridization after *in situ* amplification of bacteriophage plaques.[12] All materials and procedures not described

[6] Y. Takahashi, K. Kato, Y. Hayashizaki, T. Wakabayashi, E. Ohtsuka, S. Matsuki, M. Ikehara, and K. Matsubara, *Proc. Natl. Acad. Sci. U.S.A.* **82,** 1931 (1985).
[7] F. H. Martin and M. M. Castro, *Nucleic Acids Res.* **13,** 8927 (1985).
[8] D. Guerini and C. Klee, *Proc. Natl. Acad. Sci. U.S.A.* **86,** 9183 (1989).
[9] R. B. Wallace and C. G. Miyada, this series, Vol. 152, p. 432.
[10] T. V. Huynh, R. A. Young, and R. W. Davis, *in* "DNA Cloning: A Practical Approach" (D. M. Glover, ed.), Vol. 1, p. 49. IRL Press, Oxford, 1985.
[11] U. Gubler and B. J. Hoffman, *Gene* **25,** 263 (1983).
[12] S. L. C. Woo, this series, Vol. 68, p. 389.

in detail below can be found in Maniatis *et al.*[13] The following modifications proved to be effective.

1. To encourage preservation of poorly growing clones, recombinant λgt10 phage and host *Escherichia coli* C600 Hfl cells (Clontech) are plated on LB plates containing 10 mM MgSO$_4$, since Mg^{2+} helps stabilize phage heads. In addition, ≤0.12 ml of plating cells [and ~30,000 plaque-forming units (pfu)] are used per 138-mm plate (plating C600 Hfl cells are grown overnight to an A_{600} of 2.0, centrifuged and the cells resuspended in 0.4 vol of 10 mM MgSO$_4$). Addition of more plating cells causes some slow-growing clones to be lost.

2. To obtain sufficient *in situ* amplification, nitrocellulose filters are submerged for 2 min in a 1 : 5 dilution of an overnight culture of C600 Hfl cells in LB medium, and allowed to dry thoroughly before use.

3. To eliminate high backgrounds, after *in situ* amplification of bacteriophage plaques on nitrocellulose filters, the phage DNA on the filters is denatured, neutralized and air dried, placed under vacuum, then heated for 2 hr at 80° in the presence of silica gel desiccant. Damp filters at this stage invariably lead to high backgrounds. Cell debris, another cause of high backgrounds, is removed by washing the filters in 3× SSC, 1 mM EDTA, 0.1% SDS for 1 hr at 37°, placing them between thin plastic sheets and rubbing the top with a scrubbing brush, and finally rewashing them.

Prehybridization and hybridization are carried out in 0.9 M NaCl, 90 mM Tris-HCl, pH 7.5, 6 mM EDTA, 1 mM sodium pyrophosphate, 0.5% (v/v) Nonidet P-40, 0.2% (w/v) SDS, 0.04% (w/v) Ficoll, 0.04% (w/v) polyvinylpyrrolidone, 0.02% (w/v) bovine serum albumin, 0.1 mg/ml denatured herring sperm DNA. The filters are incubated in plastic bags on an orbital shaker, at 37° for 21–23 base oligonucleotides, 42–45° for 24–39 base oligonucleotides, and 55° for 40–60 base oligonucleotides. The oligonucleotide probe concentration is 0.8 pmol/ml and washes are in 6× SSC, initially at the hybridization temperature (1× SSC is 0.15 M NaCl, 0.015 M sodium citrate, pH 7.0). The best final wash (15 min) is determined empirically by gradually increasing the temperature to 60° and then lowering the salt concentration.

Ideally, clones that are positive with two or more oligonucleotides are chosen for further analysis. If no such clones are found with the available oligonucleotides, clones positive with only a single oligonucleotide are analyzed and in some cases turn out to be correct. Standard methods are used for plaque purification of positive clones, CsCl density gradient

[13] T. Maniatis, E. F. Fritsch, and J. Sambrook, "Molecular Cloning: A Laboratory Handbook." Cold Spring Harbor Lab., Cold Spring Harbor, New York, 1982.

purification of the phage[13] and extraction of phage DNA.[14] Subcloning of cDNA is performed without prior separation of the cDNA insert from the phage DNA. The recombinant phage DNA is digested with EcoRI, phenol extracted, ethanol precipitated, and ligated in a 2 : 1 molar ratio of phage DNA : sequencing vector with T4 DNA ligase.[15] The ligation mixture is used directly to transform E. coli JM109.[16] Since small fragments subclone more readily than large fragments, the plasmid recombinant DNA virtually always contains the cDNA insert. The Bluescript plasmid vector (Stratagene) is convenient for double-stranded sequencing by the dideoxy chain-termination method.[17] Oligonucleotides used for screening can often be used as sequencing primers and the DNA sequence obtained with them used to construct a complementary oligonucleotide which, in turn, could be used as a primer to read the initial sequence. The use of 7-deaza-2'-dGTP,[18] instead of dGTP, is essential to resolve compressions with most of the protein phosphatases (PP1 is highly GC rich).

Although as yet, no protein phosphatases have been cloned from cDNA libraries in the bacteriophage vector λ ZAP[19] (Stratagene, San Diego, CA), use of this cloning vector, which allows the integrated phagmid pBluescript to be excised by helper phage, has the advantage that it eliminates the need for subcloning.

Procedures for Cloning of Protein Phosphatases 2B and 2C

The Ca^{2+}/calmodulin-dependent protein phosphatase (PP2B) is present at much higher levels in the brain than in other mammalian tissues, and cDNA cloning has so far only been carried out using brain libraries. A partial clone encoding PP2B has been isolated from a mouse brain library in λgt11 by immunoscreening techniques.[20] Full-length clones encoding PP2B have also been isolated from rat[21] and human[8] brain libraries, using

[14] R. W. Davis, D. Botstein, and J. R. Roth, in "Advanced Bacterial Genetics: A Manual for Genetic Engineering," p. 106. Cold Spring Harbor Lab., Cold Spring Harbor, New York, 1980.

[15] J. R. Greene and L. Guarnte, this series, Vol. 152, p. 512.

[16] D. Hanahan, in "DNA Cloning: A Practical Approach" (D. M. Glover, ed.), Vol. 1, p. 109. IRL Press, Oxford, 1985.

[17] F. Sanger, S. Nicklen, and A. R. Coulson, Proc. Natl. Acad. Sci. U.S.A. 74, 5463 (1977).

[18] S. Mizusawa, S. Nishimura, and F. Seela, Nucleic Acids Res. 14, 1319 (1986).

[19] J. M. Short, J. M. Fernandez, J. A. Sorge, and W. D. Huse, Nucleic Acids Res. 16, 7583 (1988).

[20] R. L. Kincaid, M. S. Nightingale, and B. M. Martin, Proc. Natl. Acad. Sci. U.S.A. 85, 8983 (1988).

[21] T. Kuno, T. Takeda, M. Hirai, A. Ito, H. Mukai, and C. Tanaka, Biochem. Biophys. Res. Commun. 165, 1352 (1989).

similar methods to those described above for PP1 and PP2A. PP2C has been cloned from a rat liver cDNA library by oligonucleotide screening.[22]

Identification of Novel Protein Phosphatases, Isoforms, and Homologs by Screening with PP1 cDNA Probes

Screening of cDNA Libraries at Low Stringencies

Libraries are screened with a PP1 clone at low stringencies (hybridization at 50–55° in 0.05 M sodium phosphate, pH 7.4, 0.75 M NaCl, 5 mM EDTA, 0.1% Ficoll, 0.1% polyvinylpyrrolidone, 0.1% bovine serum albumin, 0.5% SDS, 0.1 mg/ml denatured nonhomologous DNA, followed by washing in 2× SSC at 50–55°). Use of the *in situ* amplification method (see Screening of Bacteriophage Libraries) enables weak positives to be picked up easily. The probe, labeled by the method of Feinberg and Vogelstein,[23] can be the entire cDNA insert of the clone, providing that no poly(A) tail is present, but for optimum results the coding region only is preferable. High-stringency screening of duplicate filters with the PP1 cDNA or specific oligonucleotides is used to prevent the isolation of further identical PP1 clones.

This approach has led to the isolation of several novel protein phosphatases related in structure to PP1 and PP2A. PPX was identified from a rabbit liver library,[24] PPV[25] and PPY[26] from *Drosophila* libraries, while PPZ[25] and PP2B$_w$[27] are yeast protein phosphatases, which were identified in a library supplied by Clontech as a rabbit brain cDNA library, but which is probably a *Saccharomyces cerevisiae* library.[28] PPX, PPV and PPPP2B$_w$ were picked out with a PP1 probe, despite the fact that the overall nucleotide identity was only 50–60% in the catalytic domain.

[22] S. Tamura, K. R. Lynch, J. Larner, J. Fox, A. Yasui, K. Kikuchi, Y. Suzuki, and S. Tsuiki, *Proc. Natl. Acad. Sci. U.S.A.* **86**, 1796 (1989).

[23] A. P. Feinberg and B. Vogelstein, *Anal. Biochem.* **132**, 6 (1983); addendum: **137**, 266 (1984).

[24] O. B. da Cruz e Silva, E. F. da Cruz e Silva, and P. T. W. Cohen, *FEBS Lett.* **242**, 105 (1988).

[25] P. T. W. Cohen, N. D. Brewis, V. Hughes, and D. J. Mann, *FEBS Lett.* **268**, 355 (1990).

[26] V. Dombrádi, J. M. Axton, D. M. Glover, and P. T. W. Cohen, *FEBS Lett.* **247**, 391 (1989).

[27] E. F. da Cruz e Silva and P. T. W. Cohen, *Biochim. Biophys. Acta* **1009**, 293 (1989).

[28] P. T. W. Cohen, unpublished observations.

[29] H. M. Barker, T. A. Jones, E. F. da Cruz e Silva, N. K. Spurr, D. Sheer, and P. T. W. Cohen, *Genomics* **7**, 159 (1990).

[30] V. Dombrádi, J. M. Axton, N. D. Brewis, E. F. da Cruz e Silva, L. Alphey, and P. T. W. Cohen, *Eur. J. Biochem.*, **194**, 739 (1991).

The same approach can be used to isolate isoforms of protein phosphatases and homologs from different species. For example, the coding region of the rabbit PP1 cDNA was used to isolate both human PP1[29] and three isoforms of PP1 from *Drosophila*,[30] while the coding region of rabbit PP2A cDNA was used to isolate both the *Drosophila* PP2A[31] and yeast PP2A.[32] This method was successful because both PP1 and PP2A show extraordinarily high evolutionary conservation.[33]

Searching Databases

One of the most commonly employed searches is the FASTP search program,[34] which has been successful in many cases. However, this method has the drawback that it may miss sequence similarities occurring over short regions. Thus when the sequences of PP1 and PP2A were used to search the National Biomedical Research Foundation databases with this algorithm, no proteins with similar sequences were identified. The use of a more exhaustive searching program,[35] which locates all locally strong regions of similarity, identified homology between PP1/PP2A and an open reading frame (ORF221) of bacteriophage λ.[36] The overall sequence identity of the first 115 residues of ORF221 to either PP1α (amino acids 45–164) or PP2Aα (amino acids 38–158) was 35%. Subsequently it was demonstrated that ORF221 does indeed encode a protein phosphatase.[37]

Relationships between Different Protein Phosphatases Discerned by cDNA Cloning

Evolutionary Conservation and Isoforms

Complementary DNA cloning of PP1 from mammals, *Drosophila*, yeast, and *Aspergillus*[1] and PP2A from mammals, *Drosophila*,[31,38] and yeast[32] has revealed that these enzymes are among the most evolutionarily conserved of all known proteins, showing greater conservation of structure than the protein kinases.[33] Cloning of PP2B from mammals[8,20,21] and

[31] S. Orgad, N. D. Brewis, L. Alphey, J. M. Axton, Y. Dudai, and P. T. W. Cohen, *FEBS Lett.* **275**, 414 (1990).

[32] A. Sneddon, P. T. W. Cohen, and M. J. R. Stark, *EMBO J.* **9**, 4339 (1990).

[33] P. T. W. Cohen, *in* "Genetics and Human Nutrition" (P. J. Randle, ed.), pp. 27–39. Libbey, London (1990).

[34] D. J. Lipman and W. R. Pearson, *Science* **227**, 1435 (1985).

[35] A. F. W. Coulson, J. F. Collins, and A. Lyall, *Comput. J.* **30**, 420 (1987).

[36] P. T. W. Cohen, J. F. Collins, A. F. W. Coulson, N. Berndt, and O. B. da Cruz e Silva, *Gene* **69**, 131 (1988).

[37] P. T. W. Cohen and P. Cohen, *Biochem. J.* **260**, 931 (1989).

[38] P. T. W. Cohen and V. Dombrádi, *Adv. Protein Phosphatases* **5**, 447 (1989).

yeast[27,28] shows that the structure of this phosphatase is also conserved, but not as highly conserved as PP1 and PP2A.

cDNA cloning is much more powerful than enzymatic and protein chemical techniques in demonstrating the existence of isoforms. Thus at least two isoforms of PP1 (PP1α and PP1β) and PP2A (PP2Aα and PP2Aβ) have been detected in mammalian tissues.[1] Although encoded by distinct genes, the isoforms of PP1 are extremely similar (>85% identity), as are the isoforms of PP2A (>95% identity), virtually all the differences being confined to the N and C termini. In *Drosophila,* three isoforms of PP1 have been identified from their cDNAs.[30] Two of the *Drosophila* PP1 isoforms are very similar to mammalian PP1α, while the other would appear to be the homolog of mammalian PP1β. An α-like PP1 (*dis2*), as well as a further isoform (*sds21*), are also present in *Schizosaccharomyces pombe.*[39]

cDNA cloning of PP2B from mammalian brain has revealed two isoforms (PP2Bα and PP2Bβ) which are the products of separate genes and show 82% overall identity.[21] Two further forms of PP2Bα[40] and three of PP2Bβ[8,41] can be generated by alternative splicing at the 3' end, and, in the case of one PP2Bβ form, additional splicing near the 5' end.[8]

The structure of one isoform of PP2C (termed PP2Cα) has been determined by cDNA cloning.[22] PP2Cα and PP2Cβ have been purified from both rabbit skeletal muscle and rabbit liver and showed 80% amino acid sequence identity over the 62 residues where they could be compared directly.[42]

Two Distinct Protein-Serine/Threonine Phosphatase Gene Families

The amino acid sequences deduced from the cDNAs of mammalian PP1 and PP2A show 41% overall identity, indicating that they are members of the same gene family. The most divergent regions are the N and C termini, and the identity is 49% in the catalytic domain (Table I). The novel protein phosphatases PPX,[25] PPY,[26] PPZ[25] and PPV,[25] and a yeast phosphatase, termed SIT4,[43] have similar domain structures to PP1 and PP2A and their identities to PP1 and PP2A are summarized in Table I. PPY and PPZ are more similar to PP1, while PPX, PPV, and SIT4 resemble PP2A more closely. PP2B is a member of the same gene family, but a more

[39] H. Ohkura, N. Kinoshita, S. Migatani, T. Toda, and M. Yanagida, *Cell (Cambridge, Mass.)* **57,** 997 (1989).

[40] R. L. Kincaid, R. Rathna Giri, S. Higuchi, J. Tamura, S. C. Dixon, C. A. Marietta, D. A. Amorese, and B. M. Martin, *J. Biol. Chem.* **265,** 11312 (1990).

[41] A. E. McPartlin, H. M. Barker, and P. T. W. Cohen, *Biochim. Biophys. Acta,* **1088,** 308 (1991).

[42] C. H. McGowan, Ph.D. Thesis, University of Dundee (1988).

[43] K. T. Arndt, C. A. Styles, and G. R. Fink, *Cell (Cambridge, Mass.)* **56,** 527 (1989).

TABLE I
PROTEIN-SERINE/THREONINE PHOSPHATASES IDENTIFIED FROM DNA SEQUENCES

Phosphatase	Source	Amino acid residues	Molecular mass (kDa)	Identity (%) in catalytic domain[a] to		Refs.
				PP1	PP2A	
PP1α	Rabbit skeletal muscle, liver	330	37.5	100	49	1
PPZα	*Saccharomyces cerevisiae*	348 or 373		69	46	25
PPY	*Drosophila*	314	36.0	66	44	28
PP2Aα	Rabbit skeletal muscle	309	35.6	49	100	1
PPX	Rabbit liver	307	35.0	49	69	24,25
SIT4	*Saccharomyces cerevisiae*	311	35.5	46	60	43
PPV	*Drosophila*	303	34.5	45	57	25
PP2Bβ	Human brain	524	59	39	42	8
λ ORF221	Bacteriophage	221	25.2	20	19	35,36

[a] Catalytic domain used for the calculation of percentage identity is amino acids 29–299 of rabbit PP1α and 22–292 of rabbit PP2Aα. cDNA clones for PP1α, PP2Aα, and PP2Bβ have also been isolated from other tissues and species (see text). PP2Bβ has two insertions of seven and six residues in the catalytic domain and a long C-terminal extension which contains a calmodulin-binding domain.

```
         60                                                         

PP1α     L K I C G D I H G Q Y Y D L L R L F E - Y G G F P P E S N Y L F L G D Y V D R G
PPY      Y H I I V G D V H G Q F H E Y G Q Y L F R L F E - K A C G G F P P K T N N F L F G G D Y V D R G
PPZα     Y K I V C G D V H G Q Y H D L L R E L F K I - G G D P P S D V P D T H N R L F L G D Y V D R G
PP2Aα    Y T V V C G D V H G Q F H D L M E L F R I - G G K S P D T N Y L F M G D Y V D R G
PPX      Y T I V C G D I H G Q F F D L M K L F E V - G G S P A N T R Y L F M G D F V D R G
PPV      Y T V V C G D V H G Q F Y D L L E L F R I - G G F P P D D T N Y L F L G D F V D R G
SIT4     I I Y I C G D I H G Q F F D L M K L F E T - G G F P P E D N Y L I L S G D Y V D R G
PP2Bβ    L L T V V G D I H G Q L H G C Y T N D L L M N K L D T I G F D N K K D L L S Y G D L V D R G
ORF221   A E N V C L E L I H I T C L L L H I T E P - W F R A Y R G N H E Q M M L G D L V D R G


PP1α     K Q S L E T I C L L L A Y K I K Y P E N F F L L R G N H E C A S I
PPY      K Q S L E T I C L L L A Y K V K Y P L N F F L L R G N H E C A S I
PPZα     K Q S L E T I L L L L A Y C Y I R Y P E E R R G N H E S R Q V I
PP2Aα    Y Y S V E T V F L L L A Y K R R R P D R I H R G N H E S R Q H I
PPX      F Y S V E T F F L L L R L A K V P A R I P D R I T L L R G N H E S R Q Q I
PPV      Y Y S L E E F L L M C L I W V L K I H I T E P - W F R A Y R G N H E S R H L
SIT4     Y Y S I E C I L L R L Y L - - - - - R G N H E C R H L
PP2Bβ    Y F S H E C L E L - - - - - R G N H E C R H L
ORF221   A E N V C L E L I H I T E P - W F R A Y R G N H E Q M M L
```

Fig. 1. The most conserved region of protein-serine/threonine phosphatases. Identities (boxed) and most conservative replacements (underlined) in all nine enzymes are highlighted. The section shown is amino acids 59–130 of rabbit PP1α. The sources of the clones from which each protein phosphatase sequence was derived are listed in Table I. Single-letter abbreviations are used for amino acids. Alanine, A; arginine, R; asparagine, N; aspartic acid, D; cysteine, C; glutamic acid, E; glutamine, Q; glycine, G; histidine, H; isoleucine, I; leucine, L; lysine, K; methionine, M; phenylalanine, F; proline, P; serine, S; threonine, T; tryptophan, W; tyrosine, Y; valine, V.

distantly related one. Its catalytic domain shows less similarity to PP1 and PP2A, and in addition it possesses two insertions in the catalytic domain as well as a long C-terminal extension that is involved in interaction with calmodulin.

Bacteriophage λ ORF221 is the shortest protein phosphatase known (221 amino acids) and the most divergent member of this family (Table I).[36,37] Although the overall sequence identity is low (Table I), the N-terminal half of ORF221 shows 35% identity to PP1 or PP2A and includes the three highly conserved sections, corresponding to residues 63–67, 90–97, and 121–126 of PP1α (Fig. 1). It is presumed that these regions are part of the active site. Oligonucleotides synthesized to these most highly conserved regions may prove useful for cloning protein phosphatases in this family that are highly divergent from PP1 and PP2A. This strategy has been successful for protein kinases.[44,45]

PP2C, investigated both by peptide sequencing[42] and cDNA cloning,[22] has no discernable sequence similarities to the PP1/PP2A/PP2B/λORF221 family. Thus it is a member of a quite distinct protein-serine/threonine phosphatase gene family.

[44] S. K. Hanks and R. A. Lindberg, this series, Vol. 200, p. 525.
[45] A. E. Wilks, this series, Vol. 200, p. 533.

[35] Reactivation of Protein Phosphatase 1 Expressed at High Levels from Recombinant Baculovirus

By Patricia T. W. Cohen and Norbert Berndt

The baculovirus/insect cell system compares favorably with bacterial, yeast, and mammalian expression systems in that it can be used to express eukaryotic genes as active nonfusion proteins that for the most part undergo posttranslational processing, are targeted to the correct organelle, and can be obtained in significant levels, even though they may be toxic to the cell.[1] In this chapter we describe the application of this system for the high-level expression of protein phosphatase 1 (PP1), one of the major protein-serine/threonine phosphatases of eukaryotic cells.[2] Although the expressed protein was mostly present as inactive, insoluble aggregates, complete reactivation could be achieved by a novel procedure that may be applicable to other protein-serine/threonine phosphatases.

[1] V. A. Luckow and M. D. Summers, Bio/Technology 6, 47 (1988).
[2] P. Cohen, this volume [33].

Principle

cDNA comprising the whole coding region of PP1, but truncated in the 5' noncoding region, is ligated into the baculovirus plasmid pAcYM1,[3] so that it is placed under the control of the polyhedrin promoter, with only a few bases between the promoter and the initiating ATG of the cDNA. The recombinant plasmid is cotransfected with the wild-type baculovirus (*Autographa californica*) DNA into *Spodoptera frugiperda* cells. *In vivo* recombination leads to the production of recombinant virus expressing PP1 instead of polyhedrin, which can be isolated by standard procedures.[4] After infection of a culture of *S. frugiperda* cells with the pure recombinant virus, the cells are harvested when production of PP1 is maximal. The cells are lysed in the presence of salt and detergent, and after centrifugation the supernatant containing active PP1 is decanted. The pellet, containing almost pure (but insoluble) PP1, is dissolved in 6 M guanidinium chloride, renatured by dilution into buffers containing Mn^{2+} and sulfhydryl reagents, and concentrated by chromatography on DEAE-Sepharose.[5]

Procedure[5]

1. Linearize the Bluescript plasmid (Stratagene, La Jolla, CA) containing PP1 cDNA from rabbit skeletal muscle with *Hind*III, and then partially digest with *Nar*I, since three *Nar*I restriction sites are present, one being just upstream of the initiation ATG. Following separation of the *Nar*I/*Hind*III fragments by agarose gel electrophoresis, electroelute (Biotrap, Schleicher and Schuell, Dassel, Germany) the fragment containing the whole coding region. Remove the overhanging ends with mung bean nuclease and, after addition of *Bam*HI linkers, ligate the cDNA into the *Bam*HI site of plasmid pAcYM1. This procedure places the initiating ATG of PP1 cDNA 15 bases 3' to the polyhedrin prometer.

2. Transfect the recombinant pAcYM1–PP1 plasmid and wild-type baculovirus DNA into *S. frugiperda* cells. Isolate and amplify recombinant virus.[4]

3. Grow 10 ml of *S. frugiperda* cells to a density of 10^6 cells/ml at 28° in suspension culture and infect at 1 plaque-forming unit (pfu)/cell with recombinant virus containing PP1 cDNA.

[3] Y. Matsuura, R. D. Possee, H. A. Overton, and D. H. L. Bishop, *J. Gen. Virol.* **68**, 1233 (1987).
[4] M. D. Summers and G. E. Smith, eds., "A Manual of Methods for Baculovirus Vectors and Insect Cell Culture Procedures," Tex. Agric. Exp. Stn. Bull. No. 1555. Texas A&M Agric. Coll., College Station, 1987.
[5] N. Berndt and P. T. W. Cohen, *Eur. J. Biochem.* **190**, 291 (1990).

4. Harvest the cells 42 hr postinfection by centrifugation for 20 min at 1500 g at room temperature.

5. Wash the cells with 50 mM Tris-HCl, pH 7.0, 1 mM dithiothreitol, 250 mM sucrose at room temperature.

6. Resuspend the cells in 1.0 ml of 50 mM Tris-HCl, pH 7.0, 1 mM dithiothreitol, 0.5 M NaCl and lyse them by four freeze-thaw cycles in a dry ice–ethanol bath. The resulting lysate can be stored for at least 2 months at −20° without any effect on PP1 activity (Fig. 1).[6]

7. Make the lysate 1% (v/v) in Triton X-100, incubate for 1 hr at room temperature, centrifuge for 10 min at 15,000 g, and remove the supernatant for analysis of active PP1.

8. Wash the pellet with 50 mM Tris-HCl, pH 7.0, 1 mM dithiothreitol, 0.5 M NaCl, and 1% Triton X-100. Centrifuge for 10 min at 15,000 g and discard the supernatant.

9. The pellet contains PP1 of >80% purity (Fig. 2).[7] Solubilize in 1.0 ml of 50 mM Tris-HCl, pH 8.0, 1 mM dithiothreitol, 6 M guanidinium chloride. Incubate for ≥1 hr at room temperature and pass the solution through a 0.22-μm filter (Flow Laboratories, London, UK).

10. Dilute the solution by rapidly mixing with 1.0 liter of 50 mM Tris-HCl, pH 7.0, 1 mM MnCl$_2$, 50 mM 2-mercaptoethanol, 0.5 M NaCl, 0.02% Tween 20. Maximum reactivation of PP1 is regained within 60 min of dilution and remains stable for >3 hr at room temperature (Fig. 3).

11. Dilute the solution containing active PP1 to 12.5 liters to reduce NaCl to 0.04 M.

12. Equilibrate 50 g of DEAE-Sepharose Fast Flow (Pharmacia, Uppsala, Sweden) in 25 mM Tris-HCl, pH 8, 15 mM 2-mercaptoethanol, 0.02% Tween 20, 0.25 mM EDTA and add to the solution. Stir for 1 hr at room temperature.

13. Pour through Whatman (Clifton, NJ) 541 filter paper on a Büchner funnel. Transfer the DEAE-Sepharose to a 20 × 3 cm glass column and elute PP1 with 25 mM Tris-HCl, pH 8, 15 mM 2-mercaptoethanol, 0.3 M NaCl.

Comments on the Procedure

The maximum production of PP1 in *S. frugiperda* cells was found to occur 42 hr postinfection as judged by both assay of cellular supernatants (Fig. 1)[6] and SDS–polyacrylamide gel electrophoresis of pellets (not shown). At this time, expressed PP1 composed ~25% of the total protein in the insect cells. About 5% was active, representing a 15-fold increase

[6] P. Cohen, S. Alemany, B. A. Hemmings, T. J. Resink, P. Strålfors, and H. Y. L. Tung, this series, Vol. 159, p. 390.

[7] U. K. Laemmli, *Nature (London)* **227,** 680 (1970).

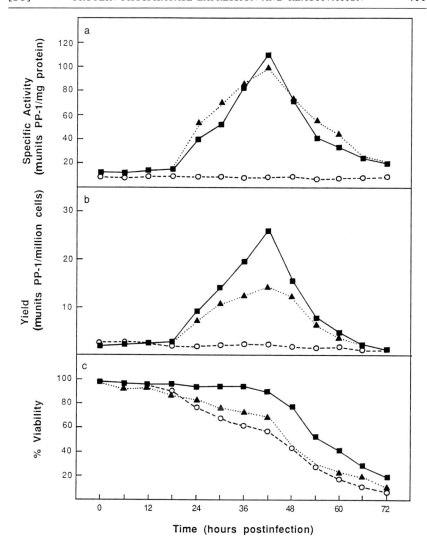

FIG. 1. Time course for the expression of PP1 from recombinant baculovirus in *S. frugiperda* cells. Suspension cultures of *S. frugiperda* cells (10^6 cells/ml) were infected with recombinant virus at a multiplicity of infection of 1 (■) and 10 (▲) and wild-type virus multiplicity of infection of 1 (○) and incubated at 28°. The specific activity (a) and yield (b) of PP1 in the cellular supernatants is shown. One unit of activity is that amount which catalyzes the dephosphorylation of 1.0 μmol of phosphorylase *a* in 1 min in the standard assay.[6] (c) Percentage of viable *S. frugiperda* cells.

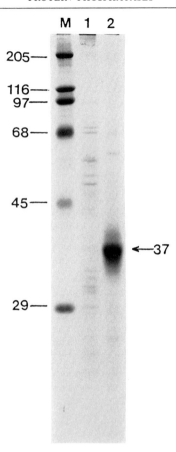

FIG. 2. SDS–polyacrylamide gel analysis of the Triton X-100 and salt-washed particulate fraction from *S. frugiperda* cells containing recombinant baculovirus expressing PP1. *Spodoptera frugiperda* cells (2×10^6) were harvested 48 hr postinfection, lysed, and centrifuged at 15,000 *g*. The pellet was washed with Triton X-100 and NaCl, then dissolved in 1% (w/v) SDS and heated for 5 min at 100°. An aliquot (10%) was loaded onto the 12.5% gel. Electrophoresis was carried out according to Laemmli[7] and the gels were stained with Coomassie Blue. Lane M, molecular mass markers, with values (kDa) indicated; lane 1, Triton/NaCl wash; lane 2, pellet.

over background PP1, while ~95% was in an inactive, particulate form. The latter was not solubilized by washing with either detergent or salt, allowing nearly all contaminating proteins to be removed. The insoluble PP1, which was >80% pure at this stage (Fig. 2),[7] could then be redissolved in 6 *M* guanidinium chloride, other compounds, such as 8 *M* urea, being less effective.[5]

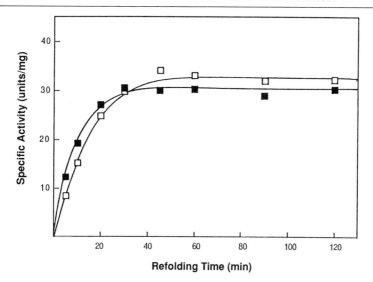

FIG. 3. Time course for reactivation of PP1 expressed from recombinant baculovirus in the particulate fraction of *S. frugiperda* cells. The PP1, solubilized in 6 *M* guanidinium chloride, was diluted to a concentration of 0.15 μg/ml in 50 m*M* Tris-HCl, pH 7.0, 1 m*M* MnCl$_2$, 50 m*M* 2-mercaptoethanol, 0.5 *M* NaCl, and 0.02% (w/v) Tween 20 and assayed for phosphorylase phosphatase activity in the presence (■) and absence (□) of 1 m*M* Mn^{2+}.[6]

Reactivation from 6 *M* guanidinium chloride was achieved by dilution in the presence of MnCl$_2$. No reactivation occurred if MnCl$_2$ was omitted.[5] The rationale for adding Mn^{2+} was that this divalent cation was well known to reactivate PP1 and PP2A that had been inactivated during storage, or by addition of EDTA, fluoride, pyrophosphate, or ATP.[8,9] Reactivation from guanidinium chloride was maximal at 1 m*M* Mn^{2+}, a 2 × 10^5-fold excess over the enzyme, suggesting that Mn^{2+} may stabilize a conformation important in refolding the active site, rather than being bound stoichiometrically as an essential cofactor. A protein phosphatase purified from rabbit liver,[8] which was probably a mixture of PP1 and PP2A,[10] did not contain bound Mn^{2+}.[8] Once refolded, the enzyme is independent of Mn^{2+} for activity.[5]

Thiols, salt, nonionic detergent, and high dilution aid the recovery of PP1 activity at step 10. It is critical to dilute the inactive PP1 (~0.2 mg/ml) at least 50-fold from 6 *M* guanidinium chloiride by rapidly adding the dilution buffer to the solution and mixing. Dialysis, or even dilution by

[8] S. C. B. Yan and D. J. Graves, *Mol. Cell. Biochem.* **42**, 21 (1989).
[9] A. Burchell and P. Cohen, *Biochem. Soc. Trans.* **6**, 220 (1978).
[10] T. S. Ingebritsen, J. G. Foulkes, and P. Cohen, *FEBS Lett.* **119**, 9 (1980).

adding the inactive enzyme to the appropriate volume of dilution buffer, leads to precipitation.[5]

The method yields 0.1 mg of active PP1 from 10 ml of cultured *S. frugiperda* cells within a few hours, the cell culture and infection with recombinant virus requiring minimal man hours. The active, expressed enzyme and renatured enzyme are indistinguishable from PP1 isolated from rabbit skeletal muscle by a number of criteria. These include specific activity toward glycogen phosphorylase, and sensitivity to inhibition by okadaic acid and inhibitor 1 and 2.[5] The expression of PP1 in this system will be useful for studies of structure/function relationships using site-directed mutagenesis. Although the method can, in principle, be used to generate the large amounts of enzyme required for crystallization and analysis of the three-dimensional structure of PP1, the volumes involved in scaling up the procedure are still a problem.

The procedure should also be useful for the expression of other related protein-serine/threonine phosphatases, such as PP2A, PP2B, and other enzymes of this gene family (e.g., PPX, PPY, and PPZ) that have been identified by cDNA cloning.[11] Expression of the latter enzymes will be critical to elucidate their substrate specificities and regulatory behavior.

[11] P. T. W. Cohen, this volume [34].

[36] Targeting Subunits for Protein Phosphatases

By MICHAEL J. HUBBARD and PHILIP COHEN

Protein phosphatase 1 (PP1), one of the major protein-serine/threonine phosphatase catalytic subunits in eukaryotic cells, does not exist as a monomer *in vivo*, but as complexes with other proteins that target it to particular subcellular locations, modify its substrate specificity, and appear to be the key to its regulation (reviewed in Ref. 1). The best characterized situation is in rabbit skeletal muscle, where PP1 is found associated with glycogen particles, the sarcoplasmic reticulum (SR), and the myofibrils, as well as in the cytosol. The glycogen-associated enzyme, termed $PP1_G$, is extremely similar, if not identical, to the form that is associated with the SR.[2] $PP1_G$ is a heterodimer composed of the catalytic C subunit and a G subunit, which is responsible for targeting the enzyme to both the

[1] P. Cohen, *Annu. Rev. Biochem.* **58**, 453 (1989).
[2] M. J. Hubbard, P. Dent, C. Smythe, and P. Cohen, *Eur. J. Biochem.* **189**, 243 (1990).

glycogen particles[3-5] and the SR.[2] The G subunit is phosphorylated by cyclic AMP-dependent protein kinase (A-kinase) at two serine residues, site 1 and site 2,[6] and by an insulin-stimulated protein kinase (ISPK) at site 1.[7] Phosphorylation of site 1 increases the rate at which $PP1_G$ dephosphorylates (activates) glycogen synthase and dephosphorylates (inactivates) phosphorylase kinase and appears to underlie the effects of insulin on glycogen metabolism.[7] In contrast, phosphorylation of site 2, which is located 19 residues C terminal to site 1,[6] causes dissociation of the C subunit from the G subunit, resulting in translocation of the former from glycogen particles[6] and SR[2] to the cytosol. Conversely, dephosphorylation of site 2 permits rebinding of the C subunit to the G subunit.[6] At physiological ionic strength, in the presence of glycogen, and under conditions where both the phosphatase and its substrates are largely bound to glycogen, $PP1_G$ is at least five- to eightfold more active in dephosphorylating glycogen phosphorylase and glycogen synthase than the released C subunit.[8] In contrast, $PP1_G$ and the released C subunit are equally effective in dephosphorylating substrates that do not bind to glycogen, such as myosin.[8] These observations, together with in vivo studies,[9-11] indicate that phosphorylation of the G subunit at site 2 represents a mechanism by which adrenalin (acting via cyclic AMP) overides the effect of insulin and prevents PP1 from dephosphorylating glycogen-bound substrates. The phosphorylation of site 1 by A-kinase may be a device for accelerating the rate of reactivation of glycogen synthesis after adrenergic stimulation has terminated and site 2 has been dephosphorylated.[7] Two further serines, four and eight residues N terminal to site 1, are phosphorylated by glycogen synthase kinase-3, but only after prior phosphorylation of site 1 by A-kinase.[12,13] The role of these phosphorylations is unknown. In view of the multiple phosphorylation sites and role of A-kinase and ISPK phosphorylation, this important region of the G subunit is termed the phosphoregulatory domain.[6]

[3] P. Stralfors, A. Hiraga, and P. Cohen, Eur. J. Biochem. 149, 295 (1985).
[4] A. Hiraga, B. E. Kemp, and P. Cohen, Eur. J. Biochem. 163, 253 (1987).
[5] M. J. Hubbard and P. Cohen, Eur. J. Biochem. 180, 457 (1989).
[6] M. J. Hubbard and P. Cohen, Eur. J. Biochem. 186, 701 (1989).
[7] P. Dent, A. Lavoinne, S. Nakielny, F. B. Caudwell, P. Watt, and P. Cohen, Nature (London) 348, 302 (1990).
[8] M. J. Hubbard and P. Cohen, Eur. J. Biochem. 186, 711 (1989).
[9] A. Hiraga and P. Cohen, Eur. J. Biochem. 161, 763 (1986).
[10] C. MacKintosh, D. G. Campbell, A. Hiraga, and P. Cohen, FEBS Lett. 234, 189 (1988).
[11] P. Dent, D. G. Campbell, F. B. Caudwell, and P. Cohen, FEBS Lett. 259, 281 (1990).
[12] P. Dent, D. G. Campbell, M. J. Hubbard, and P. Cohen, FEBS Lett. 248, 67 (1989).
[13] C. J. Fiol, J. H. Haseman, Y. Wang, P. J. Roach, R. W. Roeske, M. Kawalczuk, and A. A. DePaoli-Roach, Arch. Biochem. Biophys. 267, 797 (1988).

The form of PP1 associated with myofibrils, termed $PP1_M$, is composed of the C subunit complexed to a protein, distinct from the G subunit, which appears to enhance the ability of $PP1_M$ to dephosphorylate myosin.[14,15] The major form of PP1 in the cytosol is inactive and therefore termed $PP1_I$. It is a heterodimer composed of the C subunit and a thermostable protein, inhibitor 2, and can be activated *in vitro* through a mechanism that involves the reversible phosphorylation of inhibitor-2.[1]

Here, we detail methods that have been used to characterize the structure, localization, and regulation of $PP1_G$, several of which may be applicable to the analysis of other targeting subunits of PP1, as they are identified and purified.

Buffer Solutions

Buffer A: 50 mM Tris-HCl, pH 7.5 (22°), 0.1 mM EDTA, 0.1 mM EGTA, 10% (v/v) glycerol, 0.1% (v/v) 2-mercaptoethanol, 1 mM benzamidine, 4 μg ml^{-1} leupeptin, 0.1 mg ml^{-1} tosylphenyl chloromethyl ketone (TPCK), 0.2 mM phenylmethylsulfonyl fluoride (PMSF)

Buffer B: 25 mM Tris-HCl, 25 mM bis-Tris-HCl, pH 7.4 (4°), 0.2 mM EGTA, 1 mM benzamidine, 2 mg ml^{-1} bovine serum albumin, with freshly added 1 mM dithiothreitol

Buffer C: 50 mM Tris-HCl, pH 7.5 (22°), 5% (v/v) glycerol, 0.1% (v/v) 2-mercaptoethanol, 0.05% (v/v) Brij 35, and the proteinase inhibitors of buffer A

Buffer D: 50 mM Tris-HCl, pH 7.5 (22°), 5% (v/v) glycerol, 0.1% 2-mercaptoethanol, 0.1 mM EDTA, 0.1 mM EGTA, 1 mM benzamidine, 0.01% (w/v) Triton X-100

Buffer E: 50 mM Tris-HCl, pH 7.0, 0.1 mM EGTA, 0.1% (v/v) 2-mercaptoethanol, 1 mg ml^{-1} ovalbumin, and the proteinase inhibitors present in buffer A

Buffer F: 10 mM Tris, 75 mM glycine, pH 8.3

Purification of Protein Phosphatase 1_G

The purification of glycogen-associated $PP1_G$ was described earlier in this series[16] and only minor modifications have been introduced subsequently.[5,6] These procedures result in purification to near homogeneity of $PP1_G$, in which the native 161-kDa G subunit is partially proteolyzed to

[14] A. A. K. Chisholm and P. Cohen, *Biochim. Biophys. Acta* **968**, 392 (1988).
[15] A. A. K. Chisholm and P. Cohen, *Biochim. Biophys. Acta* **971**, 163 (1988).
[16] P. Cohen, S. Alemany, B. A. Hemmings, T. J. Resink, P. Stralfors, and H. Y. L. Tung, this series, Vol. 159, p. 390.

the 103-kDa G' fragment (this occurs principally at the chromatographic step on DEAE-cellulose[5]). Both the G' subunit and activity of purified $PP1_G$ are stable at $-80°$ in solutions containing 10% (v/v) glycerol for >1 year. The same procedure has also been used to purify $PP1_G$ from the SR following its extraction from the myofibrillar pellet with Triton X-100.[2] For studies of the *in vivo* phosphorylation state of $PP1_G$, the phosphatase inhibitors NaF and inorganic pyrophosphate are included.[10,11] A more rapid procedure for isolating glycogen-associated $PP1_G$ described below takes only 26 hr and yields nearly homogenous enzyme in which some intact (161 kDa) G subunit is still present.[5]

1. Isolate glycogen–protein particles from the skeletal muscle of three New Zealand White rabbits as described[16] and resuspend them in 210 ml of 50 mM Tris-HCl, pH 7.5, 0.1 mM EDTA, 10% (v/v) glycerol, 0.1% (v/v) 2-mercaptoethanol, 1 mM benzamidine, 8 μg ml^{-1} leupeptin, 0.1 mg ml^{-1} TPCK, 1 mM PMSF.

2. Add α-amylase (5 μg ml^{-1}) to degrade the glycogen. Incubate the suspension for 45 min at 30° (fresh 0.1 mM PMSF being added at 20 min), chill on ice, and add EGTA (0.1 mM) and fresh PMSF.

3. Centrifuge the suspension and chromatograph the supernatant, containing 75% of the $PP1_G$ activity, on a 60-ml column of DEAE-cellulose as described,[16] but with 0.1 mM EDTA and 0.1 mM EGTA in the solutions.

4. Apply the 0.15 M Tris-HCl eluate directly to an 8-ml column of poly(L-lysine)-Sepharose, prepared as described.[16] Wash with buffer A + 0.2 M NaCl, and elute $PP1_G$ with buffer A + 0.5 M NaCl.

5. Dilute the active fractions with an equal volume of buffer A and apply to a 3-ml column of aminohexyl-Sepharose. Elute with a 36-ml linear gradient of 0.25–1.0 M NaCl in buffer A.

6. Concentrate the active fractions by ultrafiltration (YM30 membrane; Amicon, Danvers, MA), dilute with buffer A to <2.5 mS conductivity, and apply to a 2-ml column of heparin-Sepharose. After washing with buffer A, elute $PP1_G$ with buffer A + 0.4 M NaCl and store at $-80°$. A subsequent gel-permeation chromatography on Superose 12 (30 × 1 cm) equilibrated in buffer A resolves $PP1_G$ containing intact 161-kDa G subunit (which elutes at the void volume, V_0) from $PP1_G$ containing the 103-kDa G' subunit (which elutes slightly later, $V_e/V_0 = 1.2$).

Glycogen Binding Assays

1. Purify glycogen from rabbit skeletal muscle extracts by centrifugation and washing sequentially with SDS, trichloroacetic acid, and water[17]

[17] C. Smythe, F. B. Caudwell, M. Ferguson, and P. Cohen, *EMBO J.* **7**, 2681 (1988).

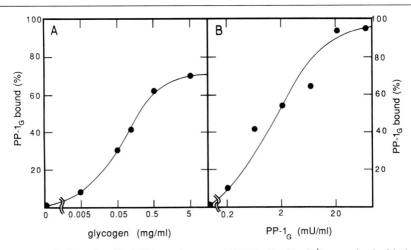

FIG. 1. Binding of purified PP1$_G$ to glycogen. (A) PP1$_G$ (5 mU ml^{-1}) was mixed with the indicated concentrations of glycogen and binding was assessed as described in the text. (B) The indicated concentrations of PP1$_G$ were mixed with glycogen (5 mg ml^{-1} final) and binding was determined as above. The data indicated a high-affinity bimolecular interaction between PP1$_G$ and glycogen, in near-physiological conditions. (Reproduced from Hubbard and Cohen.[5])

and deionize by passing over mixed cation/anion-exchange resin. Quantitate glycogen by the phenol/H$_2$SO$_4$ procedure with glucose as a standard.[18]

2. Mix 50 μl PP1$_G$ (up to 50 mU ml^{-1} final, in buffer B + 0.1 M NaCl) at 4° with 5 μl glycogen (up to 10 mg ml^{-1} final) in an Airfuge tube (20 × 5 mm; Ultraclear, Beckman, Palo Alto, CA), incubate on ice for 5 min, then centrifuge for 10 min at 200 kPa (~120,000 g_{av}) in an Airfuge A100 rotor (Beckman) at 4°.

3. Withdraw 25 μl of supernatant and measure PP1 activity with the standard assay (below). Correct activity for carryover of NaCl and glycogen if necessary.

Binding is routinely assessed on the basis of depletion of PP1 activity from the supernatant. Recovery of activity (total supernatant plus pellet) is >90% that of noncentrifuged controls. With a fixed concentration of PP1$_G$ (5 mU ml^{-1}), binding is dependent on glycogen concentration and approaches saturation at 5 mg ml^{-1} glycogen (Fig. 1A). With 5 mg ml^{-1} glycogen, binding is dependent on PP1$_G$ concentration and approaches saturation at 50 mU ml^{-1} (Fig. 1B). These data indicate a bimolecular

[18] M. Dubois, K. A. Giles, J. K. Hamilton, P. A. Rebers, and F. Smith, *Anal. Chem.* **28**, 350 (1956).

process for PP1$_G$ binding to glycogen, with K_{app} values of about 0.1 mg ml^{-1} glycogen and 2 mU ml^{-1} PP1$_G$. These values (for purified PP1$_G$ and glycogen) correspond to a K_{app} of 4–8 nM and are in good agreement with the value (K_{app} 10 nM) obtained for dissociation of intact PP1$_G$ from isolated glycogen-protein particles.[5] No binding of free C subunit to glycogen is detectable in these conditions. Binding is not significantly affected by increasing ionic strength to 0.5 M NaCl or by variation of pH from 6 to 8.[5] These properties can be used to distinguish glycogen-associated PP1$_G$ from SR-associated PP1$_G$, which does not dissociate from these membranes, even on dilution to 0.1 mU ml^{-1}.[2]

Procedures for Dissociating G and C Subunits

Gel-Permeation Chromatography[4]

1. Dissociate the subunits by incubating PP1$_G$ (0.3 ml, ≤ μM) for 20 hr at 4° in buffer C containing 2 M NaCl.
2. Separate the subunits by applying dissociated PP1$_G$ to a column of Sephadex G-100 Superfine (Pharmacia, Piscataway, NJ) (16 × 1 cm) equilibrated at 4° in buffer C containing 1 M NaCl. Collect 0.25-ml fractions at 2 ml hr^{-1}. The G subunit elutes at the void volume, and the C subunit at $V_e/V_0 \approx 1.5$ with a recovery of ≥80%.

G and C subunits prepared by this procedure can be reconstituted to PP1$_G$, following concentration and dialysis to remove NaCl.[3]

Hydrophobic Interaction Chromatography[18a]

1. Dissociate the subunits by diluting the PP1$_G$ sample (~5 μM, ≤0.5 ml) 5- to 10-fold in buffer D containing 1.2 M (NH$_4$)$_2$SO$_4$.
2. Separate the subunits of applying to a column of phenyl-Superose (HR5/5, 5 × 0.5 cm; Pharmacia, Piscataway, NJ) equilibrated in the same buffer at room temperature. Collect 0.5-ml fractions at 0.5 ml min^{-1}.
3. Elute the subunits with a linear gradient to buffer D over 40 min. The C subunit is eluted first at ~1.1–0.9 M (NH$_4$)$_2$SO$_4$, followed by G subunit at ~0.2 M (NH$_4$)$_2$SO$_4$. Recovery of activity is usually >80%.
4. Desalt samples by ultrafiltration, or quick freeze the fractions in liquid nitrogen and store at −80°. Thawed fractions are desalted by dialysis, since precipitation may occur during ultrafiltration.

[18a] C. MacIntosh, unpublished observations (1990).

Phosphorylation with A-Kinase[6]

1. Prepare a solution comprising buffer B + 0.1 M NaCl (pH 7.4 at 22°), 5 mM magnesium acetate and 10 mU ml^{-1} A-kinase catalytic subunit.[19] One unit of A-kinase activity is that amount which catalyzes the incorporation of 1.0 μmol of phosphate/min into mixed histones (type IIA; Sigma, St. Louis, MO).

2. Equilibrate 50 μl PP1$_G$ in this solution at 30° and commence phosphorylation by adding 5 μl ATP (0.1 mM final). Incubate for 7–10 min at 30°.

3. Place tubes on ice, add 5 μl glycogen (5 mg ml^{-1} final), then centrifuge for 10 min at 200 kPa to pellet the glycogen. Phosphorylation in these conditions releases $\geq$95% of the C subunit which remains in the supernatant following centrifugation. Phosphorylated G subunit is almost exclusively in the glycogen pellet.[6] This is a simple and rapid procedure for separating the G and C subunits, and unlike the other procedures described above is appropriate for microliter-scale preparations.[6] This procedure can also be used to dissociate the C subunit from SR-associated PP1$_G$.[2] Phosphorylation results in a >4000-fold decrease in the affinity of G subunit for C subunit.[6]

Phosphatase Assays

Standard Assay

The standard assay, with [^{32}P]phosphorylase (10 μM) as substrate, was described earlier in this series.[16] This procedure, employing low phosphatase concentrations (<0.3 mU ml^{-1}) and a low ionic strength, is used to monitor PP1$_G$ purification and for all other routine activity measurements.

Modified ("Physiological") Assay[8]

1. Dilute PP1, NaCl (or KCl), glycogen, and substrate in buffer B. For phosphorylase phosphatase assays include 5 mM caffeine.

2. Mix 10 μl enzyme (~45 nM, 30 mU/ml), 5 μl glycogen (30 mg ml^{-1}), 5 μl NaCl (0.6–1.2 M), and equilibrate at 30°.

3. Start the reaction by adding 10 μl substrate (30°) and, for assays with [^{32}P]phosphorylase, stop at 15 to 30 sec with 200 μl ice-cold 10% (w/v) trichloroacetic acid. To ensure complete precipitation of other ^{32}P-labeled substrates (e.g., glycogen synthase and phosphorylase kinase, both 1 μM)

[19] E. M. Reimann and R. A. Beham, this series, Vol. 99, p. 51.

in the presence of glycogen, add 10 μl of 100 mg ml^{-1} albumin as a carrier, before trichloroacetic acid.

4. Quantitate released [^{32}P]phosphate as in the standard assay.[16]

A relatively high concentration of PP1$_G$ is used so that >80% of the phosphatase is bound to glycogen. Physiological levels of PP1$_G$ (the inferred average concentration in muscle cytosol is ~200 nM[6]) give essentially quantitative binding but unfeasibly short reaction times. Using this procedure, the activity of glycogen-bound PP1$_G$ toward glycogen-bound substrates (glycogen phosphorylase and glycogen synthase) at physiological ionic strength is five to eight times higher than that of the free C subunit or PP1$_G$ that is not bound to glycogen. Note that by starting the reaction with substrate, effects resulting from the binding of substrate to glycogen may be underestimated due to incomplete binding during the brief duration of the assays.[8]

Phosphorylation and Phosphopeptide Mapping

Phosphorylation with A-Kinase[4,5]

1. Prepare solution containing 35 mM Tris-HCl, pH 7.0, 2 mM magnesium acetate, 5% (v/v) glycerol, 1 mM dithiothreitol, 10 mU ml^{-1} A-kinase catalytic subunit, and PP1$_G$ ($\leq$2 μM). Equilibrate at 30° and start the reaction by adding 0.1 vol of 1 mM [γ-^{32}P]ATP (1–2 × 10^6 cpm nmol^{-1}).

2. Terminate the reaction after 10 min by adding trichloroacetic acid, SDS,[4] EDTA, or A-kinase inhibitor peptide,[5,6] as appropriate for the subsequent steps.

Phosphorylation is rapid, with half-times <0.5 min for site 1 and <1 min for site 2 and essentially stoichiometric for site 1.[5] However, phosphorylation of site 2 is opposed by autodephosphorylation of that site[5] and net incorporation into site 2 is markedly reduced at >5 μM PP1$_G$.[6] Stoichiometries of >0.9 mol phosphate/mol for site 2 can be obtained if much lower PP1$_G$ concentrations (~100 nM) and higher ionic strengths are employed,[6] probably due to decreased rates of autodephosphorylation. Autodephosphorylation of site 2 can be prevented by including 25 mM NaF[12] or 0.1 μM okadaic acid[7] in the phosphorylation mixture. Site 1 is resistant to autodephosphorylation.[5]

Phosphorylation with the ISPK.[7] This is carried out as described above except that MgOAc is increased to 4 mM and ISPK (10 U/ml) replaces A-kinase. One unit of ISPK is that amount which catalyzes the incorporation of 1 nmol of phosphate into a synthetic peptide corresponding to the C-terminal 32 residues of ribosomal protein S6 in 1 min.[7]

Measurement of Phosphorylation Stoichiometries.[6] The G subunit in purified $PP1_G$ comprises a mixture of proteolytic fragments ranging from 161 to 103 kDa, which complicates precise determination of the phosphorylation stoichiometry. $PP1_G$ is therefore quantitated on the basis of phosphorylase phosphatase activity,[5] assuming 0.5 U $nmol^{-1}$ or by measuring ^{32}P radioactivity incorporated into site 1 by phosphopeptide mapping, since phosphorylation of site 1 is essentially stoichiometric under the standard phosphorylation conditions.[3,5] The two methods agree to ±10%.

Phosphopeptide Mapping[4-6]

1. Following phosphorylation by A-kinase in the presence of 0.2 mg ml^{-1} bovine serum albumin, precipitate ^{32}P-labeled $PP1_G$ by addition of 5 vol of 12% (w/v) trichloroacetic acid, and, after incubation for 2 min on ice, centrifuge for 3 min at 14,000 g.

2. Aspirate the supernatant and wash the pellet three times with 300 μl of 10% (w/v) trichloroacetic acid to remove $[^{32}P]ATP$ and twice with diethyl ether to remove trichloroacetic acid.

3. Add 0.25 ml of 0.25 M NH_4HCO_3 to the precipitate, followed by 20 μg trypsin (treated with tosylphenyl chloromethyl ketone) and incubate the suspension for 16 hr at 37°.

4. Concentrate the digest to 10–20 μl under vacuum, but rigorously avoid taking the solution to dryness. Add 0.2 ml 25% (v/v) acetonitrile in 0.2% (v/v) trifluoroacetic acid, sonicate for 5 to 10 min at room temperature, then dilute to 1 ml with 0.2% (v/v) trifluoroacetic acid.

5. Centrifuge for 5 min at 14,000 g, then inject the supernatant directly onto a C_{18} high-performance liquid chromatography (HPLC) column equilibrated in 0.1% (v/v) trifluoroacetic acid. Develop the column at 1 ml min^{-1} with a 0–40% acetonitrile gradient in 0.1% trifluoroacetic acid (increasing at 0.3 or 0.6% acetonitrile/min), and measure radioactivity (Cerenkov radiation) in the eluate with an online monitor.

After trypsinolysis, transfer steps are minimized and drying avoided, to prevent irretrievable loss of the hydrophobic site 2 peptide to surface.[5] This procedure gives reproducible and clean separation of the site 1 (eluting at ~15% acetonitrile) and site 2 (~30% acetonitrile) tryptic phosphopeptides (Fig. 2), with near-quantitative recoveries of both, thereby permitting quantitative analysis of G subunit phosphorylation and dephosphorylation.[5,6] The characteristic phosphopeptide map (Fig. 2) was used to demonstrate that protein staining bands in $PP1_G$ preparations with molecular masses between 40 and 160 kDa were fragments of the G subunit,[4,5]

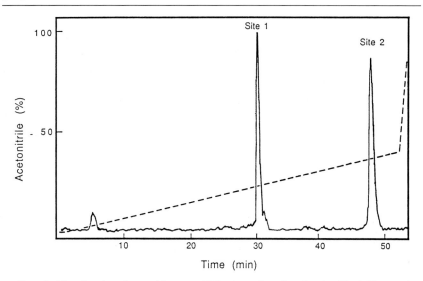

FIG. 2. Tryptic phosphopeptide map. PP1$_G$ was phosphorylated with A-kinase plus Mg[γ-^{32}P]ATP, digested with trypsin, and subjected to reversed-phase HPLC on a Vydac 218TP54 column (The Separations Group, Hesperia, CA) developed with a water/acetonitrile gradient (broken line) containing 0.1% trifluoroacetic acid. The solid line shows ^{32}P radioactivity in arbitrary units measured with an on-line monitor. The site 1 and site 2 phosphopeptides are indicated. The integrated areas under the two peaks are identical (Reproduced from Hubbard *et al.*[2])

and provided evidence that SR-associated PP1 was very similar or identical to PP1$_G$.[2]

Immunological Procedures

Preparation of Anti-Site 1 Peptide Antibodies[4]

1. Synthesize the peptide SPQPSRRGSESSEE, corresponding to the region in the G subunit surrounding the A-kinase phosphorylation site 1.

2. Prepare a peptide–albumin conjugate by mixing synthetic peptide (15 mg in 1.5 ml H$_2$O) with bovine serum albumin (20 mg in 0.5 ml 0.4 M sodium phosphate, pH 7.5) at room temperature. Add 1 ml of 20 mM glutaraldehyde dropwise with stirring over 5 min, then stir for a further 30 min. Quench free aldehyde groups by adding 0.3 ml of 1 M glycine and stirring for 30 min. Dialyze the conjugate at 4° versus phosphate-buffered saline and store in aliquots at −20°.

3. To hyperimmunize a New Zealand White rabbit, emulsify 1.5 mg

peptide–albumin conjugate in 1.5 ml Freund's complete adjuvant and inject intradermally at multiple sites in the back. Repeat after 10 days with 1 mg conjugate emulsified in incomplete adjuvant, plus a further 0.5 mg emulsified conjugate injected intramuscularly in the hind limbs. At day 21 repeat with 0.5 mg conjugate subcutaneously in the back and 1 mg intramuscularly in the legs. At day 31, inject 100 μg conjugate (no adjuvant) into the marginal ear vein. After a further 7 days, test bleed the animal and verify reactivity of the serum. Obtain the blood, allow to clot, centrifuge for 10 min at 10,000 g, and store the antiserum in aliquots at $-20°$.

4. Prepare monospecific antibodies by affinity chromatography on immobilized synthetic peptide as follows. Couple 7.5 mg site 1 peptide to 2.5 ml Affi-Gel 15 (Bio-Rad, Richmond, CA) following the manufacturer's procedure.[4] Mix 25 ml antiserum with the affinity resin for 2 hr at room temperature, pack into a 5-ml column, then wash the bed with phosphate-buffered saline until the absorbance of the eluate at 280 nm is <0.02. Elute the anti-peptide antibodies with 50 mM glycine hydrochloride, pH 2.0. Immediately neutralize each fraction with the addition of 1 M Na$_2$HPO$_4$. Pool the major 280 nm-absorbing fractions, dialyze against phosphate-buffered saline, and store in aliquots at $-20°$.

Anti-site 1 antibodies (IgG, Ref. 6) are used to quantitatively immunoprecipitate PP1$_G$[4,5,14] and to immunoblot the G subunit.[5] Proteolyzed forms of the G subunit (G' $\geq$ 40 kDa) are recognized as well as the intact G subunit. The sensitivity of detection during immunoblotting is severalfold less than with the antibody preparations described below, consistent with low epitope density resulting from the use of a peptide immunogen. In addition, the anti-site 1 antibodies do not recognize the phosphorylated G-subunit nearly as well as the dephosphorylated protein.

Preparation of Anti-Native G Subunit Antibodies[20]

1. To hyperimmunize an adult female sheep, absorb 0.16 mg homogeneous PP1$_G$ to a piece (3 × 1 cm) of nitrocellulose membrane filter (e.g., BA85; Schleicher and Schull), then quench the filter by incubating with 10 mg ml^{-1} keyhole limpet hemocyanin for 2 hr at room temperature. Rinse the filter in phosphate-buffered saline and implant it in the peritoneal cavity through a small (~1 cm) stab incision in the shaved right side. Close the wound with intramuscular and dermal sutures and spray with plastic skin. The operation is done under extensive local analgesia.

2. Emulsify 0.4 mg homogeneous PP1$_G$ in 2 ml Freund's complete

[20] M. J. Hubbard, unpublished observations (1990).

adjuvant and inject at multiple sites in the neck. Give three boost injections of 0.16 mg $PP1_G$ in 1 ml Freund's incomplete adjuvant every 2 weeks.

3. After 7 weeks (from the time of implantation), test bleed the animal and verify the presence of immunoreactivity. Obtain the antiserum, isolate the γ-globulin fraction by $(NH_4)_2SO_4$ fractionation, dialyze against Tris-buffered saline, and store at $-20°$.

Anti-$PP1_G$ antibodies obtained by this procedure react strongly with $PP1_G$ on dot blots and with the G subunit on immunoblots.

Preparation of Anti-Denatured G′ Subunit Antibodies[10]

1. Denature 2 mg $PP1_G$ by boiling for 2 min in solubilization buffer containing 2% (w/v) SDS, then subject to preparative scale SDS–polyacrylamide gel electrophoresis.
2. Lightly stain the gel with aqueous 0.1% (w/v) Coomassie Blue, locate and excise the band corresponding to the major 103-kDa G′ fragment of the G subunit, then lyophilize and powder it.
3. To hyperimmunize a sheep, emulsify the powdered gel with Freund's incomplete adjuvant and inject one-third of the material at multiple subcutaneous sites. Repeat at 2 to 3 week intervals. After 9 weeks, collect the antiserum and store at $-20°$.

Anti-G′ subunit antibodies obtained by this procedure reacted strongly with G subunit and proteolytic fragments ($\geq$30 kDa) on immunoblots (the phospho and dephospho forms stain equally well),[6] and with $PP1_G$ on dot blots,[2,8,10] but do not immunoprecipitate significant amounts of $PP1_G$.

Immunoprecipitation[4,20]

1. Dilute $PP1_G$, or an unknown phosphatase, to ~1 mU ml^{-1} and antibodies (anti-site 1, or anti-native G subunit) to 0.2 mg ml^{-1} in buffer E. An equivalent amount of nonimmune IgG is used routinely as a control. For anti-site 1 antibodies, specificity is illustrated by preincubating the antibodies with excess synthetic site 1 peptide to prevent immunoprecipitation.[4]
2. Mix 10 μl of diluted phosphatase with 10 μl antibody or control IgG and incubate for 10 min at room temperature or 30 min on ice.
3. For rabbit antibodies, precipitate immune complexes by adding 5 μl of 10% (v/v) heat-treated, formalinized *Staphylococcus aureus* suspension (e.g., Pansorbin; Calbiochem, San Diego, CA), and after 3 min at room temperature or 10 min on ice centrifuge for 2 min at 14,000 g. For sheep antibodies, use Pansorbin pretreated with an excess of rabbit anti-sheep IgG, and double the incubation time. In both cases, wash Pansorbin imme-

diately before use by centrifuging and resuspending three times in buffer E.

4. Assess specific immunoprecipitation by measuring PP1 activity in the supernatant of test control samples, using the standard phosphatase assay.

Immunoprecipitation of $PP1_G$ by antibodies to the G subunit was used to establish that the dephosphorylated G subunit is tightly associated with the C subunit[4] and showed that extensive proteolytic fragmentation of the G subunit did not destroy the site(s) of interaction with the C subunit.[4,5] Immunoprecipitation experiments also indicated that dissociation of PP1 from glycogen at high dilution results from release of the $PP1_G$ holoenzyme and not dissociation of the C subunit from the G subunit.[5] Immunoprecipitation of SR-associated PP1 was one of the experiments which indicated that this enzyme was very similar or identical to $PP1_G$,[2] while the failure to immunoprecipitate myofibrillar PP1 suggested that $PP1_M$ was a distinct species.[14]

Immunoblotting the G Subunit[5]

1. Soak SDS–polyacrylamide gel in electrotransfer buffer F for 30 min at room temperature.

2. Prepare the usual sandwich of gel with 0.45-μm nominal pore size nitrocellulose and subject to transverse electrophoresis in a Transblot apparatus (Bio-Rad) configured for high field intensity (4 cm between electrode plates). Transfer for at least 200 V hr cm^{-1} with stirred water cooling.

3. For immunodetection, quench the nitrocellulose with albumin or skim milk proteins and expose to primary antibodies and an amplified chromogenic detection system in the normal way.[5]

The 161-kDa G subunit is slow to move from the gel, being only partially transferred under conditions that give quantitative transfer of the 205-kDa myosin heavy chain.[5] The above procedure uses half the normal electrotransfer buffer concentration, and with methanol omitted to reduce heating and facilitate transfer of high molecular weight species. To avoid distortion artifacts, it is essential to equilibrate the gel thoroughly in electrotransfer buffer prior to transfer. Many low-molecular-weight species are poorly retained on the nitrocellulose under these high field transfer conditions.[5] Quantitative comparisons between different molecular weight species must therefore be approached with caution. Immunoblotting experiments are used to demonstrate the intact form (161 kDa) and degradation of the G subunit,[5] to monitor association of G subunit with glycogen under various conditions,[5,6,10] and to identify SR-associated $PP1_G$.[2]

Concluding Remarks

The procedures described here have led to increased understanding of the structure, localization, and regulation of glycogen-associated $PP1_G$. The observed effects of $PP1_G$ activity of ionic strength, pH, G–C subunit interaction, and binding of glycogen to $PP1_G$ and substrates emphasizes the importance of using near-physiological conditions for assessment of phosphatase function. Together with associated studies of myofibrillar and SR-associated PP1, this work has led to the concept of PP1 regulation in skeletal muscle by targeting subunits. It remains to be seen whether the control mechanisms now being elucidated for $PP1_G$ are general ones, utilized by PP1 in other tissues and perhaps by other phosphatases.

Acknowledgment

This work was supported by a Programme Grant and Group Support from the Medical Research Council, London, the British Diabetic Association, and the Royal Society. Michael Hubbard was a Postdoctoral Fellow of the Juvenile Diabetes Foundation International. We thank our colleagues for useful suggestions during the development of several of the methods described here.

[37] Purification of Protein-Tyrosine Phosphatases from Human Placenta

By NICHOLAS K. TONKS, CURTIS D. DILTZ, and EDMOND H. FISCHER

There is now a substantial body of evidence implicating the phosphorylation of proteins on tyrosyl residues as an essential element in the control of normal and neoplastic cell growth.[1] The phosphorylation state of a protein obviously reflects the relative activities of the kinase that phosphorylates it and the phosphatase that removes the phosphate. Thus, while considerable progress has been made in the characterization of the protein-tyrosine kinases, such a focus of attention furnishes an incomplete picture of this dynamic and reversible process. Characterization of the protein-tyrosine phosphatases (PTPases) provides a necessary complementary perspective from which to achieve an overall understanding of the control

[1] Y. Yarden and A. Ullrich, *Annu. Rev. Biochem.* **57**, 443 (1988).

of cellular function by tyrosine phosphorylation. This chapter describes procedures for the preparation of substrates and assay of PTPases, as well as the purification of a major low-molecular-weight PTPase from human placenta.

Buffers

Buffer A: 10 mM imidazole hydrochloride, pH 7.2, 5 mM EDTA, 0.25 M sucrose, 1 mM benzamidine, 0.002% (w/v) phenylmethylsulfonyl fluoride (PMSF), 2 μg/ml leupeptin, 50 U/ml aprotinin, 0.1% (v/v) 2-mercaptoethanol

Buffer B: 40 mM imidazole hydrochloride, pH 7.2, 0.2 M NaCl, 1 mM benzamidine, 0.002% (w/v) PMSF, 2 μg/ml leupeptin, 50 U/ml aprotinin

Buffer C: 40 mM imidazole hydrochloride, pH 7.2, 0.5 M NaCl, 1 mM benzamidine, 10% (v/v) glycerol, 0.05% (v/v) Triton X-100

Buffer D: 25 mM imidazole hydrochloride, pH 7.2, 1 mg/ml fatty acid- and globulin-free bovine serum albumin (BSA), 0.1% (v/v) 2-mercaptoethanol

Buffer E: 10 mM imidazole hydrochloride, pH 7.2, 2 mM EDTA, 0.1% (v/v) 2-mercaptoethanol, 1 mM benzamidine, 0.002% (w/v) PMSF

Buffer F: 5 mM imidazole hydrochloride, pH 7.2, 1 mM EDTA, 0.1% (v/v) 2-mercaptoethanol, 1 mM benzamidine, 0.002% (w/v) PMSF

Choice of Substrate

A major technical difficulty associated with the study of protein phosphatases, as compared to the kinases, is the requirement for suitably purified phosphorylated substrates. In the case of the PTPases this is even more pronounced since, in comparison to Ser and Thr phosphorylation, tyrosine phosphorylation is a rare event and the physiological substrates have yet to be clearly identified and characterized.[1] The best studied substrates are the protein-tyrosine kinases themselves, which undergo autophosphorylation; however, the large quantities of substrate required prohibit their routine use during purification procedures. Consequently a variety of artificial substrates have been utilized. Throughout our experiments reduced carboxamidomethylated and maleylated (RCM) lysozyme has been used as the standard substrate.

Preparation of RCM Lysozyme

The preparation follows a modification of the procedure of Crestfield *et al.*[2] Two grams of chicken egg white lysozyme (three times crystallized; Sigma, St. Louis, MO) is dissolved in 12.5 ml of 0.5 M Bicine–NaOH, pH 8.5, then 14.3 g of guanidine hydrochloride (ultra-pure enzyme grade) is added to denature the protein and the volume made up to 25 ml for a final concentration of 0.25 M Bicine–NaOH, pH 8.5, 6 M guanidine hydrochloride. The 4 disulfide bonds are reduced by addition of 3.8 ml of 3 M dithiothreitol (20-fold excess over protein thiol) and incubation at 50° for 90 min under nitrogen in the dark. The reaction is set up in a 40-ml screw cap glass septum vial (Pierce, Rockford, IL) fitted with a silicon/Teflon seal. Penetrate the seal with two hypodermic needles, one for nitrogen inlet, the other as an outlet and, while flushing with nitrogen, stir the reaction components with a magnetic stirring bar to facilitate removal of O_2. On removing the needles, the silicon is self-sealing. The vessels can then be wrapped in aluminum foil for incubation at 50°. After cooling on ice, 1.17 g of iodoacetamide (5.5-fold molar excess over protein thiol) is added and the solution is incubated under nitrogen at room temperature in the dark for 15 min to carboxamidomethylate the SH residues. This reaction is quenched by addition of 0.15 ml of 2-mercaptoethanol (14.3 M standard solution) and subjected to dialysis against water. At this point the protein may precipitate and can be collected by centrifugation at 15,000 g for 10 min. The pellet should be washed once with distilled water (4°) and recentrifuged as above; this helps to remove any remaining reducing agent. Following washing, the pellet is redissolved by addition of 6 M urea, 50 mM Bicine–NaOH, pH 8.5, in a final volume of 100 ml. The amino groups are maleylated by addition of 1.64 g maleic anhydride (20-fold molar excess over lysyl residues) dissolved in ~20 ml of dioxane. This is added in small aliquots, stirring the reaction mixture on ice and maintaining the pH at ~8.7 by addition of 1 N NaOH (the pH optimum for the reaction is 8.5–9). After adding all of the maleic anhydride, the solution is left on ice overnight. Due to the presence of urea the solution may become turbid; it can be reclarified by warming to room temperature.

The modified protein is desalted at room temperature on Sephadex G-25 Superfine (Pharmacia, Piscataway, NJ) equilibrated in 10 mM NH$_4$OH. Routinely, aliquots of 40 ml are applied to a 5 × 20 cm column at a flow rate of ~200 ml/hr, although larger aliquots can be processed over bigger columns. Protein is detected by UV absorption at 280 nm (A_{280}), pooled,

[2] A. M. Crestfield, S. Moore, and W. H. Stein, *J. Biol. Chem.* **238,** 622 (1963).

and lyophilized. It may be stored as a lyophilized powder; however, routinely, we redissolve it in 0.5 M imidazole hydrochloride, pH 7.2 ($\sim$20 mg/ml) and dialyze it against the same buffer at 50 mM concentration to complete the solubilization. The resulting solution is then stored in $\sim$5-ml aliquots at $-20°$.

Preparation of Protein-Tyrosine Kinase

Human placenta is chosen as the tissue source because it is rich in insulin and epidermal growth factor (EGF) receptor protein-tyrosine kinases and is relatively easily available. The receptors are partially purified as follows.

Placentas are obtained by arrangement with local hospitals, transported to the laboratory on ice, and normally processed as far as the first homogenization step within 1–2 hr after delivery. Normal standard precautions for the handling of human tissue (rubber gloves, goggles, etc.) should be observed. Routinely, the preparation begins from one placenta (250–300 g). After scraping away the supporting membranes, the tissue is washed extensively in 10 mM imidazole hydrochloride, pH 7.2, 150 mM NaCl to remove blood clots, then homogenized for three periods of 15 sec each at high speed in a Waring blender in 2 vol of buffer A minus 2-mercaptoethanol. After centrifugation at 6000 g for 30 min at 4°, the pellet is discarded and the supernatant further centrifuged at 27,000 g for 60 min. This second pellet is resuspended in $\sim$200 ml of buffer B and centrifuged at 45,000 g for 30 min. The resulting washed membrane pellet is resuspended in $\sim$30 ml of buffer B containing 10% (v/v) glycerol (final volume $\approx$ 50 ml, including membranes) and can either by stored in aliquots at $-70°$, or further processed directly. Normally the subsequent stages are carried out with membranes equivalent to $\sim$100 g of placental tissue.

Leupeptin, lima bean trypsin inhibitor, and Triton X-100 are added to final concentrations of 6 μg/ml, 20 μg/ml, and 2% (v/v), respectively, and the suspension is left to stir on ice for 45 min, then centrifuged at $\sim$150,000 g for 30 min at 4°. The supernatant is mixed with $\sim$30 ml of wheat germ lectin-Sepharose equilibrated in buffer C and gently rotated end over end at 4° overnight. The slurry is poured into a 2.5-cm diameter column and the breakthrough collected and stored separately. After washing the gel with buffer C, bound receptor protein-tyrosine kinase is eluted with buffer C containing 0.3 M N-acetylglucosamine. Fractions of $\sim$3 ml are collected and the peak of protein, estimated by Bradford,[3] is pooled. The column can be regenerated by washing with $\sim$200 ml of 1 mM acetic acid, then reequilibrated in buffer C. Routinely, the breakthrough from the first pass

[3] M. M. Bradford, *Anal. Biochem.* **72**, 248 (1976).

over the column is recombined with the wheat germ lectin-Sepharose, rotated end over end at 4° for ~3 hr, and eluted as above. Generally, this second step recovers less than 30% of the protein obtained in the first run. The two pools are combined and concentrated to ~3 ml in an Amicon (Danvers, MA) stirred cell equipped with a YM10 membrane.

When protein-tyrosine kinase assays are conducted on this pool following the phosphorylation procedure described below for RCM lysozyme [with the exception that the 4 : 1 random poly(Glu : Tyr) copolymer is used as substrate at 2 mg/ml final concentration and ATP is present at 0.2 mM] one routinely recovers ~30 units of kinase activity from a 300-g placenta. One unit equals 1 nmol phosphate incorporated into 4 : 1 poly(Glu : Tyr)/min. Note that PTPase inhibitors such as vanadate and molybdate must be included to overcome the problem of phosphatases contaminating the kinase preparation.

Phosphorylation of RCM Lysozyme

The phosphorylated substrate is routinely prepared in aliquots of ~50 mg. RCM lysozyme (2 mg/ml final concentration) is phosphorylated by the partially purified kinase preparations in a reaction mixture that contains the following (at final concentrations): 40 mM imidazole hydrochloride, pH 7.2, 50 mM NaCl, 12 mM magnesium acetate, 4 mM MnCl$_2$, 0.1 mM Na$_3$VO$_4$, 0.2 mM EGTA, 0.05% (v/v) Triton X-100, and 3% (v/v) glycerol (added as a cocktail in which each component is at 10 times the final concentration). Insulin and epidermal growth factor (EGF) are added to 300 and 200 nM, respectively. The receptor kinases are routinely used at 0.3–0.6 U/ml, resulting in a final concentration of ~30 mM N-acetylglucosamine.

Dithiothreitol is added to 2 mM and [γ-^{32}P]ATP (~200 cpm/pmol) is usually included at 4 mM. Ammonium molybdate (an additional PTPase inhibitor), is included at 100 μM and deoxycholate is added to 0.2% (w/v). The reaction is initiated by addition of RCM lysozyme and allowed to proceed at 30° overnight. A white precipitate, most likely a complex between deoxycholate and metal ions, is detectable after prolonged incubation. Incorporation of phosphate is assayed by spotting an aliquot of the phosphorylation mix (usually 10 μl) onto a P81 phosphocellulose paper square (Whatman, Clifton, NJ), followed by washing three times in 75 mM H$_3$PO$_4$ and counting in scintillant.[4] A parallel control incubation in the absence of RCM lysozyme indicates that autophosphorylation of the kinase routinely accounts for <5% of the total counts incorporated.

[4] R. Roskoski, this series, Vol. 99, p. 3.

RCM Lysozyme: T N R N T D G S T D <u>Y</u> G I L Q I N S R W W

MBP: D G H H A A R T T H <u>Y</u> G S L P Q K A Q G H

FIG. 1. Phosphorylation of RCM lysozyme and myelin basic protein (MBP) by insulin and EGF receptor protein-tyrosine kinases: identity of phosphorylation sites. RCM lysozyme and MBP were phosphorylated using $[\gamma\text{-}^{32}P]$ATP and a wheat germ lectin-Sepharose-purified preparation of insulin and EGF receptors, as described in the text. Only phosphotyrosine was detected on phosphoamino acid analysis (data not shown). The proteins were digested with trypsin and the tryptic peptides separated by reversed-phase high-performance liquid chromatography. In both cases >80% of the ^{32}P radioactivity was recovered in a single peak, which was pooled and the phosphotyrosine-containing peptide identified by amino acid sequence determination using a Beckman model 890C protein sequencer. The phosphorylated residues, which did not extract as the phenylthiohydantoin, are highlighted. Single-letter abbreviations are used for the amino acids. A, Alanine; R, arginine; N, asparagine; D, aspartic acid; C, cysteine; Q, glutamine; E, glutamic acid; G, glycine; H, histidine; I, isoleucine; L, leucine; K, lysine; M, methionine; F, phenylalanine; P, proline; S, serine; T, threonine; W, tryptophan; Y, tyrosine; V, valine.

The reaction is terminated by addition of 100% (w/v) trichloroacetic acid to a final concentration of 10%; the advantage of this step is that it will also denature contaminating PTPases. After ~30 min on ice, the suspension is centrifuged at 27,000 g for 15 min at 4°, the supernatant discarded, and the protein pellet washed three times further with 20% (w/v) trichloroacetic acid to remove $[\gamma\text{-}^{32}P]$ATP. The pellet is disrupted in ~3 ml of 2 M Tris base, using a glass rod, and then left on ice overnight; any remaining particles are broken up in a glass Dounce homogenizer. After ~2 hr further on ice, the sample is dialyzed against 50 mM imidazole hydrochloride, pH 7.2 (four changes of 1 liter each) and stored at 4°.

Stoichiometries of up to 0.5 mol ^{32}P incorporated/mol protein can be achieved with ~1 U/ml protein-tyrosine kinase. At the kinase concentrations routinely used, the stoichiometry is generally in the range 0.2–0.3 mol/mol. Labeling occurs exclusively on tyrosine and the site of phosphorylation has been identified (Fig. 1).

Assay of PTPases

Principle of Assay

The assay is based on the release of $[^{32}P]P_i$ from phosphorylated protein substrate. That inorganic phosphate is being released, and not phosphopeptides, can be confirmed by applying the molybdate/isobutanol/benzene extraction method described in Antoniw and Cohen.[5] One unit of activity

[5] J. F. Antoniw and P. Cohen, Eur. J. Biochem. 68, 45 (1976).

is defined as the amount of enzyme that catalyzes the release of 1 nmol of phosphate/min from the substrate. Routinely, the extent of dephosphorylation is kept below 30% to ensure linear kinetics.

Procedure

The reaction is carried out in a 1.5-ml plastic microcentrifuge tube. A mixture of 0.02 ml of sample containing·PTPase (diluted to less than 1 U/ml in buffer D) and 0.02 ml of buffer D (or whatever "additions" are being tested for their effect on activity) are preincubated at 30°, for ~5 min. The assay is initiated by addition of 0.02 ml substrate (routinely at a final concentration of 5 μM phosphotyrosine) and allowed to proceed at 30° for up to 10 min. The reaction is terminated by the addition of 0.18 ml of 20% (w/v) trichloroacetic acid and 0.02 ml of 25 mg/ml BSA added as carrier protein. Samples are normally frozen briefly to facilitate the precipitation of proteins, thawed, then centrifuged at 12,500 g for 5 min in a microcentrifuge. A 0.2-ml aliquot of the supernatant is counted in 1 ml of scintillant. Blank incubations in which the PTPase is replaced by buffer D should be included; total radioactivity in the assay is determined by counting 0.02 ml of the substrate. Counts released in the blank are routinely less than 2% of the total radioactivity in the assay. If the specific activity of ATP used in the phosphorylation of RCM lysozyme is 200 cpm/pmol, a final substrate concentration of 5 μM phosphotyrosine will be equivalent to 60,000 cpm in the assay. Substrate prepared in this way may be used at this concentration for approximately two half-lives of ^{32}P. One should set a lower limit of 10,000–15,000 cpm in the assay in order to detect significant release of [^{32}P]P$_i$ above background within the linear range of dephosphorylation.

Comments on Substrate Preparation and Assay of PTPases

RCM lysozyme offers many advantages over other chemically modified proteins that have been used previously. A problem frequently encountered is that phosphorylation of potential substrates occurs to low stoichiometry and at multiple sites, resulting in nonlinear kinetics of dephosphorylation. However, RCM lysozyme is phosphorylated on a single tyrosyl residue (Y^{53}), to a relatively high stoichiometry; linear rates of dephosphorylation are observed up to ~50% ^{32}P released. Furthermore, maleylation of the lysine side chains in the modified lysozyme significantly improves its solubility, thus expanding the range of concentrations at which it can be used in an assay; it also eliminates the need for harsh conditions to solubilize the modified protein that may lead to its hydrolysis.

Bovine brain myelin basic protein (MBP) is also phosphorylated to relatively high levels (~0.3 mol/mol) under the same conditions. Again, a single site (Y^{67})(is modified (Fig. 1). However, in this case, no prior chemical modification of the protein is required. While the low-molecular-weight PTPases dephosphorylate both of these substrates with similar high efficiency,[6] integral membrane PTPases such as CD45 greatly favor MBP.[7] The preparation of MBP is detailed elsewhere in this volume.[8]

Purification of PTPases

Human placenta was selected as a tissue source because of the inherent interest attached to the study of a human enzyme, the ready availability of the tissue, and the fact that it contains high levels of PTPase activity, comparable to those found in the kidney, which was previously identified as one of the richest sources of the enzyme.[9]

Classification of Low-Molecular-Weight PTPases

PTPase activity distributes roughly equally between the soluble (aqueous buffer extract) and particulate (Triton X-100-containing buffer extract) fractions of human placenta, with at least four distinct low-molecular-weight forms observed in each. In addition, Brautigan's group (Brown University, Providence, RI) has presented evidence for a cytoskeletal form that resists Triton extraction.[10] Current understanding of the characteristics of these enzymes is somewhat limited and thus it would be premature to introduce a system of nomenclature at this time. However, our method of classification has been based on the pattern of elution of the enzymes from ion-exchange columns. On passage of the placenta extract over DEAE-cellulose equilibrated in buffer E, 70–80% of the PTPase activity is retained on the column and is designated type 1. When the breakthrough fraction is subsequently passed over phosphocellulose equilibrated in the same buffer, the activity retained is designated as type 2 and that emerging in this second breakthrough, as type 3. After a number of additional purification steps, PTPases 2 and 3 each seem to compose a single major species (J. A. Lorenzen, N. K. Tonks, and E. H. Fischer, unpublished observations). The type 1 activity can be resolved into two peaks by gradient elution from DEAE-cellulose; these are termed 1A and 1B, based on their order of elution at ~50 mM and ~100 mM NaCl,

[6] N. K. Tonks, C. D. Diltz, and E. H. Fischer, *J. Biol. Chem.* **263**, 6731 (1988).

[7] N. K. Tonks, C. D. Diltz, and E. H. Fischer, *J. Biol. Chem.* **265**, 10674 (1990).

[8] N. K. Tonks, C. D. Diltz, and E. H. Fischer, this volume [38].

[9] C. L. Shriner and D. L. Brautigan, *J. Biol. Chem.* **259**, 11383 (1984).

[10] J. Roome, T. O'Hare, P. F. Pilch, and D. L. Brautigan, *Biochem. J.* **256**, 498 (1988).

respectively (Fig. 2). PTPase 1B constitutes the major activity toward RCM lysozyme. In order to illustrate the principles of the purification procedure, the subsequent discussion will be largely restricted to this form.

Construction of Substrate Affinity Column

The key feature of the purification procedure is the application of a thiophosphorylated substrate affinity-chromatography step. Many kinases, including the insulin and EGF receptors, can utilize ATPγS as a phosphate donor. While the thiophosphorylated product displays a high affinity for the phosphatase, it is resistant to dephosphorylation. Therefore by immobilizing thiophosphorylated RCM lysozyme on CNBr-Sepharose, an affinity support is generated on which the PTPase can be retained and subsequently eluted with a salt gradient. This technique has been successfully applied to the purification of several protein-Ser/Thr phosphatases.[11,12]

RCM lysozyme is thiophosphorylated with [γ-^{35}S]ATP so that the stoichiometry of reaction (~0.25 mol/mol), the efficiency of coupling to CNBr-Sepharose, and the extent of any dethiophosphorylation can all be easily monitored. The derivative is dialyzed against coupling buffer (0.1 M sodium borate, pH 8.2, 0.5 M NaCl), mixed with CNBr-Sepharose (Pharmacia, Piscataway, NJ) (1 g/4 mg thiophosphoryl protein) that had been first swelled in 1 mM HCl, then equilibrated in the same buffer, and rotated end over end at 4° overnight. Excess protein is removed by washing the gel with coupling buffer in a Büchner funnel, then any remaining active groups are blocked by rotating end over end at 4° overnight in 1 M ethanolamine hydrochloride, pH 8.0. The product is then washed with 0.1 M sodium acetate, pH 5.0, 1 M NaCl, followed by 0.1 M sodium borate, pH 8.2, 1 M NaCl; this cycle of washing is performed three times to remove noncovalently bound protein. The gel is then equilibrated in buffer F and stored at 4°. Routinely, the coupling efficiency with RCM lysozyme is ~80%.

Purification Procedure

All steps are performed at 4° (see Table I). Placenta tissue, washed in imidazole-buffered saline as described for the isolation of the receptor protein-tyrosine kinases, is homogenized in 2 vol of buffer A for three 15-sec bursts in a Waring blender set at high speed. After centrifugation at

[11] M. D. Pato and R. S. Adelstein, J. Biol. Chem. 255, 6535 (1980).
[12] N. K. Tonks and P. Cohen, Biochim. Biophys. Acta 747, 191 (1983).

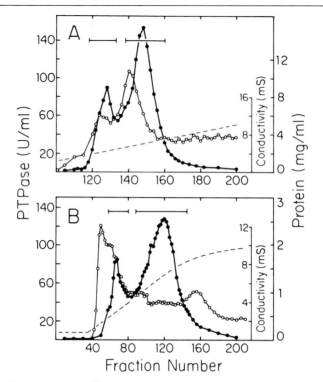

FIG. 2. Chromatography of human placenta extracts on DE-52. Aqueous buffer (soluble) and Triton X-100-containing buffer (particulate) extracts from 6.5 kg of human placenta were prepared and subjected to batchwise chromatography on DE-52 as described in the text. Each was rechromatographed on DE-52 eluting with a linear salt gradient as follows: (A) Soluble fraction: Column dimensions, 7.5 × 65 cm; gradient, 0–0.3 M NaCl in 8 liters of buffer E. Fractions of 35 ml were collected at 525 ml/hr. (B) Particulate fraction: Column dimensions, 5 × 27 cm; gradient, 0–0.25 M NaCl in 4 liters of buffer E. Fractions of 13 ml were collected at 200 ml/hr. (●), PTPase activity; (○), protein; (----), conductivity. The horizontal bars indicate the fractions that were pooled. Two peaks of activity, assayed with RCM lysozyme, were resolved; further purification of the second, quantitatively larger peak, PTPase 1B, is described in the text. (Reproduced from N. K. Tonks et al.[15])

27,000 g for 60 min, the supernatant is filtered through glass wool and stored on ice. The pellet is rehomogenized as above in 2 vol of buffer A plus 0.5% (v/v) Triton X-100. After stirring for 60 min, the suspension is centrifuged at 27,000 g for 60 min and the supernatant filtered through glass wool, then combined with the aqueous extract. The pellet is discarded.

If the extract is to be stockpiled while sufficient placentas are collected for a bulk preparation, the following steps are included. DE-52 cellulose equilibrated in buffer E is added to the extract (~30 ml of slurry/g protein)

TABLE I
PURIFICATION OF PLACENTA PTPASE 1B[a]

| | Soluble | | Particulate | |
Step	Specific activity (U/mg)	Purification (-fold)	Specific activity (U/mg)	Purification (-fold)
Extract	2	1	10	1
DEAE-Cellulose	14	7	119	12
Sephacryl S200	33	17	255	26
Affi-Gel Blue	141	70	613	61
Polylysine-Sepharose	371	186	675	68
FPLC gel filtration	6,300	3,150	2,450	245
RCM lysozyme- Sepharose affinity chromatography	46,780	23,390	42,710	4,270

[a] Data taken from Ref. 15.

and the suspension stirred overnight. The DE-52 is then allowed to settle out. The supernatant is decanted and the DE-52 resuspended in 2 vol of buffer, then collected in a 25-cm diameter Büchner funnel and washed further with 5 vol of buffer. The breakthrough and wash fractions containing PTPases 2 and 3 will not be considered further. Bound PTPase 1 is eluted stepwise by removing the DE-52 from the funnel and stirring in two bed volumes of buffer E plus 0.3 M NaCl for 20 min. The slurry is returned to the funnel, the eluate collected under mild suction, and the process is then repeated. This procedure minimizes the elution volume. Solid ammonium sulfate is added to 70% saturation (470 g/liter), the suspension stirred for 60 min, and centrifuged at 5000 g for 60 min. The pellet is resuspended in a minimum volume of buffer E, dialyzed against the same buffer, and stored at $-70°$ after a final dialysis versus buffer E plus 50% (v/v) glycerol. Routinely the next steps are not initiated until at least 3 kg of placenta has been thus processed.

The frozen pools are allowed to thaw overnight at 4°, then dialyzed against three changes of buffer E. The dialysate is clarified by centrifugation at 16,000 g for 20 min and the conductivity of the supernatant adjusted to ~0.5 mS by addition of 0.1% (v/v) 2-mercaptoethanol, 1 mM benzamidine, 0.002% (w/v) PMSF. One liter of a packed slurry of DE-52, equilibrated in buffer E, is added, the mixture stirred for ~60 min, then poured onto a 2-liter bed of DE-52, equilibrated in the same buffer prepacked in a 7.5-cm diameter column. The column is washed with ~4 liters of equilibration buffer, then developed with a linear salt gradient from 0 to

0.3 M NaCl in a total volume of 14 liters of buffer E. Fractions of 35 ml
are collected at a flow rate of ~500 ml/hr. Activity is assayed across the
profile and the peak containing PTPase 1B is pooled.

The sample is diluted with buffer E until the conductivity has been
reduced to ~3 mS. To this is added ~450 ml of Affi-Gel Blue (at least 1
ml/10 mg protein) equilibrated in buffer E, and the mixture stirred for 2
hr. The Affi-Gel Blue is collected on a Büchner funnel, washed with 7–10
bed volumes of buffer E, then further washed with a final volume of 1500
ml of buffer E containing 0.8 M NaCl. This final step is performed by
removing the gel from the funnel and stirring for 2 hr in ~750 ml of elution
buffer, then returning the slurry to the funnel and collecting the eluate
under mild suction. This process is repeated and the two eluate fractions
are combined and dialyzed versus three changes of buffer E.

The dialysate is adsorbed onto a 2.5 × 16 cm column of polylysine-
Sepharose [prepared by covalently cross-linking 0.5 g of 16-kDa polylysine
(Sigma) to 15 g CNBr-Sepharose, as described earlier], equilibrated in
buffer E plus 10% (v/v) glycerol. The column is washed with ~5 bed
volumes of equilibration buffer, then developed with a linear salt gradient
from 0 to 0.5 M NaCl in a total volume of 1 liter buffer E containing
10% (v/v) glycerol. Fractions of ~12 ml are collected at a flow rate of
120 ml/hr.

The peak of activity is pooled and dialyzed versus buffer E, then
adsorbed onto a 2-ml pad of DE-52 equilibrated in the same buffer. Bound
protein is eluted with buffer E containing 0.35 M NaCl. This manipulation
is performed in a Bio-Rad (Richmond, CA) disposable Poly Prep column,
the sample being loaded and eluted by gravity. The eluate is collected in
0.5-ml fractions and pooled according to protein estimated by Bradford.[3]
Samples of ~2.5 ml are applied to a Pharmacia FPLC (fast protein liquid
chromatography) gel-filtration system comprising one Superose 6 and two
Superose 12 columns (each 1.6 × 50 cm) connected in series and equilibrated in buffer E containing 50 mM NaCl. Fractions of 2 ml are collected
at a flow rate of 90 ml/hr. If an FPLC system were not available, Sephadex
G-75 superfine would be a suitable alternative. The peak of activity is
pooled so as to minimize contamination by higher molecular weight protein
present in the accompanying peak, and the material obtained dialyzed
against buffer F.

The dialysate is adsorbed onto a 2.5 × 15 cm column of thiophosphorylated RCM lysozyme-Sepharose in the same buffer, washed with three to
four bed volumes, and then eluted with a linear salt gradient from 0 to 0.25
M NaCl in 300 ml of buffer F. Fractions of 3 ml are collected at a flow rate
of 120 ml/hr. The peak fractions can either be pooled or stored separately
at −70° following addition of glycerol to ~30% (v/v). In this form the

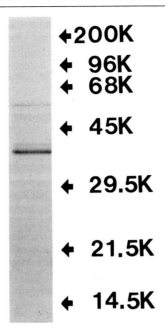

+200K
+ 96K
+ 68K

+ 45K

+ 29.5K

+ 21.5K

+ 14.5K

FIG. 3. SDS–polyacrylamide gel electrophoresis of PTPase 1B. One microgram of protein from the pooled peak of activity eluted from the thiophosphorylated substrate affinity column was subjected to electrophoresis on a 12% polyacrylamide gel, according to Laemmli.[12a] The gel was stained with Coomassie Brilliant Blue and the arrows denote the position of the marker proteins: myosin (200 kDa), phosphorylase b (96 kDa), bovine serum albumin (68 kDa), ovalbumin (45 kDa), carbonic anhydrase (29.5 kDa), trypsin inhibitor (21.5 kDa), and lysozyme (14.5 kDa).

enzyme is very stable, with negligible loss of activity over a 6-month period. From 3 kg of placenta one should obtain a total of ~0.5 mg of PTPase 1B. SDS–PAGE of the purified product is depicted in Fig. 3.[12a]

Comments on Purification

A major difficulty encountered is the "stickiness" of the purified protein. Therefore, in preparation for its storage, it is advisable to add glycerol rather than dialyze against a glycerol-containing buffer. Attempts to concentrate the pure enzyme by either ultrafiltration in Amicon stirred cells or dialysis procedures will lead to substantial losses of activity. If the

[12a] U. K. Laemmli, *Nature (London)* **227**, 680 (1970).

protein is required in a more concentrated form,[13,14] a microscale DE-52 concentration step similar to that described above can be applied. The enzyme also sticks to glass, so plastic test tubes, measuring cylinders, etc., should be utilized when handling the affinity column eluate.

The procedure described here represents a streamlined version of that in Tonks et al.[15] The early gel-filtration step on Sephacryl S-200 has been eliminated and gradient elution from Affi-Gel Blue has been replaced by a batchwise chromatography step on the same support. Furthermore, since determination of the complete amino acid sequence of the soluble and particulate forms of the enzymes revealed no difference between the molecules,[16] the aqueous and Triton X-100 extracts can be combined at the beginning of the preparation. Although the resulting product is only ~80–90% pure (Fig. 3), mixing the two extracts makes the purification far more convenient.

Many of the substrate affinity columns utilized in the purification of Ser/Thr phosphatases have been hampered by a low binding capacity. However, a 1.5 × 15 cm column of the RCM lysozyme derivative is not saturated even with ~0.5 mg of PTPase 1B. This improvement may be due to increased accessibility of the substrate: by maleylation of the lysine side chains in RCM lysozyme, the number of free amino groups available for cross-linking to CNBr-Sepharose will be dramatically reduced. Thus the chance of immobilizing the protein in a manner that "masks" the phosphorylated residue will be decreased. Perhaps such procedures may be applicable to other similar substrate affinity columns.

Characteristics of PTPase 1B

When oligonucleotide probes, based on the sequence of PTPase 1B, were used to screen a human peripheral T cell cDNA library, a clone was isolated which encoded a protein of 48 kDa that displayed 85% sequence similarity to PTPase 1B within the conserved core structure[17]; however, this species bore an ~11-kDa extension at the C terminus. Subsequently, several groups have isolated cDNAs for PTPase 1B that all encode proteins

[13] N. K. Tonks, M. F. Cicirelli, C. D. Diltz, E. G. Krebs, and E. H. Fischer, Mol. Cell. Biol. 10, 458 (1990).

[14] M. F. Cicirelli, N. K. Tonks, C. D. Diltz, E. H. Fischer, and E. G. Krebs, Proc. Natl. Acad. Sci. U.S.A. 87, 5514 (1990).

[15] N. K. Tonks, C. D. Diltz, and E. H. Fischer, J. Biol. Chem. 263, 6722 (1988).

[16] H. Charbonneau, N. K. Tonks, S. Kumar, C. D. Diltz, M. Harrylock, D. E. Cool, E. G. Krebs, E. H. Fischer, and K. A. Walsh, Proc. Natl. Acad. Sci. U.S.A. 86, 5252 (1989).

[17] D. E. Cool, N. K. Tonks, H. Charbonneau, K. A. Walsh, E. H. Fischer, and E. G. Krebs, Proc. Natl. Acad. Sci. U.S.A. 86, 5257 (1989).

of ~50 kDa, also bearing C-terminal extensions relative to the placenta protein.[18-20] Thus, PTPase 1B appears to have been isolated from human placenta as a truncated species; whether this represents a physiologically relevant processing event or an artifact of purification remains to be established. At least for the T cell PTPase, the C-terminal tail appears to be involved in modulation of enzyme activity and control of intracellular localization.[21,22]

A feature exhibited by all the PTPases characterized to date is their total dependence on sulfhydryl compounds for activity. Without 2-mercaptoethanol or dithiothreitol in the assay buffer, no activity will be detected. The enzymes are also irreversibly inhibited by alkylating agents such as iodoacetic acid, suggesting the presence of at least one reactive cysteinyl residue that is essential for catalysis.[6,7] Two cysteines (residues 121 and 215 in PTPase 1B[16]) are conserved in all PTPase sequences identified thus far, and evidence has been presented indicating that site-directed mutagenesis to convert at least one of these to Ser inactivates the enzyme.[23]

There is no inhibitor at present that is absolutely specific for the PTPases. While all PTPases are inhibited by micromolar concentrations of vanadate and molybdate, PTPase 1B is the most sensitive to inhibition by heparin (IC_{50} 20 nM) or poly(Gly : Tyr) (1 : 1 and 4 : 1, K_I 50 nM) of those so far tested: for instance, PTPase 1A requires a two-orders-of-magnitude higher concentration for inhibition.[6] However, when added *in vivo* these reagents will exhibit pleiotropic effects not restricted to phosphatases.

PTPase 1B is absolutely specific for phosphotyrosyl residues and dephosphorylates several artificial substrates with similar high affinity and specific activity.[6] It can also dephosphorylate proteins known to be phosphorylated on tyrosyl residues *in vivo,* such as the insulin receptor[6] or p56[lck].[7] Furthermore, on microinjection into *Xenopus* oocytes, PTPase 1B is capable of antagonizing the action of insulin and dephosphorylating tyrosyl residues in a molecule of the same apparent molecular weight as

[18] K. Guan, R. S. Haun, S. J. Watson, R. L. Geahlen, and J. E. Dixon, *Proc. Natl. Acad. Sci. U.S.A.* **87,** 1501 (1990).
[19] J. Chernoff, A. R. Schievella, C. A. Jost, R. L. Erikson, and B. G. Neel, *Proc. Natl. Acad. Sci. U.S.A.* **87,** 2735 (1990).
[20] S. Brown-Shimer, K. A. Johnson, J. B. Lawrence, C. Johnson, A. Bruskin, N. R. Green, and D. E. Hill, *Proc. Natl. Acad. Sci. U.S.A.* **87,** 5148 (1990).
[21] D. E. Cool, N. K. Tonks, H. Charbonneau, E. H. Fischer, and E. G. Krebs, *Proc. Natl. Acad. Sci. U.S.A.* **87,** 7280 (1990).
[22] N. K. Tonks, *Curr. Opin. Cell Biol.* **2,** 1114 (1990).
[23] M. Streuli, N. X. Krueger, A. Y. M. Tsai, and H. Saito, *Proc. Natl. Acad. Sci. U.S.A.* **86,** 8698 (1989).

the β subunit of the insulin receptor.[13,14] Clearly, the microinjection of PTPases into cells or overexpression of their cDNAs may be a useful technique for defining roles for protein-tyrosine phosphorylation in signal transduction processes. In addition, the overexpression of PTPases may counteract the action of oncogenic protein-tyrosine kinases and confer resistance to transformation or reverse a transformed phenotype.

The determination of the primary amino acid sequence of PTPase 1B demonstrated that, unlike the protein kinases that are all derived from a common ancestor, the PTPases compose a unique family of enzymes distinct from the protein-Ser/Thr phosphatases. However, extensive homology has been detected between PTPase 1B and the intracellular segment of CD45, the leukocyte common antigen.[16,24] Isolated CD45 possesses PTPase activity,[7,25] providing evidence for a novel class of receptor-linked enzymes with the potential to initiate signal transduction processes by the dephosphorylation of tyrosyl residues in target proteins.[26] The purification and assay of CD45 are described in [38] in this volume.[8]

Acknowledgments

The amino acid sequence analysis described in Fig. 1 was performed in collaboration with Ken Walsh, Department of Biochemistry, University of Washington. This work was supported by the National Institutes of Health Grant DK07902 from the National Institute of Diabetes and Digestive and Kidney Diseases, GM42508 from the National Institute of General Medical Sciences, and by a grant from the Muscular Dystrophy Association of America. We thank Carmen Westwater for typing the manuscript.

[24] H. Charbonneau, N. K. Tonks, K. A. Walsh, and E. H. Fischer, *Proc. Natl. Acad. Sci. U.S.A.* **85**, 7182 (1988).
[25] N. K. Tonks, H. Charbonneau, C. D. Diltz, E. H. Fischer, and K. A. Walsh, *Biochemistry* **27**, 8695 (1988).
[26] N. K. Tonks and H. Charbonneau, *Trends Biochem. Sci.* **14**, 497 (1989).

[38] Purification and Assay of CD45: An Integral Membrane Protein-Tyrosine Phosphatase

By Nicholas K. Tonks, Curtis D. Diltz, and Edmond H. Fischer

The determination of the amino acid sequence of protein-tyrosine phosphatase (PTPase) 1B, a major low-molecular-weight PTPase isolated from human placenta (see [37] in this volume) demonstrated that this enzyme is not structurally related to the protein-Ser/Thr phosphatases.[1,2] This is unlike the protein kinases that are all derived from a common ancestor.[3] However, a homologous relationship was established between PTPase 1B and each of the two, internally homologous domains of CD45, the leukocyte common antigen.[1,2]

CD45 represents a family of high-molecular-weight (180k–220k) integral membrane proteins, the expression of which is restricted to cells of the hematopoietic lineage.[4] The molecule can be defined in terms of 3 segments. There is a heavily O- and N-glycosylated extracellular segment, that varies in size from ~400–550 residues and bears a cysteine-rich putative ligand-binding motif. The variable molecular weight of members of this family is associated with the differential expression of three exons encoding sequences in this segment, at the extreme N terminus of the molecule. The intracellular segment is highly conserved between isoforms and comprises two homologous domains of ~300 residues, each structurally equivalent to one PTPase molecule. The intracellular and extracellular segments are connected by a single transmembrane hydrophobic stretch of 22 residues. Considering its receptor-like configuration and following the demonstration of intrinsic PTPase activity,[5] CD45 may be regarded as a prototype for a new class of receptor-linked molecules with the potential to initiate novel pathways of signal tranduction via the dephosphorylation of tyrosyl residues in proteins.[6] A simple procedure for the purification and assay of CD45 is described.

Antibodies to CD45

A number of antibodies directed against the extracellular segment of CD45 are available.[7] Some recognize restricted epitopes and are specific for particular isoforms, e.g., 2H4 and UCHL-1, whereas others are categorized as pan-anti-CD45 antibodies, recognizing all CD45 family members in a particular species.[4,7] Our studies utilize an antibody termed 9.4,[8] which

[1] H. Charbonneau, N. K. Tonks, K. A. Walsh, and E. H. Fischer, Proc. Natl. Acad. Sci. U.S.A. 85, 7182 (1988).
[2] H. Charbonneau, N. K. Tonks, S. Kumar, C. D. Diltz, M. Harrylock, D. E. Cool, E. G. Krebs, E. H. Fischer, and K. A. Walsh, Proc. Natl. Acad. Sci. U.S.A. 86, 5252 (1989).
[3] S. K. Hanks, A. M. Quinn, and T. Hunter, Science 241, 42 (1988).
[4] M. L. Thomas, Annu. Rev. Immunol. 7, 339 (1989).
[5] N. K. Tonks, H. Charbonneau, C. D. Diltz, E. H. Fischer, and K. A. Walsh, Biochemistry 27, 8695 (1988).
[6] N. K. Tonks and H. Charbonneau, Trends Biochem. Sci. 14, 497 (1989).
[7] A. J. McMichael, ed., "Leukocyte Typing," Vol. 3. Oxford Univ. Press, Oxford, 1987.
[8] S. Cobold, G. Hale, and H. Waldmann, in "Leukocyte Typing" (A. J. McMichael, ed.), Vol. 3, p. 788. Oxford Univ. Press, Oxford, 1987.

is from this latter category; however, any pan-anti-CD45 antibody should be appropriate.

Tissue Source

The choice of antibody will influence the source of biological material from which CD45 is to be purified. Thus 9.4 is species specific, recognizing human CD45, but not the murine molecule. Nevertheless, similar strategies can be applied to the purification of these molecules from mouse or rat. Furthermore, since CD45 is restricted to cells of the hematopoietic lineage and because of the severe limitations in the quantity of starting material that would result from the use of cultured cells, human spleen was chosen as the tissue source. This material was obtained fresh following surgical excision by arrangement with local hospitals.

Preparation and Phosphorylation of Myelin Basic Protein as Substrate for CD45

The fundamentals of the assay for CD45 are the same as described for PTPase 1B in [37] in this volume.[9] Activity is measured by following the release of $[^{32}P]P_i$ from a phosphorylated substrate. In this case the substrate of choice is myelin basic protein (MBP), isolated according to the procedure of Diebler et al.[10]

Twenty grams of bovine brain acetone powder (purchased from Sigma, St. Louis, MO) is stirred with 400 ml of 30 mM HCl for 30 min. At this time, the pH is adjusted to 3.0 with HCl and the mixture is stirred for a further 60 min at 4°, then centrifuged at ~13,000 g for 30 min at 4°. The supernatant is filtered by gravity through Whatman (Clifton, NJ) 114 paper and sufficient 8 M urea added to give a final concentration of 2 M. Whatman DE-52 cellulose is then added with stirring to titrate the pH to 9.0. Stirring is continued for a further 30 min. The slurry is centrifuged at 13,000 g for 30 min and the supernatant (~350 ml) is concentrated in an Amicon (Danvers, MA) stirred cell equipped with a PM10 membrane, dialyzed versus 10 mM acetic acid, and lyophilized for prolonged storage.

Alternatively, the preparation can begin from bovine brain tissue. Three hundred-gram aliquots are homogenized in ~12 vol of a 2 : 1 mix of chloroform : methanol and left to stir at 4° overnight. The suspension is filtered through Whatman #114 paper under suction, and the insoluble material resuspended twice more with 2 liters of 2 : 1 chloroform : metha-

---

[9] N. K. Tonks, C. D. Diltz, and E. H. Fischer, this volume [37].
[10] G. E. Diebler, L. F. Boyd, and M. W. Keis, Prog. Clin. Biol. Res. 146, 149 (1984).

nol, once with 2 liters of acetone, filtering each time, and finally stirred for 1 hr in water). After filtration the delipidated tissue is resuspended in 30 mM HCl (2 ml/g starting material) and processed as described above.

Before phosphorylation, aliquots of the lyophilized powder are resuspended in 25 mM imidazole, pH 7.2, at ~20 mg/ml by weight and dialyzed versus at least two changes of the same buffer. The phosphorylation reaction is performed with insulin and epidermal growth factor (EGF) receptor kinase, purified over wheat germ lectin-Sepharose, as described in Tonks et al.,[9] with the exception that deoxycholate is not included in the buffers. Routinely, stoichiometries of phosphorylation of ~0.3 mol/mol are obtained, with labeling at a single site.[9]

Purification of CD45

Buffers

Buffer A: 10 mM imidazole hydrochloride, pH 7.2, 5 mM EDTA, 1 mM EGTA, 0.1% (v/v) 2-mercaptoethanol, 1 mM benzamidine, 0.002% (w/v) phenylmethylsulfonyl fluoride (PMSF), 2 μg/ml leupeptin, 50 U/ml aprotinin
Buffer B: Buffer A containing only 2 mM EDTA instead of 5 mM
Buffer C: 20 mM imidazole hydrochloride, pH 7.2, 1 mM EGTA, 0.1% (v/v) 2-mercaptoethanol, 50 U/ml aprotinin, 1 mM benzamidine, 0.002% (w/v) PMSF, 2 μg/ml leupeptin, 0.1% (v/v) Triton X-100
Buffer D: 50 mM triethylamine, pH 11, 0.1% (v/v) Triton X-100, 0.1% (v/v) 2-mercaptoethanol

Reagents

DE-52 was purchased from Whatman, protein A-Sepharose from Sigma. Other chemicals were from standard sources.

The monoclonal antibody (MAb) 9.4–protein A-Sepharose column was constructed by incubating 56 mg of the antibody with 10 ml of protein A-Sepharose and covalently cross-linking with dimethyl pimelimidate, according to the procedure of Schneider et al.[11]

Procedures

All steps are performed at 4°. Care must be taken to process and assay samples as quickly as possible, particularly in the early stages of the preparation, to minimize the problem of contaminating proteases. Normal

[11] C. Schneider, R. A. Newman, D. R. Sutherland, U. Asser, and M. F. Greaves, J. Biol. Chem. 257, 10766 (1982).

precautions (rubber gloves, goggles, etc.) should be taken when handling human tissue.

Human spleen, obtained as rapidly as possible following surgical excision and transported from the hospital on ice, is divided into aliquots of ~50 g and washed extensively with ice-cold 10 mM imidazole hydrochloride, pH 7.2, 150 mM NaCl to remove blood clots. Sequential washes in the buffered saline, squeezing out blood clots with the fingers, will efficiently clean the tissue. Washed spleen is then homogenized in a Waring blender at high speed for four intervals of 15 sec each in 2.5 vol of buffer A containing 0.25 M sucrose. After a low-speed spin to pellet out large debris, nuclei, etc. (6000 g for 20 min) the supernatant is centrifuged for 60 min at 48,000 g. The high-speed supernatant is discarded and the pellet washed with buffer A (~50 ml/100 g of starting material), resuspending with five strokes in a glass Dounce homogenizer. At this stage, the membrane pellet may be resuspended, in buffer A containing 20% (v/v) glycerol (~20 ml/100 g of starting material), again using the homogenizer, and stored in aliquots at −70° for further processing.

Fresh or thawed frozen membranes are washed once more with buffer A as above and then resuspended in buffer B containing 0.5% (v/v) Triton X-100 (~50 ml/100 g starting material) with five strokes in the homogenizer, stirred gently on ice for 60 min, and then centrifuged at 48,000 g for 60 min. The pellets are discarded and the supernatant is filtered through glass wool and applied to a column of DE-52 equilibrated in buffer B containing 0.1% (v/v) Triton X-100. After washing with ~two column volumes, the bound material is eluted with a linear salt gradient from 0 to 0.3 M NaCl in equilibration buffer. The total volume of the gradient is ~10 times that of the column.

An example of an elution profile is shown in Fig. 1. The two major peaks of activity correspond to the observed elution positions of low-molecular-weight PTPases 1A and 1B from human placenta.[12] Superimposed on these is the elution of CD45, which is best detected by dot-blot analysis, using the 9.4 antibody and following standard protocols. Perhaps not unexpectedly, in view of the differences in the state of glycosylation of the various isoforms,[4] CD45 is detected over a fairly broad section of the profile. The peak eluting between a conductivity of 6 and 10 mS is pooled and passed over a protein A-Sepharose preclearing column (~5 ml) equilibrated in buffer C. The flow-through is then applied to the anti-CD45 MAb 9.4–protein A-Sepharose column equilibrated in the same buffer. The flow-through from this step is collected and stored separately; the column is washed with buffer C containing 0.3 M NaCl to remove

---

[12] N. K. Tonks, C. D. Diltz, and E. H. Fischer, *J. Biol. Chem.* **263**, 6722 (1988).

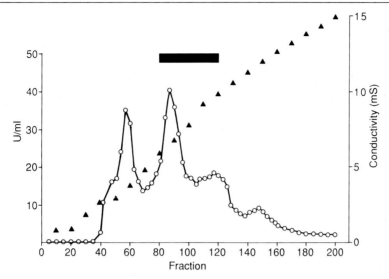

Fig. 1. Ion-exchange chromatography of a Triton X-100 extract of human spleen on DE-52. After loading, the column (4 × 20 cm) was washed with ~500 ml of equilibration buffer and bound PTPase was eluted with a linear salt gradient from 0 to 0.3 M NaCl in a total volume of 2 liters of buffer B containing 0.1% (v/v) Triton X-100. Fractions of 10 ml were collected at a flow rate of 120 ml/hr. Activity was measured using reduced carboxamido-methylated and maleylated (RCM) lysozyme as substrate (○) as described in Tonks et al.[9] The closed triangles denote conductivity and the solid bar indicates the CD45-containing fractions that were pooled.

protein adhering nonspecifically. After washing with ~two column volumes of a solution of 0.1% (v/v) Triton X-100, 0.1% (v/v) 2-mercaptoethanol, bound CD45 is eluted with buffer D. In order to rapidly neutralize the high pH eluate, five aliquots of 10 ml are collected, each into 0.55 ml of 2 M 3-[N-morpholino] propane sulfonic acid (MOPS), pH 6.0. In this way, the final pH of each aliquot is ~7.2. The neutralized eluate fractions are combined, and immediately desalted by gel filtration on a 4 × 16.4 cm column of Sephadex G-25 Superfine (Pharmacia, Piscataway, NJ) equilibrated in buffer C. Collection of the eluate begins as the sample is loaded. The first 60 ml is discarded; then the next 75 ml, containing CD45 fraction, is retained and concentrated to ~0.5 ml in an Amicon stirred cell equipped with an XM100A (100 kDa cut-off) membrane. Glycerol is added to a final concentration of 20% (v/v) and the CD45 pool is stored in aliquots at −70°. A sodium dodecyl sulfate (SDS) gel of the purified product is depicted in Fig. 2.[13] The desalting column can be regenerated by washing with a further ~100 ml of buffer C.

[13] U. K. Laemmli, *Nature* (*London*) **227**, 680 (1970).

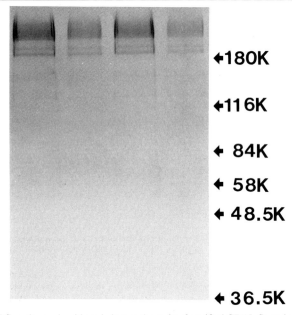

←180K

←116K

← 84K

← 58K

← 48.5K

← 36.5K

FIG. 2. SDS–polyacrylamide gel electrophoresis of purified CD45. Samples of CD45 from two separate preparations are illustrated. In each case, 1 and 0.5 μg (from left to right) of eluate from the antibody affinity chromatography column were subjected to electrophoresis on a 7.5% polyacrylamide gel according to Laemmli.[13] The gel was stained with Coomassie Blue and arrows denote the position of marker proteins [α_2-macroglobulin (180k), β-galactosidase (116k), fructose-6-phosphate kinase (84k), pyruvate kinase (58k), fumarase (48.5k), and lactate dehydrogenase (36.5k) (Sigma, St. Louis, MO, prestained molecular weight standards kit)].

Routinely the preparation begins from ~500 g of human spleen. Under these conditions, the amount of CD45 in the pool from DE52 is sufficient to saturate an anti-CD45 antibody column of this size. Consequently multiple passes are made over the column, each treated as discussed above, storing the breakthrough from each run for further processing. In three such preparations, each involving four passages over the antibody affinity column, 1.5–2 mg of CD45 (estimated by Bradford[14] with bovine serum albumin as standard) were obtained.

Characteristics of CD45

A comparison of the properties of CD45 and PTPase 1B is shown in Table I. One common feature between the low-molecular-weight and receptor-like PTPases is that *all* tested thus far are absolutely dependent

[14] M. M. Bradford, *Anal. Biochem.* **72**, 248 (1976).

TABLE I
PROPERTIES OF CD45 AND PTPASE 1B

Property	Human spleen CD45	Human placenta PTPase 1B
M_r	180k–220k	37.5k
SH dependence	Yes	Yes
Substrate specificity	Tyrosyl residues	Tyrosyl residues
RCML[a]		
Specific activity[b]	~1,000 U/mg	~20,000 U/mg
K_m	~15 μM	280 nM
MBP		
Specific activity[b]	~21,000 U/mg	~17,000 U/mg
K_m	1–4 μM	250 nM
Modulators of RCML dephosphorylation		
Vanadate and Molybdate	Inhibitory	Inhibitory
Zn^{2+} and Mn^{2+}	Activating	Inhibitory
Spermine	12-fold activation	1.5-fold activation
EDTA	~2-fold activation	~2- to 3-fold activation
Heparin and poly(Glu : Tyr)	$IC_{50} > 200$ nM	IC_{50} 20–50 nM

[a] RCML, Reduced carboxamidomethylated and maleylated lysozyme.
[b] 1 Unit = 1 nmol phosphate released/min.

on the presence of sulfhydryl compounds for activity.[15,16] Thus it is essential to include 2-mercaptoethanol or dithiothreitol in the assay buffer, to avoid inactivation of the enzymes. This phenomenon is apparently reversible and at least some activity can be restored by adding back the reducing agent. However, alkylating agents such as iadoacetic acid or N-ethylmaleimide irreversibly inactivate the enzymes, implying that at least one cysteinyl residue is an indispensable component in the catalytic mechanism. Interestingly, comparison of protein sequences between members of the family indicates two totally conserved cysteinyl residues (numbers 121 and 215 in PTPase 1B[2]). Support for the idea that one of these is essential for activity has been provided by Streuli et al.[17]

At the present moment, there is no specific inhibitor of the PTPases. Vanadate and molybdate inhibit both the low-molecular-weight and integral membrane forms; however, these agents exhibit pleiotropic effects upon addition to cells. Zinc ion (at micromolar concentrations) has been

[15] N. K. Tonks, C. D. Diltz, and E. H. Fischer, *J. Biol. Chem.* **263**, 6731 (1988).
[16] N. K. Tonks, C. D. Diltz, and E. H. Fischer, *J. Biol. Chem.* **265**, 10674 (1990).
[17] M. Streuli, N. X. Krueger, A. Y. M. Tsai, and H. Saito, *Proc. Natl. Acad. Sci. U.S.A.* **86**, 8698 (1989).

established as an inhibitor of the low-molecular-weight enzymes,[18] but CD45 is much less sensitive; in fact, with RCM lysozyme as substrate, Zn^{2+} actually stimulates the activity of CD45.[16]

The low-molecular-weight enzymes dephosphorylate all the artificial substrates tested with similar high specific activity and high affinity, suggesting a lack of selectivity for the nature of the protein bearing the phosphorylated tyrosyl residue.[15] By comparison, CD45 seems to display some elements of substrate specificity. Thus with RCM lysozyme, CD45 displays ~5% of the activity of PTPase 1B.[5] However, when an appropriate substrate is utilized in an *in vitro* assay, CD45 exhibits an even more powerful phosphatase activity than the low-molecular-weight enzyme. Thus with phosphotyrosyl MBP, the turnover numbers are 4260 min^{-1} and 1690 min^{-1} for CD45 and PTPase 1B, respectively.[16] Interestingly, the dephosphorylation of RCM lysozyme by CD45 is stimulated more than 10-fold by polyamines such as spermine: nonphosphorylated MBP is also stimulatory. Furthermore, microtubule-associated protein (MAP) kinase, a protein-Ser/Thr kinase that becomes phosphorylated on tyrosyl and threonyl residues in response to a variety of mitogenic stimuli and is active in its phospho form,[19] is dephosphorylated and concomitantly inactivated *in vitro* by CD45 (at ~10 U/ml vs RCM lysozyme).[20] However, PTPase 1B at ~25 U/ml vs the same substrate was ineffective.[20] Perhaps this observation may reflect a difference in the specificity of CD45 and PTPase 1B toward proteins that are phosphorylated on tyrosyl residues *in vivo*, in addition to that observed toward the substrates of practical rather than physiological relevance described above. Such a phenomenon, in conjunction with the obvious potential for direct control of receptor-linked PTPases by extracellular ligands, may be important in delineating whether a low-molecular-weight or integral membrane PTPase is involved in the control of a particular cellular function.

Acknowledgments

On the basis of extensive sequence identity between the human placenta enzyme, PTPase 1B, and the C-terminal homologous domains of CD45, it was proposed that CD45 possesses intrinsic PTPase activity. Without these studies, performed in collaboration with Harry

---

[18] D. L. Brautigan, P. Bornstein, and B. Gallis, *J. Biol. Chem.* **256,** 6519 (1981).
[19] L. B. Ray and T. W. Sturgill, *Proc. Natl. Acad. Sci. U.S.A.* **85,** 3753 (1988).
[20] N. G. Anderson, J. L. Maller, N. K. Tonks, and T. W. Sturgill, *Nature (London)* **343,** 651 (1990).

Charbonneau and Ken Walsh, this manuscript could not have been written. We thank them for many helpful discussions. This work was supported by National Institutes of Health Grant DK07902 from the National Institute of Diabetes and Digestive and Kidney Diseases and by a grant from the Muscular Dystrophy Association of America. We thank Carmen Westwater for typing the manuscript.

[39] Resolution and Characterization of Multiple Protein-Tyrosine Phosphatase Activities

By Thomas S. Ingebritsen

Introduction

Reversible phosphorylation of proteins on tyrosine residues is one of the earliest events in signal transduction pathways leading to the stimulation of cell proliferation. This regulatory mechanism also appears to play a critical role in the transformation of animal cells to a tumor-like phenotype[1,2] and it may also participate in the control of other cell functions (e.g., insulin action, neural function, and platelet activation).[3-5] The extent of phosphorylation of these proteins is dependent on the balance of the protein kinase and protein phosphatase activities toward a particular substrate protein and there is growing evidence that protein-tyrosine phosphatases (PTPs) play a critical role in regulating tyrosine phosphorylation reactions.[6-9]

Protein-tyrosine phosphatases, like protein-tyrosine kinases, can be grouped into two categories depending on whether they are part of receptor-like molecules.[6-9] The sequences of three nonreceptor PTPs are known and they share extensive homology (30–70%) with the receptor-like PTPs

[1] T. Hunter and J. A. Cooper, *Annu. Rev. Biochem.* **54,** 897 (1985).
[2] T. Hunter and J. A. Cooper, *in* "The Enzymes" (P. D. Boyer and E. G. Krebs, eds.), Vol. 17, p. 191. Academic Press, Orlando, Florida, 1986.
[3] O. M. Rosen, *Science* **237,** 1452 (1987).
[4] J. Brugge, P. Cotton, A. Lustig, W. Yonemoto, L. Lipsich, P. Coussens, J. N. Barrett, D. Nonner, and R. W. Keane, *Genes Dev.* **1,** 287 (1987).
[5] A. Golden, S. P. Nemeth, and J. S. Brugge, *Proc. Natl. Acad. Sci. U.S.A.* **83,** 852 (1986).
[6] T. Hunter, *Cell (Cambridge, Mass.)* **58,** 1013 (1989).
[7] T. S. Ingebritsen, S. K. Lewis, V. M. Ingebritsen, B. P. Jena, K. T. Hiriyanna, S. W. Jones, and R. L. Erikson, *Adv. Protein Phosphatases* **5,** 121 (1989).
[8] T. S. Ingebritsen and K. T. Hiriyanna, *in* "Bioinformatics: Information Transduction and Processing Systems from Cell to Whole Body" (O. Hatase and J. H. Wang, eds.), p. 117. Elsevier, Amsterdam, 1989.
[9] N. K. Tonks and H. Charbonneau, *Trends Biochem. Sci.* **14,** 497 (1989).

(CD45 and LAR) within a core (perhaps catalytic) domain.[6,9-13] None of the PTP sequences exhibits homology with those of the type 1 and type 2 protein-Ser/Thr phosphatases, indicating that the PTPs represent a distinct family of protein phosphatases. The properties of receptor-like PTPs are reviewed elsewhere in this volume.[14]

Six nonreceptor PTPs have been purified to homogeneity or near homogeneity from the cytoplasmic fraction of various mammalian tissues. The enzymes are the human placental 1A PTP (M_r 35,000),[15] human placental 1B PTP (M_r 37,000),[15] rabbit kidney type I PTP (M_r 34,000),[16] rabbit kidney type II PTP (M_r 37,000),[16] bovine spleen PTP (M_r 52,000),[17] and bovine brain PTP5 (M_r 48,000).[18] Two additional PTPs have been identified via cDNA cloning, the human T cell PTP (M_r 48,000)[11] and rat brain PTP (M_r 50,000).[12]

Although the relationships among these PTPs are not fully defined, the available information indicates that there are a minimum of three distinct nonreceptor PTP catalytic subunits. The complete amino acid sequences of the human placental 1B PTP,[10] human T cell PTP,[11] and rat brain PTP[12] are known. The rat brain PTP sequence is 97% identical to the human placental PTP over the first 321 amino acids but extends for an additional 111 amino acids at the C terminus. This indicates that the two enzymes correspond to the human and rat forms of the same PTP gene product. The C-terminal domain of the PTP may have been selectively lost via limited proteolysis during purification of the enzyme from human placenta. The human T cell PTP is the product of a gene that is distinct from but related to that encoding the rat brain/human placental 1B PTP.

This chapter focuses on methods for the assay and purification of bovine brain PTPs and on methods for the assay and purification of two regulatory proteins, inhibitor H and inhibitor L, that function as PTP inhibitors. Bovine brain PTP5 is distinct from the human placental 1A PTP

[10] H. Charbonneau, N. K. Tonks, S. Kumar, C. D. Diltz, M. Harrylock, D. E. Cool, E. G. Krebs, E. H. Fischer, and K. A. Walsh, *Proc. Natl. Acad. Sci. U.S.A.* **86**, 5252 (1989).
[11] D. E. Cool, N. K. Tonks, H. Charbonneau, K. A. Walsh, E. H. Fischer, and E. G. Krebs, *Proc. Natl. Acad. Sci. U.S.A.* **86**, 5257 (1989).
[12] K. Guan, R. S. Haun, S. J. Watson, R. L. Geahlen, and J. E. Dixon, *Proc. Natl. Acad. Sci. U.S.A.* **87**, 1501 (1990).
[13] M. Streuli, N. X. Krueger, A. Y. M. Tsai, and H. Saito, *Proc. Natl. Acad. Sci. U.S.A.* **86**, 8698 (1989).
[14] N. K. Tonks, C. D. Diltz, and E. H. Fischer, this volume [38].
[15] N. K. Tonks, C. D. Diltz, and E. H. Fischer, *J. Biol. Chem.* **263**, 6722 (1988).
[16] C. L. Shriner and D. L. Brautigan, *J. Biol. Chem.* **259**, 11383 (1984).
[17] H. Y. L. Tung and L. J. Reed, *Anal. Biochem.* **161**, 412 (1987).
[18] S. W. Jones, R. L. Erikson, V. M. Ingebritsen, and T. S. Ingebritsen, *J. Biol. Chem.* **264**, 7747 (1989).

and rabbit kidney type II PTP.[18] The relationship of PTP5 to the rat brain/ human placental 1B PTP, human T cell PTP, rabbit kidney type I PTP, and bovine spleen PTP is unknown. In addition to PTP5, six other PTP activities, termed PTP1A, PTP1B, PTP2, PTP3, PTP4, and PTP6, have been partially purified from the cytosolic fraction from bovine brain.[18] Like PTP5 these activities are distinct from protein-Ser/Thr phosphatases but differ from PTP5 in their chromatographic properties on DEAE-cellulose, phosphocellulose, and/or gel filtration. The seven PTPs also differ in their sensitivity to inhibitor proteins (see below)[19] and in their sensitivity to inhibition by macromolecular polyanions [heparin, $poly(Glu^{80},Tyr^{20})$, and the synthetic polynucleotide, $poly(G,I)$].[7] It is not yet clear whether the seven bovine brain activities correspond to distinct gene products, different forms of the same gene product, or some combination of these two possibilities.

The turnover number of bovine brain PTP5[18] is at least an order of magnitude greater than those of the known protein-tyrosine kinases. This suggests that the activity of PTP5 may need to be closely regulated in order to allow tyrosine phosphorylation reactions to occur in intact cells. Pursuing this idea we have identified two regulatory proteins, inhibitor H (>500 kDa) and inhibitor L (38 kDa) in bovine brain, that potently and preferentially inhibit PTP5.[19] Inhibitor H and inhibitor L also inhibit bovine brain PTP4 but the IC_{50} values for inhibition of PTP4 are 10- and 2-fold greater, respectively, than those for the inhibition of PTP5. The two inhibitor proteins only inhibit the other five bovine brain protein-tyrosine phosphatases at very high concentrations ($IC_{50} \geq$ 100-fold higher than those for PTP5). This suggests that PTP4 and PTP5 may be closely related enzymes.

While the precise functions of PTPs are poorly understood, the physiological importance of these enzymes is underscored by several types of observations. First, addition of vanadate, a potent PTP inhibitor, to normal cells results in a dramatic rise in tyrosine phosphorylation,[6] although it should be noted that vanadate may also have a stimulatory effect on some protein-tyrosine kinases. Second, the existence of receptor-like PTPs and of protein inhibitors of nonreceptor PTPs suggests that PTPs may be tightly regulated and that this regulation in some cases may initiate signal transduction events. Third, PTP4 and/or PTP5 like activities are present in a wide range of vertebrate cell types and they are particularly enriched in proliferating cell populations.[7] Fourth, the similarity of receptor-like PTP LAR to neural adhesion molecules suggests that it may have a role

[19] T. S. Ingebritsen, *J. Biol. Chem.* **264**, 7754 (1989).

in cell–cell or cell–matrix interactions.[20] Fifth, the receptor-like PTP, CD45, is required for antigen-induced T lymphocyte proliferation[21] and influences signal transduction events mediated via the CD2, CD3, and CD4 cell surface antigens of T cells.[22] Sixth, microinjection of the placental PTP into *Xenopus* oocytes delays the insulin-induced maturation response.[9]

Assay Methods

Protein-Tyrosine Phosphatase Assay

Principle. Protein-tyrosine phosphatases are assayed by following the time-dependent release of trichloroacetic acid-soluble radioactivity from [^{32}P]casein.

Reagents

Bovine serum albumin (BSA): Ultrapure grade from Boehringer-Mannheim (Cat. #238 031)

Assay buffer: 50 mM Tris-HCl (pH 7.0 at 25°)–0.05 mM EDTA–1 mg/ml BSA–0.3% (v/v) 2-mercaptoethanol

Tris–Brij buffer: 50 mM Tris-HCl (pH 7.0 at 25°)–0.01% (w/v) Brij 35

[^{32}P]Casein: The stock solution of casein phosphorylated by the insulin receptor kinase (see below) is diluted to 300 nM with assay buffer. Between assays the diluted [^{32}P]casein is stored at 4°

Procedure

Mix 20 μl of protein phosphatase diluted in assay buffer with 20 μl of Tris–Brij buffer and preincubate for 5 min at 30°. Start the assay by adding 20 μl of [^{32}P]casein to the reaction mixture. Continue the incubation for 1–30 min and then stop the reaction with 100 μl of 20% (w/v) trichloroacetic acid (TCA). Vortex the mixture and hold on ice for 10 min. Vortex again, centrifuge at room temperature for 2 min at 12,000 g in a microfuge, and count 120 μl of the supernatant.

The protein phosphatase assays are linear with respect to enzyme concentration and time up to 30% release of inorganic phosphate from casein. One unit of protein-tyrosine phosphatase is the amount that re-

[20] M. Streuli, N. X. Krueger, L. R. Hall, S. F. Schlossman, and H. Saito, *J. Exp. Med.* **168**, 1523 (1988).

[21] J. T. Pingel and M. L. Thomas, *Cell (Cambridge, Mass.)* **58**, 1055 (1989).

[22] J. A. Ledbetter, N. K. Tonks, E. H. Fischer, and E. A. Clark, *Proc. Natl. Acad. Sci. U.S.A.* **85**, 8628 (1988).

leases 1 nmol of inorganic phosphate from casein per minute under the standard assay conditions. When initially characterizing a new protein-tyrosine phosphatase using this assay it is important to establish that the trichloroacetic acid-soluble counts represent inorganic phosphate rather than small phosphopeptides that may also be trichloroacetic acid soluble. This can readily be done by treating the TCA supernatant with molybdate and extracting the resulting inorganic phosphate–molybdate complexes into organic solvent.[23] Phosphate esters or anhydrides do not form complexes with molybdate.

Protein-Tyrosine Phosphatase Inhibitor Assay

Principle. Inhibitor H and inhibitor L are assayed based on their ability to inhibit the activity of PTP5.

Reagents

PTP5: The enzyme is partially or highly purified from bovine brain as described below and diluted to a concentration of 10 mU/ml in assay buffer

Other reagents: The preparation of these are described above (see Protein-Tyrosine Phosphatase Assay)

Procedure

Fractions containing inhibitor activity are incubated at 95° for 5 min to inactivate any endogenous protein-tyrosine phosphatase activity. The mixture is cooled on ice, centrifuged for 2 min at 15,000 g to remove denatured protein, and diluted in Tris–Brij buffer. Then, mix 20 μl of protein phosphatase diluted in assay buffer with 20 μl of diluted inhibitor H or inhibitor L, preincubate for 5 min at 30°, and initiate the protein-tyrosine phosphatase assay by adding 20 μl of [^{32}P]casein to the reaction mixture. After 10 min the reaction is terminated and further processed as described above (see Protein-Tyrosine Phosphatase Assay). It is important to use high-quality BSA (Boehringer Mannheim, ultrapure grade) in the assay buffers since fraction V BSA contains contaminants that interfere with the inhibition of PTP5 by the two inhibitor proteins.

Inhibitor activity is quantitated based on the concentration-dependent decrease in PTP5 compared to a control incubation lacking inhibitor (Fig.

---

[23] S. Shenolikar and T. S. Ingebritsen, this series, Vol. 107, p. 102.

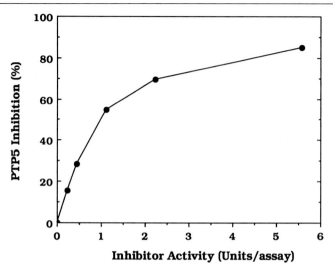

FIG. 1. Standard curve for the inhibition of PTP5. The standard inhibitor assay was used and the amounts of inhibitor (DEAE-cellulose fraction) added are indicated. See text for definition of units. (Reprinted with permission from Ingebritsen.[19])

1). Concentrations producing up to 50–60% PTP5 inhibition can be estimated with little loss of accuracy. The sensitivity of PTP5 to inhibition by inhibitor H and inhibitor L decreases as the concentration of PTP5 in the assay increases.[19] The standard assay contains 0.2 mU PTP5, which is equivalent to a PTP5 concentration of 20 pM based on a molecular weight of 46,000, and K_m and V_{max} values of 130 nM and 10,000 U/mg, respectively. This concentration of PTP5 gives near-maximal sensitivity of the enzyme to the two inhibitor proteins. One unit of inhibitor is the amount that decreases the activity of 0.2 mU PTP5 by 50% under the standard assay conditions.[24]

Phosphorylation of Casein by Insulin Receptor Kinase

Principle. [^{32}P]Casein, phosphorylated exclusively on tyrosine residues to a stoichiometry of 1–2 nmol/mg, is prepared by incubation with the insulin receptor kinase and [γ-^{32}P]ATP.

---

[24] The definition of inhibitor units given here is equivalent to that given in Ref. 19. The apparent difference in the amount of PTP5 in the standard assay reflects a PTP5 activity unit definition based on a substrate concentration of 100 nM vs 1 nM in Ref. 19. The actual PTP5 concentation (pM) is approximately the same as in Ref. 19.

Reagents

Buffer A: 50 mM HEPES, pH 7.4–0.1% (v/v) Triton X-100–10 mM MgCl$_2$

Buffer B: 166.7 mM HEPES, pH 7.4–16.7 mM MgCl$_2$–1.7 mM MnCl$_2$–3.3 mM EDTA

Casein: This is a mixture of 70% α_{s1}- + α_{s2}-caseins and 30% β- + κ-caseins. It is purchased from Sigma (Cat. #C-7891; St. Louis, MO)

Insulin receptor kinase: The enzyme is partially purified from NIH 3T3 HIR3.5 cells as described by Treadwell *et al.*[25] and stored at −80° in 50 mM Tris-HCl (pH 8.0 at 25°)–0.3 M N-acetylglucosamine–0.05% (v/v) Triton X-100–10% (v/v) glycerol at a concentration of 0.4–0.8 U/ml. Insulin receptor kinase activity is assayed essentially as described by Tonks *et al.*[15] using poly(Glu^{80},Tyr^{20}) as substrate (2 mg/ml) and 0.1 mM [γ-^{32}P]ATP (1000 cpm/pmol). One unit of insulin receptor kinase is the amount that incorporates 1 nmol of ^{32}P into poly(Glu^{80},Tyr^{20}) in 1 min under the standard assay conditions [γ-^{32}P]ATP (5 mM, 10,000–20,000 cpm/pmol) in H$_2$O

Phosphorylation Procedure

Mix 320 μl of buffer A containing 12.5 mg/ml casein and 7 μg/ml insulin with 240 μl buffer B plus 80 μl insulin receptor. Preincubate the mixture for 15 min at 30°. Add 160 μl [γ-^{32}P]ATP and incubate overnight at 30°. Remove a 5-μl aliquot and process to determine phosphate incorporation (see below). Terminate incubation by adding 160 μl of 100% TCA. Hold on ice for 60 min and centrifuge for 2 min at 12,000 g in a microfuge. Wash the pellet six times with 0.5 ml ice-cold 20% TCA (centrifuge 1 min after each wash). Add 0.5 ml of 0.5 M Tris, pH 8.5 to the washed pellet and allow the pellet to redissolve overnight in the refrigerator. Gel filter the solution on a Sephadex G-50 column (0.7 × 17 cm) equilibrated with 50 mM Tris (pH 7.0 at 25°)–0.05 mM EDTA. Pool the ^{32}P-labeled protein peak, divide into aliquots containing 300 pmol of [^{32}P]casein, and store at −20°.

Measurement of Phosphate Incorporation into Casein

Dilute 5 μl of the reaction mixture to 100 μl with H$_2$O. Transfer four 5-μl aliquots of the diluted reaction mixture to 2 × 2 cm squares of Whatman (Clifton, NJ) P-81 phosphocellulose paper. Wash the filter papers three times with 200 ml of 0.5% (w/v) H$_3$PO$_4$ (5 min for each wash). Wash the filter papers once with 200 ml of ethanol, dry, and count in scintillant.

[25] J. L. Treadwell, J. Whittaker, and J. E. Pessin, *J. Biol. Chem.* **264,** 15136 (1989).

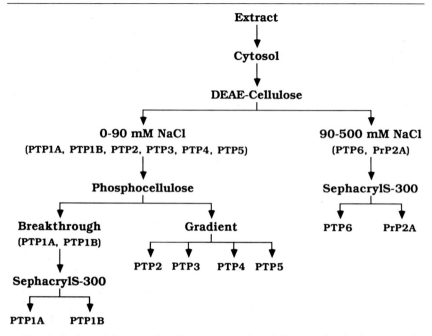

FIG. 2. Outline of the procedure for the separation of the seven bovine brain protein-tyrosine phosphatases (PTPs) from each other and from protein-Ser/Thr Phosphatase 2A (PrP2A).

Protein-Tyrosine Phosphatase and Phosphatase Inhibitor Preparations

Separation of Bovine Brain Protein-Tyrosine Phosphatases

Principle. Most of the protein-tyrosine phosphatase activity in bovine brain extracts (70%) is in the cytosolic fraction. Seven protein-tyrosine phosphatases (PTP1A, PTP1B, PTP2, PTP3, PTP4, PTP5, and PTP6) account for 90–95% of the cytosolic activity. These activities are resolved by successive chromatographies on DEAE-cellulose, phosphocellulose, and Sephacryl S-300. The remaining 5–10% of the activity is due to protein-Ser/Thr phosphatase 2A. The separation protocol is shown in schematic form in Fig. 2.

Reagents

Homogenization solution: 4.0 m*M* EDTA, pH 7.0–250 m*M* sucrose–0.1 m*M* phenylmethylsulfonyl fluoride (PMSF)–0.1 m*M* benzamidine–0.2% (v/v) 2-mercaptoethanol

Buffer C: 20 mM Tris-HCl (pH 7.0 at 25°)–0.1 mM EDTA–0.1 mM PMSF–0.1 mM benzamidine–0.2% (v/v) 2-mercaptoethanol

Buffer D: 50 mM Tris-HCl (pH 7.0 at 25°)–50 mM NaCl–0.1 mM EDTA–0.1 mM PMSF–0.1 mM benzamidine–0.01% (w/v) Brij 35–0.2% (v/v) 2-mercaptoethanol

Buffer E: 50 mM Tris-HCl (pH 7.0 at 25°)–0.1 mM EDTA–0.1 mM PMSF–0.1 mM benzamidine–50% (v/v) glycerol–0.2% (v/v) 2-mercaptoethanol

Procedure. The entire procedure is performed at 4°. Two fresh bovine brains are deveined, passed through a meat grinder, and homogenized in a Waring blender (2 min at low speed) in 2 vol of homogenization solution. The homogenate is centrifuged at 12,000 g for 20 min and the resulting supernatant (extract) is further centrifuged at 100,000 g for 60 min.

DEAE-Cellulose Chromatography Step. The supernatant (cytosol) is applied to a DEAE-cellulose column (7.5 × 14 cm) equilibrated with buffer C at a flow rate of 300–400 ml/hr. The column is washed with buffer C containing 90 mM NaCl until the A_{280} of the effluent is <0.2 (~2 liters). The breakthrough and 90 mM NaCl wash fractions are combined (DEAE-cellulose 0–90 mM NaCl fraction). This fraction contains PTP1A, PTP1B, PTP2, PTP3, PTP4, and PTP5. The column is then washed with buffer C containing 200 mM NaCl and the resulting protein peak (DEAE-cellulose 90–200 mM NaCl fraction), which contains PTP6 and protein-Ser/Thr phosphatase 2A, is pooled. Ammonium sulfate (472 g/liter) (70% saturation) is added to the two DEAE-cellulose fractions and the mixtures are stirred for 30 min, adjusted to pH 7.0 with ammonium hydroxide, and centrifuged at 8000 g for 35 min. The pellets are resuspended in a minimal volume (~120 ml) of buffer C and dialyzed overnight against three 2-liter volumes of buffer C.

Phosphocellulose Chromatography Step. The DEAE-cellulose 90 mM NaCl fraction is applied at a flow rate of 100 ml/hr to a phosphocellulose column (3.6 × 10 cm) equilibrated with buffer C. The column is washed with buffer C until the A_{280} approaches zero (~300 ml). The protein peak from the breakthrough and wash step (PTP1A plus PTP1B) is pooled, an aliquot is dialyzed overnight against buffer E, and stored at −20°. The phosphocellulose column is developed with a 1-liter linear gradient of 0–600 mM NaCl in buffer C. The PTP2, PTP3, PTP4, and PTP5 peaks (eluting at 160, 260, 320, and 420 mM NaCl, respectively) are separately pooled, dialyzed overnight against buffer E, and stored at −20°. Protein-tyrosine phosphatase activity is stable for at least 1 year under these conditions.

Gel-Permeation Chromatography. A 2-ml aliquot of the phosphocellulose breakthrough and wash fraction (PTP1A plus PTP1B) (see above)

TABLE I
PROPERTIES OF BOVINE BRAIN PROTEIN-TYROSINE PHOSPHATASES[a]

PTP	Molecular weight	Elution with NaCl (mM) from		Sensivity to	
		DEAE-cellulose	Phospho-cellulose	Inhibitor proteins	Polyanions
PTP5	46,000	0–90	420	+ + + +	+ + + +
PTP4	80,000	0–90	320	+ + +	+ + +
PTP3	90,000	0–90	260	+ +	+ + +
PTP2	88,000	0–90	160	+ +	+ +
PTP1A	86,000	0–90	0	+	+ +
PTP1B	24,000	0–90	0	+	+ +
PTP6	104,000	90–200	0		+

[a] Reprinted with permission from Ingebritsen and Hiriyanna.[8] Pluses indicate degree of inhibitor sensitivity.

stored in buffer E is dialyzed overnight against 1 liter of buffer D and then applied to a Sephacryl S-300 (2.2 × 45 cm) column equilibrated in buffer D. The column is developed with buffer D at a flow rate of 24 ml/hr. The PTP1A (M_r 86,000) and PTP1B (M_r 24,000) peaks are separately pooled, dialyzed overnight against 1 liter of buffer E, and stored at $-20°$.

A 2-ml aliquot of the ammonium sulfate-concentrated DEAE-cellulose 90–200 mM NaCl fraction is applied to a Sephacryl S-300 column equilibrated with buffer D and eluted as described above. The PTP6 (M_r 104,000) and protein-Ser/Thr phosphatase 2A (M_r 201,000) peaks are separately pooled, dialyzed overnight against 1 liter of buffer E, and stored at $-20°$.

Properties of Bovine Brain Protein-Tyrosine Phosphatases

The physical, chromatographic, and regulatory properties of the seven bovine brain protein-tyrosine phosphatases are summarized in Table I. The apparent molecular weights of PTP5 and PTP1B on gel filtration are distinctive whereas molecular weight is not a useful criterion for distinguishing among the other five protein-tyrosine phosphatases. The properties that are most useful for distinguishing among the protein-tyrosine phosphatases are their chromatographic properties on phosphocellulose (see Table I), their sensitivity to inhibition by inhibitors H and L, and their sensitivity to inhibition by macromolecular polyanions.

Inhibitor H and inhibitor L inhibit the seven bovine brain protein-tyrosine phosphatases with the following order of potencies: PTP5 > PTP4 > PTP3 ≈ PTP2 > PTP1 > PTP6.[19] PTP5 is 2- and 10-fold more sensitive to inhibition by inhibitor L and inhibitor H, respectively, than PTP4 and

≥50- to 100-fold more sensitive than the other five protein-tyrosine phosphatases.

PTP5 is potently inhibited by three classes of macromolecular polyanions: polynucleotides (e.g., DNA, RNA, and synthetic oligonucleotides), acidic polyamino acids [e.g., poly(Glu^{80},Tyr^{20}), poly(Glu)], and the acidic polysaccharide heparin.[7] The IC$_{50}$ values for inhibition of PTP5 by heparin, poly(Glu^{80},Tyr^{20}), and poly(G,I) are ~0.2 μg/ml. The three types of polyanions inhibit the bovine brain protein-tyrosine phosphatases with the following order of potencies: PTP5 > PTP4 ≈ PTP3 > PTP2 ≈ PTP1 > PTP6.

The separation procedure can also be applied to other tissues. Using this procedure and the sensitivity to inhibition by inhibitor H, we have identified PTP4- and PTP5-like activities in seven rabbit tissues (brain, spleen, kidney, liver, heart, adipose tissue, and skeletal muscle) as well as several vertebrate cell types in culture (chicken embryo fibroblasts, a mouse T cell hybridoma, and a mouse B cell hybridoma).[7]

Purification of PTP5 to Near Homogeneity

Principle. Highly purified PTP5 is prepared from bovine brain using a seven-step procedure employing successive chromatographies of the cytosolic fraction on DEAE-cellulose, phosphocellulose, Affi-Gel Blue, heparin-agarose, Mono S, and TSK G3000 SW. Early steps in the procedure are similar to those used for the separation of the seven bovine brain PTPs; however, they have been shortened by using a batch adsorption procedure at the DEAE-cellulose step and step NaCl cuts at the phosphocellulose step. It is critical to maintain the protease inhibitor, PMSF, in all buffers used in the purification procedure to avoid limited proteolysis of the enzyme. Starting with 750 g of bovine brain, approximately 6 μg of protein is obtained after the final purification step with the M_r 48,000 PTP5 peptide accounting for about 10–25% of the total protein. The overall yield based on total activity in the brain extracts is about 1.5% with a 13,000-fold purification.

Reagents

Homogenization solution: See Separation of Bovine Brain Protein-Tyrosine Phosphatases
Buffer F: 10 mM Tris-HCl (pH 7.0 at 25°)–0.1 mM EDTA–0.2 mM PMSF–0.2% (v/v) 2-mercaptoethanol
Buffer G: 10 mM potassium phosphate, pH 7.0–0.1 mM EDTA–0.2 mM PMSF–0.2 (v/v) 2-mercaptoethanol
Buffer H: 10 mM potassium phosphate, pH 7.0–0.1 mM EDTA–5 mM DTT–0.05% (w/v) Brij 35–10% (v/v) glycerol

Buffer I: 50 mM potassium phosphate, pH 7.0–150 mM NaCl–0.1 mM
EDTA–0.05% (w/v) Brij 35–10% (v/v) glycerol–0.2% (v/v) 2-mercaptoethanol

PTP5 storage buffer: 50 mM Tris-HCl (pH 7.0 at 25°)–50 mM NaCl–0.1
mM EDTA–5 mM DTT–50% (v/v) glycerol

Purification Procedure. All procedures are carried out at 4° unless otherwise specified. Brain (750 g) is homogenized (see above) in 1.5 liters of homogenization solution. The homogenate is centrifuged at 7700 g, the supernatant (extract) is decanted through glass wool, and the pH adjusted to 7.5 with NaOH. The extract is centrifuged at 53,000 g for 5.5 hr.

DEAE-Cellulose Chromatography. The supernatant is added to 500 ml (packed volume) of DEAE-cellulose equilibrated with buffer F. The mixture is stirred for 30 min, allowed to settle for 20 min, and decanted through a sintered glass funnel. The DEAE-cellulose is washed in the funnel with buffer F until the volume of flow through plus wash totals about 2 liters. Solid ammonium sulfate (472 g/liter) is added to the combined flow through and wash fractions to give a 70% saturation solution and the mixture is stirred for 20 min. Precipitated proteins are collected by centrifugation at 7700 g for 20 min, dissolved in buffer F, dialyzed for 8 hr against 20 liters of buffer F with two changes of buffer, and fresh PMSF is added to give a final concentration of 1 mM.

Phosphocellulose Chromatography. The resulting solution (DEAE-cellulose fraction) is applied at a flow rate of 2.5 ml/min to a phosphocellulose column (50-ml bed volume) equilibrated in buffer F. The column is washed with 250 mM NaCl in buffer F (~4 liters) at a flow rate of 3.6 ml/min and PTP5 is eluted with 350 mM NaCl in buffer F at a flow rate of 0.8 ml/min. Fractions (15 ml) containing the activity peak are pooled.

Affi-Gel Blue Chromatography. The pooled phosphocellulose fraction is applied at a flow rate of 1 ml/min to an Affi-Gel Blue column (10-ml bed volume) equilibrated with buffer F containing 350 mM NaCl. PTP5 is eluted from the column at a flow rate of 0.75 ml/min with a 500-ml linear gradient of 0.5–1.0 M NaCl in buffer F (adjusted to pH 7.5 at 25°). Fractions (15 ml) containing the activity peak are pooled.

Heparin-Agarose Chromatography. The pooled fractions from the Affi-Gel Blue chromatography step are dialyzed for 8 hr against 20 liters of buffer F with two changes of buffer and applied at a flow rate of 1 ml/min to a heparin-agarose column (5-ml bed volume) equilibrated in buffer F. PTP5 is eluted at a flow rate of 0.5 ml/min with a 250-ml linear gradient of 0–1 M NaCl in buffer G. Fractions (5 ml) containing the activity peak are pooled.

Mono S Chromatography. The pooled fractions from the heparin-agarose step are dialyzed overnight against 2 liters of buffer F and applied to a Pharmacia (Piscataway, NJ) Mono S cation-exchange column (1-ml bed volume) equilibrated with buffer H. PTP5 is eluted with a 28-ml gradient of 0–600 mM NaCl in buffer H using a Pharmacia FPLC (fast protein liquid chromatography) system (flow rate of 0.8 ml/min). Fractions (1 ml) containing the PTP5 activity peak are pooled.

Gel-Permeation Chromatography. The pooled fractions from the Mono S chromatography step are dialyzed against 1 liter of buffer I for 4 hr and concentrated to about 0.5 ml by ultrafiltration using an Amicon (Danvers, MA) YM10 membrane. The resulting fraction is applied to a TSK G3000 SW column (0.75 × 60 cm) equilibrated with buffer I. The procedure is carried out at room temperature at a flow rate of 0.5 ml/min. Fractions (0.25 ml) are collected, immediately placed on ice, and assayed for protein-tyrosine phosphatase activity. The peak of activity is pooled, dialyzed against PTP5 storage buffer, and stored at −20°.

Properties of Highly Purified PTP5

PTP5 is a monomeric protein with a rather broad pH optimum of 7.0. The enzyme dephosphorylates four of six substrates that have been tested to date, including casein, RCM lysozyme (phosphorylated by the insulin receptor kinase), pp60$^{v\text{-}src}$ (Tyr-416 autophosphorylation site), and insulin receptor kinase (autophosphorylation sites).[7] In the latter case, the cloned cytoplasmic domain of the insulin receptor kinase expressed in insect cells using the baculovirus expression system[26] was used as substrate after autophosphorylation *in vitro.* Caseins phosphorylated by the insulin receptor kinase, pp60$^{v\text{-}src}$, or pp60$^{c\text{-}src}$ are equivalent substrates for PTP5.[27] Kinetic parameters (K_m, 100–130 nM; k_{cat}, 6–8 sec^{-1}) for dephosphorylation of casein and lysozyme are similar. At a fixed substrate concentration, pp60$^{v\text{-}src}$ and insulin receptor kinase are dephosphorylated at 10% of the rate with casein or RCM lysozyme as substrates. Calpactin I (phosphorylated by pp60$^{v\text{-}src}$) and mixed histones [phosphorylated by epidermal growth factor (EGF) receptor kinase] are not dephosphorylated at significant rates. *p*-Nitrophenyl phosphate is also a substrate for PTP5, although the enzyme is not a major alkaline phosphatase activity in bovine brain.[18]

Partial Purification of Protein-Tyrosine Phosphatase Inhibitors

Principle. Inhibitor H and inhibitor L are partially purified by successive chromatographies of the bovine brain cytosolic fraction on DEAE-cellulose and Sephacryl S-300. The DEAE-cellulose chromatography step

[26] R. Herrera, D. Lebwohl, A. G. Herreros, R. G. Kallen, and O. M. Rosen, *J. Biol. Chem.* **263,** 5560 (1988).
[27] S. K. Jakes, K. L. Hippen, and T. S. Ingebritsen, unpublished observations.

gives a 40-fold purification with a 300–400% yield and separates the two inhibitor proteins from the seven bovine brain PTP activities. The reason for the increase in activity at this step is not known. The Sephacryl S-300 step separates inhibitor H ($M_r > 500,000$) from inhibitor L (M_r 38,000).

Reagents

Buffers C and D: See Separation of Bovine Brain Protein-Tyrosine Phosphatases

Buffer J: 50 mM Tris-HCl (pH 7.0 at 25°)–0.05 mM EDTA–0.1 mM PMSF–0.1 mM benzamidine–0.2% (v/v) 2-mercaptoethanol

Procedure. The starting point for the preparation is the unboiled 12,000 *g* supernatant (extract) obtained from a single bovine brain (see Separation of Bovine Brain Protein-Tyrosine Phosphatases). The extract is centrifuged at 100,000 *g* for 60 min at 4° and the resulting supernatant (cytosol) is applied to a DEAE-cellulose (DE-52) column (5.5 × 16 cm) equilibrated with buffer C at a flow rate of 300 ml/hr. The column is washed with 400 ml buffer C and then further washed with buffer C containing 200 mM NaCl until the A_{280} of the effluent is <0.04 (2–3 liters). Inhibitor activity is eluted from the column with buffer C containing 700 mM NaCl. During this last wash, 10-ml fractions are collected and the A_{280} of each fraction is measured. The protein peak is pooled, dialyzed overnight against two changes (2 liters each) of buffer J, and stored at −20°.

The DEAE-cellulose pool is concentrated 100-fold by ultrafiltration using an Amicon PM10 membrane, dialyzed overnight against buffer D, and aliquots (equal to 50% of a one-brain preparation) are applied to a Sephacryl S-300 column (2.2 × 42 cm) equilibrated with the same buffer. The column is developed at a flow rate of 24 ml/hr. The inhibitor H and inhibitor L peaks are separately pooled, dialyzed overnight against two changes (2 liters each) of buffer J, and stored at −20°.

Properties of Inhibitor H and Inhibitor L

The two protein inhibitors are inactivated by treatment with protease K, *Staphylococcus aureus* V-8 protease, and subtilisin although, interestingly, they are not inactivated by treatment with high concentrations of trypsin, chymotrypsin, thermolysin, or pepsin. It is important to check inhibitor preparations for protease sensitivity since polynucleotides (i.e., DNA fragments and RNA) also inhibit PTP5. After the gel-filtration step, the inhibitor H and inhibitor L preparations are free of contaminating protein-tyrosine phosphatases. The distinctive heat stability of inhibitors H and L ($t_{1/2}$ 50–120 min) is a very convenient property that allows their assay even in the presence of contaminating protein-tyrosine phosphatases. However, it should be noted that the activity of the two inhibitor proteins is not a consequence of the heat-treatment step.[19] The inhibition of PTP5 is rapid (maximal inhibition in <1 min at 30°) and is readily reversed upon dilution of mixtures of PTP5 and the inhibitor proteins. The potency and specificity for inhibition of protein-tyrosine phosphatases by inhibitor H and inhibitor L are not dependent on the substrate (casein, RCM lysozyme, or insulin receptor kinase) used to assay the protein-tyrosine phosphatases. This suggests that the inhibitors act by binding with different affinities to the bovine brain protein-tyrosine phosphatases rather than by binding to the substrate. Inhibitor H and inhibitor L are not competing substrates for PTP5. The two inhibitor proteins are distinct from the three protein inhibitors (inhibitor 1, inhibitor 2, and DARPP-32) that act on protein-Ser/Thr phosphatase 1.

Acknowledgments

Supported by Grants GM33431 and GM38415 from the National Institutes of Health, Grant NP608A from the American Cancer Society, and a grant from the Office of Biotechnology at Iowa State University. T.S.I. is an Established Investigator of the American Heart Association.

Section IV

Protein Phosphatase Inhibitors

[40] Use of Okadaic Acid to Inhibit Protein Phosphatases in Intact Cells

By D. Grahame Hardie, Timothy A. J. Haystead, and Alistair T. R. Sim

Okadaic acid is a marine toxin originally isolated from the black sponge, *Halichondria okadaii*.[1] It is now known to be one of a family of related toxins, including dinophysistoxin 1 and acanthifolicin[2] (Fig. 1). These toxins are synthesized by dinoflagellates, especially of the genus *Dinophysis*, but the toxins accumulate in organisms further up the food chain, including sponges, shellfish, and, ultimately, humans. In humans, consumption of contaminated shellfish causes diarrhetic shellfish poisoning, and this is a particular problem at certain times of year when there is a proliferation of plankton containing the dinoflagellates.

Okadaic acid is a complex fatty acid derivative containing numerous polyether linkages[1] (Fig. 1). As well as causing acute diarrhea, it has been shown to act as a tumor promoter in the mouse skin bioassay, but unlike many other tumor promoters does not activate protein kinase C.[3] The first clue to its mechanism of action came from observations that it enhanced the contraction of skinned smooth muscle,[4] suggesting that it either stimulated myosin light-chain kinase, or inhibited myosin light-chain phosphatase. The latter proved to be correct, and it was subsequently shown that it inhibited both protein phosphatase 1 and 2A.[5] Because the structure of okadaic acid is very hydrophobic, it seemed that it should be possible to use it as a protein phosphatase inhibitor in intact cells, and this has proved to be the case.[6] It is now apparent that okadaic acid is an extremely valuable tool with which to test the physiological role of protein phosphorylation in any physiological response in intact cells. The discovery that

[1] K. Tachibana, P. J. Scheuer, Y. Tsukitani, H. Kikuchi, D. Van Enden, J. Clardy, Y. Gopichand, and F. J. Schmitz, *J. Am. Chem. Soc.* 103, 2469 (1981).
[2] F. J. Schmitz, R. S. Prasad, Y. Gopichand, M. B. Houssain, D. Van der Helm, and P. Schmidt, *J. Am. Chem. Soc.* 103, 2467 (1981).
[3] M. Suganuma, H. Fujiki, H. Suguri, S. Yoshizawa, M. Hirota, M. Nakayasu, M. Ojika, K. Wakamatsu, K. Yamado, and T. Sugimura, *Proc. Natl. Acad. Sci. U.S.A.* 85, 1768 (1988).
[4] S. Shibata, Y. Ishida, H. Kitano, Y. Ohizumi, J. Habon, Y. Tsukitani, and H. Kikuchi, *J. Pharmacol. Exp. Ther.* 223, 135 (1982).
[5] C. Bialojan and A. Takai, *Biochem. J.* 256, 283 (1988).
[6] T. J. Haystead, A. T. R. Sim, D. Carling, R. C. Honnor, Y. Tsukitani, P. Cohen, and D. G. Hardie, *Nature (London)* 337, 78 (1989).

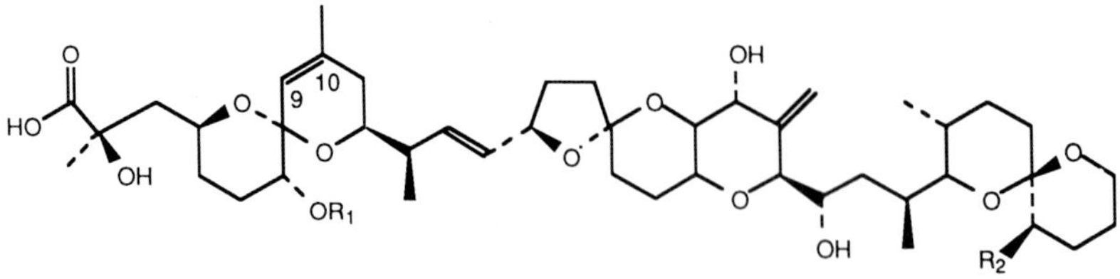

OKADAIC ACID: R₁ = H, R₂ = H

DINOPHYSISTOXIN: R₁= H, R₂ = CH₃

ACANTHIFOLICIN: C9-C10 = $\overset{C}{\underset{C}{|}}$ S

FIG. 1. Structures of marine toxins.

okadaic acid inhibits protein phosphatases in intact cells is analogous to the finding that phorbol esters activate protein kinase C; but because protein phosphatases 1 and 2A reverse the effects of many protein kinases, the toxin should have even more widespread applicability.

Availability and Properties

Okadaic acid used in our studies was a gift from Dr. Y. Tsukitani (Fujisawa Pharmaceutical, Tokyo, Japan). Okadaic acid is commercially available from Moana Bioproducts (Honolulu, HI). The toxin (M_r 804) is a white powder which is sparingly soluble in water, but can be dissolved to at least 5 mM in dimethyl sulfoxide, in which form it can be stored indefinitely at $-20°$.

Specificity of Okadaic Acid Inhibition

In the standard phosphatase assay, 50% inhibition of the catalytic subunits of protein phosphatases 1 and 2A occurs at about 10 and 0.1 nM, respectively (see [33] in this volume). Total inhibition of protein phosphatase 2A occurs at 1 nM okadaic acid, which is equivalent to the molar concentration of the phosphatase in the assay, emphasizing the extreme potency of the effect. The concentration of okadaic acid needed to inhibit protein phosphatase 2A is therefore critically dependent on phosphatase concentration. Using suitable dilutions of okadaic acid and the phosphatase activities, the toxin can be used as a specific inhibitor to discriminate between protein phosphatase 1, protein phosphatase 2A, and other phosphatases in cell-free assays (see [33] in this volume). Because both protein phosphatases 1 and 2A are present in cells at concentrations approaching the micromolar range, 1 μM okadaic acid is required to totally block these phosphatases in intact cells, and selective inhibition of protein phosphatases 1 and 2A is not possible. Nevertheless it may be possible to establish that one or the other of the two protein phosphatases is predominant in a particular system from cell-free assays. In the case of acetyl-CoA carboxylase, experiments using inhibitor 2 and okadaic acid as selective inhibitors of protein phosphatases 1 and 2A, respectively, in dilute crude extracts have established that protein phosphatase 1 accounts for less than 5% of the observed dephosphorylating activity.[7] The finding that okadaic acid converts acetyl-CoA carboxylase completely into the phosphorylated form in adipocytes[6] therefore shows that protein phosphatase 2A is responsible for dephosphorylation in the intact cells.

Protein phosphatase 2B, a Ca^{2+}-dependent protein phosphatase whose catalytic subunit is homologous to those of protein phosphatases 1 and 2A, is also inhibited by okadaic acid, but with much lower potency (half-maximal effect at 5 μM[5]). Protein phosphatase 2C, pyruvate dehydrogenase phosphatase, protein-tyrosine phosphatases, and acid and alkaline phosphatases are not inhibited.[5,6] Okadaic acid up to 5 μM does not inhibit protein kinase C, cyclic AMP-dependent protein kinase, the AMP-activated protein kinase, phosphorylase kinase, glycogen synthase kinase 3, or casein kinases 1 and 2.[3,6]

Effects of Okadaic Acid on Protein Phosphorylation in Isolated Cells

Experimental protocols for isolation of cells and their ^{32}P labeling are described elsewhere in this series (e.g., Garrison, Vol. 99 [4]). In our experiments,[6] rat hepatocytes or adipocytes are isolated as described previously,[8,9] and prelabeled at 100 mg wet wt/ml (hepatocytes) or 10^6 cells/ml (adipocytes) in 0.2 mM [^{32}P]phosphate (500 Ci/mol) for 60 min. Okadaic acid [0.4 mM in 10% (v/v) dimethyl sulfoxide (DMSO)] is then added to final concentrations up to 1 μM. The final concentration of DMSO is 0.025% (v/v) in all incubations.

After incubation of hepatocytes for 15 min at 37°, cells are harvested by centrifugation, and resuspended in ice-cold homogenization buffer containing fluoride ions to inhibit protein-serine/threonine phosphatases (0.25

[7] T. A. J. Haystead, F. P. Moore, P. Cohen, and D. G. Hardie, *Eur. J. Biochem.* **187,** 199 (1990).
[8] R. Holland, L. A. Witters, and D. G. Hardie, *Eur. J. Biochem.* **140,** 325 (1984).
[9] T. A. J. Haystead and D. G. Hardie, *Biochem. J.* **234,** 279 (1986).

M mannitol, 2 mM EDTA, 1 mM EGTA, 50 mM NaF, 50 mM Tris-HCl, pH 7.4; 2.5 ml/g wet wt). The cells are lysed using 40 strokes in a Dounce homogenizer (hepatocytes) or by using a Ystral T1500 homogenizer (Döttingen, Germany) for 3–5 sec at setting 3 (adipocytes). A postmitochondrial supernatant is prepared by centrifugation (10,000 g for 15 min at 4°). In the case of hepatocytes this is then separated into cytosol, microsomal, and glycogen fractions by further centrifugation (100,000 g for 60 min at 4°). The supernatant (cytosol fraction) is poured off and the upper, fluffy microsomal pellet is resuspended separately from the lower, white glycogen pellet, both in homogenization buffer. All three fractions are analyzed by electrophoresis in 5–15% (w/v) polyacrylamide gels in the presence of sodium dodecyl sulfate (SDS). The gels are dried and subjected to autoradiography, and the results for the cytosol fraction are shown in Fig. 2.

Within 15 min, 0.1 or 1 μM okadaic acid causes a marked stimulation of ^{32}P labeling of a very large number of hepatocyte phosphoproteins, including several that could be tentatively identified by comigration with purified markers (see Fig. 2 and Haystead et al.[6]). Analysis of trichloroacetic acid-precipitable radioactivity in the cytosol fraction indicates that the toxin stimulates total ^{32}P labeling of protein by ~2.5-fold, with a maximal effect at ~1 μM and a half-maximal effect at ~200 nM. Analysis of cellular nucleotides by ion-exchange high-performance liquid chromatography (HPLC) shows that this experimental protocol does not affect the specific radioactivity of ATP or ATP : ADP ratios.

Very similar results are obtained with isolated adipocytes.[6] In these cells, time course experiments show that there is a marked effect of okadaic acid on protein phosphorylation after only 1 min (T. A. Haystead, unpublished results), indicating that inhibition of protein phosphatases is very rapid. In isolated bovine adrenal chromaffin cells 1 μM okadaic acid increases total protein phosphorylation 2.7-fold, and phosphorylation of tyrosine hydroxylase (monooxygenase) by 3.1-fold.[10] Using lower concentrations of the toxin and longer incubation times (93 nM for 90 min), okadaic acid has also been shown to stimulate phosphorylation of a nucleolar protein in primary human fibroblasts.[11]

Effects of Okadaic Acid on Metabolic Parameters

Along with the original observations that it caused contraction of smooth muscle,[4] okadaic acid has been shown to modulate dramatically several metabolic parameters in intact cells. This confirms that the eleva-

[10] J. Haavik, D. L. Schelling, D. G. Campbell, K. K. Andersson, T. Flatmark, and P. Cohen, FEBS Lett. 251, 36 (1989).

[11] O. G. Issinger, T. Martin, W. R. Richter, M. Olson, and H. Fujiki, EMBO J. 7, 1621 (1988).

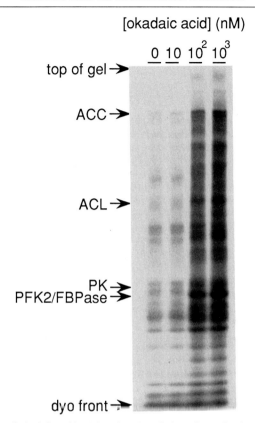

FIG. 2. Effects of okadaic acid on the phosphorylation of proteins in the cytosol fraction of ^{32}P-labeled rat hepatocytes. Hepatocytes were incubated for 15 min with the indicated concentration of okadaic acid. Phosphoproteins were tentatively identified by reference to other studies in the literature and by comigration with purified markers as follows: ACC, acetyl-CoA carboxylase; ACL, ATP-citrate lyase; PK, pyruvate kinase (L type); PFK2/ FBPase, 6-phosphofructo-2-kinase/fructose-2,6-bisphosphatase.

tion in ^{32}P labeling is due to real increases in protein phosphorylation, rather than being merely due to effects on phosphate turnover. The toxin stimulates glucose output and gluconeogenesis in hepatocytes, and stimulates lipolysis and inhibits fatty acid synthesis in adipocytes.[6] These are the effects predicted if the toxin acted exclusively to increase protein phosphorylation, strengthening the view that it has no nonspecific toxic effects, at least in these short-term incubations (15 min). Given the dramatic effects of okadaic acid on protein phosphorylation, it might be expected that all manner of toxic effects secondary to increases in protein phosphorylation would occur in long-term incubations. Okadaic acid has

TABLE I

EFFECTS OF OKADAIC ACID ON BIOCHEMICAL PARAMETERS IN INTACT CELLS AND
CELL-FREE EXTRACTS

Experimental system	Effect of okadaic acid	Ref.
Intact cells		
Human and rabbit smooth muscle	Initiates contraction	4
Mouse skin	Promotes induction of tumors by dimethylbenzanthracene	3
Mouse skin	Induces ornithine decarboxylase	13
Rat hepatocytes	Stimulates glucose output	6
Rat hepatocytes	Stimulates gluconeogenesis	6
Rat adipocytes	Mimics effect of insulin on deoxyglucose uptake	6
Rat adipocytes	Inhibits fatty acid synthesis	6
Rat adipocytes	Stimulates lipolysis	6
Rat peritoneal macrophages	Increases production of prostaglandin E_2	14
Molluskan neuron (*Helix aspersa*)	Potentiates 5-hydroxytryptamine-induced suppression of K^+ current	15
Paramecium tetraurelia	Prolongs backward swimming due to delayed inactivation of ciliary Ca^{2+} channels	16
Cardiac myocytes	Increases L-type inward Ca^{2+} current	17
Tracheal myocytes	Increases open state probability of Ca^{2+}-dependent K^+ channels	18
NIH 3T3 cells	Causes reversion of transformed phenotype induced by *raf-1* and *ret-2* oncogenes	12
Cell-free systems		
Reticulocyte lysate	Abolishes protein synthesis due to phosphorylation of elongation factor-2	19
SV40 virus	Inhibits stimulation of DNA replication by large T antigen	20
Xenopus oocyte extract	Increases activation of protein kinase activity (cdc2) of maturation-promoting factor	21

been incubated with NIH 3T3 cells at 10 nM for up to 10 days, and has intriguing effects (see Table I), but at higher concentrations it killed the cells.[12]

Clearly the most exciting use of okadaic acid will be in obtaining evidence for novel roles for protein phosphorylation, and many such studies have now been initiated. A summary of publications is presented

[12] R. Sakai, I. Ikeda, H. Kitani, H. Fujiki, F. Takaku, U. Rapp, T. Sugimura, and M. Nagao, *Proc. Natl. Acad. Sci. U.S.A.* **86,** 9946 (1989).

in Table I,[13-21] which illustrates the great potential of the toxin for studies of diverse processes in many different types of cell. As shown in Table I, okadaic acid has also been very useful as a probe in cell-free systems in which the complexity of the process (e.g., protein synthesis or DNA replication) has as yet prevented complete purification of all components.

One point to bear in mind in this type of study is that okadaic acid can only be effective if the kinase whose effect it reverses is at least partially active. This is illustrated by the fact that inhibition of protein synthesis by okadaic acid in reticulocyte lysates occurs only in the presence of Ca^{2+}, because elongation factor-2 kinase is Ca^{2+}/calmodulin dependent.[19] Another example may be the effect of the toxin on glucose transport in adipocytes. Although okadaic acid quantitatively mimicks the effect of insulin, it only does so after a lag of several minutes.[6] Since inhibition of protein phosphatases occurs within 1 min (see above), this time lag may indicate that the kinase responsible for activating glucose transport is almost inactive in the absence of insulin, so that it takes some time for the phosphorylation to reach a new steady state level.

Another toxin, calyculin A, has been isolated from sea sponges.[22] Although not obviously related in structure to okadaic acid, it also causes contraction of skinned muscle fibres. It inhibits protein phosphatase 2A (IC_{50} 1 nM) with similar potency to okadaic acid, but was found to inhibit protein phosphatase 1 at concentrations (IC_{50} 2 nM) much lower than did okadaic acid (IC_{50} 60 nM). Its specificity and efficacy on intact cells have not yet been tested, but these results suggest that comparative experiments with the two toxins might enable one to obtain evidence for the relative importance of the two protein phosphatases in some systems. After the original submission of this chapter, two other natural toxins have been found to be potent protein phosphatase inhibitors. Tautomycin is related

[13] H. Fujiki, M. Suganuma, H. Suguri, S. Yoshizawa, M. Ojika, K. Wakamatsu, K. Tamada, and T. Sugimura, Proc. Jpn. Acad. Ser. B, 63, 51 (1987).

[14] K. Ohuchi, T. Tamuri, M. Ohashi, M. Watanabe, N. Hirasaura, S. Tsurufiji, and H. Fujiki, Biochim. Biophys. Acta 1013, 86 (1989).

[15] P. Cohen, G. Cottrell, and C. Hill-Venning, J. Physiol. (London) 415, 33P (1989).

[16] S. Klumpp, P. Cohen, and J. Schultz, EMBO J. 9, 685 (1990).

[17] J. Hescheler, G. Mieskes, J. C. Ruegg, A. Takai, and W. Trautwein, Pfluegers Arch. 412, 248 (1988).

[18] H. Kume, A. Takai, H. Tokuno, and T. Tomita, Nature (London) 341, 152 (1989).

[19] N. T. Redpath and C. G. Proud, Biochem. J. 262, 69 (1989).

[20] R. E. Lawson, P. Cohen, and D. P. Lane, J. Virol. 64, 2380 (1990).

[21] M. A. Felix, P. Cohen, and E. Karsenti, EMBO J. 9, 675 (1990).

[22] H. Ishihara, B. L. Martin, D. L. Brautigan, H. Karaki, H. Ozaki, H. Y. Kato, N. Fusetani, S. Watabe, K. Hashimoto, D. Uemura, and D. J. Hartshorne, Biochim. Biophys. Res. Commun. 159, 871 (1989).

in structure to okadaic acid, but is produced by soil bacteria, i.e., *Streptomyces griseochromogenes* and *S. verticillatus*. Microcystins (e.g., Microcystin-LR) are unrelated in structure to okadaic acid, being cyclic hepta-
peptides containing unusual amino acids, produced by the cyanobacterial
genera *Microcystis, Oscillatoria,* and *Anabaena.* Both tautomycin and
Microcystin-LR inhibit protein phosphatase 2A with similar potency to
okadaic acid, but are even more potent inhibitors of protein phosphatase
1 than okadaic acid.[23,24] Despite their diversity of structure, there is evidence that okadaic acid, tautomycin, Microcystin-LR, and inhibitors 1
and 2 interact at the same site on protein phosphatase 1.[23,24] The effects
of these compounds on protein phosphorylation in intact cells are not yet
as well characterized as those of okadaic acid, but it seems likely that they
will be very useful experimental tools, particularly Microcystin-LR, which
is already commercially available (Calbiochem).

Acknowledgments

Studies in this laboratory were supported by the Medical Research Council (U.K.) and
the British Heart Foundation. We thank Philip Cohen for helpful discussion and Y. Tsukitani
for samples of okadaic acid.

[23] C. MacKintosh and S. Klumpp, *FEBS Lett.* **277**, 137 (1990).
[24] C. MacKintosh, K. A. Beattie, S. Klumpp, P. Cohen, and G. A. Codd, *FEBS Lett.* **264**,
187 (1990).

[41] Use of Vanadate as Protein-Phosphotyrosine Phosphatase Inhibitor

By Julius A. Gordon

Introduction

Vanadium ions have found widespread use as an inhibitor of protein-phosphotyrosine phosphatases.[1-4] Almost routinely, some form of vanadium ion is added to preserve the phosphotyrosine content of cells, cell lysates, and tyrosine kinase assays. Protein-phosphotyrosine phosphatase (PPTPase) activity is found in most mammalian and avian cells and tissues as well as in bacterial and yeast cultures.[1]

Earlier observations suggested that PPTPase activity might account for the rapid loss and turnover of phosphate in phosphotyrosine-containing proteins in membrane preparations and in cells containing temperature-sensitive tyrosine kinase activity.[5-8]

Current evidence suggests that many different phosphotyrosine phosphatases coexist in the same cell, either embedded in the cytoplasmic surface of the membrane or in the cytosol. A specific regulatory or modulating function of the phosphotyrosine phosphatases in cells seems reasonable.[9] PPTPases have been purified to partial homogeneity from the cytoplasm of several cell types.[10-13] A problem which plagues the attempts to characterize the activity of the many PPTPases is of concern in studies on the inhibition of protein-phosphotyrosine phosphatases. The physiological

[1] K.-H. William Lau, J. K. Farley, and D. J. Baylink, *Biochem. J.* **257**, 23 (1989).
[2] D. L. Brautigan and C. L. Shriner, this series, vol. 159, p. 339.
[3] G. Swarup, S. Cohen, and D. L. Garbers, *Biochem. Biophys. Res. Commun.* **107**, 1104 (1982).
[4] J. F. Leis and N. O. Kaplan, *Proc. Natl. Acad. Sci. U.S.A.* **79**, 6507 (1982).
[5] G. Carpenter, L. King, and S. Cohen, *J. Biol. Chem.* **254**, 4884 (1979).
[6] D. L. Brautigan, P. Bornstein, and B. Gallis, *J. Biol. Chem.* **256**, 6519 (1981).
[7] K. Radke and G. S. Martin, *Cold Spring Harbor Symp. Quant. Biol.* **44**, 975 (1979).
[8] B. M. Sefton, T. Hunter, K. Beemon, and W. Eckhart, *Cell (Cambridge, Mass.)* **20**, 807 (1980).
[9] T. Hunter, *Cell (Cambridge, Mass.)* **58**, 1013 (1989).
[10] D. Horlein, B. Gallis, D. L. Brautigan, and P. Bornstein, *Biochemistry* **27**, 5577 (1982).
[11] G. Swarup, K. V. Speeg, S. Cohen, and D. L. Garbers, *Biochem. Biophys. Res. Commun.* **107**, 1104 (1982).
[12] J. G. Foulkes, E. Erikson, and R. L. Erikson, *J. Biol. Chem.* **258**, 431 (1983).
[13] B. Gallis, P. Bornstein, and D. L. Brautigan, *Proc. Natl. Acad. Sci. U.S.A.* **78**, 6689 (1981).

substrates of the specific PPTPases are largely unknown; the assays for PPTPase activity depend on the utilization of nonspecific substrates which have been generated from off-the-shelf proteins or small polypeptides.[1] This uncertainty in substrate suitability may influence the effectiveness of inhibitors. Also, the multiple phosphotyrosine phosphatases may be more or less sensitive to the same inhibitor. Zinc chloride is a general PPTPase inhibitor, yet a family of PPTPases have been purified and separated on zinc-affinity chromatography which suggests a differential level of zinc inhibition.[1,2,6,10] As vanadate fails to inhibit the assay of nickel-activated calcineurin phosphotyrosine phosphatase activity,[14] caution also must be observed in reliance on vanadate as a complete inhibitor. Most published studies report a residual (5–10%) phosphotyrosine phosphate activity in inhibitor concentrations as high as 1 mM vanadate.

The protein-phosphotyrosine phosphatase activities sensitive to inhibition by zinc chloride and sodium orthovanadate are conveniently separable from the phosphoseryl and phosphothreonyl phosphatase activities inhibited by EDTA and fluoride.[2] However, the casual inclusion of fluoride in lysates may inhibit phosphotyrosine phosphatase activity while EDTA may stimulate PPTPase activity.[12] Therefore the judicious and trial-by-error mixing of vanadium ions with other inhibitors should be carefully considered in any study. EDTA may even inactivate vanadium ions effects by complex formation.[15]

Tabulated in Table I are some effects of vanadate that may not be immediately apparent as a consequence of the inhibition of PPTPase activity. These effects should be particularly considered when intact cells are incubated for long periods in vanadate.

Chemistry of Vanadium under Physiological Conditions

Vanadium is a Group 5b transition metal in aqueous solution. Its oxidized forms in dilute aqueous solution, tetravalent vanadyl ($+4$ oxidation state, VO_2^+) and pentavalent vanadate ($+5$ oxidation state, generally written as HVO_4^-, VO_3^-, or $H_2VO_4^-$), have been shown to have a variety of biological and biochemical effects (see Table I).[16–34] The underlying

[14] C. J. Pallen, K. A. Valentine, J. H. Wang, and M. D. Hollenberg, *Biochemistry* **24**, 4727 (1985).

[15] K. A. Rubinson, *Proc. R. Soc. London, Ser. B* **212**, 65 (1981).

[16] H. S. Earp, R. A. Rubin, K. S. Austin, and R. C. Dy, *FEBS Lett.* **161**, 180 (1983).

[17] D. J. Brown and J. A. Gordon, *J. Biol. Chem.* **259**, 9540 (1984).

[18] R. Gherzi, C. Caratti, S. Andraghetti, A. Bertolini, G. Montemurro, G. Sesti, and R. Cordera, *Biochem. Biophys. Res. Commun.* **152**, 1474 (1988).

[19] J. K. Klarlund, S. Latini, and J. Forchhammer, *Biochim. Biophys. Acta* **971**, 112 (1988).

[20] D. Cassel, Y.-X. Zhuang, and L. Glaser, *Biochem. Biophys. Res. Commun.* **118**, 675 (1984).

TABLE I

DIVERSE EFFECTS OF MONOMERIC VANADIUM IONS
OTHER THAN DIRECT PPTPASE INHIBITION

Effect	Refs.
Examined in intact cells or membrane preparations	
Possible direct stimulation of tyrosine kinase activity	16–19
Increase in intracellular pH and Ca^{2+}	20
Effect on intracellular degradation of proteins	21–23
Mitogenic effects and increased DNA synthesis	24,25
Action as an intracellular redox system	26,27
Effect on phosphoinositide metabolism	28
Examined in cell lysates or purified systems	
Binding to F-actin	29
Inhibition of protein degradation by binding to substrate	30
Formation of covalent esters with hydroxyl groups (i.e., glucose and tyrosine)	31–33
Inhibition of RNase and other enzymes	34

mechanisms or mechanism of these biological effects is not clearly elucidated although a structural similarity of vanadate to phosphate intermediates has been implicated in phosphoryl transferase reactions.[34–36] The complete chemistry of vanadium is complex and need not be considered under most aqueous physiological conditions.[15] In the presence of oxidizing agents vanadium ions exist as the hydrated monomer of vanadate (HVO_4^{2-} or $H_2VO_4^{-}$) at micromolar concentations near neutral pH. In the presence of extracellular reducing agents or intracellular glutathione,

[21] M. Desautels and A. L. Goldberg, *Proc. Natl. Acad. Sci. U.S.A.* **79**, 1869 (1982).
[22] K. Tanaka, L. Waxman, and A. L. Goldberg, *J. Biol. Chem.* **259**, 2803 (1984).
[23] J. L. Varga, F. Anieto, J. Cervera, and E. Knecht, *Biochem. J.* **258**, 33 (1989).
[24] J. B. Smith, *Proc. Natl. Acad. Sci. U.S.A.* **80**, 6162 (1983).
[25] G. Carpenter, *Biochem. Biophys. Res. Commun.* **102**, 1115 (1981).
[26] J. E. Benabe, L. A. Echegoyen, B. Pastrana, and M. Martinez-Maldonado, *J. Biol. Chem.* **262**, 9555 (1987).
[27] J. C. Cantley, Jr., J. H. Ferguson, and K. Kustin, *J. Am. Chem. Soc.* **100**, 5210 (1978).
[28] S. Paris, J.-C. Chambard, and J. Pouyssegur, *J. Biol. Chem.* **262**, 1977 (1987).
[29] C. Combeau and M.-F. Carlier, *J. Biol. Chem.* **263**, 17429 (1988).
[30] S. Pillai and J. E. Zull, *J. Biol. Chem.* **260**, 8384 (1985).
[31] P. P. Layne and V. A. Najjar, *Proc. Natl. Acad. Sci. U.S.A.* **76**, 5010 (1979).
[32] A. F. Nour-Eldeen, M. M. Craig, and M. J. Gresser, *J. Biol. Chem.* **260**, 6836 (1985).
[33] A. S. Tracey and M. J. Gresser, *Proc. Natl. Acad. Sci. U.S.A.* **83**, 609 (1986).
[34] B. R. Nechay, L. B. Nanninga, P. S. E. Nechay, R. L. Post, J. J. Grantham, I. S. Macara, L. F. Kubena, T. D. Phillips, and H. Forest, *Fed. Proc., Fed. Am. Soc. Exp. Biol.* **45**, 123 (1986).
[35] T. J. B. Simons, *Nature (London)* **281**, 337 (1979).
[36] D. W. Boyd and K. Kustin, *Adv. Inorg. Biochem.* **6**, 311 (1984).

the oxyanion is rapidly reduced to the oxycation vanadyl (VO_2^+) within 30 min or less.[15] Vanadate also begins to polymerize at concentrations greater than 0.1 mM at neutral pH. The yellow–orange solutions of decavanadate can be converted to the colorless solutions of monomeric vanadate by dilution after a period of many hours. The process is hastened by boiling at pH 10, which encourages the kinetically sluggish depolymerization process.[15]

Coordination complexes of vanadium compounds with sulfhydryls, disulfides, hydroxyls, and carboxylate compounds are usually important only if they are multidentate. Relevant examples are the sequestration of vanadyl from oxidation when in a coordination complex with EDTA, ATP, or other multidentate partners.[15,34] Less obvious was the observation that bovine serum albumin and transferrin may "chelate" one or more ionic forms of vanadium.[34] With longer incubation periods and higher vanadate concentrations, the role of nonenzymatic esters of vanadate with hydroxyl compounds (i.e., glucose and tyrosine) should not be overlooked in evaluating the effects of vanadium ions in biological systems.[32,33]

Preparation of Vanadate Solutions

Stock solutions of sodium orthovanadate (Na_3VO_4) may be constituted in sterile and distilled water at a concentration of 1 mM adjusted to approximately pH 10. To ensure the presence of monomers the solution is heated to boiling until translucent and the pH readjusted to pH 10. We routinely store a stock solution in flint glass for several months. In our laboratory the preparation seems to improve with age during this time. We know of no rationale to explain this effect nor the mysticism attached to finding a single supplier and lot number for vanadate. Vanadate solutions can also be aliquoted, stored in plastic, and frozen; we thaw the frozen solution overnight prior to sue. Vanadyl or decavanadate solutions may be prepared for comparison to the effects of vanadate, but remember that added vanadyl, metavanadate, orthovanadate, or decavanadate will interconvert in aqueous solution without suitable precautions (i.e., pH, oxidation state, complexing compounds, and concentration).

Sodium orthovanadate ($Na_3VO_4 \cdot nH_2O$) may be purchased from Fisher Scientific Company (Fairlawn, NJ). As the hydration number of commercial vanadate preparations is rarely given, the precise concentration at pH 10.5 of a dilute aqueous solution may be determined from the extinction coefficient of 3.55×10^3 at 260 nm.[37] Alternatively, a dye-dependent

---

[37] L. Newman, W. J. LaFleur, F. J. Brousaides, and A. M. Ross, *J. Am. Chem. Soc.* **80**, 4491 (1958).

spectrophotometric assay may be used.[38] For less precise work we assume a hydration number of 10 in preparing stock solutions of sodium orthovanadate.

Practical Considerations

The following is a rough guide for the investigator first undertaking studies in which Na_3VO_4 will be used to inhibit PPTPases. Considerations for use may be divided into those situations where the cell membrane is disrupted or intact.

To preserve the protein phosphotyrosines during cell disruption, the lysis buffer should be made about 1.0 mM in vanadate (higher only if necessary to account for extraordinary complex formation). Subsequently, the concentration of vanadate may be reduced to provide the necessary inhibition of PPTPases. Note that the addition of vanadate from stock solution (pH 10) will change the pH of the lysate or assay if the buffering capacity of the medium is not adequate.

With intact cells in culture where metabolic studies are to be done, the situation is more complicated. Toxicity effects of vanadate must be considered. Our experience with normal and transformed chicken embryo fibroblasts indicates that toxicity occurs beyond a 6-hr incubation in 100 μM vanadate. Toxicity was judged by a decrease in the intracellular transport of ortho[^{32}P]phosphate and ^{32}P incorporation into intracellular proteins. Morphological changes and cell detachment usually occur at later and at somewhat higher vanadate concentrations. In our experience, longer incubation periods require a reduction in vanadate concentration below 100 μM when either using avian or murine cells. We have never seen the "transformation" of normal avian and murine cells by vanadate,[39] in agreement with others.[40]

Two points should be emphasized concerning the use of vanadium compounds. First, we reemphasize that vanadyl, metavanadate, and orthovanadate, as well as decavanadate, are interconverted in aqueous solution depending on concentration, pH, and the state of the redox potential. Many published studies on the effect of vanadyl on cells and assays were probably following the effect of vanadate.

Second, when vanadate is added with other reagents it is not easy to predict whether its effect will be synergistic or antagonistic. We have incubated Rous sarcoma virus-tranformed cells with genistein, vanadate,

[38] C. C. Goodno, this series, Vol. 85, p. 116.
[39] J. K. Klarlund, Cell (Cambridge, Mass.) 41, 707 (1985).
[40] J. E. Van Wart-Hood, M. E. Linder, and J. G. Burr, Oncogene 4, 1267 (1989).

or in both. Genistein was presumably acting as a tyrosine kinase inhibitor.[41] The incorporation of ortho[^{32}P]phosphate into the phosphotryosines of pp60^{v-src} and calpactin I was followed for up to 6 hr. We observed no increase of phosphotyrosine content in vanadate- and genistein-treated cells compared to genistein-treated cells. The expectation was that the effect of partial inhibition of protein-phosphotyrosine kinases by moderate genistein concentrations would be masked in the presence of vanadate, which would act as an inhibitor of PPTPases.

From other studies (Table I) and our own experience, we have found that the results obtained with intact cells may not easily be explained by the simple inhibition of PPTPases by vanadate. The interpretation of the results obtained with vanadate in cell lysates and assays is more convincing as an inhibitor of PPTPases.

---

[41] T. Akiyama, J. Ishida, S. N. Nakagawa, H. Ogawara, S. I. Watanabe, N. Itoh, M. Shibuya, and Y. Fukami, *J. Biol. Chem.* **262,** 5592 (1987).

Author Index

Numbers in parentheses are footnote reference numbers and indicate that an author's work is referred to although the name is not cited in the text.

Subject Index

A

isotopic equilibrium of, 153
location of, 159
number of, determination of, 153, 260
at or near pseudosubstrate regions, 293
^{32}P-labeled, identification of, 187
prediction of, 186
with sequencing, 266–267
Phosphorylation site sequences, for protein
kinases, 293–294
Phosphorylation state. *See also* Phosphorylation stoichiometry
definition of, 251
Phosphorylation state-specific antibodies.
See also Dephospho-specific antibodies; Phospho-specific antibodies
characterization of, 276–280
limitations, 282
measurement of phosphorylation stoichiometry with, 276–279, 282–283
monoclonal, 265
production of, 269, 271
polyclonal, production of, 269–271
production of, 264–283
with synthetic peptides as phosphorylation state-specific epitopes, 266–
271
screening for, 271–276
utility of, 266, 280–283
Phosphorylation state-specific epitopes,
synthetic peptides as, 266–271
Phosphorylation stoichiometry
direct measurement of, 282
measurement of, 154, 160, 251–252, 260,
264
by biosynthetic labeling, 245–251
by FABMS, 160–163
with immunoblotting, 276–278
methods for, 142, 248, 265
with nonequilibirum radioimmunoassay, 277
one-dimensional procedures for, 258–
259
peptide mapping for, 260–261
two-dimensional, 252–256
at phosphorylation site, 153–154
with phosphorylation state-specific
antibodies, 276–279
by ^{32}P incorporation, 277–278
by separation of phosphorylated isoforms, 251–261

with two-dimensional electrophoresis,
252–256
two-dimensional procedures for, 258
Phosphoryl groups, hydrolysis of, in sequencing, 198
Phosphorylhistidine, 199
Phosphoserine, 128, 144, 169. *See also* O-
Phosphates
ability to block antibody binding, 36–37,
49–50
acidic protons of, 256
in anti-phosphotyrosine-immunoprecipi-
tated proteins, percentage of, 38
β elimination of, 169, 180
charge on, 134
chemical properties of, 12–13
conversion to dithiothreitol adduct of
dehydroalanine, 183
conversion to phenylthiocarbamyl-*S*-
ethylcysteine derivative, 186
conversion to *S*-ethylcysteine, 155, 163,
169–185
with performic acid oxidation, 174–
175
without performic acid oxidation, 174–
175
determination of, 155, 163, 169–185
in cardiac troponin I, 171–172
by double chromatography, 18
by double electrophoresis, 20–21
by Edman degradation, 166, 180
by electrophoresis and chromatography, 19–20
by high-performance liquid chromatography, 3–4, 6–9
by one-dimensional thin-layer systems, 15–16
by phosphoamino acid analysis, 130
in proteins and peptides, 182–183
by sequence analysis, 182
S-ethylcysteine method for, 179–180
by sequence analysis without *S*-ethyl-
cysteine modification, 183
distribution of, after alkali treatment, 26
S-ethylcysteine derivatives of, sequence
analysis of, 165
indicators of, 138
measurement of
acid hydrolysis for, 3–4
base hydrolysis for, 4–5